Microbes in Agri-Forestry Biotechnology

This book explores recent advances on the use of microbes for agri-forestry biotechnological applications. It provides technical concepts and discussions on the use of microorganisms for processes such as bioprocessing, bioremediation, soil enhancement, aquaponics advances, and plant-host symbiosis. The book provides an overview of the microbial approach to the tools and processes used in agriculture and forestry that make or modify products, improve plants for specific uses, and make use of livestock in agricultural systems. The authors discuss the main process conditions that enhance agri-forestry applications with the use of microbes and introduce the use of genetically modified (GM) microbes in agrobiotechnology. Finally, the authors explore the main technological advances in the production of secondary metabolites with potential applications in agri-forestry. This book is intended for biotechnologists, biologists, bioengineers, biochemists, microbiologists, food technologists, enzymologists, and related researchers.

Advances and Applications in Biotechnology

Series Editors:
Dr. Gustavo Molina
Dr. Vijai Kumar Gupta

Plant-Microbe Interactions: Harnessing Next-Generation Molecular Technologies for Sustainable Agriculture
Edited by Jagajjit Sahu, Anukool Vaishnav, and Harikesh Bahadur Singh

Microbes in Agri-Forestry Biotechnology
Edited by Gustavo Molina, Zeba Usmani, Minaxi Sharma, Abdelaziz Yasri, and Vijai Kumar Gupta

For more information about this series, please visit: https://www.routledge.com/ Advances-and-Applications-in-Biotechnology/book-series/AABT

Microbes in Agri-Forestry Biotechnology

Edited by
Gustavo Molina
Zeba Usmani
Minaxi Sharma
Abdelaziz Yasri
Vijai Kumar Gupta

CRC Press
Taylor & Francis Group
Boca Raton London New York

CRC Press is an imprint of the
Taylor & Francis Group, an **informa** business

First edition published 2023
by CRC Press
6000 Broken Sound Parkway NW, Suite 300, Boca Raton, FL 33487-2742

and by CRC Press
4 Park Square, Milton Park, Abingdon, Oxon, OX14 4RN

CRC Press is an imprint of Taylor & Francis Group, LLC

© 2023 selection and editorial matter, Gustavo Molina, Zeba Usmani, Minaxi Sharma, Abdelaziz Yasri, Vijai Kumar Gupta; individual chapters, the contributors

ISBN: 978-0-367-62426-2 (hbk)
ISBN: 978-0-367-62713-3 (pbk)
ISBN: 978-1-003-11047-7 (ebk)

DOI: 10.1201/9781003110477

Typeset in Times
by KnowledgeWorks Global Ltd.

Contents

Editors

Gustavo Molina, PhD, graduated in Food Engineering with master's degree and a PhD at the University of Campinas – Unicamp (Campinas – Brazil), and part of his doctoral research was developed at the Laboratoire de Génie Chimique et Biochimique at the Université Blaise Pascal (Clermont-Ferrand – France). He is an Associate Professor at the Institute of Science and Technology at UFVJM (Diamantina – Brazil) and the Head of the Laboratory of Food Biotechnology. He developed his postdoctoral research at the Institut Polytechnique de Grenoble (Grenoble – France) in the area of biorefinery and development of the enzymatic hydrolysis process of lignocellulosic materials.

Zeba Usmani, PhD, holds a doctorate from the reputed Indian Institute of Technology (IIT-ISM), Dhanbad, India and serves as a visiting Assistant Professor in the Department of Applied Biology, University of Science and Technology, Meghalaya, India. Her research focuses on the intersection of environmental chemistry and microbiology, wherein she brings a diverse understanding of developing novel green methodologies for sustainable degradation of organic and inorganic chemicals. She is an accomplished researcher with 47 peer-reviewed publications, 13 book chapters, and 1 book.

Minaxi Sharma, PhD, is currently working as a Senior Researcher at Haute Ecole Provinciale de Hainaut – Condorcet (Belgium). She earned her doctoral degree in Dairy Chemistry from ICAR-National Dairy Research Institute, India. She is also working as a Visiting Assistant Professor in the Department of Applied Biology, University of Science and Technology, Meghalaya, India. Dr. Sharma has 12 years of enriched experience in the area of food science and technology, nanoencapsulation, extraction of plant bioactives, food waste valorization, food chemistry, food microbiology, and microbial biotechnology, and has written 62 research/review articles, 7 book chapters, and 3 books with reputed publishers.

Abdelaziz Yasri, PhD, earned a PhD in Biophysics at Montpellier Medical School in France and also did MBA at Montpellier Business School. He is an Affiliated Professor at Mohammed VI Polytechnic University in Morocco and General Secretary of the National Institut of Agronomic Research (INRA-Morocco). He directed the School of Agriculture, Fertilizers and Environment Sciences at University of Mohamed VI Polytechnic in Benguerir – Morocco. Prof. Yasri has more than 27 years of experience in academic research and in start-ups and biotechnology. He worked in biotech companies in France, Belgium, United States, and Morocco, where he occupied several positions such as Principal Scientist and Director.

Vijai Kumar Gupta, PhD, holds a doctoral degree in Microbiology from Dr. RML Avadh University, India. Currently, he is working at *Center for Safe and Improved Foods & Biorefining and Advanced Materials Research Center*, SRUC, Edinburgh, United Kingdom. Dr. Gupta is an expert in *Biomass Valorisation, Microbial Engineering Biotechnologies, Bioactive Natural Products, and Functional Microbiome.* He has 231 peer-reviewed publications, 43 book chapters, and 39 books in his hands.

1 Multi-Disciplinary Nature of Microbes in Agricultural Research
Current Status

Zengwei Feng[1,2], Honghui Zhu[1], and Qing Yao[2]
[1]Guangdong Provincial Key Laboratory of Microbial Culture Collection and Application, Key Laboratory of Agricultural Microbiomics and Precision Application, Ministry of Agriculture and Rural Affairs, State Key Laboratory of Applied Microbiology Southern China, Guangdong Microbial Culture Collection Center (GDMCC), Institute of Microbiology, Guangdong Academy of Sciences, Guangzhou, China
[2]College of Horticulture, Guangdong Province Key Laboratory of Microbial Signals and Disease Control, Guangdong Engineering Research Center for Litchi, South China Agricultural University, Guangzhou, China

CONTENTS

DOI: 10.1201/9781003110477-1

1

1.1   INTRODUCTION

Agriculture, as the primary industry, is of paramount importance to human society, although the proportion of its production value is far less than that of other industries. To sustain the demand for foods on Earth, agricultural industry has been striving to provide adequate, safe, and clean food across the world. Along with this process, agricultural science and technology have increasingly proven the key factors promoting the crop production. Among them, fertilizer nutrients contribute 30–50% to the crop yield in the USA (Stewart et al. 2005), and crop improvement contributes 21–50% to the growth in major crop yields in developing countries (Beddington 2010). The contribution of pest control to crop yield can be high as well because the losses in production due to pests are very high (Beddington 2010). Microbes are the ubiquitous organisms in all ecosystems on Earth, and microbial technology involves most aspects of agricultural industry, such as soil improvement and control of pests. Recently, microbiome has been proposed to be one of the most important breakthroughs to advance food and agricultural research in the future (National Academies of Sciences, Engineering, and Medicine 2019), which highlights the essential roles of agricultural microbes.

Agricultural microbes comprise diverse groups both taxonomically and functionally. To facilitate the increase in production, many studies on agricultural microbes have been conducted, including but not confined to microbial resources, environmental adaptation, action mechanisms, and application techniques. Accordingly, research on the agricultural microbes covers multiple disciplines, from taxonomy, biology (e.g., physiology, biochemistry, molecular biology, and multi-omics) through to ecology (Figure 1.1). During the past decade, many novel species and strains have been isolated and identified, which greatly deepens our understanding of the agricultural microbial resource and their taxonomy. For example, most of recently reported novel strains of legume rhizobia belong to β-rhizobia, while the traditionally reported strains are normally α-rhizobia (Hassen et al. 2020). Considering the application in field conditions, the adaptation of agricultural microbes to environments and the underlying mechanisms have been highlighted, for example, acidic soil, drought condition, and heavy metal contaminated soil (Satapute et al. 2019; Feng et al. 2020a; Pang et al. 2020). In this scenario, exploring the effects of abiotic factors on microbial inocula with multi-omics (meta-genomics, meta-transcriptomics, meta-proteomics, etc.) technology attracts more and more attention.

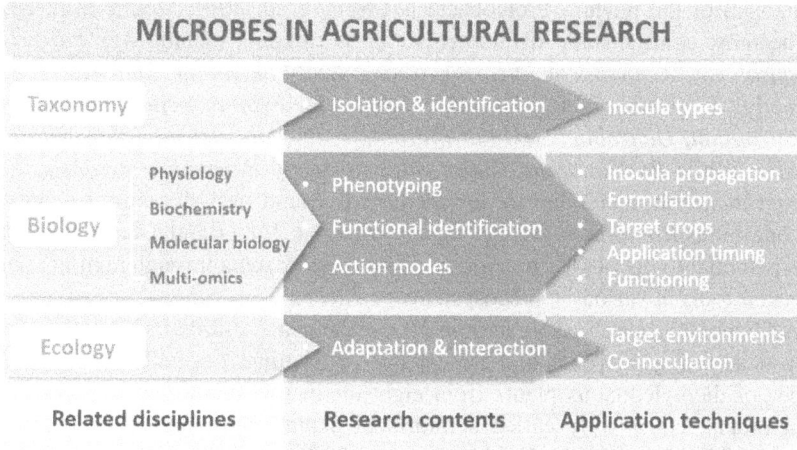

FIGURE 1.1 Multi-disciplinary nature of microbes in agricultural research.

1.2 RESOURCE AND TAXONOMY

Since agricultural microbes comprise diverse microbes in terms of their classification, it is necessary to probe into the resource and taxonomy. Recently, with increasingly abundant resource and accurate classification, more and more potential isolates of high efficiency can be applied in agricultural production.

1.2.1 NITROGEN-FIXING MICROBES

As an essential element for all living organisms, nitrogen (N) is a requisite for the biosynthesis of important cellular components, such as nucleic acids and proteins (Kuypers et al. 2018; Yu et al. 2020). Dinitrogen (N_2), however, which comprises up to approximately 78% of the Earth's atmosphere, is a non-bioavailable form of N for most organisms (Han et al. 2020). In agricultural ecosystems, N_2 can be converted into ammonia (NH_3) and further be used by plants. The first and key process is called nitrogen fixation, which is mainly conducted in soils by N-fixing bacteria and archaea (Mus et al. 2016; Han et al. 2020). Biological nitrogen fixation (BNF) plays an extremely important role in agricultural ecosystems due to its high-efficiency, low-cost, and non-polluting traits (Moura et al. 2020). Its contribution to total N uptake by plants is estimated to reach as high as 65–95% (Bolger et al. 1995).

Among N-fixing microbes, rhizobia are a predominant group and also have the highest species diversity. They are comprised of gram-negative bacteria that can form symbiotic N-fixing root and/or stem nodules in association with leguminous plants (Shamseldin et al. 2017). These symbiotic associations have been estimated to fix approximately 80% of the biologically fixed N_2 in agricultural areas (Moura et al. 2020). Rhizobia can live in soil as saprophytes or in the root nodules of their host legumes as symbionts (Han et al. 2020). Currently, rhizobia have more than 400 species in over 20 genera, which belong to three phylogenetic classes α-proteobacteria (α-rhizobia), β-proteobacteria (β-rhizobia), and γ-proteobacteria (γ-rhizobia) of the phylum Proteobacteria (Wang et al. 2019; Moura et al. 2020). As the most common and diverse rhizobia, α-rhizobia include *Agrobacterium, Allorhizobium, Aminobacter, Azorhizobium, Bradyrhizobium, Devosia, Ensifer* (formerly *Sinorhizobium*), *Mesorhizobium, Methylobacterium, Microvirga, Neorhizobium, Ochrobactrum, Pararhizobium, Phyllobacterium, Rhizobium*, and *Shinella*. While *Burkholderia, Cupriavidus* (formerly *Ralstonia*), *Herbaspirillum, Paraburkholderia*, and *Trinickia* are frequently found in β-rhizobia. In contrast, only one genus *Pseudomonasis* belongs to γ-rhizobia. Importantly, there is still enormous potential to dig out the previously unreported rhizobia through high-efficiency molecular tools in various soil ecosystems.

As filamentous aerobic soil actinobacteria, the genus *Frankia* are another major group of N-fixing microbes and are capable of forming root nodules in over 200 species of dicotyledonous plants from eight distinctive families within three different orders, collectively called actinorhizal plants (Normand et al. 2014; Van Nguyen and Pawlowski 2017; Van Nguyen et al. 2019). Similar to the symbiotic status of rhizobia, *Frankia* strains can live as vesicles during saprophytic growth and in symbiosis. When vesicles live in soil as saprophytes, they are always septate and spherical. Interestingly, in symbiosis, the shape of vesicles and their position in the infected root depends on the identity of host plants (Pawlowski and Demchenko 2012). Phylogenetically, *Frankia* strains can be grouped in four clusters, including three clusters with distinct host ranges (Pawlowski and Demchenko 2012; Van Nguyen et al. 2019). Cluster I strains are associated with host plants classified in the order Fagales, including Betulaceae, Casuarinaceae (apart from *Gymnostoma*), and Myricaceae (apart from *Morella*). Cluster II strains form nodules with host plants classified in the order Cucurbitales, including Coriariaceae, Datiscaceae, and some families in the order Rosales (Rosaceae and Rhamnaceae [*Ceanothus*]). Cluster III strains infect host plants mainly belonging to the order Rosales, including Elaeagnaceae, Rhamnaceae (apart from *Ceanothus*), and *Gymnostoma* and *Morella* in the order Fagales. The fourth cluster contains noninfective or noneffective strains (Pawlowski and Demchenko 2012; Van Nguyen et al. 2019; Nouioui et al. 2019).

1.2.2 MYCORRHIZAL FUNGI

Mycorrhizal fungi, which are ubiquitous and multifunctional fungi of the plant rhizospheric microbiome, can enhance their host plant access to soil nutrients (e.g., N and P) and water, improving the tolerance of their host plants to a variety of biotic and abiotic stresses, and improving ambient soil structure and properties (Smith and Read 2008; van der Heijden et al. 2015). By far, four major types of mycorrhizal fungi have been reported in terrestrial ecosystems based on their structure and/or the identity of their host plant, namely arbuscular mycorrhizal (AM) fungi, ectomycorrhizal (EcM) fungi, orchid mycorrhizal fungi, and ericoid mycorrhizal fungi. However, two predominant functional groups, AM fungi and EcM fungi, are potentially applied and function in agricultural ecosystems.

AM fungi, classified in the subphylum Glomeromycotina, are capable of forming mutualistic symbiotic associations with more than 80% of terrestrial plant species (approximately 200,000 plant species), which include most herbs and grasses, many trees, many hornworts, and liverworts (Smith and Read 2008; van der Heijden et al. 2015; Spatafora et al. 2016; Feng et al. 2020a). Once AM fungal hyphae enter the root, they can form a highly branched hyphal structure (i.e., arbuscule) in root cortical cells in which the exchange of nutrients occurs (Luginbuehl and Oldroyd 2017; Feng et al. 2020b). To sustain their growth and function, AM fungi rely exclusively on the carbon source in the form of sugars and lipids delivered from their host plants (Bago et al. 2003; Helber et al. 2011; Jiang et al. 2017; Luginbuehl et al. 2017; Feng et al. 2020b). In return, AM fungi can help their host plants grow vigorously by increasing nutrient levels (especially P), optimizing water balance, enhancing photosynthesis, and increasing reactive oxygen species (ROS) scavenging activity in host plants (Li et al. 2019a; Zhang et al. 2019; Chu et al. 2020; Feng et al. 2020b). Some isotopic labeling experiments have revealed that the contribution of AM fungi to P and N uptake by their host plants, namely the mycorrhizal pathway of P and N uptake, was predominant in plants (Smith et al. 2004; Tanaka and Yano 2005; Chu et al. 2020). Although AM fungi are ubiquitous in terrestrial ecosystems, to date, only 334 AM fungal species have been validly described and belong to 36 genera within 12 families of 4 orders according to Schüßler's website (http://www.amf-phylogeny.com/; last updated 17 February 2020).

EcM fungi can establish symbioses with approximately 6,000 plant species in 145 genera of 26 families (approximately 5,600 angiosperms and 285 gymnosperms), most of which are trees or shrubs (Brundrett 2009; van der Heijden et al. 2015). Once EM fungal hyphae attach to the surface of the root, the hyphae proliferate and differentiate to form a fungal mantle around the root surface. After that, a typical hyphal network of EcM fungi, namely the Hartig net, forms around epidermal and/or cortical root cells (Martin et al. 2016). The Hartig net is considered the main functional structure of the symbiosis due to its abundant hyphal branching and large surface area, which is necessary for the bidirectional transport of nutrients between these two symbiont partners (Peterson and Massicotte 2004; Martin et al. 2016). In contrast to AM fungi, most EcM fungi are saprotrophic as they can be grown on artificial agar media without host plants (van der Heijden et al. 2015). However, in terms of physiological and ecological functions of AM and EcM symbioses, they play a similar role in their host plants (van der Heijden et al. 2015). It is noteworthy that 20,000–25,000 EcM fungal species, classified in basidiomycetes and ascomycetes, are estimated at the global scale (Brundrett 2009; van der Heijden et al. 2015). Furthermore, a recent study revealed that ectomycorrhizal plant species were more nutrient use-conservative than arbuscular mycorrhizal plant species, an effect that was robust to controlling for plant growth form and evolutionary history (Averill et al. 2019).

1.2.3 PGPRs

Plant growth-promoting rhizobacteria (PGPRs) are a group of soil bacteria that can colonize the rhizosphere of plants and are characterized by promoting plant growth. In agricultural ecosystems, the PGPRs that have been widely concerned mainly include phosphate-solubilizing bacteria (PSB), potassium-solubilizing bacteria (KSB), siderophore-producing bacteria, indole acetic acid (IAA)-producing bacteria, and 1-aminocyclopropane-1-carboxylate (ACC) deaminase-producing bacteria. Interestingly, many PGPRs possess more than one of these functional characteristics, enabling a strong promotion of plant growth.

The accumulated P in agricultural soils is far sufficient to sustain crop yields worldwide (Zhu et al. 2018). However, these P sources, mainly in the form of insoluble organic/inorganic phosphate, need to be further activated by PSB in soil before plants take them up. PSB have a large phylogenetic composition. By far, more than 40 genera have been reported as PSB, such as *Acinetobacter, Acromobacter, Aerobacter, Agrobacterium, Alicyclobacillus, Arthrobacter, Azotobacter, Bacillus, Beijerinckia, Bradyrhizobium, Burkholderia, Clostridium, Enterbacter, Enterococcus, Erwinia, Flavobacterium, Kitasatospora, Klebsiella, Micrococcus, Microbacterium, Paenibacillus, Pantoea, Paraburkholderia, Pseudomonas, Rahnella, Rhizobium, Serratia, Stenotrophomonas,* and *Streptomyces* (Mander et al. 2012; Ahemad and Kibret 2014; Shameer and Prasad 2018). Insoluble potassium (K) sources (e.g., biotite, feldspar, illite, muscovite, orthoclase, and mica) in soils also need to be activated by KSB and further used by plants (Etesami et al. 2017). However, different from PSB, fewer KSB species and genera have been reported, such as *Achromobacter, Acidothiobacillus, Agrobacterium, Aminobacter, Arthrobacter, Bacillus, Burkholderia, Enterobacter, Klebsiella, Kosakonia, Leclercia, Microbacterium, Myroides, Paenibacillus, Pantoea, Pseudomonas, Rahnella, Ralstonia, Raoultella,* and *Sphingomonas* (Zhang and Kong 2014; Etesami et al. 2017; Khanghahi et al. 2018; Sun et al. 2020). In soils with low available iron, the absorption of iron by plants often requires the participation of siderophore-producing bacteria. So far, some siderophore-producing bacteria have been identified in a wide range of genera, including *Acinetobacter, Azotobacter, Bacillus, Bradyrhizobium, Enterobacter, Klebsiella, Kluyvera, Mesorhizobium, Paenibacillus, Proteus, Pseudomonas, Ralstonia, Rhizobium,* and *Stenotrophomonas* (Ahemad and Kibret 2014; Shameer and Prasad 2018). These processes play a significant role in reducing application of chemical fertilizers, protecting soil structure and ecological environment.

In addition to above-mentioned PGPRs functioning in nutritional mechanisms, other PGPRs, such as IAA-producing bacteria and ACC deaminase-producing bacteria, promote plant growth by directly regulating phytohormones (IAA and ethylene) in plants. Similarly, these types of PGPRs also have been found mainly in above-mentioned genera (Ahemad and Kibret 2014; Shameer and Prasad 2018).

1.2.4 ENDOPHYTES

Endophytes are an endosymbiotic group of bacteria and fungi that colonize the internal tissues of plants, including roots, stems, and leaves. However, most studies have focused on endophytes residing within root, such as dark septate endophytes (DSE).

DSE are characterized by intense dark pigmentation, the formation of septate hyphae and occasionally microsclerotia (Dos Santos et al. 2017). They are sterile or conidial Ascomycetes and are common in roots (e.g., rice and wheat) in agricultural ecosystems. DSE have been reported in nearly 600 plant species representing about 320 genera of 114 families (Jumpponen and Trappe 1998). By far, an increasing number of DSEs with different functions have been reported in diverse genera, including *Acephala, Alternaria, Ascochyta, Aspergillus, Aquilomyces, Cadophora, Cladosporium, Cochliobolus, Coniothyrium, Curvularia, Cryptosporiopsis, Darksidea, Drechslera, Embellisia, Exophiala, Flavomyces, Harpophora, Leptodontidium, Microdiplodia, Microdochium, Nigrospora, Ophiosphaerella, Paraphoma, Periconia, Phaeomollisia, Phialocephala, Phialophora, Phoma,* and *Setosphaeria* (Knapp et al. 2015; Berthelot et al. 2016; Bonfim et al. 2016; Vergara et al. 2018; Li et al. 2019b). A tremendous number of studies have revealed that DSE have an overall positive effect on their host plant performance by enhancing nutritional status and improving the tolerance of their host plants to various abiotic stresses, especially heavy metal stress (Newsham 2011; Berthelot et al. 2016; He et al. 2017; Zhu et al. 2018). Nonetheless, compared with mycorrhizal fungi, studies on DSE are still in its infancy. Overall, DSE show potential for application as biofertilizers to maintain sustainable agricultural ecosystems.

1.3   ECOLOGY AND ENVIRONMENTAL ADAPTATION

Environmental adaptation of agricultural microbes is the precondition for their functioning because edaphic factors strongly shape the microbial community structure and functionality (Zhou et al. 2019). Moreover, the isolates with high abundance can be more effective than the isolates with low abundance. In this circumstance, understanding how agricultural microbes interact with their environments is critical for their application in field condition. In most cases, the promoting effects of the abundant isolates in their native environments can be expected.

1.3.1   NITROGEN-FIXING MICROBES

Globally, nodulated legumes have been found in all ecosystems except open seas and are arguably more significant at high than low latitudes (Sprent et al. 2017). There are also longitudinal differences in their occurrence, especially at low latitudes (Sprent et al. 2017). However, it is obviously deficient and one-sided to study the biogeographic distribution and diversity of N-fixing microbes based only on leguminous plants, nodule morphology, and culturable strains. BNF is catalyzed by the nitrogenase whose multiple subunits are encoded by three genes *nifH, nifD,* and *nifK* (Zehr et al. 2003; Gaby and Buckley 2012). The gene *nifH*, which encodes nitrogenase reductase, is the most widely sequenced marker gene for studying the phylogeny, diversity, and abundance of N-fixing microbes (Gaby and Buckley 2012).

In recent years, with the rapid development of next-generation sequencing technologies, increasing studies based on the marker gene *nifH* have showed a considerable geographical distribution and diversity of diazotrophs in various terrestrial ecosystems, including farmland (Kumar et al. 2017; Ouyang et al. 2018; Han et al. 2019; Vadakattu et al. 2019), forest (Tu et al. 2016a; Lee et al. 2019; Meng et al. 2019), grassland (Tu et al. 2016b), mangrove (Zhang et al. 2017), desert (Köberl et al. 2016), badland (Dahal et al. 2017), and permafrost (Penton et al. 2016). Tu et al. (2016a), revealed that the diazotrophic communities across six North American forests follow traditional biogeographic patterns similar to plant and animal communities, including the taxa-area relationship and latitudinal diversity gradient; specifically, significantly higher diazotrophic community diversities and lower microbial spatial turnover rates were found for rainforests than temperate forests. A recent study along the longitude within China by Han et al. (2019) found that diazotrophs were mainly dominated by *Bradyrhizobium* (29.15%), *Azospirillum* (9.95%), *Myxobacter* (6.95%), *Desulfovibrio* (5.22%), and *Methylobacterium* (4.42%) in three farmland soil types (black soil, fluvo-aquicsoil, and red soil); moreover, *Bradyrhizobium* was the core group in all soils. Additionally, Meng et al. (2019) showed that the three most abundant N-fixing genera were *Bradyrhizobium* (45.96%), *Desulfovibrio* (25.66%), and *Methylobacterium* (12.77%), respectively, in five acidic forest soils. Similarly, these genera, *Methylobacterium*, *Bradyrhizobium*, and *Rhizobium*, are also predominant in an Alaskan boreal soil along a permafrost thaw gradient and a desert region of Egypt, suggesting a strong ecological adaptability of diazotrophs in extreme environments (Köberl et al. 2016; Penton et al. 2016).

In terms of adaptability, a global meta-analysis revealed that N addition inhibited terrestrial BNF by 19.0%, micronutrient addition stimulated terrestrial BNF by 30.4%, and P addition had an inconsistent effect on terrestrial BNF, that is, inhibiting free-living N fixation by 7.5% and stimulating symbiotic N fixation by 85.5% (Zheng et al. 2019a). Furthermore, Tu et al. (2016a) indicated that the diazotrophic community diversity was strongly correlated with latitude, annual mean temperature, plant species richness and precipitation, and weakly correlated with pH and soil moisture. Compared with those in black soil and fluvo-aquic soil, the most abundant diazotrophs were found in red soil; a network analysis further revealed that the interaction relationships between N-fixing genera levels in red soil were more complex compared to those two soils (Han et al. 2019). A meta-analysis comprising 47 field studies in agricultural ecosystems revealed that N fertilization had no effect on the abundance of *nifH* (Ouyang et al. 2018). However, N fertilizer form (inorganic versus organic) and soil pH strongly affected the response of *nifH* abundance to N fertilization (Ouyang et al. 2018). Soil pH was considered the most important soil property driving the community composition of diazotrophs (Han et al. 2019). Tu et al. (2016a) found that the diazotrophic richness was significantly positively correlated with soil ammonium (NH_4^+) and nitrate (NO_3^-), but significantly negatively correlated with total soil carbon. However, these contrary results were found in the study by Han et al. (2019), indicating complicated effects of soil properties on diazotrophic community and diversity in different ecosystems.

1.3.2 Mycorrhizal Fungi

The rapid development of DNA sequencing technology has been breaking some cognitive limitations in AM fungal species richness caused by unculturable AM fungi and showed a faster and deeper understanding about ecological distribution and species diversity of AM fungi. A global scale assessment of AM fungal diversity in natural ecosystems has revealed that 93% of 236 AM fungal taxa were present on at least two continents and 34% were present on all six continents, providing a robust evidence of their very low endemism (Davison et al. 2015). The species richness of AM fungi on these six continents, including Europe, Africa, South America, North America, Oceania, and Asia, is decreasing in order (Davison et al. 2015). Generally, the species richness of AM fungi in forest, grassland, and cropland decrease in order and the declining AM fungal diversity follow land use intensity (Xiang et al. 2014; Bowles et al. 2016; Oehl et al. 2017; Rodríguez-Echeverría et al. 2017). On the other hand, there are still some clear differences in community composition of AM fungi among major continents and ecosystems. A more similar community composition of AM fungi is usually present in these continents formerly in Laurasia (Asia, Europe, and North America) or Gondwanaland (Africa, Australia, and South America) (Kivlin et al. 2011). Furthermore, although the community composition of AM fungi in the four ecosystems (agriculture areas, grasslands, forests, and wetlands) was significantly different, the community composition in agriculture areas and grasslands is more similar (Kivlin et al. 2011). There are many key drivers that affect the community composition of AM fungi, but the decisive drivers usually depend on the study scale. In general, geology and climate type shape AM fungal communities at the global scale, whereas soil type, soil pH, and ecosystem type are key drivers at regional scales (Rodríguez-Echeverría et al. 2017). Van Geel et al. (2018) indicated that abiotic conditions, rather than biotic filtering through host plants specificity, were the most important drivers in shaping AM fungal communities in European seminatural grasslands. These studies revealed that there are extremely complex regulatory mechanisms on AM fungal communities in various ecosystems. Overall, AM fungi exhibit a very wide geographical distribution, suggesting a strong ecological adaptability of AM fungi and a further help for their host plants to adapt extreme environmental conditions.

In natural ecosystems, EcM fungi have a surprising number of species richness, most of which belong to Basidiomycota, whereas the rest belong to Ascomycota (Brundrett 2009; Matsuoka et al. 2016; Boeraeve et al. 2018; van der Linde et al. 2018). However, in contrast to AM fungi, EcM fungi have a significantly smaller proportion of host plant species (approximately 2%), which are mainly distributed in temperate, boreal, and some subtropical forests (Smith and Read 2008; van der Heijden et al. 2015). This trait indicates a limited biogeographical distribution of EcM fungi at least as symbiotic fungi. However, the distribution of host plants is not the only decisive factor driving community composition of EcM fungi. Intriguingly, Wen et al. (2015) indicated that EcM fungal communities were determined more by geographical location than host monophyly on this spatial (approximately 770–1,600 km) and geological time scale (approximately 20–25 My). A large-scale research in 20 European countries across strong environmental gradients revealed that host factor, soil factor, geographical distance, and climatic variables explained 23%, 21%, 14%, and 12% of the variance in community composition of EcM fungi, respectively (van der Linde et al. 2018). Similarly, Matsuoka et al. (2016) found that the community composition of EcM fungi along an elevation gradient in northern Japan explained by the host community, environmental, and spatial fractions were 32.6%, 26.0%, and 30.6%, respectively. Bai et al. (2020) revealed that EcM fungal communities in a temperature broad-leaved mixed forest were more abundant in the nutrient-poor and low-moisture environments than in the nutrient-rich and high-moisture environments. In addition to these environmental factors, forest age can also affect EcM fungal community composition by an interaction with host plant and soil variables (Boeraeve et al. 2018).

Remarkably, based on observing arbuscules or coils for AM status and Hartig net or similar structures for EcM status within the same plant species, 238 plant species of 89 genera within 32 families have been confirmed to be dual-mycorrhizal plants (Teste et al. 2020). Additionally, they are widespread globally, generally covering most areas where EcM plants occur (Steidinger et al. 2019; Teste et al. 2020).

1.3.3   PGPRs

PGPRs, as the name suggests, exist in the rhizosphere of almost all plants. Considering that the global distribution of plants spans various harsh environments, it is obvious that PGPRs have an extremely wide ecological distribution and strong environmental adaptation.

So far, PGPRs have been investigated in various ecosystems, including farm-land, forest, grassland, wetland, and desert. In grassland soils, Mander et al. (2012) found a very rich PSB community with 39 genera, spanning 24 families in 4 phyla (Actinobacteria, Bacilli, Bacteroidetes, and Proteobacteria); these PSB isolates were predominantly represented by five key families: Pseudomonadaceae (32%), Micrococcaceae (13%), Enterobacteriaceae (9%), Sphingobacteriaceae (8%), and Microbacteriaceae (5%) (Mander et al. 2012). Different from traditional strain iso-lation and identification methods, these two marker genes, *phoD* for solubilizing organic P and *pqqC* for solubilizing inorganic P, are considered high-efficiency molecular tools for studying PSB communities (Fraser et al. 2015; Ragot et al. 2017; Long et al. 2018). Long et al. (2018) indicated that these PSB affiliated to Rhizobiales and Actinomycetales played a dominant role in paddy soil, and the copy number of the *phoD* gene was significantly higher than that of *pqqC* gene; further-more, these *phoD*-harboring bacteria were affiliated to two phyla: Proteobacteria and Actinobacteria, whereas *pqqC*-harboring bacteria were divided into three phyla: Proteobacteria, Actinobacteria, and Verrucomicrobia. A large-scale investigation across three types of land uses (arable, forest, and grassland) located in arid-to-temperate climates on two continents by Ragot et al. (2017) revealed that the *phoD* gene presented in 13 phyla, of which Actinobacteria, Cyanobacteria, Deinococcus-Thermus, Firmicutes, Gemmatimonadetes, Planctomycetes, and Proteobacteria were common and dominant *phoD*-harboring phyla in all environments. In addition, Ragot et al. (2017) found that climate, soil group, land use, and soil nutrient contents were the common environmental drivers of PSB community. Apart from this, other studies also revealed that soil pH and P availability content were dominant factors affecting PSB community. Specifically, there was a prominent inhibiting effect of low pH rang-ing pH 4–8 and high-P level on the abundance of PSB, and low P input promoted the growth of PSB of *Alicyclobacillus*, *Bacillus*, and *Clostridium* (Mander et al. 2012; Long et al. 2018; Zheng et al. 2019b).

However, there are fewer and lagging research on other PGPRs. Zhang and Kong (2014) isolated 27 KSB strains from tobacco fields, which belong to *Klebsiella* (17 strains), *Enterobacter* (5 strains), *Agrobacterium*, *Burkholderia*, *Microbacterium*, *Myroides*, and *Pantoea*, respectively. Sun et al. (2020) isolated 18 efficient KSB strains from rhizospheric soil of *Mikania micrantha* in abandoned orchards and found that the *Burkholderia* genus had the highest solubilizing ability. Remarkably, Bouffaud et al. (2018) developed a direct molecular tool to quantify the ACC deam-inase gene (*acdS*) in soil microbe communities and found that Proteobacteria and

Actinobacteria from the rhizosphere of maize played a dominant role in cropped and meadow soils. Furthermore, the abundance of ACC deaminase-producing bacteria in soils is related to the identity and genotype of their host plants (Bouffaud et al. 2018).

1.3.4 ENDOPHYTES

DSE are widely distributed worldwide as root-colonizing fungi and have been reported in various ecosystems, such as rain forest, semiarid area, extreme arid area, metal mine tailing area, mineral soil, mire, and a mostly unvegetated and high-elevation landscape (Terhonen et al. 2014; Knapp et al. 2015; Lugo et al. 2015; Xu et al. 2015a; Bonfim et al. 2016; Xie et al. 2017; de Mesquita et al. 2018). Terhonen et al. (2014) isolated 87 DSE strains in spruce-dominated pristine mires, most of which consisted of *Acephala* and *Phialocephala* of Helotiales. Xu et al. (2015a) found that DSE strains isolated from metal mine tailing areas belonged to the genera *Cladosporium, Exophiala, Leptodontidium,* and *Phialophora.* Bonfim et al. (2016) isolated 251 DSE isolates from 27 species in the Atlantic rain forest along an altitudinal gradient and found that the most abundant genera were *Alternaria, Ascochyta, Cladosporium, Coniothyrium, Nigrospora, Microdiplodia,* and *Phoma.* Furthermore, de Mesquita et al. (2018) revealed that DSE colonization was the highest in areas with less snowpack and higher inorganic nitrogen levels, suggesting an important role for these fungi in mineralizing organic nitrogen. Overall, the host plant species is the most important driving factor of the community of DSE, whereas altitude, soil parameters, or season are of minor importance (Bonfim et al. 2016; de Mesquita et al. 2018). More notably, the presence of microsclerotia in roots of *Hedysarum scoparium* was positively correlated with soil ammonium, and negatively correlated with soil organic matter and pH (Xie et al. 2017).

1.4 ACTION MODES: PHYSIOLOGICAL, MOLECULAR, AND GENOMIC ASPECTS

Due to the high diversity of these agricultural microbes in both taxonomy and function, the action modes for their functioning can vary depending on taxa and environments. Meanwhile, the physiological, molecular, and genomic mechanisms underlying their functioning have been uncovered for better application.

1.4.1 NITROGEN-FIXING MICROBES

Nitrogen fixation is the most prominent feature of N-fixing microbes and also the core mechanism for them to promote growth of host plants. It is catalyzed by one of the nitrogenase metalloenzymes, including iron-iron (FeFe), vanadium-iron (VFe), and molybdenum-iron (MoFe) nitrogenases, which are cooperatively regulated by the electron transfer protein (known as nitrogenase reductase or iron protein and encoded by *nifH*) and the α-subunit (encoded by *nifD*) and β-subunit (encoded by *nifK*) of the catalytic enzyme in the root and/or stem nodules (Eady 1996; Kuypers et al. 2018). BNF can effectively offset N limitation, especially in N-depleted soil, resulting in a release of plant growth potential. For instance, inoculation with

Klebsiella variicola strain DX120E significantly increased the dry weight, N, P, and K contents of sugarcane plants at 52%, 80%, 69%, and 67%, respectively (Wei et al. 2014). A field experiments using ^{15}N natural abundance or ^{15}N-enrichment assessments over five years indicated that atmospheric N fixation contributed 29–82% of the N nutrition of maize (van Deynze et al. 2018).

In addition to N contribution, they are also capable of regulating physiological, biochemical, cytological, and molecular processes to help their host plants grow better under a series of abiotic stress conditions, such as salt (Qu et al. 2016; Wang et al. 2016; Stritzler et al. 2018), drought (Staudinger et al. 2016; Mohammadi et al. 2019), heavy metal (Guo and Chi 2014; Rizvi and Khan 2018), acidity, and aluminum stress (Ferreira et al. 2012; Mendoza-Soto et al. 2015). These stresses usually cause ROS accumulation and membrane lipid peroxidation, further resulting in an imbalance homeostasis and decreased membrane integrity in plant cells. However, effective N-fixing symbiont can mitigate these symptoms by improving photosynthetic efficiency, N-fixing ability, the activity of antioxidant enzymes as well as osmotic adjustment capacity.

For instance, Wang et al. (2016) found that active nodules significantly increased survival rate of alfalfa (*Medicago sativa*) under salt stress; higher survival rate was associated with reduced lipid peroxidation (malondialdehyde, i.e., MDA), higher activities of peroxidase (POD), superoxide dismutase (SOD), catalase (CAT), and ascorbate peroxidase (APX) as well as higher concentrations of soluble sugar and reduced glutathione (GSH), especially in roots under salt stress. These results were supported by Qu et al. (2016), beyond that, they revealed some responses to salt stress at molecular level in soybean seedling (*Glycine max*). Several key genes involved in flavonoid biosynthesis (cytochromeP450 monooxygenase, chalcone synthase, and chalcone isomerase) and other salt-responsive genes (stress-induced protein SAM22, PR10-like protein, and phosphatidylinositol-specific phospholipase C) were significantly up-regulated with inoculation of *Sinorhizobium meliloti* 1021 under salt stress (Qu et al. 2016); overall, the regulation of this rhizobium in flavonoids metabolism was considered the key process for soybean's ability to adapt to saline soil (Qu et al. 2016). Independent of rhizobial strain and uncoupled from initial leaf N content, Staudinger et al. (2016) observed a significant delay in drought-induced leaf senescence in nodulated relative to non-nodulated plants. Additionally, a comprehensive omics (i.e., ionomic, metabolomic, and proteomic) data suggested that phytohormone interactions (ethylene and jasmonic acid) and enhanced translational regulation played a role in increased leaf maintenance in nodulated plants during drought; furthermore, increasing K^+ and decreasing Na^+ in nodulated plants are universal strategies for plants response to salt and drought stresses (Qu et al. 2016; Staudinger et al. 2016). Restrictions on the transfer and accumulation of heavy metals to plants is one of the core mechanisms of N-fixing bacteria to mitigate heavy metal stress in plants (Guo and Chi 2014; Hassan et al. 2016; Rizvi and Khan 2018). Guo and Chi (2014) revealed that *Bradyrhizobium* sp. strain YL-6 decreased cadmium (Cd) accumulation in *G. max* by increasing Fe availability. Moreover, *Azotobacter chroococcum* strain CAZ3 showed a metal-chelating ability and the roots of inoculated plants accumulated greatest amounts of metals compared to other organs (shoots and kernels) (Rizvi and Khan 2018). More intriguingly, several Al toxicity-responsive miRNAs and their target genes were reported to be relevant for symbiotic N-fixing common bean response/adaptation to Al stress, including miR164/NAC1 (NAM/ATAF/CUC transcription factor) and miR393/TIR1 (transport inhibitor response 1-like protein) in auxin, and miR170/SCL (SCARECROW-like protein transcription factor) in gibberellin signaling (Mendoza-Soto et al. 2015).

In general, the adaptability of N-fixing microbes is determined by the original environmental traits of the isolated strains. A *Rhizobium leguminosarum* isolate was recovered from the root nodule of the faba bean (*Vicia faba*) grown in sludge-contaminated fields and showed resistance to heavy metal stress in decreasing order Ba^{2+} (80 mg/kg), Zn^{2+} (70 mg/kg), Co^{2+} (50 mg/kg), Al^{3+} (40 mg/kg), Ni^{2+} (30 mg/kg), and Cd^{2+} (10 mg/kg) (Abd-Alla et al. 2012). Several *Rhizobium* strains isolated from Amazonian soils were highly tolerant to acidity and aluminum toxicity mainly due to exopolysaccharides production and a resistant cell outer membrane (Ferreira et al. 2012). Lipa and Janczarek (2020) suggested that phosphorylation systems played a considerable role in the physiology of symbiotic N-fixing microbe cells and adaptation to various environmental conditions.

1.4.2 MYCORRHIZAL FUNGI

P-acquiring is the core mechanism of AM fungi to promote growth of their host plants, especially in P-limited soils (Table 1.1). AM fungal extraradical hyphae (EH) continuously extend and then form a huge hypha network in the soil, which can efficiently absorb inorganic phosphate (Pi) via high-affinity fungal phosphate transporters such as GintPT and GigmPT from P-depleted soils (Fiorilli et al. 2013; Xie et al. 2016). Subsequently, Pi transported into the cytosol is incorporated into ATP in mitochondria and then is accumulated in the vacuoles as polyphosphate (polyPi), which is a high-efficiency form for translocating rapidly to arbuscules along EH and intraradical hyphae (IH) (Ezawa et al. 2003; Kikuchi et al. 2014). After that, polyPi is hydrolyzed into Pi by AM fungal phosphatase and exopolyphosphatase. Pi is exported from arbuscules to the periarbuscular space via fungal transporters (Xie et al. 2016), whereafter, it is transported across the periarbuscular membrane via mycorrhiza-specific and high-affinity phosphate transporters of their host plant and finally enters root cortical cells to meet plant P-requirement (Loth-Pereda et al. 2011; Yang et al. 2012). AM fungi made major contributions to P uptake of close to 100% to *Solanum lycopersicum* and *Linum usitatissimum*, and somewhat lower contributions to *M. truncatula* (60–80%) (Smith et al. 2004). According to the latest study, the maximum contribution of *Rhizophagus intraradices* to P uptake of *Zea mays* was up to 60% (Chu et al. 2020). These studies suggest that AM fungi play a vital important role in improving P nutrient of their host plants.

Additionally, numerous studies have proven the roles of AM fungi in the acquisition of N, K, S, other metal elements and water (Table 1.1). Some key transporters and their functional genes involved in these processes have been identified, including nitrate transporters (Tian et al. 2010), ammonium transporters (Pérez-Tienda et al. 2011), K uptake transporters and shaker-like K channel (Garcia and Zimmermann 2014; Garcia et al. 2016), and aquaporins (Li et al. 2013). In addition, physiological-biochemical changes induced by AM fungi are one of the most notable plant

TABLE 1.1
The Action Modes of Mycorrhizal Fungal Functionality

Function Category	AM Fungi	EcM Fungi	References
Promotion of N uptake	Nitrate transporter: GiNT of *Rhizophagus irregularis* (namely *Glomus intraradices*); Ammonium transporters: GintAMT1-3 of *R. irregularis*; Amino acid permease: GmosAAP1 of *Funneliformis mosseae* (namely *Glomus mosseae*); Dipeptide transporter: RiPTR2 of *R. irregularis*	Nitrate transporter: LbNrts of *Laccaria bicolor*; Peptide transporters: HcPTR2A and HcPTR2B of *Hebeloma cylindrosporum*; Utilization of nitrate and organic nitrogen sources: fHANT-ACs of *L. bicolor*; Extracellular proteases and surface-reactive metabolites of *Paxillus involutus*	López-Pedrosa et al. 2006; Cappellazzo et al. 2008; Kemppainen et al. 2010; Tian et al. 2010; Pérez-Tienda et al. 2011; Belmondo et al. 2014; Calabrese et al. 2016; Müller et al. 2020; Wang et al. 2020
Promotion of P uptake	Phosphate transporters: GvPT of *Glomus versiforme*; GiPT, GintPT and RiPT5 of *R. irregularis*; GmosPT of *F. mosseae*; GigmPT of *Gigaspora margarita*	Phosphate transporters: HcPT1 and HcPT2 of *H. cylindrosporum*; BePT of *Boletus edulis*; RlPT of *Rhizopogon luteolus*; LbPT of *Leucocortinarius bulbiger*	Harrison and van Buuren 1995; Maldonado-Mendoza et al. 2001; Benedetto et al. 2005; van Aarle et al. 2007; Tatry et al. 2009; Fiorilli et al. 2013; Wang et al. 2014; Walder et al. 2016; Xie et al. 2016; Zheng et al. 2016; Becquer et al. 2018
Promotion of K uptake	Potassium transporters: RiHAK of *R. irregularis*	Potassium transporters: HcTrk1, HcTrk2 and HcHAK of *H. cylindrosporum*; K⁺ channels: HcTOKs and HcSKC of *H. cylindrosporum*	Garcia and Zimmermann 2014; Garcia et al. 2014; 2016; Guerrero-Galán et al. 2018
Promotion of other elements uptake	Ferric reductase and Fe permeases: RiFRE1 and RiFTRs of *R. irregularis*; Thirty putative metal transporters involved in copper, iron, and zinc transfer of *R. irregularis*	Zinc transporter: SlZRT1 and SlZRT2 of *Suillus luteus*; SlZnT1 of *S. luteus*; RaZIP1 of *Russula atropurpurea*; Two metallothionein (MT) coding genes: SlMTa and SlMTb of *S. luteus*; Sulfate transporter: TbSul1 of *Tuber borchii*	Zeppa et al. 2010; Tamayo et al. 2014; 2018; Coninx et al. 2017; 2019; Nguyen et al. 2017; Ruytinx et al. 2017; Leonhardt et al. 2018; Xie and Tang 2019

(Continued)

TABLE 1.1 *(Continued)*
The Action Modes of Mycorrhizal Fungal Functionality

Function Category	Action Modes		References
	AM Fungi	EcM Fungi	
Promotion of water uptake	Aquaporins: GintAQPF1 and GintAQPF2 of *R. irregularis*; RcAQP3 of *R. clarus*	Promoted soil aggregation and soil water repellency; Aquaporins: LbAQP1 and JQ585595 of *L. bicolor*	Li et al. 2013; Zheng et al. 2014; Navarro-RóDenas et al. 2015; Xu et al. 2015b; Kikuchi et al. 2016; Xu and Zwiazek 2020
Increased resistance to abiotic stresses (drought, salinity, heavy metals, extreme temperatures, aluminum toxicity, low pH, acid rain, and pollutants)	Optimized water balance, enhanced photosynthesis and related metabolic processes, more active ROS scavenging processes, and lower membrane damage	Improving mineral nutrition and water balances, reducing of metal ion uptake towards photosynthetic organs, diluting heavy metal concentrations; A transporter (Mte1) for abiotic stress and plant metabolite resistance in *Tricholoma vaccinum*; Glutathione biosynthesis	Calonne et al. 2014; Danielsen and Polle 2014; Yang et al. 2014; Zheng et al. 2014; Chen et al. 2015; 2017; Schlunk et al. 2015; Canton et al. 2016; Aguilera et al. 2018; Li et al. 2019a; 2019c; Khullar and Reddy 2018; Mathur et al. 2018; Coninx et al. 2019; Guerrero-Galán et al. 2019; Rask et al. 2019; Tang et al. 2019; Feng et al. 2020a; 2020b; Khullar and Reddy 2020; Kilpeläinen et al. 2020

responses to abiotic stresses, including enhanced photosynthesis and related metabolic processes, more active ROS-scavenging processes, and lower membrane damage (Yang et al. 2014; Chen et al. 2017; Li et al. 2019a).

Similar to AM fungi, EcM fungi can also perform various physiological and ecological functions (Table 1.1). Recently, Wang et al. (2020) found that EcM fungi had an ability to acquire N from the mineral-associated proteins by simultaneously generating extracellular proteases and surface-reactive metabolites, meaning a large N pool of mineral-associated proteins to trees in forests. Additionally, the molecular pathways involved in P and K uptake from the soil to host plant roots have been identified, including *BePT* and *HcTOK1* (Wang et al. 2014; Guerrero-Galán et al. 2018). Remarkably, EcM fungi enhance host plant photosynthesis through N and water economy, not only through increased belowground carbon demand (Heinonsalo et al. 2015). Not only that, EcM fungi in association with *Pinus sylvestris* seedlings promoted soil aggregation and soil water repellency (Zheng et al. 2014). In addition to the direct uptake of nutrients and water, EcM fungi can directly promote the growth of host plants by improving their tolerance to abiotic stress. For instance, EcM fungi enhanced a tolerance of their host plants to salt stress by reducing sodium (Na^+) uptake towards photosynthetic organs and improving mineral nutrition and water balances (Guerrero-Galán et al. 2019). Other studies showed that EcM fungi could enhance the tolerance of their host plants to acid rain (Li et al. 2019c), drought (Danielsen and Polle 2014; Kilpeläinen et al. 2020), pollutant (Khullar and Reddy 2020), heavy metal stresses (Chen et al. 2015; Canton et al. 2016) to varying degrees. The inoculation of EcM fungi *Pisolithus tinctorius* could alleviate the deleterious effects of acid rain (pH 3.5) on nitrogen cycling microbes in Masson pine forest soils (Li et al. 2019c). A key gene *mte1* of *Tricholoma vaccinum* was found to be induced during symbiotic interaction and mediated detoxification of xenobiotics and metal ions such as Al, Cu, Li, and Ni, as well as secondary plant metabolites (Schlunk et al. 2015). Intriguingly, Tang et al. (2019) revealed that EcM fungi did not act as a barrier inhibiting the absorption of heavy metals by host plants, but rather promoted this absorption; improving the plant's nutritional status and promoting growth, diluting heavy metal concentrations, thereby reducing the toxic effects of heavy metals on host plants. Importantly, glutathione biosynthesis in EcM fungi is considered the key molecular mechanism responsible for their tolerance to heavy metal and pollutant stresses (Khullar and Reddy 2020).

1.4.3 PGPRs

PGPRs are a group of rhizobacteria that are completely classified according to their function, that is, promotion in plant growth driven by a variety of mechanisms. P is an essential component of cell membrane and nucleic acid and is of great significance for plant growth and development. However, 95–99% of P exists in insoluble, immobilized, or precipitated forms; therefore, it is difficult for plants to utilize it (Gouda et al. 2018). On one hand, these insoluble inorganic phosphates in soils, such as tricalcium phosphate, aluminum phosphate, iron phosphate, can be converted to orthophosphate by PSB via secreting low molecular weight organic acids (e.g., gluconic acid, oxalic acid, and citric acid) and released H^+ (Long et al. 2018). On the other hand, PSB can secrete phosphomonoesterases, such as acid and alkaline phosphatases, phytases, and nucleotidases, which are capable of catalyzing mineralization of insoluble organic phosphate (sugar phosphates, phytate, and nucleotides) in soils (Fraser et al. 2015). After successful colonization in the rhizosphere, PSB can significantly regulate soil P status and promote their host plant growth, especially in low-P soils. For instance, inoculation with two PSB strains significantly increased soil available P by 35–38% and plant biomass by 22–26% for maize, moreover, increased soil available P by 38–41% and plant biomass by 18–24% for wheat (Kaur and Reddy 2015). Analogous to the study by García-López et al. (2016), they found that *Bacillus subtilis* QST713 increased P uptake by 107% and dry matter by 73% for cucumber.

More than 90% of potassium (K) exists in the form of insoluble rock and silicate minerals (Gouda et al. 2018). Generally, it is believed that KSB can dissolve silicate minerals and release K through the production of inorganic and organic acids (gluconic acid, oxalic acid, and citric acid), acidolysis, polysaccharides, complexolysis, chelation, and exchange reactions (Etesami et al. 2017). Zhang and Kong (2014) showed that four KSB isolates enhanced tobacco growth by 31–69% and K uptake by 8–20%. A *Bacillus pseudomycoides* strain O-5 increased soil K availability content by 47.0 mg kg^{-1} after 105 days incubation, while the strain released 104.9 mg K kg^{-1} in muscovite type mineral-treated soil (Pramanik et al. 2019).

Furthermore, plants can assimilate iron deriving from iron-limiting environments by some siderophore-producing bacteria, which can secrete low-molecular mass iron chelators called siderophores in response to low Fe levels in rhizosphere (Ahmed and Holmström 2014; Etesami and Maheshwari 2018). Meanwhile, siderophores also form stable complexes with other heavy metals, resulting in alleviating the heavy metal stress of plants (Ahemad and Kibret 2014).

In addition to these nutritional mechanisms, some PGPRs have an ability to synthesize and secrete IAA and ACC deaminase. For instance, IAA-producing bacteria can change root system architecture of their host plants and stimulate the formation of lateral roots and root hair, suggesting more active sites to colonization of other PGPRs and a more high-efficiency uptake of nutrients and water (Vacheron et al. 2013; Etesami et al. 2015). When plants are exposed to various abiotic stress conditions, endogenous ethylene level of plants is induced dramatically (Ahemad and Kibret 2014; Gouda et al. 2018). Ethylene at high level, however, is able to induce defoliation and cell senescence that may lead to the inhibition in plant growth. ACC deaminase is capable of hydrolyzing ACC, which is the biosynthetic precursor for ethylene in plants, into ammonia and α-ketobutyrate, resulting in reduced ethylene levels and enhanced plant growth (Ullah et al. 2015; Parray et al. 2016). For example, an ACC deaminase-producing PGPR strain, *Enterobacter* sp. UPMR18, inoculated okra plants exhibited higher germination percentage (increased by 60%), growth parameters (increased by 84%), and chlorophyll content (increased by approximately 60%) than non-inoculated plants (Habib et al. 2016). In addition, increased antioxidant enzyme activities (APX, CAT, and SOD) and up-regulation of ROS pathway genes (APX, CAT, DHAR, and GR) were observed in this strain-inoculated okra plants under salinity stress (Habib et al. 2016).

Although above-mentioned PGPRs have different direct mechanisms of plant growth promotion, they have similar effects on their host plants under abiotic stress conditions, including alteration in root morphology, decrease in absorption of excess heavy metals, accumulation of osmolytes, induction of plant antioxidative enzymes, and up-regulation of abiotic stress resistance genes (Vacheron et al. 2013; Hassan et al. 2016; Vurukonda et al. 2016; Mukhtar et al. 2020).

1.4.4 ENDOPHYTES

Overall, DSE show a positive effect on plant performance, especially under various abiotic stress conditions. A meta-analysis revealed that inoculation with DSE increased the biomass and nutrient content of their host plants by 26–103%; this positive effect would be more prominent when all or the majority of N was available in organic form, suggesting a preference for organic N utilization in DSE (Newsham 2011). Not only that, several DSE strains have different potential to solubilize inorganic P (tricalcium phosphate, aluminum phosphate, and iron phosphate) as well as organic P (Della Mónica et al. 2015; Spagnoletti et al. 2017). Additionally, producing auxin and releasing volatile organic compounds (e.g., 2-methyl-1-butanol, 3-methyl-1-butanol, ethyl acetate, and styrene) by DSE are other mechanisms of stimulating plant growth (Berthelot et al. 2016). Li et al. (2019b) suggested that improved root biomass and length were the key mechanisms of DSE for promoting growth of host and non-host plants under drought stress. Under Cd stress, DSE promoted plant growth achieved by a decrease in Cd content, the regulation on phytohormone balance, and photosynthetic activities in maize leaves (He et al. 2017). Shen et al. (2015) found that glutathione S-transferase was involved in heavy metal detoxification for DSE. A comprehensive RNA-seq analysis revealed an all-sided molecular response of DSE to excessive heavy metals, including organic acid metabolism and transportation, metal ion binding and transportation, sulfate assimilation, redox homeostasis, ROS scavenging, transcription factors production, DNA repair, and cell wall integrity maintenance (Zhao et al. 2015). These responses benefit the survival of DSE and alleviate the heavy metal stress for plants under heavy metal conditions.

1.5   APPLICATION TECHNIQUES

Application of agricultural microbes in field conditions has been practiced intentionally or unintentionally for more than hundreds of years. Beneficial taxa can be employed by applying the inocula to target environments or by proliferating them in situ with promoting agents. Some novel application techniques have been developed to enable the field application more efficient and user-friendly.

1.5.1   Nitrogen-Fixing Microbes

Co-inoculation of N-fixing microbes and P-related microbes (e.g., AM fungi and PSB) can meet simultaneously the requirements of plants for N and P. Püschel et al. (2017) indicated that inoculation with AM fungi stimulated BNF in two *Medicago* spp. through improved P acquisition. Additionally, the combination of rhizobium and AM fungi has an overall positive effect on plants when exposed to various abiotic stresses (Hao et al. 2019; Ren et al. 2019; Razakatiana et al. 2020). Importantly, the exogenous application of signaling molecules, such as hydrogen sulfide (H_2S) and nitric oxide (NO), could enhance the resistance of legume-rhizobium symbiosis under metal toxicity by regulating the biochemical response of the plant-soil system, thereby minimizing potential health risks (Fang et al. 2020). In addition, H_2S and rhizobia synergistically regulated N assimilation and remobilization during N deficiency-induced senescence in soybean (Zhang et al. 2020). These signaling molecules are industrially available and low-cost, meaning an application prospect in agriculture. Recently, Tewari and Sharma (2020) revealed that exopolysaccharides (EPS) as supplement could promote rhizobial growth and the supplementation of EPS in the form of talc-based bioformulation with rhizobial strains enhanced nodulation and growth attributes of *Cajanus cajan* under multi-stress conditions. As a metabolite produced by rhizobia, EPS can assist in infection process, nodule formation, and growth enhancement of crops (Castellane et al. 2015; Tewari and Sharma

2020). Overall, EPS has potential to be an additive in fertilizers to enhance the ability of biological nitrogen fixation in the future.

In natural ecosystems, these combination, rhizobia/legumes and *Frankia/* actinorhizal plants, are ubiquitous. However, most plants cannot form nodule with N-fixing microbes for biological nitrogen fixation. Remarkably, some scientists have been identified how binding specificities in lysine motif (LysM) receptors can be altered to facilitate Nod factor recognition and signaling from a chitin receptor, advancing the prospects of engineering rhizobial symbiosis into nonlegumes (Bozsoki et al. 2020).

1.5.2 Mycorrhizal Fungi

In recent years, the co-application of AM fungi and some additives, such as short-chain chitin oligomers, gibberellin (GA$_3$), biochar, compost, humic substances, and nano-TiO$_2$, has been proven to enhance AM fungal functions (Khalloufi et al. 2017; Pinos et al. 2019; Ogunkunle et al. 2020; Volpe et al. 2020; Raklami et al. 2021). For instance, exogenous short-chain chitin oligomers derived from crustacean exoskeleton could trigger symbiotic signaling in all host plant species and increase AM colonization in *M. truncatula* (Volpe et al. 2020). In addition, there is an interactive positive effect between AM fungi and exogenous GA$_3$ which alleviates growth impairment under salinity conditions by modifying the hormonal balance of the plant (Khalloufi et al. 2017). Recently, application of nano-TiO$_2$ had a stimulatory effect on the activity of AM fungi in alleviating Cd stress in pre-flowering cowpea (Ogunkunle et al. 2020). Among these additives, short-chain chitin oligomers, biochar, and nano-TiO$_2$ are likely to be promising candidate materials due to their low-cost, high-efficiency, and non-polluting traits.

1.5.3 PGPRs

Some high performance PGPR strains isolated in the laboratory usually cannot colonize successfully or perform their functions fully when applied into fields, mainly due to their inadaptability to the applied environment. To avoid this result, the indigenous PGPR strains should be the first choice to be isolated and applied. For instance, a KSB strain (*Bacillus pseudomycoides* strain O-5) isolated from tea-growing soil enhanced K uptake in tea plants by increasing K availability in soil (Pramanik et al. 2019). Similar to the rhizosphere, the hyphosphere of mycorrhizal fungi is also an ideal niche for recruiting some PGPRs, especially PSB, which can form a stable cooperative relationship with these fungi (Fontaine et al. 2016; Zhang et al. 2016). Fontaine et al. (2016) found that several PSB strains isolated from ectomycorrhizal mycelium (*Wilcoxina* sp.) of *Picea glauca* were highly efficient at fluorapatite weathering. Co-inoculation with PSB and their cooperative partner (AM fungi and EcM fungi) may have a greater potential to promote nutrient uptake of plants.

Co-inoculation with different functional PGPRs can make up for multiple environmental disadvantages (e.g., infertile soil) (Han and Lee 2006; Korir et al. 2017). Korir et al. (2017) showed that co-inoculation of PGPRs and rhizobia significantly increased nodule weight and growth of common bean in a low-P soil. In addition, introduction of micro predator also accelerates bacterial turnover, increase the quantity of bacteria, and promote bacterial activity (Jiang et al. 2020). Jiang et al. (2020) found that interactions of bacteria-feeding nematodes and IAA-producing bacteria promote growth of *Arabidopsis thaliana* by regulating soil auxin status. Biochar is a carbon-rich product with high degree of porosity and extensive surface area and has been proven to fix heavy metals. The combination of biochar and PSB is a promising candidate material for heavy metal remediation (Chen et al. 2019). More notably, Pour et al. (2019) revealed that nano-encapsulation of PGPRs (using silica nanoparticles and carbon nanotubes) and their metabolites improved the commercial rootstock UCB1 of pistachio micropropagation. These modern approaches and techniques, such as micro-encapsulation and nano-encapsulation, have already been developed in the area of pesticides, fertilizers, and genetic material for plant transformation; as a matter of fact, they also show potential to cover these shortcomings of inadaptation by enhancing applied strains service life, dispersion in fertilizer formulation, and allowing the controlled release (Vejan et al. 2016; Gouda et al. 2018).

1.5.4 ENDOPHYTES

Compared with other soil microbes, fewer applications have been reported on DSE, most of which is involved to co-inoculation with AM fungi or EcM fungi. For instance, the interactions between DSE (*Cadophora* sp.) and AM fungi (*Funneliformis mosseae*) involved a decrease of oxidative stress and Cd accumulation in shoots of ryegrass (*Lolium perenne*) (Berthelot et al. 2018). Consortia of DSE and/or EcM fungi strains have potential to function in bioaugmentation strategies for the phytomanagement of marginal lands, which are usually characterized by poor nutrient availability and metal contamination conditions (Berthelot et al. 2019).

1.6 CONCLUDING REMARKS

In limited arable lands, conventional agricultural food production that rely on chemical fertilizers and pesticides are hard to feed the global increasing population in the future. In addition, abiotic stresses that are widely distributed in agricultural ecosystems become increasingly severe because of dramatic changes in the global climate, environmental pollution, and unscientific agricultural management during the past several decades. Therefore, food production and food safety have been receiving extensive attention in these years. From an applicative point of view, application of soil microbial resources can replace main functions of chemical fertilizers and pesticides, furthermore, they have been proven to be high-efficiency, low-cost, and sustainable practices in agricultural ecosystems. However, how to explore and utilize these beneficial soil microbial resources is still the largest challenge in the next few decades.

These beneficial soil microbes are usually composed of N-fixing microbes (e.g., rhizobia and *Frankia*), mycorrhizal fungi (e.g., AM fungi and EcM fungi), PGPRs (e.g., PSB, KSB, IAA-producing bacteria, ACC deaminase-producing producing bacteria, and siderophore-producing bacteria), and endophytes (e.g., DSE). They have abundant species richness and diversity and can perform various physiological and ecological functions in a variety of agricultural ecosystems which span moderate and harsh climate and soil environments, suggesting a strong ecological adaptability. A tremendous number of studies have highlighted that these beneficial soil microbes are able to improve the nutrient uptake (e.g., N, P, K, S, Fe) and water status of plants, which has a main contribution to the survival of plants under abiotic stress conditions (e.g., drought, salinity, heavy metals, acidic soils, aluminum toxicity, and pollutants). Additionally, the physiological, biochemical, and molecular changes of plants afforded by beneficial soil microbes are also the main mechanisms of plants response to abiotic stresses, including improved root system, enhanced photosynthesis, improved osmotic adjustment capacity, increased ROS-scavenging activity, reduced lipid peroxidation, optimized phytohormone balance, and up-regulated abiotic stress resistance genes.

In recent years, some application techniques of beneficial soil microbes have been raised and found to be effective, such as co-inoculation between different functional microbes, utilization of indigenous microbial resources, application of their metabolites, exogenous application of signaling molecules, nano-encapsulation, and micro-encapsulation. Most of them are promising in future agriculture due to their low-cost, high-efficiency, and sustainable traits.

These studies revealed the infinite potential of beneficial soil microbes in agricultural applications and put forward challenges in current agricultural system. More notably, although many scientists have tried to apply these microbial resources to practical food production, this still cannot meet the needs of agricultural development. It's because the contradiction between sufficient theoretical research and limited application technologies is still the main obstacle hindering the utilization of microbial resources. In addition, popularization of soil microbial knowledge and scientific management practices are indispensable for most farmers. Overall, more studies, especially those on the side of application techniques of beneficial soil microbes, are needed to provide deeper insights into practical food production and safety.

ACKNOWLEDGEMENTS

This work was financially supported by GDAS' Project of Science and Technology Development (2022GDASZH-2022010101), Natural Science Foundation of Guangdong (2022A1515010868), Guangdong Province Science and Technology Innovation Strategy Special Fund (No. 2018B020205001) and Natural Science Foundation of China (42077040).

REFERENCES

Abd-Alla, M. H., Morsy, F. M., El-Enany, A. W. E., and Ohyama, T. 2012. Isolation and characterization of a heavy-metal-resistant isolate of *Rhizobium leguminosarum* bv. viciae potentially applicable for biosorption of Cd^{2+} and Co^{2+}. *Int. Biodeter. Biodegr.* 67: 48–55.

Aguilera, P., Larsen, J., Borie, F., Berríos, D., Tapia, C., and Cornejo, P. 2018. New evidences on the contribution of arbuscular mycorrhizal fungi inducing Al tolerance in wheat. *Rhizosphere.* 5: 43–50.

Ahemad, M., and Kibret, M. 2014. Mechanisms and applications of plant growth promoting rhizobacteria: Current perspective. *J. King Saud Univ. Sci.* 26: 1–20.

Ahmed, E., and Holmström, S. J. 2014. Siderophores in environmental research: Roles and applications. *Microb. Biotechnol.* 7: 196–208.

Averill, C., Bhatnagar, J. M., Dietze, M. C., Pearse, W. D., and Kivlin, S. N. 2019. Global imprint of mycorrhizal fungi on whole-plant nutrient economics. *P. Natl. Acad. Sci. USA.* 116: 23163–23168.

Bago, B., Pfeffer, P. E., Abubaker, J., et al. 2003. Carbon export from arbuscular mycorrhizal roots involves the translocation of carbohydrate as well as lipid. *Plant Physiol.* 131: 1496–1507.

Bai, Z., Yuan, Z. Q., Wang, D. M., et al. 2020. Ectomycorrhizal fungus-associated determinants jointly reflect ecological processes in a temperature broad-leaved mixed forest. *Sci. Total Environ.* 703: 135475.

Becquer, A., Garcia, K., Amenc, L., et al. 2018. The *Hebeloma cylindrosporum* HcPT2 Pi transporter plays a key role in ectomycorrhizal symbiosis. *New Phytol.* 220: 1185–1199.

Beddington, J. 2010. Food security: Contributions from science to a new and greener revolution. *Philos. T. R. Soc. B.* 365: 61–71.

Belmondo, S., Fiorilli, V., Perez-Tienda, J., Ferrol, N., Marmeisse, R., and Lanfranco, L. 2014. A dipeptide transporter from the arbuscular mycorrhizal fungus *Rhizophagus irregularis* is upregulated in the intraradical phase. *Front. Plant Sci.* 5: 436.

Benedetto, A., Magurno, F., Bonfante, P., and Lanfranco, L. 2005. Expression profiles of a phosphate transporter gene (*GmosPT*) from the endomycorrhizal fungus *Glomus mosseae*. *Mycorrhiza.* 15: 620–627.

Berthelot, C., Blaudez, D., Beguiristain, T., Chalot, M., and Leyval, C. 2018. Co-inoculation of *Lolium perenne* with *Funneliformis mosseae* and the dark septate endophyte *Cadophora* sp. in a trace element-polluted soil. *Mycorrhiza.* 28: 301–314.

Berthelot, C., Leyval, C., Chalot, M., and Blaudez, D. 2019. Interactions between dark septate endophytes, ectomycorrhizal fungi and root pathogens in vitro. *FEMS Microbiol. Lett.* 366: fnz158.

Berthelot, C., Leyval, C., Foulon, J., Chalot, M., and Blaudez, D. 2016. Plant growth promotion, metabolite production and metal tolerance of dark septate endophytes isolated from metal-polluted poplar phytomanagement sites. *FEMS Microbiol. Ecol.* 92: fiw144.

Boeraeve, M., Honnay, O., and Jacquemyn, H. 2018. Effects of host species, environmental filtering and forest age on community assembly of ectomycorrhizal fungi in fragmented forests. *Fungal Ecol.* 36: 89–98.

Bolger, T. P., Pate, J. S., Unkovich, M. J., and Turner, N. C. 1995. Estimates of seasonal nitrogen fixation of annual subterranean clover-based pastures using the ^{15}N natural abundance technique. *Plant Soil.* 175: 57–66.

Bonfim, J. A., Vasconcellos, R. L. F., Baldesin, L. F., Sieber, T. N., and Cardoso, E. J. B. N. 2016. Dark septate endophytic fungi of native plants along an altitudinal gradient in the Brazilian Atlantic forest. *Fungal Ecol.* 20: 202–210.

Bouffaud, M. L., Renoud, S., Dubost, A., Moënne-Loccoz, Y., and Muller, D. 2018. 1-Aminocyclopropane-1-carboxylate deaminase producers associated to maize and other Poaceae species. *Microbiome.* 6: 114.

Bowles, T. M., Jackson, L. E., Loeher, M., and Cavagnaro, T. R. 2016. Ecological intensification and arbuscular mycorrhizas: A meta-analysis of tillage and cover crop effects. *J. Appl. Ecol.* 54: 1785–1793.

Bozsoki, Z., Gysel, K., Hansen, S. B., et al. 2020. Ligand-recognizing motifs in plant LysM receptors are major determinants of specificity. *Science.* 369: 663–670.

Brundrett, M. C. 2009. Mycorrhizal associations and other means of nutrition of vascular plants: Understanding the global diversity of host plants by resolving conflicting information and developing reliable means of diagnosis. *Plant Soil.* 320: 37–77.

Calabrese, S., Pérez-Tienda, J., Ellerbeck, T., et al. 2016. GintAMT3—a low-affinity ammonium transporter of the arbuscular mycorrhizal *Rhizophagus irregularis*. *Front. Plant Sci.* 7: 679.

Calonne, M., Fontaine, J., Debiane, D., Laruelle, F., Grandmougin-Ferjani, A., and Sahraoui, A. L. H. 2014. The arbuscular mycorrhizal *Rhizophagus irregularis* activates storage lipid biosynthesis to cope with the benzo[a] pyrene oxidative stress. *Phytochemistry.* 97: 30–37.

Canton, G. C., Bertolazi, A. A., Cogo, A. J., et al. 2016. Biochemical and ecophysiological responses to manganese stress by ectomycorrhizal fungus *Pisolithus tinctorius* and in association with *Eucalyptus grandis*. *Mycorrhiza.* 26: 475–487.

Cappellazzo, G., Lanfranco, L., Fitz, M., Wipf, D., and Bonfante, P. 2008. Characterization of an amino acid permease from the endomycorrhizal fungus *Glomus mosseae*. *Plant Physiol.* 147: 429–437.

Castellane, T. C. L., Persona, M. R., Campanharo, J. C., and de Macedo Lemos, E. G. 2015. Production of exopolysaccharide from rhizobia with potential biotechnological and bioremediation applications. *Int. J. Biol. Macromol.* 74: 515–522.

Chen, H., Zhang, J., Tang, L., et al. 2019. Enhanced Pb immobilization via the combination of biochar and phosphate solubilizing bacteria. *Environ. Int.* 127: 395–401.

Chen, S., Zhao, H., Zou, C., et al. 2017. Combined inoculation with multiple arbuscular mycorrhizal fungi improves growth, nutrient uptake and photosynthesis in cucumber seedlings. *Front. Microbiol.* 8: 2516.

Chen, Y., Nara, K., Wen, Z., et al. 2015. Growth and photosynthetic responses of ectomycorrhizal pine seedlings exposed to elevated Cu in soils. *Mycorrhiza.* 25: 561–571.

Chu, Q., Zhang, L., Zhou, J., et al. 2020. Soil plant-available phosphorus levels and maize genotypes determine the phosphorus acquisition efficiency and contribution of mycorrhizal pathway. *Plant Soil.* 449: 357–371.

Coninx, L., Smisdom, N., Kohler, A., et al. 2019. *SlZRT2* encodes a ZIP family Zn transporter with dual localization in the ectomycorrhizal fungus *Suillus luteus*. *Front. Microbiol.* 10: 2251.

Coninx, L., Thoonen, A., Slenders, E., et al. 2017. The *SlZRT1* gene encodes a plasma membrane-located ZIP (Zrt-, Irt-Like Protein) transporter in the ectomycorrhizal fungus *Suillus luteus*. *Front. Microbiol.* 8: 2320.

Dahal, B., NandaKafle, G., Perkins, L., and Brözel, V. S. 2017. Diversity of free-living nitrogen fixing *Streptomyces* in soils of the badlands of South Dakota. *Microbiol. Res.* 195: 31–39.

Danielsen, L., and Polle, A. 2014. Poplar nutrition under drought as affected by ectomycorrhizal colonization. *Environ. Exp. Bot.* 108: 89–98.

Davison, J., Moora, M., Öpik, M., et al. 2015. Global assessment of arbuscular mycorrhizal fungus diversity reveals very low endemism. *Science.* 349: 970–973.

Della Mónica, I. F., Saparrat, M. C., Godeas, A. M., and Scervino, J. M. 2015. The co-existence between DSE and AMF symbionts affects plant P pools through P mineralization and solubilization processes. *Fungal Ecol.* 17: 10–17.

de Mesquita, C. P. B., Sartwell, S. A., Ordemann, E. V., et al. 2018. Patterns of root colonization by arbuscular mycorrhizal fungi and dark septate endophytes across a mostly-unvegetated, high-elevation landscape. *Fungal Ecol.* 36: 63–74.

Dos Santos, S. G., da Silva, P. R. A., Garcia, A. C., Zilli, J. É., and Berbara, R. L. L. 2017. Dark septate endophyte decreases stress on rice plants. *Braz. J. Microbiol.* 48: 333–341.

Eady, R. R. 1996. Structure-function relationships of alternative nitrogenases. *Chem. Rev.* 96: 3013–3030.

Etesami, H., Alikhani, H. A., and Hosseini, H. M. 2015. Indole-3-acetic acid (IAA) production trait, a useful screening to select endophytic and rhizosphere competent bacteria for rice growth promoting agents. *MethodsX* 2: 72–78.

Etesami, H., Emami, S., and Alikhani, H. A. 2017. Potassium solubilizing bacteria (KSB): Mechanisms, promotion of plant growth, and future prospects: A review. *J. Soil Sci. Plant Nut.* 17: 897–911.

Etesami, H., and Maheshwari, D. K. 2018. Use of plant growth promoting rhizobacteria (PGPRs) with multiple plant growth promoting traits in stress agriculture: Action mechanisms and future prospects. *Ecotox. Environ. Safe.* 156: 225–246.

Ezawa, T., Cavagnaro, T. R., Smith, S. E., Smith, F. A., and Ohtomo, R. 2003. Rapid accumulation of polyphosphate in extraradical hyphae of an arbuscular mycorrhizal fungus as revealed by histochemistry and a polyphosphate kinase/luciferase system. *New Phytol.* 161: 387–392.

Fang, L., Ju, W., Yang, C., et al. 2020. Exogenous application of signaling molecules to enhance the resistance of legume-rhizobium symbiosis in Pb/Cd-contaminated soils. *Environ. Pollut.* 265: 114744.

Feng, Z., Liu, X., Feng, G., Zhu, H., and Yao, Q. 2020a. Linking lipid transfer with reduced arbuscule formation in tomato roots colonized by arbuscular mycorrhizal fungus under low pH stress. *Environ. Microbiol.* 22: 1036–1051.

Feng, Z., Liu, X., Zhu, H., and Yao, Q. 2020b. Responses of arbuscular mycorrhizal symbiosis to abiotic stress: A lipid-centric perspective. *Front. Plant Sci.* 11: 1754.

Ferreira, P. A. A., Bomfeti, C. A., Soares, B. L., and de Souza Moreira, F. M. 2012. Efficient nitrogen-fixing Rhizobium strains isolated from Amazonian soils are highly tolerant to acidity and aluminium. *World J. Microb. Biot.* 28: 1947–1959.

Fiorilli, V., Lanfranco, L., and Bonfante, P. 2013. The expression of *GintPT*, the phosphate transporter of *Rhizophagus irregularis*, depends on the symbiotic status and phosphate availability. *Planta.* 237: 1267–1277.

Fontaine, L., Thiffault, N., Paré, D., Fortin, J. A., and Piché, Y. 2016. Phosphate-solubilizing bacteria isolated from ectomycorrhizal mycelium of *Picea glauca* are highly efficient at fluorapatite weathering. *Botany.* 94: 1183–1193.

Fraser, T. D., Lynch, D. H., Bent, E., Entz, M. H., and Dunfield, K. E. 2015. Soil bacterial *phoD* gene abundance and expression in response to applied phosphorus and long-term management. *Soil Biol. Biochem.* 88: 137–147.

Gaby, J. C., and Buckley, D. H. 2012. A comprehensive evaluation of PCR primers to amplify the *nifH* gene of nitrogenase. *PloS One* 7: e42149.

Garcia, K., Delteil, A., Conéjéro, G., et al. 2014. Potassium nutrition of ectomycorrhizal *Pinus pinaster*: overexpression of the *Hebeloma cylindrosporum* HcTrk1 transporter affects the translocation of both K+ and P in the host plant. *New Phytol.* 201: 951–960.

Garcia, K., Doidy, J., Zimmermann, S. D., Wipf, D., and Courty, P. 2016. Take a trip through the plant and fungal transportome of mycorrhiza. *Trends Plant Sci.* 21: 937–950.

Garcia, K., and Zimmermann, S. D. 2014. The role of mycorrhizal associations in plant potassium nutrition. *Front. Plant Sci.* 5: 337.

García-López, A. M., Avilés, M., and Delgado, A. 2016. Effect of various microorganisms on phosphorus uptake from insoluble Ca-phosphates by cucumber plants. *J. Plant Nutr. Soil Sc.* 179: 454–465.

Gouda, S., Kerry, R. G., Das, G., Paramithiotis, S., Shin, H. S., and Patra, J. K. 2018. Revitalization of plant growth promoting rhizobacteria for sustainable development in agriculture. *Microbiol. Res.* 206: 131–140.

Guerrero-Galán, C., Calvo-Polanco, M., and Zimmermann, S. D. 2019. Ectomycorrhizal symbiosis helps plants to challenge salt stress conditions. *Mycorrhiza.* 29: 291–301.

Guerrero-Galán, C., Garcia, K., Houdinet, G., and Zimmermann, S. D. 2018. *Hc*TOK1 participates in the maintenance of K+ homeostasis in the ectomycorrhizal fungus *Hebeloma cylindrosporum*, which is essential for the symbiotic K+ nutrition of *Pinus pinaster*. *Plant Signal. Behav.* 13: e1480845.

Guo, J., and Chi, J. 2014. Effect of Cd-tolerant plant growth-promoting rhizobium on plant growth and Cd uptake by *Lolium multiflorum* Lam. and *Glycine max* (L.) Merr. in Cd-contaminated soil. *Plant Soil*. 375: 205–214.

Habib, S. H., Kausar, H., and Saud, H. M. 2016. Plant growth-promoting rhizobacteria enhance salinity stress tolerance in okra through ROS-scavenging enzymes. *Biomed Res. Int.* 2016: 6284547.

Han, H. S., and Lee, K. D. 2006. Effect of co-inoculation with phosphate and potassium solubilizing bacteria on mineral uptake and growth of pepper and cucumber. *Plant Soil Environ.* 52: 130–136.

Han, L. L., Wang, Q., Shen, J. P., et al. 2019. Multiple factors drive the abundance and diversity of the diazotrophic community in typical farmland soils of China. *FEMS Microbiol. Ecol.* 95: fiz113.

Han, Q., Ma, Q., Chen, Y., et al. 2020. Variation in rhizosphere microbial communities and its association with the symbiotic efficiency of rhizobia in soybean. *ISME J.* 14: 1915–1928.

Hao, Z., Xie, W., Jiang, X., Wu, Z., Zhang, X., and Chen, B. 2019. Arbuscular mycorrhizal fungus improves rhizobium-glycyrrhiza seedling symbiosis under drought stress. *Agronomy.* 9: 572.

Harrison, M. J., and van Buuren, M. L. 1995. A phosphate transporter from the mycorrhizal fungus *Glomus versiforme*. *Nature.* 378: 626–629.

Hassan, W., Bashir, S., Ali, F., Ijaz, M., Hussain, M., and David, J. 2016. Role of ACC-deaminase and/or nitrogen fixing rhizobacteria in growth promotion of wheat (*Triticum aestivum* L.) under cadmium pollution. *Environ. Earth. Sci.* 75: 267.

Hassen, A. I., Lamprecht, S. C., and Bopape, F. L. 2020. Emergence of β-rhizobia as new root nodulating bacteria in legumes and current status of the legume-rhizobium host specificity dogma. *World J. Microbiol. Biotech.* 36: 1–13.

Heinonsalo, J., Juurola, E., Linden, A., and Pumpanen, J. 2015. Ectomycorrhizal fungi affect Scots pine photosynthesis through nitrogen and water economy, not only through increased carbon demand. *Environ. Exp. Bot.* 109: 103–112.

Helber, N., Wippel, K., Sauer, N., Schaarschmidt, S., Hause, B., and Requena, N. 2011. A versatile monosaccharide transporter that operates in the arbuscular mycorrhizal fungus *Glomus* sp is crucial for the symbiotic relationship with plants. *Plant Cell.* 23: 3812–3823.

He, Y., Yang, Z., Li, M., et al. 2017. Effects of a dark septate endophyte (DSE) on growth, cadmium content, and physiology in maize under cadmium stress. *Environ. Sci. Pollut. R.* 24: 18494–18504.

Jiang, Y., Wang, W., Xie, Q., et al. 2017. Plants transfer lipids to sustain colonization by mutualistic mycorrhizal and parasitic fungi. *Science.* 356: 1172–1175.

Jiang, Y., Wu, Y., Hu, N., Li, H., and Jiao, J. 2020. Interactions of bacterial-feeding nematodes and indole-3-acetic acid (IAA)-producing bacteria promotes growth of *Arabidopsis thaliana* by regulating soil auxin status. *Appl. Soil Ecol.* 147: 103447.

Jumpponen, A. R. I., and Trappe, J. M. 1998. Dark septate endophytes: A review of facultative biotrophic root-colonizing fungi. *New Phytol.* 140: 295–310.

Kaur, G., and Reddy, M. S. 2015. Effects of phosphate-solubilizing bacteria, rock phosphate and chemical fertilizers on maize-wheat cropping cycle and economics. *Pedosphere.* 25: 428–437.

Kemppainen, M. J., Alvarez Crespo, M. C., and Pardo, A. G. 2010. fHANT-AC genes of the ectomycorrhizal fungus *Laccaria bicolor* are not repressed by l-glutamine allowing simultaneous utilization of nitrate and organic nitrogen sources. *Env. Microbiol. Rep.* 2: 541–553.

Khalloufi, M., Martínez-Andújar, C., Lachaâl, M., Karray-Bouraoui, N., Pérez-Alfocea, F., and Albacete, A. 2017. The interaction between foliar GA_3 application and arbuscular mycorrhizal fungi inoculation improves growth in salinized tomato (*Solanum lycopersicum* L.) plants by modifying the hormonal balance. *J. Plant Physiol.* 214: 134–144.

Khanghahi, M. Y., Pirdashti, H., Rahimian, H., Nematzadeh, G., and Sepanlou, M. G. 2018. Potassium solubilising bacteria (KSB) isolated from rice paddy soil: From isolation, identification to K use efficiency. *Symbiosis.* 76: 13–23.

Khullar, S., and Reddy, M. S. 2018. Ectomycorrhizal fungi and its role in metal homeostasis through metallothionein and glutathione mechanisms. *Curr. Biotech.* 7: 231–241.

Khullar, S., and Reddy, M. S. 2020. Arsenic toxicity and its mitigation in ectomycorrhizal fungus *Hebeloma cylindrosporum* through glutathione biosynthesis. *Chemosphere.* 240: 124914.

Kikuchi, Y., Hijikata, N., Ohtomo, R., et al. 2016. Aquaporin-mediated long-distance polyphosphate translocation directed towards the host in arbuscular mycorrhizal symbiosis: Application of virus-induced gene silencing. *New Phytol.* 211: 1202–1208.

Kikuchi, Y., Hijikata, N., Yokoyama, K., et al. 2014. Polyphosphate accumulation is driven by transcriptome alterations that lead to near-synchronous and near-equivalent uptake of inorganic cations in an arbuscular mycorrhizal fungus. *New Phytol.* 204: 638–649.

Kilpeläinen, J., Aphalo, P. J., Barbero-López, A., Adamczyk, B., Nipu, S. A., and Lehto, T. 2020. Are arbuscular-mycorrhizal *Alnus incana* seedlings more resistant to drought than ectomycorrhizal and nonmycorrhizal ones?. *Tree Physiol.* 40: 782–795.

Kivlin, S. N., Hawkes, C. V., and Treseder, K. K. 2011. Global diversity and distribution of arbuscular mycorrhizal fungi. *Soil Biol. Biochem.* 43: 2294–2303.

Köberl, M., Erlacher, A., Ramadan, E. M., et al. 2016. Comparisons of diazotrophic communities in native and agricultural desert ecosystems reveal plants as important drivers in diversity. *FEMS Microbiol. Ecol.* 92: fiv166.

Korir, H., Mungai, N. W., Thuita, M., Hamba, Y., and Masso, C. 2017. Co-inoculation effect of rhizobia and plant growth promoting rhizobacteria on common bean growth in a low phosphorus soil. *Front. Plant Sci.* 8: 141.

Knapp, D. G., Kovács, G. M., Zajta, E., Groenewald, J. Z., and Crous, P. W. 2015. Dark septate endophytic pleosporalean genera from semiarid areas. *Persoonia.* 35: 87–100.

Kumar, U., Panneerselvam, P., Govindasamy, V., et al. 2017. Long-term aromatic rice cultivation effect on frequency and diversity of diazotrophs in its rhizosphere. *Ecol. Eng.* 101: 227–236.

Kuypers, M. M., Marchant, H. K., and Kartal, B. 2018. The microbial nitrogen-cycling network. *Nat. Rev. Microbiol.* 16: 263–276.

Lee, T. K., Han, I., Kim, M. S., Seong, H. J., Kim, J. S., and Sul, W. J. 2019. Characterization of a *nifH*-harboring bacterial community in the soil-limited Gotjawal forest. *Front. Microbiol.* 10, 1858.

Leonhardt, T., Sácký, J., and Kotrba, P. 2018. Functional analysis RaZIP1 transporter of the ZIP family from the ectomycorrhizal Zn-accumulating *Russula atropurpurea*. *Biometals.* 31: 255–266.

Li, J., Meng, B., Chai, H., et al. 2019a. Arbuscular mycorrhizal fungi alleviate drought stress in C_3 (*Leymus chinensis*) and C_4 (*Hemarthria altissima*) grasses via altering antioxidant enzyme activities and photosynthesis. *Front. Plant Sci.* 10: 499.

Li, T., Hu, Y. J., Hao, Z. P., Li, H., Wang, Y. S., and Chen, B. D. 2013. First cloning and characterization of two functional aquaporin genes from an arbuscular mycorrhizal fungus *Glomus intraradices*. *New Phytol.* 197: 617–630.

Li, X., He, C., He, X., et al. 2019b. Dark septate endophytes improve the growth of host and non-host plants under drought stress through altered root development. *Plant Soil.* 439: 259–272.

Li, Y., Chen, Z., He, J. Z., Wang, Q., Shen, C., and Ge, Y. 2019c. Ectomycorrhizal fungi inoculation alleviates simulated acid rain effects on soil ammonia oxidizers and denitrifiers in Masson pine forest. *Environ. Microbiol.* 21: 299–313.

Lipa, P., and Janczarek, M. 2020. Phosphorylation systems in symbiotic nitrogen-fixing bacteria and their role in bacterial adaptation to various environmental stresses. *Peer J.* 8: e8466.

López-Pedrosa, A., González-Guerrero, M., Valderas, A., Azcón-Aguilar, C., and Ferrol, N. 2006. *GintAMT1* encodes a functional high-affinity ammonium transporter that is expressed in the extraradical mycelium of *Glomus intraradices. Fungal Genet. Biol.* 43: 102–110.

Loth-Pereda, V., Orsini, E., Courty, P. E., et al. 2011. Structure and expression profile of the phosphate Pht1 transporter gene family in mycorrhizal *Populus trichocarpa. Plant Physiol.* 156: 2141–2154.

Long, X. E., Yao, H., Huang, Y., Wei, W., and Zhu, Y. G. 2018. Phosphate levels influence the utilisation of rice rhizodeposition carbon and the phosphate-solubilising microbial community in a paddy soil. *Soil Biol. Biochem.* 118: 103–114.

Luginbuehl, L. H., Menard, G. N., Kurup, S., et al. 2017. Fatty acids in arbuscular mycorrhizal fungi are synthesized by the host plant. *Science.* 356: 1175–1178.

Luginbuehl, L. H., and Oldroyd, G. 2017. Understanding the arbuscule at the heart of endomycorrhizal symbioses in plants. *Curr. Biol.* 27: 952–963.

Lugo, M. A., Reinhart, K. O., Menoyo, E., Crespo, E. M., and Urcelay, C. 2015. Plant functional traits and phylogenetic relatedness explain variation in associations with root fungal endophytes in an extreme arid environment. *Mycorrhiza.* 25: 85–95.

Maldonado-Mendoza, I. E., Dewbre, G. R., and Harrison, M. J. 2001. A phosphate transporter gene from the extra-radical mycelium of an arbuscular mycorrhizal fungus *Glomus intraradices* is regulated in response to phosphate in the environment. *Mol. Plant Microbe In.* 14: 1140–1148.

Mander, C., Wakelin, S., Young, S., Condron, L., and O'Callaghan, M. 2012. Incidence and diversity of phosphate-solubilising bacteria are linked to phosphorus status in grassland soils. *Soil Biol. Biochem.* 44: 93–101.

Martin, F., Kohler, A., Murat, C., Veneault-Fourrey, C., and Hibbett, D. S. 2016. Unearthing the roots of ectomycorrhizal symbioses. *Nat. Rev. Microbiol.* 14: 760–773.

Mathur, S., Sharma, M. P., and Jajoo, A. 2018. Improved photosynthetic efficacy of maize (*Zea mays*) plants with arbuscular mycorrhizal fungi (AMF) under high temperature stress. *J. Photochem. Photobiol. B.* 180: 149–154.

Matsuoka, S., Mori, A. S., Kawaguchi, E., Hobara, S., and Osono, T. 2016. Disentangling the relative importance of host tree community, abiotic environment and spatial factors on ectomycorrhizal fungal assemblages along an elevation gradient. *FEMS Microbiol. Ecol.* 92: fiw044.

Meng, H., Zhou, Z., Wu, R., Wang, Y., and Gu, J. D. 2019. Diazotrophic microbial community and abundance in acidic subtropical natural and re-vegetated forest soils revealed by high-throughput sequencing of *nifH* gene. *Appl. Microbiol. Biot.* 103: 995–1005.

Mendoza-Soto, A. B., Naya, L., Leija, A., and Hernández, G. 2015. Responses of symbiotic nitrogen-fixing common bean to aluminum toxicity and delineation of nodule responsive microRNAs. *Front. Plant Sci.* 6: 587.

Mohammadi, M., Modarres-Sanavy, S. A. M., Pirdashti, H., Zand, B., and Tahmasebi-Sarvestani, Z. 2019. Arbuscular mycorrhizae alleviate water deficit stress and improve antioxidant response, more than nitrogen fixing bacteria or chemical fertilizer in the evening primrose. *Rhizosphere.* 9: 76–89.

Moura, E. G., Carvalho, C. S., Bucher, C. P., et al. 2020. Diversity of rhizobia and importance of their interactions with legume trees for feasibility and sustainability of the tropical agrosystems. *Diversity.* 12: 206.

Mukhtar, S., Zareen, M., Khaliq, Z., Mehnaz, S., and Malik, K. A. 2020. Phylogenetic analysis of halophyte-associated rhizobacteria and effect of halotolerant and halophilic phosphate-solubilizing biofertilizers on maize growth under salinity stress conditions. *J. Appl. Microbiol.* 128: 556–573.

Müller, T., Neuhäuser, B., Ludewig, U., et al. 2020. New insights into HcPTR2A and HcPTR2B, two high-affinity peptide transporters from the ectomycorrhizal model fungus *Hebeloma cylindrosporum*. *Mycorrhiza*. 30: 735–747.

Mus, F., Crook, M. B., Garcia, K., et al. 2016. Symbiotic nitrogen fixation and the challenges to its extension to nonlegumes. *Appl. Environ. Microb.* 82: 3698–3710.

National Academies of Sciences, Engineering, and Medicine. 2019. *Science Breakthroughs to Advance Food and Agricultural Research by 2030*. Washington, DC: National Academies Press.

Navarro-RóDenas, A., Xu, H. A. O., Kemppainen, M., Pardo, A. G., and Zwiazek, J. J. 2015. *Laccaria bicolor* aquaporin LbAQP1 is required for Hartig net development in trembling aspen (*Populus tremuloides*). *Plant Cell Environ.* 38: 2475–2486.

Newsham, K. K. 2011. A meta-analysis of plant responses to dark septate root endophytes. *New Phytol.* 190: 783–793.

Nguyen, H., Rineau, F., Vangronsveld, J., Cuypers, A., Colpaert, J. V., and Ruytinx, J. 2017. A novel, highly conserved metallothionein family in basidiomycete fungi and characterization of two representative *SlMTa* and *SlMTb* genes in the ectomycorrhizal fungus *Suillus luteus*. *Environ. Microbiol.* 19: 2577–2587.

Normand, P., Benson, D. R., Berry, A. M., and Tisa, L. S. 2014. "Family Frankiaceae," In *The Prokaryote-Actinobacteria*, E. Rosenberg, E. F. Delong, S. Lory, E. Stackebrandt, and F. Thompson (eds.). Berlin: Springer-Verlag, 339–356.

Nouioui, I., Cortés, C. J., Carro, L., et al. 2019. Genomic insights into plant-growth-promoting potentialities of the genus *Frankia*. *Front. Microbiol.* 10: 1457.

Oehl, F., Laczko, E., Oberholzer, H. R., Jansa, J., and Egli, S. 2017. Diversity and biogeography of arbuscular mycorrhizal fungi in agricultural soils. *Biol. Fert. Soils.* 53: 777–797.

Ogunkunle, C. O., El-Imam, A. A., Bassey, E., Vishwakarma, V., and Fatoba, P. O. 2020. Co-application of indigenous arbuscular mycorrhizal fungi and nano-TiO_2 reduced Cd uptake and oxidative stress in pre-flowering cowpea plants. *Environ. Technol. Inno.* 20: 101163.

Ouyang, Y., Evans, S. E., Friesen, M. L., and Tiemann, L. K. 2018. Effect of nitrogen fertilization on the abundance of nitrogen cycling genes in agricultural soils: A meta-analysis of field studies. *Soil Biol. Biochem.* 127: 71–78.

Pang, Z., Xu, P., and Yu, D. 2020. Environmental adaptation of the root microbiome in two rice ecotypes. *Microbiol. Res.* 241: 126588.

Parray, J. A., Jan, S., Kamili, A. N., Qadri, R. A., Egamberdieva, D., and Ahmad, P. 2016. Current perspectives on plant growth-promoting rhizobacteria. *J. Plant Growth Regul.* 35: 877–902.

Pawlowski, K., and Demchenko, K. N. 2012. The diversity of actinorhizal symbiosis. *Protoplasma*. 249: 967–979.

Penton, C. R., Yang, C., Wu, L., et al. 2016. *NifH*-harboring bacterial community composition across an Alaskan permafrost thaw gradient. *Front. Microbiol.* 7: 1894.

Pérez-Tienda, J., Testillano, P. S., Balestrini, R., Fiorilli, V., Azcón-Aguilar, C., and Ferrol, N. 2011. GintAMT2, a new member of the ammonium transporter family in the arbuscular mycorrhizal fungus *Glomus intraradices*. *Fungal Genet. Biol.* 48: 1044–1055.

Peterson, R. L., and Massicotte, H. B. 2004. Exploring structural definitions of mycorrhizas, with emphasis on nutrient-exchange interfaces. *Can. J. Bot.* 82: 1074–1088.

Pinos, N. Q., Louro Berbara, R. L., Elias, S. S., van Tol de Castro, T. A., and García, A. C. 2019. Combination of humic substances and arbuscular mycorrhizal fungi affecting corn plant growth. *J. Environ. Qual.* 48: 1594–1604.

Pour, M. M., Saberi-Riseh, R., Mohammadinejad, R., and Hosseini, A. 2019. Nano-encapsulation of plant growth-promoting rhizobacteria and their metabolites using alginate-silica nanoparticles and carbon nanotube improves UCB1 pistachio micro-propagation. *J. Microbiol. Biotechnol.* 29: 1096–1103.

Pramanik, P., Goswami, A. J., Ghosh, S., and Kalita, C. 2019. An indigenous strain of potassium-solubilizing bacteria *Bacillus pseudomycoides* enhanced potassium uptake in tea plants by increasing potassium availability in the mica waste-treated soil of North-east India. *J. Appl. Microbiol.* 126: 215–222.

Püschel, D., Janoušková, M., Voříšková, A., Gryndlerová, H., Vosátka, M., and Jansa, J. 2017. Arbuscular mycorrhiza stimulates biological nitrogen fixation in two *Medicago* spp. through improved phosphorus acquisition. *Front. Plant Sci.* 8: 390.

Qu, L., Huang, Y., Zhu, C., et al. 2016. Rhizobia-inoculation enhances the soybean's tolerance to salt stress. *Plant Soil.* 400: 209–222.

Ragot, S. A., Kertesz, M. A., Mészáros, É., Frossard, E., and Bünemann, E. K. 2017. Soil *phoD* and *phoX* alkaline phosphatase gene diversity responds to multiple environmental factors. *FEMS Microbiol. Ecol.* 93: fiw212.

Raklami, A., El Gharmali, A., Ait Rahou, Y., Oufdou, K., and Meddich, A. 2021. Compost and mycorrhizae application as a technique to alleviate Cd and Zn stress in *Medicago sativa. Int. J. Phytoremediat.* 23: 190–201.

Rask, K. A., Johansen, J. L., Kjøller, R., and Ekelund, F. 2019. Differences in arbuscular mycorrhizal colonisation influence cadmium uptake in plants. *Environ. Exp. Bot.* 162: 223–229.

Razakatiana, A. T. E., Trap, J., Baohanta, R. H., et al. 2020. Benefits of dual inoculation with arbuscular mycorrhizal fungi and rhizobia on *Phaseolus vulgaris* planted in a low-fertility tropical soil. *Pedobiologia.* 83: 150685.

Ren, C. G., Kong, C. C., Wang, S. X., and Xie, Z. H. 2019. Enhanced phytoremediation of uranium-contaminated soils by arbuscular mycorrhiza and rhizobium. *Chemosphere.* 217: 773–779.

Rizvi, A., and Khan, M. S. 2018. Heavy metal induced oxidative damage and root morphology alterations of maize (*Zea mays* L.) plants and stress mitigation by metal tolerant nitrogen fixing *Azotobacter chroococcum. Ecotox. Environ. Safe.* 157: 9–20.

Rodríguez-Echeverría, S., Teixeira, H., Correia, M., et al. 2017. Arbuscular mycorrhizal fungi communities from tropical Africa reveal strong ecological structure. *New Phytol.* 213: 380–390.

Ruytinx, J., Coninx, L., Nguyen, H., et al. 2017. Identification, evolution and functional characterization of two Zn CDF-family transporters of the ectomycorrhizal fungus *Suillus luteus. Env. Microbiol. Rep.* 9: 419–427.

Satapute, P., Paidi, M. K., Kurjogi, M., and Jogaiah, S. 2019. Physiological adaptation and spectral annotation of Arsenic and Cadmium heavy metal-resistant and susceptible strain *Pseudomonas taiwanensis. Environ. Pollut.* 251: 555–563.

Schlunk, I., Krause, K., Wirth, S., and Kothe, E. 2015. A transporter for abiotic stress and plant metabolite resistance in the ectomycorrhizal fungus *Tricholoma vaccinum. Environ. Sci. Pollut. R.* 22: 19384–19393.

Shameer, S., and Prasad, T. N. V. K. V. 2018. Plant growth promoting rhizobacteria for sustainable agricultural practices with special reference to biotic and abiotic stresses. *Plant Growth Regul.* 84: 603–615.

Shamseldin, A., Abdelkhalek, A., and Sadowsky, M. J. 2017. Recent changes to the classification of symbiotic, nitrogen-fixing, legume-associating bacteria: A review. *Symbiosis.* 71: 91–109.

Shen, M., Zhao, D. K., Qiao, Q., et al. 2015. Identification of glutathione S-transferase (GST) genes from a dark septate endophytic fungus (*Exophiala pisciphila*) and their expression patterns under varied metals stress. *PloS One* 10: e0123418.

Smith, S. E., and Read, D. J. 2008. *Mycorrhizal Symbiosis*. San Diego, CA: Academic Press.

Smith, S. E., Smith, F. A., and Jakobsen, I. 2004. Functional diversity in arbuscular mycorrhizal (AM) symbioses: The contribution of the mycorrhizal P uptake pathway is not correlated with mycorrhizal responses in growth or total P uptake. *New Phytol.* 162: 511–524.

Spagnoletti, F. N., Tobar, N. E., Di Pardo, A. F., Chiocchio, V. M., and Lavado, R. S. 2017. Dark septate endophytes present different potential to solubilize calcium, iron and aluminum phosphates. *Appl. Soil Ecol.* 111: 25–32.

Spatafora, J. W., Chang, Y., Benny, G. L., et al. 2016. A phylum-level phylogenetic classification of zygomycete fungi based on genome-scale data. *Mycologia.* 108: 1028–1046.

Sprent, J. I., Ardley, J., and James, E. K. 2017. Biogeography of nodulated legumes and their nitrogen-fixing symbionts. *New Phytol.* 215: 40–56.

Staudinger, C., Mehmeti-Tershani, V., Gil-Quintana, E., et al. 2016. Evidence for a rhizobia-induced drought stress response strategy in *Medicago truncatula. J. Proteomics.* 136: 202–213.

Steidinger, B. S., Crowther, T. W., Liang, J., et al. 2019. Climatic controls of decomposition drive the global biogeography of forest-tree symbioses. *Nature.* 569: 404–408.

Stewart, W. M., Dibb, D. W., Johnston, A. E., and Smyth, T. J. 2005. The contribution of commercial fertilizer nutrients to food production. *Agron. J.* 97: 1–6.

Stritzler, M., Elba, P., Berini, C., Gomez, C., Ayub, N., and Soto, G. 2018. High-quality forage production under salinity by using a salt-tolerant *AtNXH1*-expressing transgenic alfalfa combined with a natural stress-resistant nitrogen-fixing bacterium. *J. Biotechnol.* 276: 42–45.

Sun, F., Ou, Q., Wang, N., et al. 2020. Isolation and identification of potassium-solubilizing bacteria from *Mikania micrantha* rhizospheric soil and their effect on *M. micrantha* plants. *Glob. Ecol. Conserv.* 23: e01141.

Tamayo, E., Gomez-Gallego, T., Azcon-Aguilar, C., and Ferrol, N. 2014. Genome-wide analysis of copper, iron and zinc transporters in the arbuscular mycorrhizal fungus *Rhizophagus irregularis. Front. Plant Sci.* 5: 547.

Tamayo, E., Knight, S. A. B., Valderas, A., Dancis, A., and Ferrol, N. 2018. The arbuscular mycorrhizal fungus *Rhizophagus irregularis* uses a reductive iron assimilation pathway for high-affinity iron uptake. *Environ. Microbiol.* 20: 1857–1872.

Tanaka, Y., and Yano, K. 2005. Nitrogen delivery to maize via mycorrhizal hyphae depends on the form of N supplied. *Plant Cell Environ.* 28: 1247–1254.

Tang, Y., Shi, L., Zhong, K., Shen, Z., and Chen, Y. 2019. Ectomycorrhizal fungi may not act as a barrier inhibiting host plant absorption of heavy metals. *Chemosphere.* 215: 115–123.

Tatry, M. V., El Kassis, E., Lambilliotte, R., et al. 2009. Two differentially regulated phosphate transporters from the symbiotic fungus *Hebeloma cylindrosporum* and phosphorus acquisition by ectomycorrhizal *Pinus pinaster. Plant J.* 57: 1092–1102.

Tian, C., Kasiborski, B., Koul, R., Lammers, P. J., Bücking, H., and Shachar-Hill, Y. 2010. Regulation of the nitrogen transfer pathway in the arbuscular mycorrhizal symbiosis: Gene characterization and the coordination of expression with nitrogen flux. *Plant Physiol.* 153: 1175–1187.

Terhonen, E., Keriö, S., Sun, H., and Asiegbu, F. O. 2014. Endophytic fungi of Norway spruce roots in boreal pristine mire, drained peatland and mineral soil and their inhibitory effect on *Heterobasidion parviporum* in vitro. *Fungal Ecol.* 9: 17–26.

Teste, F. P., Jones, M. D., and Dickie, I. A. 2020. Dual-mycorrhizal plants: Their ecology and relevance. *New Phytol.* 225: 1835–1851.

Tewari, S., and Sharma, S. 2020. Rhizobial exopolysaccharides as supplement for enhancing nodulation and growth attributes of *Cajanus cajan* under multi-stress conditions: A study from lab to field. *Soil Till. Res.* 198: 104545.

Tu, Q., Deng, Y., Yan, Q., et al. 2016a. Biogeographic patterns of soil diazotrophic communities across six forests in the North America. *Mol. Ecol.* 25: 2937–2948.

Tu, Q., Zhou, X., He, Z., et al. 2016b. The diversity and co-occurrence patterns of N_2-fixing communities in a CO_2-enriched grassland ecosystem. *Microb. Ecol.* 71: 604–615.

Ullah, A., Heng, S., Munis, M. F. H., Fahad, S., and Yang, X. 2015. Phytoremediation of heavy metals assisted by plant growth promoting (PGP) bacteria: A review. *Environ. Exp. Bot.* 117: 28–40.

Vacheron, J., Desbrosses, G., Bouffaud, M. L., et al. 2013. Plant growth-promoting rhizobacteria and root system functioning. *Front. Plant Sci.* 4: 356.

Vadakattu, G., Zhang, B., Penton, C. R., Yu, J., and Tiedje, J. M. 2019. Diazotroph diversity and nitrogen fixation in summer active perennial grasses in a Mediterranean region agricultural soil. *Front. Mol. Biosci.* 6: 115.

van Aarle, I. M., Viennois, G., Amenc, L. K., Tatry, M. V., Luu, D. T., and Plassard, C. 2007. Fluorescent in situ RT-PCR to visualise the expression of a phosphate transporter gene from an ectomycorrhizal fungus. *Mycorrhiza.* 17: 487–494.

van der Heijden, M. G. A., Martin, F. M., Selosse, M. A., and Sanders, I. R. 2015. Mycorrhizal ecology and evolution: The past, the present, and the future. *New Phytol.* 205: 1406–1423.

van Der Linde, S., Suz, L. M., Orme, C. D. L., et al. 2018. Environment and host as large-scale controls of ectomycorrhizal fungi. *Nature.* 558: 243–248.

van Deynze, A., Zamora, P., Delaux, P. M., et al. 2018. Nitrogen fixation in a landrace of maize is supported by a mucilage-associated diazotrophic microbiota. *PLoS Biol.* 16: e2006352.

Van Geel, M., Jacquemyn, H., Plue, J., et al. 2018. Abiotic rather than biotic filtering shapes the arbuscular mycorrhizal fungal communities of European seminatural grasslands. *New Phytol.* 220: 1262–1272.

Van Nguyen, T., and Pawlowski, K. 2017. "*Frankia* and actinorhizal plants: Symbiotic nitrogen fixation." In *Rhizotrophs: Plant Growth Promotion to Bioremediation*, S. Mehnaz (ed.). Singapore: Springer. 237–261.

Van Nguyen, T., Wibberg, D., Vigil-Stenman, T., et al. 2019. *Frankia*-enriched metagenomes from the earliest diverging symbiotic *Frankia* cluster: They come in teams. *Genome. Biol. Evol.* 11: 2273–2291.

Vergara, C., Araujo, K. E. C., Alves, L. S., et al. 2018. Contribution of dark septate fungi to the nutrient uptake and growth of rice plants. *Braz. J. Microbiol.* 49: 67–78.

Vejan, P., Abdullah, R., Khadiran, T., Ismail, S., and Nasrulhaq Boyce, A. 2016. Role of plant growth promoting rhizobacteria in agricultural sustainability-a review. *Molecules.* 21: 573.

Volpe, V., Carotenuto, G., Berzero, C., Cagnina, L., Puech-Pagès, V., and Genre, A. 2020. Short chain chito-oligosaccharides promote arbuscular mycorrhizal colonization in *Medicago truncatula*. *Carbohyd. Polym.* 229: 115505.

Vurukonda, S. S. K. P., Vardharajula, S., Shrivastava, M., and SkZ, A. 2016. Enhancement of drought stress tolerance in crops by plant growth promoting rhizobacteria. *Microbiol. Res.* 184: 13–24.

Walder, F., Boller, T., Wiemken, A., and Courty, P. E. 2016. Regulation of plants' phosphate uptake in common mycorrhizal networks: Role of intraradical fungal phosphate transporters. *Plant Signal. Behave.* 11: e1131372.

Wang, E. T., Chen, W. F., Tian, C. F., Young, J. P. W., and Chen, W. X. 2019. *Ecology and Evolution of Rhizobia*. Singapore: Springer.

Wang, J., Li, T., Wu, X., and Zhao, Z. 2014. Molecular cloning and functional analysis of a H+-dependent phosphate transporter gene from the ectomycorrhizal fungus *Boletus edulis* in southwest China. *Fungal Biol.* 118: 453–461.

Wang, T., Tian, Z., Tunlid, A., and Persson, P. 2020. Nitrogen acquisition from mineral-associated proteins by an ectomycorrhizal fungus. *New Phytol.* 228: 697–711.

Wang, Y., Zhang, Z., Zhang, P., Cao, Y., Hu, T., and Yang, P. 2016. Rhizobium symbiosis contribution to short-term salt stress tolerance in alfalfa (*Medicago sativa* L.). *Plant Soil.* 402: 247–261.

Wei, C. Y., Lin, L., Luo, L. J., et al. 2014. Endophytic nitrogen-fixing *Klebsiella variicola* strain DX120E promotes sugarcane growth. *Biol. Fert. Soils.* 50: 657–666.

Wen, Z., Murata, M., Xu, Z., Chen, Y., and Nara, K. 2015. Ectomycorrhizal fungal communitieson the endangered Chinese Douglas-fir (*Pseudotsuga sinensis*) indicating regional fungal sharing overrides host conservatism across geographical regions. *Plant Soil.* 387: 189–199.

Xiang, D., Verbruggen, E., Hu, Y., et al. 2014. Land use influences arbuscular mycorrhizal fungal communities in the farming-pastoral ecotone of northern China. *New Phytol.* 204: 968–978.

Xie, L., He, X., Wang, K., Hou, L., and Sun, Q. 2017. Spatial dynamics of dark septate endophytes in the roots and rhizospheres of *Hedysarum scoparium* in northwest China and the influence of edaphic variables. *Fungal Ecol.* 26: 135–143.

Xie, X., Lin, H., Peng, X., et al. 2016. Arbuscular mycorrhizal symbiosis requires a phosphate transceptor in the *Gigaspora margarita* fungal symbiont. *Mol. Plant.* 9: 1583–1608.

Xie, X., and Tang, M. 2019. Interactions between phosphorus, zinc and iron homeostasis in non-mycorrhizal and mycorrhizal plants. *Front. Plant Sci.* 10: 1172.

Xu, H., Kemppainen, M., El Kayal, W., et al. 2015b. Overexpression of *Laccaria bicolor* aquaporin JQ585595 alters root water transport properties in ectomycorrhizal white spruce (*Picea glauca*) seedlings. *New Phytol.* 205: 757–770.

Xu, H., and Zwiazek, J. J. 2020. Fungal aquaporins in ectomycorrhizal root water transport. *Front. Plant Sci.* 11: 302.

Xu, R., Li, T., Cui, H., et al. 2015a. Diversity and characterization of Cd-tolerant dark septate endophytes (DSEs) associated with the roots of Nepal alder (*Alnus nepalensis*) in a metal mine tailing of southwest China. *Appl. Soil Ecol.* 93: 11–18.

Yang, S. Y., Grønlund, M., Jakobsen, I., et al. 2012. Nonredundant regulation of rice arbuscular mycorrhizal symbiosis by two members of the PHOSPHATE TRANSPORTER1 gene family. *Plant Cell.* 24: 4236–4251.

Yang, Y., Tang, M., Sulpice, R., Chen, H., Tian, S., and Ban, Y. 2014. Arbuscular mycorrhizal fungi alter fractal dimension characteristics of *Robinia pseudoacacia* L. seedlings through regulating plant growth, leaf water status, photosynthesis, and nutrient concentration under drought stress. *J. Plant Growth Regul.* 33: 612–625.

Yu, F., Lin, J., Xie, D., et al. 2020. Soil properties and heavy metal concentrations affect the composition and diversity of the diazotrophs communities associated with different land use types in a mining area. *Appl. Soil Ecol.* 155: 103669.

Zehr, J. P., Jenkins, B. D., Short, S. M., and Steward, G. F. 2003. Nitrogenase gene diversity and microbial community structure: A cross-system comparison. *Environ. Microbiol.* 5: 539–554.

Zeppa, S., Marchionni, C., Saltarelli, R., et al. 2010. Sulfate metabolism in *Tuber borchii*: characterization of a putative sulfate transporter and the homocysteine synthase genes. *Curr. Genet.* 56: 109–119.

Zhang, C., and Kong, F. 2014. Isolation and identification of potassium-solubilizing bacteria from tobacco rhizospheric soil and their effect on tobacco plants. *Appl. Soil Ecol.* 82: 18–25.

Zhang, L., Xu, M., Liu, Y., Zhang, F., Hodge, A., and Feng, G. 2016. Carbon and phosphorus exchange may enable cooperation between an arbuscular mycorrhizal fungus and a phosphate-solubilizing bacterium. *New Phytol.* 210: 1022–1032.

Zhang, N. N., Zou, H., Lin, X. Y., et al. 2020. Hydrogen sulfide and rhizobia synergistically regulate nitrogen (N) assimilation and remobilization during N deficiency-induced senescence in soybean. *Plant Cell Environ.* 43: 1130–1147.

Zhang, X., Zhang, H., Lou, X., and Tang, M. 2019. Mycorrhizal and non-mycorrhizal *Medicago truncatula* roots exhibit differentially regulated NADPH oxidase and anti-oxidant response under Pb stress. *Environ. Exp. Bot.* 164: 10–19.

Zhang, Y., Yang, Q., Ling, J., et al. 2017. Diversity and structure of diazotrophic communities in mangrove rhizosphere, revealed by high-throughput sequencing. *Front. Microbiol.* 8: 2032.

Zhao, D., Li, T., Wang, J., and Zhao, Z. 2015. Diverse strategies conferring extreme cadmium (Cd) tolerance in the dark septate endophyte (DSE), *Exophiala pisciphila*: Evidence from RNA-seq data. *Microbiol. Res.* 170: 27–35.

Zheng, B. X., Zhang, D. P., Wang, Y., et al. 2019b. Responses to soil pH gradients of inorganic phosphate solubilizing bacteria community. *Sci. Rep.* 9: 25.

Zheng, M., Zhou, Z., Luo, Y., Zhao, P., and Mo, J. 2019a. Global pattern and controls of biological nitrogen fixation under nutrient enrichment: A meta-analysis. *Global Change Biol.* 25: 3018–3030.

Zheng, R., Wang, J., Liu, M., et al. 2016. Molecular cloning and functional analysis of two phosphate transporter genes from *Rhizopogon luteolus* and *Leucocortinarius bulbiger*, two ectomycorrhizal fungi of *Pinus tabulaeformis. Mycorrhiza.* 26: 633–644.

Zheng, W., Morris, E. K., and Rillig, M. C. 2014. Ectomycorrhizal fungi in association with *Pinus sylvestris* seedlings promote soil aggregation and soil water repellency. *Soil Biol. Biochem.* 78: 326–331.

Zhou, Y., Qin, Y., Liu, X., Feng, Z., Zhu, H., and Yao, Q. 2019. Soil bacterial function associated with stylo (legume) and bahiagrass (grass) is affected more strongly by soil chemical property than by bacterial community composition. *Front. Microbiol.* 10: 798.

Zhu, L., Li, T., Wang, C., et al. 2018. The effects of dark septate endophyte (DSE) inoculation on tomato seedlings under Zn and Cd stress. *Environ. Sci. Pollut. R.* 25: 35232–35241.

Zhu, J., Li, M., and Whelan, M. 2018. Phosphorus activators contribute to legacy phosphorus availability in agricultural soils: A review. *Sci. Total Environ.* 612: 522–537.

2 Advances in Microbial Molecular Biology
The Potential of the Tool for Agrobiotechnology

Deborah Catharine de Assis Leite[1]
and Naiana Cristine Gabiatti[1]
[1]Universidade Tecnológica Federal do Paraná,
Paraná, Brasil

CONTENTS

2.1 INTRODUCTION

The need for a high agricultural productivity considering the depletion of natural resources and global demand for food has become a big challenge in several countries (Tilman et al. 2002; Loevinsohn et al. 2013). One way to help reach these targets is to incorporate beneficial plant microbes and/or microbiomes into agricultural production. Once these players are in an agrosystem, they increase nutrients, use efficiency, plant productivity, tolerance to abiotic stress, and disease resistance (Busby et al., 2017). Until now, several studies have explored plant microbiome function and structure in agricultural environments (Rachid et al. 2013; Hesse et al. 2015; Hartman et al. 2018;

DOI: 10.1201/9781003110477-2

Chen et al. 2019; Monteiro et al. 2020; Singh et al. 2020). However, according to Busby et al. (2017), these ideas persist underutilized and very little has been done to consolidate and translate them into practical solutions for farmers.

Accessing this diversity is critical for agricultural microbiology, and in this sense, over the decades, several techniques have been developed with the aim of identifying and characterizing microbial isolates and/or communities. Innovations in microbes cultivation and microbiome characterization approaches produced a wealth of novel species (Epstein 2013).

Historically, the first studies involving microorganisms applied to agrobiotechnology focused on prospecting microorganisms with the potential to be used as biofertilizers or biocontrol agents. As the decades passed, next-generation DNA sequencing (NGS) techniques transformed the studies of Microbial Ecology applied to agrosystems (Weinert et al. 2011; Montanari-Coelho et al. 2018; Monteiro et al. 2020). Currently, it is possible not only to identify and characterize microorganisms in isolation but also the microbial communities as a whole (Seshadri et al. 2015; Mendes and Tsai 2018; Monteiro et al. 2020).

This chapter presents an overview of molecular approaches used in microbial ecology applied to agrosystems. Therefore, the discussion is based on three central questions:

A. *Who are they and what can they do?* Here, we intend to present the functional and taxonomic approaches on microbial ecology works in agricultural systems by the use of molecular markers.
B. *How to access microorganisms in agro-environmental systems?* This session brings information about different ways to access microorganism using culture-dependent and independent techniques considering the concepts of microbial prospection and microbial manipulation beyond these approaches.
C. *What comes after that?* Finally, we present the main advances associated with synthetic biology and genetic engineering applied to the study of microbial ecology in agricultural systems.

2.2   WHO ARE THEY AND WHAT CAN THEY DO? TAXONOMIC AND FUNCTIONAL MOLECULAR MARKERS MAY REVEAL MICROORGANISMS IN AGRI-ENVIRONMENTAL SYSTEM

In the last decades, the taxonomic and functional approaches have been widely used in studies on microbial communities in several agri-environmental systems (Delmont et al. 2011; Barret et al. 2015; Rachid et al. 2015; Mendes and Tsai 2018), and this understanding is an essential tool for increasing quality and safety of plant foods (Bokulich et al. 2016). In order to access this microbial diversity, the phylogenetic markers are commonly used to identify these microbial players (e.g., rRNA 16S, 18S, ITS); and the functional markers have been used when one is focused on accessing the roles played by these microorganisms (e.g., genes related to nitrogen cycle, such as *nir*K, *nir*S, *amo*A, *nif*H).

Used together with sequencing techniques, genetic markers have revealed the hidden treasure in microbial genetic information contained in soils and crops around the world (Schoch et al. 2012; Barret et al. 2015; Rachid et al. 2015; Thompson et al. 2017;), and the composition and function of those microorganisms can be addressed both in isolated and combined ways in the studies (Mendes and Tsai 2018; Monteiro et al. 2020).

2.2.1 TAXONOMIC APPROACHES

The taxonomic characterization of microbial isolates and communities in agri-environmental systems appear to be a key item of most studies on soil microbial ecology (Baldrian 2019). For example, identifying microbial pathogens is crucial to strategic and sustainable disease management in agricultural systems. Otherwise, recognizing the beneficial microorganisms capable of increasing soil quality and plant productivity could be useful in further studies on microbial manipulation (Aguilar-Marcelino et al. 2020).

Regarding prokaryotes (bacteria and archaea), the most widely used target is 16S rRNA, either when the objective is to identify a given microorganism previously isolated, or when the intention is to verify the composition of a given plant and/or soil microbiome (Kim et al. 2011). This gene region was first used in phylogenetic analysis by Lane et al. (1985). The ~1500 bp 16S rRNA gene sequence comprises both highly conserved regions proper to primer recognition and also hypervariable regions (V1–V9) to identify phylogenetic characteristics of prokaryotes, turning the most widely used marker gene on bacterial/archaeal communities' studies (Baker et al. 2003; Tringe and Hugenholtz 2008; Johnson et al. 2019).

However, the abundance of 16S rRNA sequences does not mean taxon abundances, once bacteria genomes have variable numbers of 16S rRNA, varying between 1 and 16 (Farrelly, Rainey, and Stackebrandt 1995). Alternatively, some studies have proposed the use of other molecular markers that have a single-copy gene in the bacterial genome, such as RNA polymerase β subunit (*rpo*B) (Case et al. 2007; Liu et al. 2012) and gyrase β subunit (*gyr*B) (Yamamoto and Harayama 1995; Kanako Watanabe et al. 2001). Nevertheless, the 16S rRNA gene is still the majorly used genetic marker in most papers on prokaryotic communities on agricultural systems.

When it comes to access to fungi, finding the ideal genetic marker is an even more complex task. Until now, the internal transcribed spacer (ITS) region seems to be the main fungi marker for offering the best taxonomy resolution (Schoch et al. 2012), even though its approach has many limitations. Unfortunately, the ITS cannot be aligned in distant taxa and, hence, it may not be used for phylogenetic studies. In this sense, there are approaches that involve the use of fungal markers based on small subunit (SSU) 18S or large subunit (LSU) 28S rRNA regions. However, these candidate regions have a low taxonomic resolution or are long for massive sequencing strategies (Andrea et al., 2014). Regarding the variation in the number of copies per fungal genome, any rRNA marker (LSU, SSU, or ITS) can present counts that are between a single copy and hundreds of copies per fungal genome (Schoch et al. 2012; Rachid et al. 2015; Baldrian 2019).

The sequences obtained by the sequencing of these regions can be easily deposited in databases as SILVA, RDP, Green genes, EMBL, NCBI, and others (Balvočiūtė and Huson 2017). For example, the SILVA database contains taxonomic information of Bacteria (16S rRNA), Archaea (16S rRNA), and Eukarya (18S rRNA) (Yilmaz et al. 2014). The RDP database is dedicated to 16S rRNA sequences from Bacteria, Archaea domains and 28S rRNA sequences from Fungi (Eukarya) (Cole et al. 2014).

Finally, for other eukaryotic groups such as soil protists, finding effective general primers is still a challenge and until now it was observed as a primer-based approach limited to some taxonomic groups (Dumack and Bonkowski 2021). In this sense, the *PR²database* (Guillou et al. 2013) consists of a useful tool, which harbors a large amount 18S rRNA of protists diversity and assigned taxonomic affiliation, allowing the characterization of plant- and soil-associated protists. Currently, this project has merged with *EukRef Initiative*, a project that provides protocols, guidelines, and other tools aimed at organizing a sequence data mining and annotation (del Campo et al. 2018).

As for viruses in agricultural systems, there are no polymerase chain reaction (PCR)-based methodologies, due to the characteristics of viral multiplication. Therefore, the prospect of this kind of organism requires a direct-sequencing approach such as metagenomic and/or metatranscriptomic data collection (Pratama and van Elsas 2018).

2.2.2 Functional Approaches

Constantly, biological diversity is used as an index that reflects an ecosystem quality. In this way, methodologies that enable the study of microbial diversity should also indicate differences among distinct sources from the same environment (i.e., soils) both with respect to their populations and functions. Therefore, when assessing microbial diversity, important considerations must be made, not only regarding the number and distribution of species but also regarding their functional diversity (Levy, Conway et al. 2018).

Speaking on microbes' role in agri-forestry systems, soil can be considered the most representative environment (Nannipieri et al. 2012). Although there is a consensus between researchers that maintaining/improving soil quality is a key element for the sustainability of agricultural systems, the selection of the best bioindicators to report soil quality is a challenge that generates extensive debate (Schloter et al. 2018).

The multiplicity of chemical, physical, and biological factors that control biogeochemical processes and their variations according to time and space, combined with the complexity of the soil, are among the factors that hinder the ability to access its quality and identify key parameters that can serve as indicators of its functioning (Rachman 2019). Despite that, it is well known that the biological soil component has a close interrelation with the physical and chemical components, and these will jointly influence both crop productivity and the sustainability of agricultural systems (Chen et al. 2019).

Because of this complex interaction among different variables, to consider only microbial taxonomy could neglect the whole macro scenario (Partida-Martínez and Heil 2011). Instead, looking for microbial traits, such as substrate utilization or growth rate, could be helpful for better understanding the biochemistry of soil and finding efficient ways to address enzymatic processes or how to adopt metabolomics approaches (Baldrian 2019). This is equivalent to describing a microbial community focusing on relevant metabolic functions or to establish a functional microbial community profile (Raes et al. 2011).

Soil microorganisms are responsible for environmental processes of fundamental importance, such as the soil formation, decomposition of organic residues (animals and plants), cycling of nutrients and formation of organic matter, as well as bioremediation of pollutants and pesticides, among others. Moreover, there are metabolic functions that are strongly coupled to certain environmental factors (Louca et al. 2018). Hence, metabolic profiling can be useful to elucidate changes in soil biological quality caused by environmental interferences as anthropogenic activities. This is possible because metabolic diversity results from genetic diversity, environmental effects in gene expression, and ecological interaction among different populations (Cardoso et al.2013). Therefore, for a given soil, diversity assessments involve genetic aspects (richness/abundance of genomes) and also functional aspects (variety of decomposition functions, nutrient transformation, promotion/suppression of plant growth, etc.; Escalas et al. 2019).

Even plant-associated microbiota are mostly originated from the soil in which the seed germinates and seedlings start to grow (Berg and Smalla 2009). Thus, the plant microbiome is not merely the collection of microorganisms associated with a plant, but the set of biotic and abiotic factors that influence the health and productivity of plants in a defined biome (Tosi et al. 2020). Precise and well-understood molecular information on functional components involved in soil quality can complement the ecological, morphological, and agronomic information of the genetic resources. In addition to that, molecular methods combined with physiological approaches may be the key to unveil gene expression, protein synthesis, and enzyme activities at the micro- and nanoscales (Bender, Wagg, and van der Heijden 2016).

It is remarkable how biotechnology has revolutionized agriculture with modern technologies that allow us to identify and select genes that encode beneficial characteristics to be used as molecular markers in assisted selection processes, or to have the expression of a certain gene in another organism by transgenics, obtaining new agronomic and nutritional characteristics desirable in plant cultivation (Montagu 2020).

The combination of relative enzyme activity in soil with enzyme-encoding genes expression (transcriptomics and proteomics) may contribute to elucidate the mechanism and pace of microbial community responses to nutrient availability, persistence, and stabilization of enzymes in the environment (Nannipieri et al. 2012).

This is a good strategy to overcome limitations of classical enzymatic methods. Hence, it can be very helpful to understand organic matter decomposition and its impact on local and global C and N cycling. Microorganisms are protagonists in decomposition of complex C compounds, once they are the most significant producers of central role enzymes, mainly the CAZymes (Boutard et al. 2014). Many studies have been investigating the genes related to CAZymes in genome and its abundance, as well as its transcription according to factors as soil diversity and seasonality (Howe et al. 2016; Montella et al. 2017; Žifčáková et al. 2017).

While searching for emphatic evidence that soil microbes perform the enzymatic activities (Trivedi et al. 2016), studies connected the variation in microbial community composition to differences in functional gene abundance. This evidence has consequences for the activity of enzymes directly related to C degradation at field to regional scales. Nevertheless, other authors (Wood et al. 2015) suggested that C mineralization process rate should be regulated by the expression of related genes, rather than the overall abundance, once they did not find a direct relationship between the two factors. These controversial statements show that parameters selected to measure ecosystem multiple functions and services need to be carefully selected in order to make molecular analysis results accurate enough to be widely applied (Xie et al. 2020).

Continuing on biogeochemical cycles, one of the oldest known functional genes is *nifH*, which has highly conserved sequences. Together with other genes, they code the formation of the FeMo protein, a component of the nitrogenase enzyme, responsible for the biological nitrogen fixation process (Hoffman et al. 2014). Although, initially, it was more applied to establish phylogenetic relationships, its use for the study of microbial populations in the environment has been more useful in the correlation between structure and function of the microbial community (Yang et al. 2019). Its use also allows the specific study of diazotrophic bacteria, enabling knowledge about the diversity of this special group of microorganisms associated with different environments (Turk-Kubo et al. 2012; Gaby and Buckley 2014).

There is a giant set of research focusing on genes involved in the N cycle. These genes are reasonable bioindicators because they can be a representative and accurate manner of function address. Besides that, they have great potential and sensitive response to catch environmental changes, mainly in soil (Pereira e Silva et al. 2013). Many studies have demonstrated that denitrifying microbial communities' specific features such as diversity, abundance, and ability to produce N_2O/N_2 could be used as indicative of denitrification in environments (Wei et al. 2014). The reduction of nitrite to oxide nitric is a key step in the denitrification process, which is catalyzed by NirK and NirS. These genes have been successfully used as marker genes in molecular studies of the ecological behavior of denitrifiers in different environments (Herold et al. 2018).

In the past few years, sequencing of thousands of bacterial isolate genomes from different plant environments allowed their comparison to identify functional genes that affect general adaptation to plants (Levy, Salas Gonzalez et al. 2018), root adaptation in relation to shoot adaptation (Bai et al. 2015), symbiosis involved in nitrogen fixation and its association to nodulation (Seshadri et al. 2015), and biological control activity (Köhl, Kolnaar, and Ravensberg 2019). Studies in microbial quorum-sensing genes expression (Schaefer et al. 2013) have been trying to elucidate the mechanisms that also allow plants to detect and respond to those signaling molecules, specifically N-acyl homoserine lactones (AHL) (Mathesius et al. 2003; Shrestha et al. 2020).

The prompt development of metabolomics has been enabling the reconstruction of dependent metabolic pathways among soil bacteria and estimation of nutrient pools (Swenson et al. 2018).The catabolic and anabolic pathways of an active prokaryotic cell are defined from biochemical processes that produce a highly complex mixture of metabolites. Endometabolomics allowed the investigation of several prokaryotic organisms, considering their substrate source and growth conditions. This was done by crossing genomic information about the metabolic potential of a cell and the patterns of metabolites produced within it (Wienhausen et al. 2017). Metabolome analysis can detect metabolites and show the size of their clusters at the time of sampling and, more specifically, exometabolome refers to exogenous metabolites that microorganisms process and produce (Paczia et al. 2012).

In this regard, soil metagenomics and metatranscriptomics can be successfully applied to portray the community-wide carbon use and even modeling potential impacts of global climate change on these ecosystems (Swenson et al. 2015; Žifčáková et al. 2017). At the same time, these tools can be used to estimate microbial contribution to ecosystem nutrient cycling or N deposition patterns (Hesse et al. 2015; Mackelprang et al. 2018) mediated by bacteria, fungi, and other members of the soil biota.

2.3 HOW TO ACCESS MICROORGANISMS IN AGRO-ENVIRONMENTAL SYSTEMS?

Soil serves as a microbial repository capable of harboring a diverse range of microorganisms, including bacteria, archaea, protozoa, fungi, and viruses (Delmont et al. 2011; Nannipieri et al. 2017). However, the distribution of these organisms in the soil is not uniform, and the soil portion capable of interacting with the root is a "hot spot" with regard to microbial abundance and activity—the *rhizosphere* (Berg and Smalla 2009; Compant et al. 2019; Kumar and Dubey 2020). Those members of the soil and plant microbiomes can play essential roles on crops such as plant growth (Li et al. 2016; Gamez et al. 2019), biomass enhancing (Farwell et al. 2006; Zhang et al. 2014), disease suppression (Shen et al. 2019), and stress tolerance (Hussain, Mehnaz and Siddique 2018). That is, the soil microbial manipulation will impact plant traits and performance (Compant et al. 2019).

This huge microbial diversity can be accessed by several amounts of techniques that includes a single organism prospect (Farwell et al. 2006; Zhang et al. 2014; Li et al. 2016) to even the whole microbiome description (Hesse et al. 2015; Vollú et al. 2018; Chen et al. 2019). These methodologies for accessing microorganisms can also be classified as culture-dependent (which involves a single microorganism isolation) and culture-independent (which provides access to the total microbial community without requiring a prior microorganism isolation). The choice of a method or another depends on the question to be answered by the researcher. However, studies on microbial ecology in agrosystems have shown a tendency towards a polyphasic approach bringing together the most diverse techniques (Nannipieri et al. 2017).

2.3.1 MICROBIAL PROSPECTION AND MANIPULATION

Traditional and advanced bioprospecting technologies have been developed and employed to enhance knowledge of the microbiological diversity of functional processes in microbial ecosystems, as well as for the identification and use of new biotechnological products (Beattie et al. 2011). It's possible to discover and characterize new genes, enzymes, bioactive metabolites, and drugs associated with the rich diversity of organisms not yet cultivated and the development of new strategies for the selection and screening of new products, targets, and assays based on knowledge of genomics and expression of different organisms (Rasheed et al. 2014; Musumeci et al. 2017; Kamble, Srinivasan, and Singh 2019).

Microbial prospecting consists of isolation, identification, assessment, and systematic exploration of the diversity of microbial life existing in natural habitats for novel biological products and activities with biotechnological applications search for genetic resources for research and/or commercial purposes (Lozada and Dionisi 2015). Considering that different strategies are applied for exploring the culture-dependent or non-cultivable microbial biodiversity.

The *in-silico* prospecting of microorganism's genomic sequences already available in the database have been gaining space among researchers (Kamble, Srinivasan, and Singh 2019). This method is mainly based on the discovery of new genes and metabolic pathways by analyzing genomic sequences deposited in a database (Gerlt 2016).

The decrease in sequencing costs has enabled the generation of a large amount of information on microbial genomes (Muir et al. 2016). Every year, plenty of new complete genomes of bacteria, fungi, and archaea are delivered by the scientific community (Land et al. 2015). In silico mining of these genomes through comparative genomics approaches and evolutionary analysis, with the help of bioinformatics tools, has allowed the rapid identification of new genes and enzymes with the most different functions and applications (Zhang et al. 2018).

The identification and validation of new enzymes related to the degradation of lignocellulosic biomass or sugar metabolism in bacteria and yeasts are good examples of the applicability of this technique. The combination of different bioprospecting strategies to identify new bioproducts has been used. In-silico bioprospecting can be used for the design of molecular probes, which are later used in prospecting strategies for cultivable microorganisms or metagenomic libraries (Xin et al. 2020; Bustamante et al. 2019; Egan et al. 2018).

Additionally, when the proposal is microbial manipulation, identifying and characterizing microorganisms is the first step in being able to manipulate them, and once soil- and plant-associated microbiomes can promote plant growth or control pathogens, manipulating microbes allow us to alter some crop processes. For example, changing the microbiome by inoculation of a microbial consortia can enhance plant productivity and resistance to pathogens. Contrarily, modulating plant holobiont by microbiome engineering is an upcoming biotechnological strategy to promote crop resilience and yields (Arif, Batool, and Schenk 2020). Thus, the success or failure of soil- and plant-microbiome manipulation may be based on suitable modifications of resource availability through soil use and management (Bell et al. 2019).

Over past years, agrosystems have been managed by focusing on individual components (e.g., nutrient utilization, pesticides, and organisms biocontrol). However, it is necessary to think of the soil-plant-microbiome-abiotic factors as an integrated system which have a greater potential to result in a sustainable ecosystem (Bell et al. 2019).

These different approaches can lead to the finding of microbes or microbiome characteristics that improve crops performance in non-ideal situations. For example, traditionally, cultivated isolates have been picked for beneficial impacts and some have been developed into agricultural products and/or processes, such as those that increase tolerance (Zhang et al. 2014). On the other hand, some studies that describe the soil microbiome altogether have shown that, during the growth of plants under stress, changes occur in soil microbial assemblies (Jurburg et al. 2018).

2.3.2 Microbial Isolation Strategies

Originally, the isolation of a microorganism supposes its cultivation. Nevertheless, traditional techniques of culturing microorganisms alone are not sufficient for a more wide exploration of the microbial community, its ecological functions, and diversity in natural environments (Stefani et al. 2015).

In any case, culture-dependent and independent techniques have unveiled an outstanding diversity of microorganisms whose most metabolisms mechanisms have yet to be well characterized (Vartoukian et al. 2016). Although culture-independent, supported by sequencing technologies and bioinformatics, have revolutionized environmental microbiology, the development of biotechnological applications from the microbiome genetic potential must also be anchored by the corresponding study of pure culture (Metzker 2010; Lagier et al. 2015).

Many researchers have been now focusing on environmental studies using molecular biology techniques. However, despite the great contribution of the new molecular tools for studies on biodiversity, the traditional enrichment and cultivation techniques are also important for the knowledge of the metabolic capacity and phenotypic characteristics of microorganisms (Kazuya Watanabe and Baker 2000; Lozada and Dionisi 2015). In this sense, as conservative cultivation methods are mainly based on nutrient-rich media, favoring only rapid growers, novel approaches are designed to get as close as possible to the real picture of an environment (Chaudhary, Khulan, and Kim 2019).

Strategies such as co-culturing with helper agents, modifying nutrients composition, simulating natural habitat, optimizing growth conditions, and prolonging the incubation time to access uncultured domains have reported good results (Stewart 2012; Chaudhary, Khulan, and Kim 2019). Advancements have been made in cultivation methods using transwell plates, diffusion chambers, optical tweezers, microbioreactors, and laser microdissection (Pham and Kim 2012).

As an example, Stevenson et al. (2004) explored an integrative approach to obtain pure cultures of formerly uncultivated members of Acidobacteria and Verrucomicrobia phylum from agricultural soil. They developed a simple, high-throughput, PCR-based procedure that facilitated distinction, detection, and the final isolation of target bacteria from several colonies on the same agar plate.

In another study, Bollmann, Lewis, and Epstein (2007) applied enrichment cultivation with diffusing bioreactors as a strategy to enrich uncultured microbial species. They were able to simulate the natural environment and allow the microbes with access to growth components and signaling compounds.

The diffusion chamber is a device in which bacteria are inoculated in an agar layer separated from the source environment by membranes. In this system, nutrients and growth factors from the environment are able to pass through and the microbial cells could isolate (Kaeberlein, Lewis, and Epstein 2002). A variation of this approach is the microbial trap, which enriches filamentous bacteria selectively (e.g., actinomycetes) by allowing the bacteria colonization of the agar through porous membranes (Gavrish et al. 2008). With a similar idea, the isolation chip is an example of second-generation high-throughput automated method. The culture and isolation device comprises several miniature diffusion chambers, and each one of those are inoculated with a single environmental cell (Nichols et al. 2010).

Currently, less traditional and more efficient selection techniques, including mimicking industrial conditions and automated process conditions, have allowed the cultivation and screening of hundreds of microorganisms for desirable, even complex, characteristics in a short time (Kato et al. 2018).

Single-cell isolation techniques are methods that allow the physical separation of cells from each other and/or from matrix materials (Zengler 2009). Many single-cell isolation techniques, including micromanipulation, dilution, micro fluidics, flow cytometry, and compartmentalization, have been developed. These techniques have been used in microbiology and biotechnology not only aiming to cultivate previously uncultured microbes but also to evaluate and monitor cell function and physiology and to screen for novel microbiological products (e.g., enzymes and antibiotics) (Ishii, Tago, and Senoo 2010).

Miniaturized tests and online detection systems can be used to search for a large number of microorganisms that produce bioactive compounds and to identify molecules of interest (Sekurova, Schneider, and Zotchev 2019). In addition, the cultivation of microorganisms previously said to be non-cultivable has been improved by expanding knowledge of physiology, biochemistry, and microbial ecology through the use of large-scale phenotyping techniques, which allow the analysis of several characteristics simultaneously and facilitate the optimization of culture media (Bollmann, Lewis, and Epstein 2007; Lewis et al. 2010; van der Meij et al. 2017; Montanari-Coelho et al. 2018).

2.3.2.1 Microbial Isolates: Identification and Characterization

Traditionally, the detection, identification, and characterization of microorganisms had been done based on carbon (or other energy sources) acquisition, their nutritional requirements, and the culture medium for their growth, in addition to microscopical direct observation. However, the use of these methodologies provided limited information and often needed further refinement (Schloter et al. 2018). As an alternative to these methods, several techniques have been developed, among which stand out those based on nucleic acids.

Efficient and reliable strain identification remains the most important objective in taxonomic studies. The techniques derived from molecular biology, mainly from recombinant DNA technology and genetic engineering, enable more enlightening studies in microbial ecology and applied microbiology (Hofstetter et al. 2019). Methods based on the cultivation of microorganisms associated with the use of molecular biology techniques for the isolate's characterization offer a great possibility for the selection of organisms that have biological functions of biotechnological interest. Moreover, metagenomics makes it possible to identify genes directly from environmental samples without the need to use culture media, probes, or specific markers for a given group of organisms (Nannipieri et al. 2017).

The first molecular techniques applied in bacterial identification were based on plasmid profiling, GC content, and compatibility to genetic transformation (Barghouthi 2011). Since the emergence of molecular markers in the 1980s, science has been following the advances of the genomic era, taking advantage of the large volume of DNA sequence information available (Hofstetter et al. 2019). In addition to the advantages presented over morphological markers, existing technologies provide a wide variety of polymorphisms randomly distributed throughout the genome. This has been achieved in an automated manner and at increasingly reduced costs. The molecular techniques progress has been closely followed by the great development in the fields of bioinformatics, statistics, and quantitative genetics (Dini-Andreote et al. 2018; Mendes and Tsai 2018). Thus, genetic information anchored to highly saturated physical and genetic maps are readily available, in addition to the countless differentially expressed genes, which are easily integrated through bioinformatics (Kumar and Dubey 2020).

The methodology for extracting environmental DNA is a key step in the representativeness of molecular microbial ecology studies (Ruppert, Kline, and Rahman 2019). The optimization of extraction procedures helps to eliminate factors of instability by focusing on the specific aims of study (Albertsen et al. 2015). The methodologies for obtaining DNA are basically divided into two types: those of direct extraction, where the cells are lysed directly from the sample; and indirect ones, where cells are initially separated from environmental samples and subsequently cell lysis is performed. The previous is considered more representative, as it would be able to lyse the cells that would be adhered to soil aggregates. Indirect extraction, on the other hand, generates better quality DNA, that is, with higher molecular weight (Delmont et al. 2011).

There have been two molecular applications being commonly utilized in bacterial identification and detection; these are based on hybridization and nucleotide sequencing. Most of the molecular tools currently used for environmental studies and analysis of microbial communities are based on the PCR technique that is the basis for molecular methods that have this purpose (Agrawal, Agrawal, and Shrivastava 2015). PCR consists of the in vitro enzymatic synthesis of DNA copies from a target sequence.

First described by Saiki et al. (1985), PCR consists of the in vitro enzymatic synthesis of DNA copies from a target sequence; it is usually the first step in many molecular approaches. For amplification, primers complementary to those found in the genome-specific sites are used. Through Taq DNA polymerase enzyme (isolated from the micro-organism Thermus aquaticus) action, the DNA fragment is extended from the primers. A PCR reaction consists of a mixture of reagents, Taq DNA polymerase, primers, free deoxynucleotide triphosphates (dNTPs), magnesium ions, and the target DNA.

The first applications of techniques based on nucleic acids in the study of microbial ecology were related to phylogenetic relationships between microorganisms determined by analyzing the sequence of the 16S rDNA (Macrae 2000), which generates a large amount of useful information for phylogenetic inferences. Currently, other techniques have been applied, based on the use of these markers (ARDRA, Hybridization in situ). Nevertheless, 16S rRNA gene sequencing is still considered one of the most powerful methods for exploring microbial diversity and molecular identification of species isolated from natural samples (Jeong et al. 2021).

This approach revealed that microorganisms isolated by traditional methods, using culture medium, may not represent the authentic strains present in a great number of environmental samples since they do not correspond to those most frequently identified through 16S rRNA gene PCR and sequencing (Baker et al. 2003).

Over the past years, a large number of studies on microbial identification and characterization have been published providing valuable insight into understanding the role of microbes in agricultural productivity (Singh et al. 2010), the role played by plant-associated microbes in global ecology and agriculture (Giangacomo et al. 2020), plant growth-promoting bacteria isolated (Park et al. 2005; Ayyaz et al. 2016), or finding relation in the beneficial traits for plants that isolated soil bacteria from extreme environments features may reveal (Gaete, Mandakovic, and González 2020).

On the other hand, molecular marker techniques have dramatically increased the number of available genetic markers, representing a powerful tool to be used in microorganism's identification (Nadeem et al. 2018). Methods such as RFLP, initially proposed by Botstein et al. (1980), are able to differentiate individuals through individual variations in nucleotides due to mutation, deletion, insertion, and inversion. The method consists in the hybridization of specific DNA sequences (probes) with the total DNA of the analyzed individuals digested by restriction enzymes. Several advantages have been pointed out for this type of marker; RFLPs are available in large numbers, exhibit great natural variability, are not influenced by the environment, and are free of epistasis effects, which allows the evaluation of several markers at the same time (Yang et al., 2013).

The Restriction Analysis of Amplified Ribosomal DNA (ARDRA) is a type of RFLP based on the conservation degree of rDNA restriction sites reflecting phylogenetic patterns. Despite ARDRA being useful for a quick analysis of the diversity of an environment (Sklarz et al. 2009), it is necessary caution when choosing the rDNA fragment to be amplified and analyzed by this method. When considering intraspecific diversity analysis or accessing microbial isolated groups with high phylogenetic affinity, the amplified fragment must include the 16S–23S rDNA intergenic spacer. This intergenic region presents greater variability in both its base composition and its size when compared to 16S or 23S rDNAs. When the study in question deals with the diversity between distant isolates, phylogenetically, the amplified fragments can be 16S or 23S rDNAs (Hoffmann et al. 2010).

Heyndrickx et al. (1996) elucidated the Bacillus genus phylogeny and taxonomy through application of ARDRA. These researchers found a considerable diversity in interspecific and intergroup phylogenetic relationships, matching with 16S rDNA sequence analysis. However, despite many attempts of using ARDRA for strain typing particularly microbial genera, differentiation have only been partially successful (Sklarz et al. 2009).

Over time, ARDRA has been used together with other techniques when concerning microbial species differentiation. *Rhizobia* species isolated from nodules of Vicia were identified using a set of molecular approaches including 16S ARDRA, RFLP of 16S–23S internally transcribed spacer (ITS), and 16S rDNA sequencing. The phylogenetic relationships assembled mostly supported the relationships estimated by the ARDRA and ITS-RFLP, although some discrepancies were detected (Lei et al. 2008).

Similarly, the randomly amplified DNA segment polymorphism (RAPD) technique, proposed by Williams et al. (1990), is characterized by the use of random primers with a size around 10 bp. Hence, whenever the genome of the individual to be analyzed has a nucleotide sequence corresponding to that of the primer, the amplification process will be initiated, and differences in the level of DNA are inferred by the presence or absence of a particular amplified fragment.

RAPD can be equivalent to both mutations and insertions or deletions at the primer pairing site. Its advantage is that, besides the convenient execution, it does not require specific sequencing and primer design. Requires low amounts of DNA and sample repetitive DNA regions (Atienzar et al. 2002). However, it presents low genetic information at each locus and low reproducibility. In addition to its application in the differentiation of close species, it is widely used for determining genetic diversity and structure, establishing phylogenetic relationships and differentiating close species (Baker et al. 2002).

Molecular phenotypes, generated from RAPD, can be used to diagnose different taxonomic levels, and may be able to discriminate even intraspecific levels. Di Cello et al. (1997) applied RAPD methodology to study changes in the population structure of *Burkholderia cepacia* during the growth of corn plants. The authors showed the high degree of genetic diversity among the isolates. In the initial stages of plant growth, the diversity presented was greater compared to the last stages at the end of maturation. The most possible reason attributed for this variation is the greater instability of an ecosystem with young plants compared to developed plants.

The presence of repetitive sequences (rep-elements) in bacterial genomes has been studied and also used for genotypic characterization. Primers were developed to amplify three families of so-called repetitive elements, resulting in the techniques of REP-PCR (repetitive extragenic palindromic sequences—PCR), ERIC-PCR (enterobacterial repetitive intergenic consensus—PCR), and BOX-PCR (Rezene et al. 2018). These strings correspond to 35–40 bp for REP; 124–127 bp for ERIC, and 154 bp for the BOX element, which consists of three subunits (boxA, boxB, and boxC). When one of these repetitive elements is detected within an amplifiable distance during the polymerase chain reaction, a PCR product of characteristic size is generated, so that the genome can generate a fingerprinting pattern on a gel. The method is a powerful tool for studying intraspecific and strain-specific genetic diversity, providing information complementary to previous characterization by other methodologies (Knutsen et al. 2006).

Although these primers were developed for repetitive elements in prokaryotic genomes, they have been successfully applied to the fingerprinting of pathogenic fungi infecting crops (Alves et al. 2007). Thus, the three primer sets BOX, ERIC, and REP have been used to distinguish variability at inter- and/or intraspecific levels of several fungal genera (Abdollahzadeh and Zolfaghari 2014).

Based on conventional (or qualitative) PCR, real-time PCR was developed, playing a technological leap, as it opened new and powerful applications for research, due to its great detection sensitivity and the possibility of monitoring in real time the generation of PCR products (Paiva-Cavalcanti, Regis-da-Silva, and Gomes 2010).

The real-time or quantitative PCR (qPCR) can be performed with initial DNA or RNA. The quantification of the specific gene or transcript generated in real time, in a very sensitive and precise way, is made possible by the use of a dye that interleaves in the chains of the fragments generated in the PCR (usually Sybr Green). This dye can be detected by thermocycler sensors, allowing the detection of the increased concentration of the generated fragment in real time. This methodology can indicate whether certain genes are too much or too little expressed in a system or in the portion of DNA present in the analyzed sample. It is a very sensitive methodology and the work with it today has a great impact. It is the methodology par excellence for evaluating studies carried out in cDNA libraries or DNA microarrays (Kralik and Ricchi 2017).

With the advent of qPCR as an important tool for DNA and RNA evaluation, a wide range of studies have been carried out in plant-microorganism interaction, enabling the identification and quantification of microorganisms in several matrices (Moustafa and Cross 2016; Xu and Wang 2019).

In plant health, this method has been helping in the management of plant disease in agriculture and forestry, once it allowed rapid, sensitive, specific, and high-throughput detection and quantification of organisms (Bilodeau 2011). Following the trend of using growth-promoting microorganisms associated with plants, the demand for methodologies able to monitor their presence within plant tissues has increased. The development of primers to perform real-time PCR based on several strains identification and quantification gained a wide space in scientific studies (Pereira e Silva et al. 2013).

Finally, the basic information about an isolated microbial strain can be quickly obtained by using high-throughput technology to determine its whole genome sequence. Gene annotation can quickly provide the sequence of relevant coding genes (Cheng et al. 2020). In this sense, whole-genome sequence (WGS) analysis has become available as a routine tool for quickly identifying, characterizing, and tracking microorganisms of interest. Increased resolution allows investigators to identify bacteria that are trusted members of a specific functional group, for instance (Hasman et al. 2014).

The first complete genome sequence of a plant-related microorganism was *Xylella fastidiosa* (Simpson 2000), a pathogenic bacterium. Since then, advances in both WGS and computational methods have played an important role in the understanding of the molecular genetics of many bacterial species (Gupta et al. 2014). Genomic information of plant bacterial pathogens can reveal the potential virulence mechanisms to be used in the design of effective disease control approaches (Xu and Wang 2019). At the same time, next generation sequencing (NGS) technologies have been employed to study genomes of several bacteria capable of stimulating and aiding plant growth from different cultures (Gupta et al. 2014; Matteoli et al. 2018; Subramaniam et al. 2020).

Over time, these genomic studies constantly assumed that the status of individual cells will be similar to that observed in the population study rather than individual cells (Chen et al., 2017). Nevertheless, reports on cell-to-cell heterogeneity at both cellular and molecular levels in the isogenic population is greater than previously thought and has been gaining space among the scientific community (Xu and Wang 2019). In this regard, studies related to the single-cell level involving NGS, flow cytometry, and microspectroscopy have been dissecting the contributions of individual cells to the biology of soil, forest, and agriculture ecosystems and organisms (Gawad, Koh, and Quake 2016).

Traditionally, the study of individual microbes are defined as *omics*, such as study of genes (genomics), transcripts (transcriptomics), proteins (proteomics), metabolites (metabolomics), lipids (lipidomics), and interactions (interactomics); however, the *meta-omics* approaches are used when referring to microbial communities, for example, metagenomics, metatranscriptomics, and metaproteomics (Wang and Bodovitz 2011).

The sustained increase in registered genomic sequences showed that in many cases the data that they provide could not be enough to elucidate all the issues raised on microbial analysis (Zuñiga, Zaramela, and Zengler 2017). Since then, the *omics* were developed as a new area that integrates different data from the functional genome. Functional genomics means the attribution to useful biological information for each gene, improving the knowledge of how the different biological molecules contained within the cell combine to turn into a viable organism (Greenbaum et al. 2001). In this way, it is possible to understand an organism at a systemic level since omics involve many components of the functional genome.

The RNA-sequencing (RNA-seq) technology has been applied on exploring the plant host responses during the disease development, showing that transcriptomic can be a robust tool to record the molecular changes for the organisms under different physiological state (Wang & Qian, 2009), as well as to understand the virulence factors and host range determinants of different pathogens (Jalan et al. 2013; Garita-Cambronero et al. 2016).

The multi-*omics* method, including genomics, transcriptomics, proteomics, metabolomics, lipidomics, and phenomics have been readily applied to characterize the diversity and metabolic potential of microbial isolates in a wide range of different environments including soil and plants (Eicher et al. 2020). This integrated -*omics* approach has the potential to reduce the understanding gap between phenotype and genotype once it opens the possibility of previously unknown features of microbial cells detection and track molecular components across multiple cellular functional states (Fondi and Liò 2015).

2.3.3 Microbial Communities Characterization

Historically, the pioneering indicators used to monitor soil microbiological quality in agrosystems have been based on indirect measures of the presence of this microorganism through parameters such as microbial biomass, microbial respiration rate and specific enzymes present in the soil (Schloter et al. 2018).

The *soil microbial biomass* is the alive part of the soil organic matter, may represent 1–5% of the total organic carbon in the soil and formed basically by bacteria, archaea, fungi, protozoa, and algae (Cardoso et al. 2013; Schloter et al. 2018). These soil microorganisms are capable of enhancing plant growth due to nutrient cycling and availability (Lange et al. 2015). In this sense, the impacts of crop rotations, soil type, soil pollutants, agricultural management, and nutrient turnover can be easily assessed by microbial biomass changes (Horwath and Paul 1994). The methods to assess microbial biomass include the chloroform fumigation incubation method (CFI) and chloroform fumigation extraction method (CFE), which aim to evaluate the microbial C stock in the soil, the substrate-induced respiration (SIR), which measure the basal rate of respiration of soil active microbes (Horwath and Paul 1994).

Soil respiration plays a fundamental role in the C cycle and its measuring allows quantifying microbial activity. When comparing the rate of microbial respiration with the amount of organic matter and microbial biomass, an overview of the soil can be obtained in relation to the survival and metabolism of the microorganisms that make up the ecosystem under study. This is possible, as microorganisms are responsive to quick changes in the soil, increasing their respiratory rate and also the use of carbon and nitrogen (Horwath and Paul 1994). These methods consider the metabolic quotient (qCO_2), an index given by the amount of C–CO_2 liberated per unit of microbial biomass in time. This quotient means the metabolic status of the soil microbes and, commonly, the higher the indices, more stressing environmental conditions and/or inputs of quickly degradable organic carbon that rapidly stimulates the microbial activity. In agrosystems, the soil management and use affect the microbial activity and biomass, and in general, the less impactful management culminates in a higher microbial biomass and activity (Cardoso et al. 2013).

Besides microbial activity and biomass, *soil enzymes* can also be helpful soil health indicators, such as dehydrogenases, asparaginase,β-glucosidases, cellulases, phosphatases, amylases, ureases, glutaminases, and aryl sulphatase (Navarrete et al. 2021). These enzymes catalyze many reactions from C and N cycles in soil, representing the metabolic level of the soil microbiome. Once the soil microbes are influenced by soil use and management, alters in soil enzyme activities are promptly expected (Cardoso et al. 2013).

Despite being used until today, these microbial quality indicators do not provide information on *who are* the main microorganisms present in the soil, and/or which are the major microbial groups in healthy or degraded soils. With regard to function, the use of individual enzymes shows a high degree of variability in response to climate, season, and/or geographic location, which may result in contradictory conclusions in the studies that describe the impact of contaminants or a soil management (Rao et al. 2017).

Direct *microscopic observation of microorganisms* could be an alternative tool for agrobiotechnology studies. These observations could be performed with fluorescence microscopy (FM), transmission electron microscopy (TEM), and scanning electron microscopy (SEM) (Singh et al. 2018; Navarrete et al. 2021). TEM and SEM are capable of providing high-resolution images that may be useful in ultrastructural studies of microorganisms and soil aggregates (Li, Dick, and Tuovinen 2004) or microbes attached to plant roots (Du et al. 2018). Liu et al. (2020) observed that root surface is a very inhomogeneous. The interaction between the bacteria *Pseudomonas fluorescens* SBW25 and *Brachypodium distachyon* on the roots was weak and limited to a few spots along the root.

For studies where lower resolution is enough, fluorescence microscopy is frequently used. For example, microbial enumeration or the measurement of the spatial distribution of soil/plant microorganisms in their microhabitats. In this approach, different fluorescent dyes have also been used isolated or in combination, to differentiate metabolically active cells from inactive cells in soil microbial populations (Romano, Ventorino, and Pepe 2020). The use of bacteria labeled with Green Fluorescent Protein (GFP) is one of the most commonly used methods for tracking inoculated bacteria in given crops. The use of GFP-tagged microbial strains may easily be found by a fluorescent microscopy and present no influence on autochthonous bacteria; however, it can also be used only in laboratory and/or greenhouse experiments since this kind of method requires production of bacterial transformants before any application (Romano, Ventorino, and Pepe 2020).

Additionally, as an alternative to specifically stain certain bacteria, it is possible to use fluorescence-labeled molecular probes for this purpose. The *Fluorescent in situ hybridization* (FISH) has already been used in several studies of soil and plant microbiomes. The method consists in using a specific oligonucleotide probe capable of targeting a given specific microbial group and allowing further *in situ* observation of those microbes in the microscope (Cardinale 2014; Romano, Ventorino, and Pepe 2020). Recently, Gamez et al. (2019) used FISH to verify the *P. fluorescens* Ps006 and *B. amyloliquefaciens* Bs006 capacity to colonize banana roots and pointed out these strains as promising bioinoculant. As the limitations of the FISH method, we can point to the long and complex preparation process; frequently it is not possible to differentiate living cells and dead cells, another point is that the increase in image noise caused by the autoflorescence of the plant cells can interfere with the capture of the fluorescence signal from microbial cells into plant tissues.

The development of molecular tools allows new types of unculturable microbes associated with agrosystems or helps to understand the ecological roles of those microorganisms (Romano, Ventorino, and Pepe 2020). Beyond, the use of DNA probes (Monteiro et al. 2020), PCR-based techniques (Fernando et al. 2019), and next-generation sequencing (Rachid et al. 2013), has greatly increased the capacity to track microorganisms in agricultural systems (Rachid et al. 2015).

The PCR-based technique revolutionized the study of microbial ecology in agrosystems, allowing the characterization of soil and plant microbiomes in a culture-independent approach (Romano, Ventorino, and Pepe 2020). The PCR reaction comprises the synthesis of copies of DNA fragments from a target sequence, but in this case the amplification of the genetic material is done from the total environmental sample, such as soil, root, leaves as well as other parts of the plant. Mainly, the PCR products amplified from soil or plant DNA may be analyzed by (i) genetic fingerprinting, (ii) DNA microarrays, (iii) labeled nucleic acids (DNA or RNA), (iv) nucleic acids quantification, and (v) sequencing methods, or also by a combination of the techniques mentioned above.

The *genetic fingerprinting techniques* are capable of generating profiles of microbial communities from direct PCR products amplified from total soil and/or plant DNA (Muyzer, De Waal, and Uitterlinden 1993). These techniques are rapid and allow the analyses of many samples during the same experiment. These approaches do not provide taxonomic information, but it is indicated when it is necessary to verify changes on microbial communities caused by soil use or management (Muyzer, De Waal, and Uitterlinden 1993). The gene of interest, typically the 16S rRNA gene (Rachid et al. 2013), but 18S rRNA, ITS region, and some functional genes may also be used (Rachid et al. 2013). The general principle of techniques that analyze DNA profiles is based on the separation by electrophoresis of a given PCR product, which can occur in different ways, such as:

A. differential denaturation of the PCR product under a chemical gradient by *Denaturing Gradient Gel Electrophoresis* (DGGE) or a thermal gradient by *Temperature Gradient Gel Electrophoresis* (TGGE) performed on polyacrylamide gels (Rychlik et al. 2017; Silva Marques et al. 2018).

B. location of restriction enzyme digestion sites by *Terminal Restriction Fragment Length Polymorphism (T-RFLP)* (Nannipieri et al. 2017) *or*

Amplified Ribosomal DNA restriction analysis (ARDRA) (Massol-Deya et al., 1995) . T-RFLP is very similar to ARDRA except for the use of a fluorescently tagged primer during the PCR reaction. The final PCR products are digested with restriction enzymes, and then terminal restriction fragments (T-RFs) are able to be segregated by a DNA sequencer, while ARDRA data is analyzed on agarose or polyacrylamide gels (Liu et al. 2012; Hashim and Al-Shuhaib 2019).

C. differences in single-stranded DNA mobility in non-denaturing gels (single-stranded conformational polymorphism—SSCP) (Nannipieri et al. 2017).

D. polymorphisms along the length of a fragment by *Length heterogeneity PCR* (LH-PCR) or automated analysis by the *Ribosomal Intergenic Space* (ARISA) (Hashim and Al-Shuhaib 2019).

Microarrays (or DNA chips) are solid plates, in which PCR products amplified from a total environmental sample are directly map by known molecular probes. This tool represents great advantages, since in a single array thousands of different oligonucleotides can be fixed in the same microchip, allowing thousands of genes to be accessed simultaneously. Microarrays can contain specific genes and, for example, those that encode the proteins ammonia monooxygenase or nitrogenase, obtaining information about the functional diversity, or they can contain patterns of samples that represent the different taxonomic groups found in soil or plant samples (Ahmad et al. 2011). In this sense, DNA microarrays used in microbial ecology studies could be classified into two major categories: (a) 16S rRNA gene microarrays and (b) functional gene arrays (FGA).

The 16S rRNA gene Microarrays (*PhyloChip*), containing several ribosomal genes labeled, are capable of differentiating 8,741 prokaryotic taxa, representing 121 archaeal and bacterial orders (Navarrete et al. 2021). PhyloChip has already been successfully applied in studies of agrosystems microbiomes (Brodie et al. 2007, Sagaram et al. 2009; Weinert et al. 2011). Recently, Sarhan et al. (2018) used this method to characterize the microbiome of maize root compartments, and also endo- and ecto-rhizospheres microbial communities, and they are capable to find 1,818 different operational taxonomic units (OTUs).

The FGA are designed mainly to detect specific metabolic groups of bacteria, and, for instance, are called *GeoChip* (He et al. 2007). This DNA chip contains approximately 28,000 probes, covering 57,000 genes, and 292 functions related to carbon, nitrogen, phosphorus and sulfur cycles, energy metabolism, resistance to antibiotics, and metals and degradation of organic contaminants. GeoChip is already used to detect the specific genes evolved on the carbon cycle that are capable of responding to different nitrogen application levels under sugarcane and soybean intercropping systems (Sarhan et al. 2018).

While the traditional approaches are focused on identifying the metabolic roles of different microbes that have involved the enrichment and/or isolation of those organisms, the *stable-isotope probing* (SIP) is a method in which particular groups of organisms capable of incorporating specific substrates are identified without requiring cultivation. So, stable-isotope-labeled nitrogen (15N) or carbon (13C) sources may be assimilated into microbial cells from environmental samples and further separation and analysis of labeled nucleic acids (DNA-SIP or RNA-SIP) reveals functional and phylogenetic information about those microorganisms evolved in the metabolism of that given substrate (Yu et al. 2020). Neufeld et al. (2007) used 15N-DNA-SIP combined with 16S rRNA (DGGE) analysis and sequencing of picked DGGE bands. In this study, Pseudonocardia sp. is prevalent on soils treated with maize residues, while *Arthrobacter* sp. and *Streptomyces* sp. were observed in the soybean residue-treated soils. Thus, they conclude that residue quality inducing contrasting N assimilation by different bacteria groups.

With regard to microbial communities, the quantitative PCR (qPCR) allows precise quantitation of phylogenetic and/or functional genes (España et al. 2011). This method may be used to verify the effect of soil use and management on the abundance of microorganisms (Schloter et al. 2018), the quantification of bioindicator microorganisms (Monteiro et al. 2020) or pathogens (Cuer et al. 2018) and also the quantification of microbial resistance genes, such as soil pollution proxies (Orlofsky et al. 2015; O'Connell and Nutman 2016).

PCR-based *sequencing methods* have become an important tool for the study of microbial ecology in agricultural systems. The first such approach was the clone libraries. In this method, PCR fragments are inserted into a cloning vector (usually a plasmid), then generating a recombinant plasmid. Competent *Escherichia coli* cells are transformed with this recombinant plasmid and their cells are plated in a selective medium to identify positive clones, that is, those that have the fragment of interest. Finally, these fragments can be sequenced. However, this is a very laborious method that results in a small number of clones, and it is not yet possible to access most of the microbial diversity (Janssen, 2006). Sagaram et al. (2009) had used 16S rRNA gene clone library sequencing to determine the microbial communities composition for symptomatic and asymptomatic citrus midribs and they verified the prevalence of *Candidatus Liberibacter asiaticus* in the symptomatic leaves.

Over the decades, there was the emergence of the massive DNA-sequencing technology—high-throughput sequencing—along with the bioinformatics tools that had assisted in the analysis of this diversity which is accessed by sequencing. This combination has provided a notable alternative to other molecular studies of microbial ecology in agrosystems which allow a high-resolution approach (Krishna, Khan, and Khan 2019). Indeed, analyzing the microbiome with the high-throughput sequencing or new generation sequencing (NGS) approach has many prospective results that could allow understanding the crop and soil microbiomes. This approach could also help to understand changes in dynamics and structure of microbiomes after microbial inoculation treatments (Erlacher et al. 2014) or pesticides usage (Cernava et al. 2019; Chen et al. 2021). High-throughput sequencing could be performed following two different approaches: (1) amplicon sequencing based on the amplification of phylogenetic marker genes; (2) shotgun sequencing based on random sequencing—metagenomics.

Amplicon sequencing allows a deep investigation of the microbial communities by the use of high-throughput amplicon-sequencing technology. This approach targets variable regions of specific phylogenetic markers (e.g., 16S rRNA and 18S rRNA genes, and ITS region), and it is commonly used to describe semi-quantitatively bacterial, archaeal, and fungal community compositions (Monteiro et al. 2020;

Chen et al. 2021). In addition, this approach could also be used for functional studies, such targeting enzyme-coding genes evolved on carbon, nitrogen, and phosphorus cycles, for example. In all of those cases, the analysis requires downstream bioinformatics analysis (Rastogi and Sani 2011; Navarrete et al. 2021).

Metagenomics consists in a direct analysis of genomes contained in a given microbiome, and currently represents a powerful tool for microbial soil/crop bioprospecting, as it allows access to the genetic potential of the soil and/or crop microbiomes (Rastogi and Sani 2011). Based on metagenome data, bacterial, fungal, and viral genomes can be reconstructed; actually, until this moment, there are several thousands of microbial genomes that were accessed without culturing the organisms behind from soil and/or crops microbiomes (Mühling et al. 2016; Stolze et al. 2016).

Additionally, Langille et al. (2013) developed the PICRUSt—Phylogenetic Investigation of Communities by Reconstruction of Unobserved States—a method to predict metagenomic functional profiles from phylogenetic (16S rRNA gene) data using previous genomic knowledge about related taxa. By this "predictive metagenomic" approach, they reinforced that phylogeny and function are sufficiently attached and also provide useful information about uncultivated microbiomes for which just phylogenetic marker genes are available. Montanari-Coelho et al. (2018) used PICRUSt to verify the impact of genetically modified (GM) soybean plants on the predictive functionality of associated microbial communities.

While in metagenomics, the extracted genetic material is the DNA, in the *metatranscriptomics*, the genetic material in question is the total RNA. After that, rRNA is removed in order to enrich the mRNA portion, and finally, cDNA synthesis is performed before sequencing (Lozada and Dionisi 2015). Metatranscriptomics represents a powerful approach for the discovery of metabolically active enzymes that are involved in specific biochemical pathways and it may also be applied in the analysis of community-specific variants of functional genes.

Metagenomics and transcriptomics can be used together in cases where the objective is to relate microbial diversity as well as function, as studied by Saminathan et al. (2018). They aimed to understand the diversity and function of microbial communities in relation to carbohydrate metabolism of ripe watermelon fruits, and they observed a high expression of genes related to infectious diseases in cultivar that presents a low microbial diversity.

Another approach that can be used for the characterization of novel enzymes is *metaproteomics*. This approach consists in the direct analysis of the proteins from a given microbial community. In this technique, the protein from an environmental sample is extracted, followed by the separation of the proteins by polyacrylamide gel electrophoresis or liquid chromatography (Wilmes and Bond 2009; Starke, Jehmlich, and Bastida 2019). Chen et al. (2019b) verified the protein profile of rhizospheric soils treated by different kinds of nitrogen application using metaproteomics and the results showed that 88.28% of the soil proteins were derived from microbes, 6.25% from animals, and 5.74% from plants.

Metabolomics has also a huge potential to better elucidate the chemical communication that results from the interaction between rhizosphere and root community members. Root exudates contain a bunch of primary and secondary plant metabolites that can inhibit or attract different kinds of microbes (Coninck et al. 2015).

The recent development of broad-spectrum and highly sensitive metabolomics platforms allows the recognition of metabolome location of the root and its exudate composition. One potential application of metabolomics can be the measurement of changes in specific metabolite levels, in response to a given treatment (Johnson, Ivanisevic, and Siuzdak 2016). With a similar approach, functional metagenomics can provide information to the identification of novel plant growth-promoting genes by heterologous expression in a root colonizer (Levy, Conway et al. 2018).

The extensive use of these meta-omics techniques remarkably increased the number of molecular data collections from environmental samples. In this context, bioinformatics arose from the initial requirement for adequate informatics tools which could be used for the organization, management, and analysis of biological data, and it proved to be fundamental in providing tools for data analysis, interpretation, and modeling (Esposito et al. 2016).

The high-throughput sequencing bioinformatics era is revolutionizing the experimental design in molecular microbiology, remarkably contributing in enhancing scientific knowledge and also impacting relevant applications in several aspects of agriculture (Esposito et al. 2016). The bioinformatics tools application has changed agriculture production and practice, giving knowledge and mechanisms for improved product quality and improved strategies of protection against environmental stress, diseases, and parasites.

In addition, bioinformatics tools do not allow us to just have a descriptive view of *who are* the microorganisms associated with a plant and/or agricultural soil; or even *what are the potential functions* performed by these microorganisms. For example, bioinformatics allows identification of the "core" microbiome—a set of microorganisms common to a particular set of environments (Shade and Handelsman 2012). These members, that are persistent features of a dataset, are hypothesized to reflect key functional relationships with the plant/and or soil (Shade and Stopnisek 2019).

Defining this core microbiome will help to identify crop-associated microbes that should be focused on further research. In this sense, researchers could be able to remove transient associations from the analysis and prioritize stable taxa that play important roles in the host (Busby et al. 2017). Moreover, the taxonomically defined core microbiome using amplicon-sequencing techniques based on 16S rRNA, it is possible identifying a functional core using metagenomic and metatranscriptomic approaches to find common predicted functions that play key roles in given agricultural system (Vandenkoornhuyse et al. 2015). Yeoh et al. (2016) identified a core sugarcane root microbiome and verified the effect of high nitrogen fertilizer rates on the presence of diazotrophs in different conditions. They suggested that these bacteria can naturally be enriched in sugarcane roots for targeted isolation and characterization and could be used for further studies in plant root microbiome manipulation.

Finally, it is important to mention that there are still many studies that neglect some basic principles of statistics such as replication and randomization, especially in works where expensive techniques such as meta-omics are used. There are two works that mention the importance of these two principles. On the one hand, Prosser (2010) highlighted that researchers need to replicate the findings observed to evaluate their characteristics. On the other hand, Webster (2017) added that they must also "randomize the selection to estimate these characteristics without bias and with known confidence." Unfortunately, it is still common to find published studies involving agrosystems microbiomes that present hypotheses and results in which such statistical principles have not been observed, which in the end are very uninformative.

2.4   WHAT COMES AFTER THAT?

Considering the predicted population growth for the next decades and the subsequent rising food demand, science and society face a huge challenge on how to produce food on a satisfactory scale, while ensuring a secure and sustainable output (Levy, Conway et al. 2018). Besides that, global climate change has already been challenging production systems and still brings over many uncertainties in this equation (FAO 2017).

The advances in science and technology led to a revision and reorientation of methodologies, allowing old and current issues to be addressed from a new perspective (Tosi et al. 2020). The paradigm shift observed in recent years in the areas of biology and biotechnology is illustrated by the focus transition from "traditional biology" to the focus of "genetic engineering" and "bioinformatics" and recently "systemic biology" (Berg et al. 2020).

The bioinformatics approach is based on the collection and storage of data in databases as well as the analysis and synthesis of information (data mining) (Burbeck and Jordan 2006). Databases for bioinformatics include genomes, proteomes, molecular structures, chemical diversity, metabolic pathways (metabolomes), biodiversity and systematics (Ong et al. 2016). Thus, innovative experiments can be carried out in silico instead of in vivo or in vitro, reducing the need to carry out experiments with biological models in the laboratory to an essential minimum (Chen, Chen, and Zhang 2017).

Together, genomics and functional genomics provide an accurate molecular map of a given cell or organism, and thereby provide information for the search for new targets for biotechnology search and discovery strategies (Oraiopoulou et al. 2018).

In this regard, we could watch the rise of systemic biology aiming for a comprehensive functional understanding due to the integration of omics sciences (Greenbaum et al. 2001). This approach was conceived as the overlap of data from transcriptomics, proteomics, metabolomics, glycomics (all carbohydrates in a cell), and lipidomics (complete set of lipids), operating in a holistic view of biological phenomena. In this way, it is expected that the complex biological phenomena will be unraveled deeply, generating new biotechnological applications for the benefit of human health, greater quality of life and sustainable use of natural resources (Pinu et al. 2019).

Nevertheless, despite the wide and rapid scientific evolution and technological advances promoted by the application of the most modern techniques of molecular biology, there is still much to be done in terms of transfer and the direct application of this knowledge to the field.

In this sense, agrosystems microbiomes may be characterized by approaches based on isolated microorganisms or microbial communities as a whole, as mentioned in this chapter and summarized in Figure 2.1.

Microbe and/or microbiome engineering can be the bridge to link the generated data and effective application. More specifically, synthetic biology (redesign and

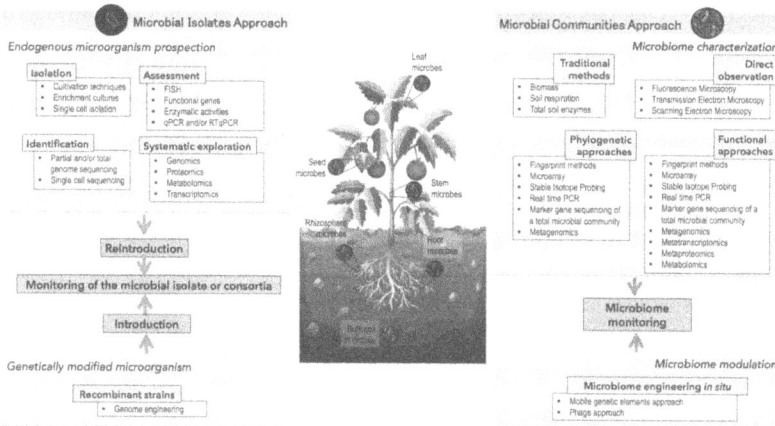

FIGURE 2.1 Overview of different approaches used to characterize microbial isolates and microbial communities in agrosystems. The crop-associated microbial communities could be found in different sites of a plant such as seeds, leaves, stem, root and rhizosphere; and also, on the bulk soil. Those endogenous microbes may be accessed individually by microbial-isolated prospection which evolve the isolation, identification, assessment, and systematic exploration of those microbes. After that, these microorganisms are capable of being reintroduced in an agrosystem for the purpose of acting in biocontrol or as a bio-fertilizer, for example, and the impact of that inoculation of a single microorganism or some of them (consortia) could be monitored by short or long term. On the other hand, the crop microbial communities may be characterized by traditional methods, direct observation, and phylogenetic and functional approaches. Those methods could be used to monitor the effects of land use and/or management of the soil. Additionally, microbe and/or microbiome engineering based on synthetic biology could be an option either by using the introduction and monitoring of genetically modified microorganisms (GMMs) or by microbiome modulation through the use of mobile genetics elements or phage approaches. (The tomato plant was designed by brgfx/Freepik and the tomato opened fruit was designed by vectorpocket/Freepik.)

repurposing of biological systems for novel applications or purposes) may be able to select and build a group of microbes able to benefit crops in various aspects (Ke, Wang, and Yoshikuni 2020).

As a strategy, microbes associated with a given plant may be isolated from an environmental microbiome, and after it is genetically engineered to carry plant-benefiting traits (Figure 2.1). So, the studies go in the direction of microbial design, which perform accurately and target specific plant-microbe interactions. This issue brief outlines current research on the topic and discusses topics that may arise if a synthetic microbe is made available for environmental release. Besides enhancing plants' growth, resistance to disease, or tolerance to environmental stressors by modifying their traits, a growing area of research is looking at how microbes influence these plant characteristics (Ryder et al. 2012; Setten et al., 2013; Hussain et al., 2018).

On the other hand, considering a modulation of microbial communities induced by a microbiome engineering (Figure 2.1), horizontal gene transfer (HGT) is used to introduce chosen traits into a broad range of plant hosts *in situ*. In this sense, a known strategy is to incorporate mobile genetic elements (MGEs), which transfer and integrate exogenous genes into a random subpopulation of microbiomes to allow the study of PGP traits (Zhang et al. 2018; Ke, Wang, and Yoshikuni 2020). Another strategy which may be cited is the development of bacteriophage systems to eliminate or engineer particular species within populations, in a microbiome (Ke, Wang, and Yoshikuni 2020), which allows disease control studies (Jalan et al. 2013; Xu and Wang 2019).

Finally, the benefits achieved with molecular biology approaches have a huge potential to contribute to the establishment of an efficient and sustainable agriculture. The scientific benefits of a deepest knowledge on microbial diversity, at functional and taxonomic levels, are leading to a better understanding of the functions performed by microbial communities in natural and farming environments.

2.5   CONCLUDING REMARKS

Therefore, considering an agrosystem, we are able to determine the microorganism's behavior, individually or as a microbiome, under different conditions. Also enlightening their interactions with other components of biodiversity, such as plants and outlining the biochemical mechanisms associated with important agrarian processes, such as plant growth. Economic and strategic benefits are related to the discovery of potentially exploitable microorganisms in biotechnological processes due to the search to provide advanced and efficient procedures in increasing food production.

REFERENCES

Abdollahzadeh, Jafar, and Sajedeh Zolfaghari. 2014. "Efficiency of Rep-PCR Fingerprinting as a Useful Technique for Molecular Typing of Plant Pathogenic Fungal Species: Botryosphaeriaceae Species as a Case Study." *FEMS Microbiology Letters* 361 (2): 144–57. https://doi.org/10.1111/1574-6968.12624.

Agrawal, Pavan Kumar, Shruti Agrawal, and Rahul Shrivastava. 2015. "Modern Molecular Approaches for Analyzing Microbial Diversity from Mushroom Compost Ecosystem." *3 Biotech* 5 (6): 853–66. https://doi.org/10.1007/s13205-015-0289-2.

Aguilar-Marcelino, Liliana, Pedro Mendoza-de-Gives, Laith Khalil Tawfeeq Al-Ani, María Eugenia López-Arellano, Olga Gómez-Rodríguez, Edgar Villar-Luna, and David Emmanuel Reyes-Guerrero. 2020. "Using Molecular Techniques Applied to Beneficial Microorganisms as Biotechnological Tools for Controlling Agricultural Plant Pathogens and Pest." *Molecular Aspects of Plant Beneficial Microbes in Agriculture*: 333–49. https://doi.org/10.1016/b978-0-12-818469-1.00027-4.

Ahmad, I., M. S. A. Khan, F. Aqil, and M. Singh. 2011. "Microbial Applications in Agriculture and the Environment: A Broad Perspective." In: Ahmad, I., Ahmad, F., and Pichtel, J. (eds.), *Microbes and Microbial Technology*. New York, NY: Springer. https://doi.org/10.1007/978-1-4419-7931-5_1.

Albertsen, Mads, Søren M. Karst, Anja S. Ziegler, Rasmus H. Kirkegaard, and Per H. Nielsen. 2015. "Back to Basics—The Influence of DNA Extraction and Primer Choice on Phylogenetic Analysis of Activated Sludge Communities." *PLoS ONE* 10 (7): 1–15. https://doi.org/10.1371/journal.pone.0132783.

Alves, Artur, Alan J. L. Phillips, Isabel Henriques, and António Correia. 2007. "Rapid Differentiation of Species of Botryosphaeriaceae by PCR Fingerprinting." *Research in Microbiology* 158 (2): 112–21. https://doi.org/10.1016/j.resmic.2006.10.003.

Arif, Inessa, Maria Batool, and Peer M. Schenk. 2020. "Plant Microbiome Engineering: Expected Benefits for Improved Crop Growth and Resilience." *Trends in Biotechnology* 38 (12): 1385–96. https://doi.org/10.1016/j.tibtech.2020.04.015.

Atienzar, Franck A., Paola Venier, Awadhesh N. Jha, and Michael H. Depledge. 2002. "Evaluation of the Random Amplified Polymorphic DNA (RAPD) Assay for the Detection of DNA Damage and Mutations." *Mutation Research—Genetic Toxicology and Environmental Mutagenesis* 521 (1–2): 151–63. https://doi.org/10.1016/S1383-5718(02)00216-4.

Ayyaz, Khadija, Ahmad Zaheer, Ghulam Rasul, and Muhammad Sajjad Mirza. 2016. "Isolation and Identification by 16S RRNA Sequence Analysis of Plant Growth-Promoting Azospirilla from the Rhizosphere of Wheat." *Brazilian Journal of Microbiology* 47 (3): 542–50. https://doi.org/10.1016/j.bjm.2015.11.035.

Bai, Yang, Daniel B. Müller, Girish Srinivas, Ruben Garrido-Oter, Eva Potthoff, Matthias Rott, Nina Dombrowski et al. 2015. "Functional Overlap of the Arabidopsis Leaf and Root Microbiota." *Nature* 528 (7582): 364–69. https://doi.org/10.1038/nature16192.

Baker, G. C., J. J. Smith, and D. A. Cowan. 2003. "Review and Re-Analysis of Domain-Specific 16S Primers." *Journal of Microbiological Methods* 55 (3): 541–55. https://doi.org/10.1016/j.mimet.2003.08.009.

Baker, Jason C., Richard E. Crumley, and Todd T. Eckdahl. 2002. "Random Amplified Polymorphic DNA PCR in the Microbiology Teaching Laboratory: Identification of Bacterial Unknowns." *Biochemistry and Molecular Biology Education* 30 (6): 394–97. https://doi.org/10.1002/bmb.2002.494030060135.

Baldrian, Petr. 2019. "The Known and the Unknown in Soil Microbial Ecology." *FEMS Microbiology Ecology* 95 (2): 1–9. https://doi.org/10.1093/femsec/fiz005.

Balvočiūtė, Monika, and Daniel H. Huson. 2017. SILVA, RDP, Greengenes, NCBI and OTT—How Do These Taxonomies Compare?. *BMC Genomics* 18 (2): 1–8. https://doi.org/10.1186/s12864-017-3501-4.

Barghouthi, Sameer A. 2011. "A Universal Method for the Identification of Bacteria Based on General PCR Primers." *Indian Journal of Microbiology* 51 (4): 430–44. https://doi.org/10.1007/s12088-011-0122-5.

Barret, Matthieu, Martial Briand, Sophie Bonneau, Anne Préveaux, Sophie Valière, Olivier Bouchez, Gilles Hunault, Philippe Simoneau, and Marie Agnès Jacquesa. 2015. "Emergence Shapes the Structure of the Seed Microbiota." *Applied and Environmental Microbiology* 81 (4): 1257–66. https://doi.org/10.1128/AEM.03722-14.

Beattie, Andrew J., Mark Hay, Bill Magnusson, de Nys, Rocky, James Smeathers, and Julian F. V. Vincent. 2011. "Ecology and Bioprospecting." *Austral Ecology* 36 (3): 341–56. https://doi.org/10.1111/j.1442-9993.2010.02170.x.

Bell, Terrence H., Kevin L. Hockett, Ricardo I. Alcalá-Briseño, Mary Barbercheck, Gwyn A. Beattie, Mary Ann Bruns, John E. Carlson et al. 2019. "Manipulating Wild and Tamed Phytobiomes: Challenges and Opportunities." *Phytobiomes Journal* 3 (1): 3–21. https://doi.org/10.1094/PBIOMES-01-19-0006-W.

Bender, S. Franz, Cameron Wagg, and Marcel G. A. van der Heijden. 2016. "An Underground Revolution: Biodiversity and Soil Ecological Engineering for Agricultural Sustainability." *Trends in Ecology and Evolution* 31 (6): 440–52. https://doi.org/10.1016/j.tree.2016.02.016.

Berg, Gabriele, Daria Rybakova, Doreen Fischer, Tomislav Cernava, Marie Christine Champomier Vergès, Trevor Charles, Xiaoyulong Chen et al. 2020. "Microbiome Definition Re-Visited: Old Concepts and New Challenges." *Microbiome* 8 (1): 1–22. https://doi.org/10.1186/s40168-020-00875-0.

Berg, Gabriele, and Kornelia Smalla. 2009. "Plant Species and Soil Type Cooperatively Shape the Structure and Function of Microbial Communities in the Rhizosphere." *FEMS Microbiology Ecology* 68 (1): 1–13. https://doi.org/10.1111/j.1574-6941.2009.00654.x.

Bilodeau, Guillaume J. 2011. "Quantitative Polymerase Chain Reaction for the Detection of Organisms in Soil." *CAB Reviews: Perspectives in Agriculture, Veterinary Science, Nutrition and Natural Resources* 6 (014): 1–14. https://doi.org/10.1079/PAVSNNR20116014.

Bokulich, Nicholas A., Zachery T. Lewis, Kyria Boundy-Mills, and David A. Mills. 2016. "A New Perspective on Microbial Landscapes within Food Production." *Current Opinion in Biotechnology* 37: 182–89. https://doi.org/10.1016/j.copbio.2015.12.008.

Bollmann, Annette, Kim Lewis, and Slava S. Epstein. 2007. "Incubation of Environmental Samples in a Diffusion Chamber Increases the Diversity of Recovered Isolates." *Applied and Environmental Microbiology* 73 (20): 6386–90. https://doi.org/10.1128/AEM.01309-07.

Botstein, David, Raymond L. White, Mark Skolnick, and Ronald W. Davis. 1980. "Botstein." *American Journal of Human Genetics* 32: 314–31.

Boutard, Magali, Tristan Cerisy, Pierre Yves Nogue, Adriana Alberti, Jean Weissenbach, Marcel Salanoubat, and Andrew C. Tolonen. 2014. "Functional Diversity of Carbohydrate-Active Enzymes Enabling a Bacterium to Ferment Plant Biomass." *PLoS Genetics* 10 (11): e1004773. https://doi.org/10.1371/journal.pgen.1004773.

Brodie, Eoin L., Todd Z. DeSantis, Jordan P. Moberg Parker, Ingrid X. Zubietta, Yvette M. Piceno, and Gary L. Andersen. 2007. "Urban Aerosols Harbor Diverse and Dynamic Bacterial Populations." *Proceedings of the National Academy of Sciences of the United States of America* 104 (1): 299–304. https://doi.org/10.1073/pnas.0608255104.

Burbeck, Steve, and Kirk E. Jordan. 2006. "An Assessment of the Role of Computing in Systems Biology." *IBM Journal of Research and Development* 50 (6): 529–43. https://doi.org/10.1147/rd.506.0529.

Busby, P. E., C. Soman, M. R. Wagner, M. L. Friesen, J. Kremer, A. Bennett, M. Morsy, J. A. Eisen, J. E. Leach, and J. L. Dangl. 2017. "Research Priorities for Harnessing Plant Microbiomes in Sustainable Agriculture." *PLoS Biology* 15 (3), 1–14. https://doi.org/10.1371/journal.pbio.2001793.

Bustamante, Daniel, Silvia Segarra, Marta Tortajada, Daniel Ramón, Carlos del Cerro, María Auxiliadora Prieto, José Ramón Iglesias, and Antonia Rojas. 2019. "In Silico Prospection of Microorganisms to Produce Polyhydroxyalkanoate from Whey: Caulobacter Segnis DSM 29236 as a Suitable Industrial Strain." *Microbial Biotechnology* 12 (3): 487–501. https://doi.org/10.1111/1751-7915.13371.

Campo, Javier del, Martin Kolisko, Vittorio Boscaro, Luciana F. Santoferrara, Serafim Nenarokov, Ramon Massana, Laure Guillou et al. 2018. "EukRef: Phylogenetic Curation of Ribosomal RNA to Enhance Understanding of Eukaryotic Diversity and Distribution." *PLoS Biology* 16 (9): 1–14. https://doi.org/10.1371/journal.pbio.2005849.

Cardinale, Massimiliano. 2014. "Scanning a Microhabitat: Plant-Microbe Interactions Revealed by Confocal Laser Microscopy." *Frontiers in Microbiology* 5 (March): 1–10. https://doi.org/10.3389/fmicb.2014.00094.

Cardoso, Elke Jurandy Bran Nogueira, Rafael Leandro Figueiredo Vasconcellos, Daniel Bini, Marina Yumi Horta Miyauchi, Cristiane Alcantara dos Santos, Paulo Roger Lopes Alves, Alessandra Monteiro de Paula, André Shigueyoshi Nakatani, Jamil de Moraes Pereira, and Marco Antonio Nogueira. 2013. "Soil Health: Looking for Suitable Indicators. What Should Be Considered to Assess the Effects of Use and Management on Soil Health?" *Scientia Agricola* 70 (4): 274–89. https://doi.org/10.1590/S0103-90162013000400009.

Case, Rebecca J., Yan Boucher, Ingela Dahllöf, Carola Holmström, W. Ford Doolittle, and Staffan Kjelleberg. 2007. "Use of 16S RRNA and RpoB Genes as Molecular Markers for Microbial Ecology Studies." *Applied and Environmental Microbiology* 73 (1): 278–88. https://doi.org/10.1128/AEM.01177-06.

Cernava, T., Chen, X., Krug, L., Li, H., Yang, M., & Berg, G. 2019. "The tea leaf microbiome shows specific responses to chemical pesticides and biocontrol applications." *Science of the Total Environment* 667, 33–40. https://doi.org/10.1016/j.scitotenv.2019.02.319.

Chaudhary, Dhiraj Kumar, Altankhuu Khulan, and Jaisoo Kim. 2019. "Development of a Novel Cultivation Technique for Uncultured Soil Bacteria." *Scientific Reports* 9(1): 1–11. https://doi.org/10.1038/s41598-019-43182-x.

Chen Chen, Han Y. H. Chen, Xinli Chen, and Zhiqun Huang. 2019a. "Meta-Analysis Shows Positive Effects of Plant Diversity on Microbial Biomass and Respiration." *Nature Communications* 10 (1): 1–10. https://doi.org/10.1038/s41467-019-09258-y.

Chen, X. D., K. E. Dunfield, T. D. Fraser, S. A. Wakelin, A. E. Richardson, and L. M. Condron. 2019b. "Soil Biodiversity and Biogeochemical Function in Managed Ecosystems." *Soil Research* 58 (1): 1–20. https://doi.org/10.1071/SR19067.

Chen, Xiaoyulong, Wisnu Adi Wicaksono, Gabriele Berg, and Tomislav Cernava. 2021. "Bacterial Communities in the Plant Phyllosphere Harbour Distinct Responders to a Broad-Spectrum Pesticide." *Science of the Total Environment* 751: 141799. https://doi.org/10.1016/j.scitotenv.2020.141799.

Chen, Zixi, Lei Chen, and Weiwen Zhang. 2017. "Tools for Genomic and Transcriptomic Analysis of Microbes at Single-Cell Level." *Frontiers in Microbiology* 8 (September): 1–12. https://doi.org/10.3389/fmicb.2017.01831.

Cheng, Wenwen, Xuanyu Yan, Jiali Xiao, Yunyun Chen, Minghui Chen, Jiayi Jin, Yu Bai, Qi Wang, Zhiyong Liao, and Qiongzhen Chen. 2020. "Isolation, Identification, and Whole Genome Sequence Analysis of the Alginate-Degrading Bacterium Cobetia Sp. Cqz5-12." *Scientific Reports* 10 (1): 1–10. https://doi.org/10.1038/s41598-020-67921-7.

Cole, James R. Wang, Qiong Fish, Jordan A Chai, Benli McGarrell, Donna M Sun, Yanni Brown, C Titus Porras-Alfaro, Andrea Kuske, Cheryl R Tiedje, James M. 2014. Ribosomal Database Project: data and tools for high throughput rRNA analysis. *Nucleic Acids Research* 42 (D1) D633–D642. https://doi.org/10.1093/nar/gkt1244.

Compant, Stéphane, Abdul Samad, Hanna Faist, and Angela Sessitsch. 2019. "A Review on the Plant Microbiome: Ecology, Functions, and Emerging Trends in Microbial Application." *Journal of Advanced Research* 19: 29–37. https://doi.org/10.1016/j.jare.2019.03.004.

Coninck, Barbara De, Pieter Timmermans, Christine Vos, Bruno P. A. Cammue, and Kemal Kazan. 2015. "What Lies Beneath: Belowground Defense Strategies in Plants." *Trends in Plant Science* 20 (2): 91–101. https://doi.org/10.1016/j.tplants.2014.09.007.

Cuer, Caroline A., Renato de A. R. Rodrigues, Fabiano C. Balieiro, Jacqueline Jesus, Elderson P. Silva, Bruno José R Alves, and Caio T. C. C. Rachid. 2018. "Short-Term Effect of Eucalyptus Plantations on Soil Microbial Communities and Soil-Atmosphere Methane and Nitrous Oxide Exchange." *Scientific Reports* 8 (1): 15133. https://doi.org/10.1038/s41598-018-33594-6.

Delmont, Tom O., Patrick Robe, Sébastien Cecillon, Ian M. Clark, Florentin Constancias, Pascal Simonet, Penny R. Hirsch, and Timothy M. Vogel. 2011. "Accessing the Soil Metagenome for Studies of Microbial Diversity." *Applied and Environmental Microbiology* 77 (4): 1315–24. https://doi.org/10.1128/AEM.01526-10.

Di Cello, F., A. Bevivino, L. Chiarini, R. Fani, D. Paffetti, S. Tabacchioni, and C. Dalmastri. 1997. "Biodiversity of a Burkholderia Cepacia Population Isolated from the Maize Rhizosphere at Different Plant Growth Stages." *Applied and Environmental Microbiology* 63 (11): 4485–93. https://doi.org/10.1128/aem.63.11.4485-4493.1997.

Dini-Andreote, Francisco, Jan Dirk Van Elsas, Han Olff, and Joana Falcão Salles. 2018. "Dispersal-Competition Tradeoff in Microbiomes in the Quest for Land Colonization." *Scientific Reports* 8 (1): 1–9. https://doi.org/10.1038/s41598-018-27783-6.

Du, Huihui, Qiaoyun Huang, Caroline L. Peacock, Boqing Tie, Ming Lei, Xiaoli Liu, and Xiangdong Wei. 2018. "Competitive Binding of Cd, Ni and Cu on Goethite Organo–Mineral Composites Made with Soil Bacteria." *Environmental Pollution* 243: 444–52. https://doi.org/10.1016/j.envpol.2018.08.087.

Dumack, K., and M. Bonkowski. 2021. "Protists in the Plant Microbiome: An Untapped Field of Research." In: Carvalhais, L. C., and Dennis P. G. (eds.), *The Plant Microbiome. Methods in Molecular Biology*, 77–84. New York: Humana Press. https://doi.org/10.1007/978-1-0716-1040-4_8.

Egan, Kevin, Des Field, R. Paul Ross, Paul D. Cotter, and Colin Hill. 2018. "In Silico prediction and Exploration of Potential Bacteriocin Gene Clusters within the Bacterial Genus Geobacillus." *Frontiers in Microbiology* 9 (September): 1–17. https://doi.org/10.3389/fmicb.2018.02116.

Eicher, Tara, Garrett Kinnebrew, Andrew Patt, Kyle Spencer, Kevin Ying, Qin Ma, Raghu Machiraju, and Ewy A. Mathé. 2020. "Metabolomics and Multi-Omics Integration: A Survey of Computational Methods and Resources." *Metabolites* 10 (5). https://doi.org/10.3390/metabo10050202.

Epstein, S. S. 2013. "The Phenomenon of Microbial Uncultivability." *Current Opinion in Microbiology* 16 (5): 636–42. https://doi.org/10.1016/j.mib.2013.08.003.

Erlacher, Armin, Massimiliano Cardinale, Rita Grosch, Martin Grube, and Gabriele Berg. 2014. "The Impact of the Pathogen Rhizoctonia Solani and Its Beneficial Counterpart Bacillus Amyloliquefaciens on the Indigenous Lettuce Microbiome." *Frontiers in Microbiology* 5 (April): 1–8. https://doi.org/10.3389/fmicb.2014.00175.

Escalas, Arthur, Lauren Hale, James W. Voordeckers, Yunfeng Yang, Mary K. Firestone, Lisa Alvarez-Cohen, and Jizhong Zhou. 2019. "Microbial Functional Diversity: From Concepts to Applications." *Ecology and Evolution* 9 (20): 12000–16. https://doi.org/10.1002/ece3.5670.

España, Mingrelia, Frank Rasche, Ellen Kandeler, Thomas Brune, Belkis Rodriguez, Gary D. Bending, and Georg Cadisch. 2011. "Identification of Active Bacteria Involved in Decomposition of Complex Maize and Soybean Residues in a Tropical Vertisol Using 15N-DNA Stable Isotope Probing." *Pedobiologia* 54 (3): 187–93. https://doi.org/10.1016/j.pedobi.2011.03.001.

Esposito, A., C. Colantuono, V. Ruggieri, and M. L. Chiusano. 2016. "Bioinformatics for Agriculture in the Next-generation Sequencing Era." *Chemical and Biological Technologies in Agriculture*, 3 (1): 1–12. https://doi.org/10.1186/s40538-016-0054-8.

FAO. 2017. *The Future of Food and Agriculture: Trends and Challenges*. Vol. 4. Rome. https://www.fao.org/3/i6583e/i6583e.pdf.

Farrelly, V., F. A. Rainey, and E. Stackebrandt. 1995. "Effect of Genome Size and Rrn Gene Copy Number on PCR Amplification of 16S RRNA Genes from a Mixture of Bacterial Species." *Applied and Environmental Microbiology* 61 (7): 2798–801. https://doi.org/10.1128/aem.61.7.2798-2801.1995.

Farwell, Andrea J., Susanne Vesely, Vincent Nero, Hilda Rodriguez, Saleh Shah, D. George Dixon, and Bernard R. Glick. 2006. "The Use of Transgenic Canola (Brassica Napus) and Plant Growth-Promoting Bacteria to Enhance Plant Biomass at a Nickel-Contaminated Field Site." *Plant and Soil* 288 (1–2): 309–18. https://doi.org/10.1007/s11104-006-9119-y.

Fernando, Eustace Y., Simon Jon McIlroy, Marta Nierychlo, Florian Alexander Herbst, Francesca Petriglieri, Markus C. Schmid, Michael Wagner, Jeppe Lund Nielsen, and Per Halkjær Nielsen. 2019. "Resolving the Individual Contribution of Key Microbial Populations to Enhanced Biological Phosphorus Removal with Raman–FISH." *ISME Journal* 13 (8): 1933–46. https://doi.org/10.1038/s41396-019-0399-7.

Fondi, Marco, and Pietro Liò. 2015. "Multi-Omics and Metabolic Modelling Pipelines: Challenges and Tools for Systems Microbiology." *Microbiological Research* 171: 52–64. https://doi.org/10.1016/j.micres.2015.01.003.

Frickmann, Hagen, Andreas Erich Zautner, Annette Moter, Judith Kikhney, Ralf Matthias Hagen, Henrik Stender, and Sven Poppert. 2017. "Fluorescence in Situ Hybridization (FISH) in the Microbiological Diagnostic Routine Laboratory: A Review." *Critical Reviews in Microbiology* 43 (3): 263–93. https://doi.org/10.3109/1040841X.2016.1169990.

Gaby, John Christian, and Daniel H. Buckley. 2014. "A Comprehensive Aligned NifH Gene Database: A Multipurpose Tool for Studies of Nitrogen-Fixing Bacteria." *Database* 2014: 1–8. https://doi.org/10.1093/database/bau001.

Gaete, Alexis, Dinka Mandakovic, and Mauricio González. 2020. "Isolation and Identification of Soil Bacteria from Extreme Environments of Chile and Their Plant Beneficial Characteristics." *Microorganisms* 8 (8): 1–13. https://doi.org/10.3390/microorganisms8081213.

Gamez, R, M. Cardinale, M. Montes, S. Ramirez, S. Schnell, and F. Rodriguez. 2019. "Screening, Plant Growth Promotion and Root Colonization Pattern of Two Rhizobacteria (Pseudomonas Fluorescens Ps006 and Bacillus Amyloliquefaciens Bs006) on Banana Cv. Williams (Musa Acuminata Colla)." *Microbiological Research* 220: 12–20. https://doi.org/10.1016/j.micres.2018.11.006.

Garita-Cambronero, Jerson, Ana Palacio-Bielsa, María M. López, and Jaime Cubero. 2016. "Draft Genome Sequence for Virulent and Avirulent Strains of Xanthomonas Arboricola Isolated from Prunus Spp. in Spain." *Standards in Genomic Sciences* 11 (1): 1–10. https://doi.org/10.1186/s40793-016-0132-3.

Gavrish, Ekaterina, Annette Bollmann, Slava Epstein, and Kim Lewis. 2008. "A Trap for in Situ Cultivation of Filamentous Actinobacteria." *Journal of Microbiological Methods* 72 (3): 257–62. https://www.ncbi.nlm.nih.gov/pmc/articles/PMC3624763/pdf/nihms412728.pdf.

Gawad, Charles, Winston Koh, and Stephen R. Quake. 2016. "Single-Cell Genome Sequencing: Current State of the Science." *Nature Reviews Genetics* 17 (3): 175–88. https://doi.org/10.1038/nrg.2015.16.

Gerlt, John A. 2016. "Tools and Strategies for Discovering Novel Enzymes and Metabolic Pathways." *Perspectives in Science* 9: 24–32. https://doi.org/10.1016/j.pisc.2016.07.001.

Giangacomo, Cecelia, Mohsen Mohseni, Lynsey Kovar, and Jason G. Wallace. 2020. "Comparing DNA Extraction and 16s Amplification Methods for Plant-Associated Bacterial Communities." *BioRxiv*, 1–32. https://doi.org/10.1101/2020.07.23.217901.

Greenbaum, Dov, Nicholas M. Luscombe, Ronald Jansen, Jiang Qian, and Mark Gerstein. 2001. "Interrelating Different Types of Genomic Data, from Proteome to Secretome: 'Oming in on Function." *Genome Research* 11 (9): 1463–68. https://doi.org/10.1101/gr.207401.

Guillou, Laure, Dipankar Bachar, Stéphane Audic, David Bass, Cédric Berney, Lucie Bittner, Christophe Boutte et al. 2013. "The Protist Ribosomal Reference Database (PR2): A Catalog of Unicellular Eukaryote Small Sub-Unit RRNA Sequences with Curated Taxonomy." *Nucleic Acids Research* 41 (D1): 597–604. https://doi.org/10.1093/nar/gks1160.

Gupta, Alka, Murali Gopal, George V. Thomas, Vinu Manikandan, John Gajewski, George Thomas, Somasekar Seshagiri, Stephan C. Schuster, Preeti Rajesh, and Ravi Gupta. 2014. "Whole Genome Sequencing and Analysis of Plant Growth Promoting Bacteria Isolated from the Rhizosphere of Plantation Crops Coconut, Cocoa and Arecanut." *PLoS ONE* 9 (8): e104259. https://doi.org/10.1371/journal.pone.0104259.

Hartman, Kyle, Marcel G. A. van der Heijden, Raphaël A. Wittwer, Samiran Banerjee, Jean Claude Walser, and Klaus Schlaeppi. 2018. "Cropping Practices Manipulate Abundance Patterns of Root and Soil Microbiome Members Paving the Way to Smart Farming." *Microbiome* 6 (1): 1–14. https://doi.org/10.1186/s40168-017-0389-9.

Hashim, Hayder O., and Mohammed Baqur S. Al-Shuhaib. 2019. "Exploring the Potential and Limitations of PCR-RFLP and PCR-SSCP for SNP Detection: A Review." *Journal of Applied Biotechnology Reports* 6 (4): 137–44. https://doi.org/10.29252/JABR.06.04.02.

Hasman, Henrik, Dhany Saputra, Thomas Sicheritz-Ponten, Ole Lund, Christina Aaby
    Svendsen, Niels Frimodt-Moller, and Frank M. Aarestrup. 2014. "Rapid Whole-
    Genome Sequencing for Detection and Characterization of Microorganisms Directly
    from Clinical Samples." *Journal of Clinical Microbiology* 52 (1): 139–46. https://doi.
    org/10.1128/JCM.02452-13.
He, Zhili et al. 2007. "GeoChip: A Comprehensive Microarray for Investigating
    Biogeochemical, Ecological and Environmental Processes." *The ISME Journal*, 1 (1):
    67–77. https://doi.org/10.1038/ismej.2007.2.
Herold, Miriam B., Madeline E. Giles, Colin J. Alexander, Elizabeth M. Baggs, and Tim
    J. Daniell. 2018. "Variable Response of NirK and NirS Containing Denitrifier
    Communities to Long-Term PH Manipulation and Cultivation." *FEMS Microbiology
    Letters* 365 (7): 1–6. https://doi.org/10.1093/femsle/fny035.
Hesse, Cedar N., Rebecca C. Mueller, Momchilo Vuyisich, La Verne Gallegos-Graves, Cheryl
    D. Gleasner, Donald R. Zak, and Cheryl R. Kuske. 2015. "Forest Floor Community
    Metatranscriptomes Identify Fungal and Bacterial Responses to N Deposition in Two
    Maple Forests." *Frontiers in Microbiology* 6 (April): 1–15. https://doi.org/10.3389/
    fmicb.2015.00337.
Heyndrickx, M., L. Vauterin, P. Vandamme, K. Kersters, and P. De Vos. 1996. "Applicability
    of Combined Amplified Ribosomal DNA Restriction Analysis (ARDRA) Patterns in
    Bacterial Phylogeny and Taxonomy." *Journal of Microbiological Methods* 26 (3): 247–59.
    https://doi.org/10.1016/0167-7012(96)00916-5.
Hoffman, Brian M., Dmitriy Lukoyanov, Zhi Yong Yang, Dennis R. Dean, and Lance C.
    Seefeldt. 2014. "Mechanism of Nitrogen Fixation by Nitrogenase: The Next Stage."
    Chemical Reviews 114 (8): 4041–62. https://doi.org/10.1021/cr400641x.
Hoffmann, Maria, Eric W. Brown, Peter Ch Feng, Christine E. Keys, Markus Fischer, and
    Steven R. Monday. 2010. "PCR-Based Method for Targeting 16S-23S RRNA Intergenic
    Spacer Regions among Vibrio Species." *BMC Microbiology* 10: 1–14. https://doi.org/
    10.1186/1471-2180-10-90.
Hofstetter, Valérie, Bart Buyck, Guillaume Eyssartier, Sylvain Schnee, and Katia Gindro.
    2019. "The Unbearable Lightness of Sequenced-Based Identification." *Fungal Diversity*
    96: 243–84. https://doi.org/10.1007/s13225-019-00428-3.
Horwath, W. R., and E. A. Paul. 1994. "Microbial Biomass." Methods of Soil Analysis: Part 2 -
    Microbiological and Biochemical Properties 5, 753–73.
Howe, Adina, Fan Yang, Ryan J. Williams, Folker Meyer, and Kirsten S. Hofmockel. 2016.
    "Identification of the Core Set of Carbon-Associated Genes in a Bioenergy Grassland
    Soil." *PLoS ONE* 11 (11): 1–14. https://doi.org/10.1371/journal.pone.0166578.
Hussain, Syed Sarfraz, Samina Mehnaz, and Kadambot H. M. Siddique. 2018a. "Harnessing
    the Plant Microbiome for Improved Abiotic Stress Tolerance." In: Egamberdieva, D., and
    Ahmad, P. (eds.), *Plant Microbiome: Stress Response. Microorganisms for Sustainability*,
    vol 5, 21–43. Singapore: Springer. https://doi.org/10.1007/978-981-10-5514-0_2.
Hussain, A., S. Ali, M. Rizwan, M. Zia ur Rehman, M. R. Javed, M. Imran, S. A. S. Chatha,
    and R. Nazir. 2018. "Zinc Oxide Nanoparticles Alter the Wheat Physiological Response
    and Reduce the Cadmium Uptake by Plants." Environmental Pollution 242: 1518–26.
    https://doi.org/10.1016/j.envpol.2018.08.036.
Ishii, Satoshi, Kanako Tago, and Keishi Senoo. 2010. "Single-Cell Analysis and Isolation for
    Microbiology and Biotechnology: Methods and Applications." *Applied Microbiology
    and Biotechnology* 86 (5): 1281–92. https://doi.org/10.1007/s00253-010-2524-4.
Jalan, Neha, Dibyendu Kumar, Maxuel O. Andrade, Fahong Yu, Jeffrey B. Jones, James
    H. Graham, Frank F. White, João C. Setubal, and Nian Wang. 2013. "Comparative
    Genomic and Transcriptome Analyses of Pathotypes of Xanthomonas Citri Subsp.
    Citri Provide Insights into Mechanisms of Bacterial Virulence and Host Range." *BMC
    Genomics* 14 (1): 551. https://doi.org/10.1186/1471-2164-14-551.

Janssen, Peter H. 2006. "Identifying the Dominant Soil Bacterial Taxa in Libraries of 16S rRNA and 16S rRNA Genes." 2006. *Applied and Environmental Microbiology* 72 (3): 1719–28. https://doi.org/10.1128/AEM.72.3.1719-1728.2006.

Jeong, Jinuk, Kyeongeui Yun, Seyoung Mun, Won Hyong Chung, Song Yi Choi, Young do Nam, Mi Young Lim et al. 2021. "The Effect of Taxonomic Classification by Full-Length 16S RRNA Sequencing with a Synthetic Long-Read Technology." *Scientific Reports* 11 (1): 1–12. https://doi.org/10.1038/s41598-020-80826-9.

Johnson, Caroline H., Julijana Ivanisevic, and Gary Siuzdak. 2016. "Metabolomics: Beyond Biomarkers and towards Mechanisms." *Nature Reviews Molecular Cell Biology* 17 (7): 451–59. https://doi.org/10.1038/nrm.2016.25.

Johnson, Jethro S., Daniel J. Spakowicz, Bo Young Hong, Lauren M. Petersen, Patrick Demkowicz, Lei Chen, Shana R. Leopold et al. 2019. "Evaluation of 16S RRNA Gene Sequencing for Species and Strain-Level Microbiome Analysis." *Nature Communications* 10 (1): 1–11. https://doi.org/10.1038/s41467-019-13036-1.

Jurburg, Stephanie D., Tiago Natal-da-Luz, João Raimundo, Paula V. Morais, José Paulo Sousa, Jan Dirk van Elsas, and Joana Falcao Salles. 2018. "Bacterial Communities in Soil Become Sensitive to Drought under Intensive Grazing." *Science of the Total Environment* 618: 1638–46. https://doi.org/10.1016/j.scitotenv.2017.10.012.

Kaeberlein, T., K. Lewis, and S. S. Epstein. 2002. "Isolating 'Uncultivabte' Microorganisms in Pure Culture in a Simulated Natural Environment." *Science* 296 (5570): 1127–29. https://doi.org/10.1126/science.1070633.

Kamble, Asmita, Sumana Srinivasan, and Harinder Singh. 2019. "In-Silico Bioprospecting: Finding Better Enzymes." *Molecular Biotechnology* 61 (1): 53–9. https://doi.org/10.1007/s12033-018-0132-1.

Kato, Souichiro, Ayasa Yamagishi, Serina Daimon, Kosei Kawasaki, Hideyuki Tamaki, Wataru Kitagawa, Ayumi Abe et al. 2018. "Isolation of Previously Uncultured Slowgrowing Bacteria by Using a Simple Modification in the Preparation of Agar Media." *Applied and Environmental Microbiology* 84 (19): 1–9. https://doi.org/10.1128/AEM.00807-18.

Ke, Jing, Bing Wang, and Yasuo Yoshikuni. 2020. "Microbiome Engineering: Synthetic Biology of Plant-Associated Microbiomes in Sustainable Agriculture." *Trends in Biotechnology* 39 (3): 1–18. https://doi.org/10.1016/j.tibtech.2020.07.008.

Kim, Ki Woo. 2019. "Plant Trichomes as Microbial Habitats and Infection Sites." *European Journal of Plant Pathology* 154 (2): 157–69. https://doi.org/10.1007/s10658-018-01656-0.

Kim, Minseok, Mark Morrison, and Zhongtang Yu. 2011. "Evaluation of Different Partial 16S RRNA Gene Sequence Regions for Phylogenetic Analysis of Microbiomes." *Journal of Microbiological Methods* 84 (1): 81–87. https://doi.org/10.1016/j.mimet.2010.10.020.

Knutsen, Eivind, Ola Johnsborg, Yves Quentin, Jean Pierre Claverys, and Leiv Sigve Håvarstein. 2006. "BOX Elements Modulate Gene Expression in Streptococcus Pneumoniae: Impact on the Fine-Tuning of Competence Development." *Journal of Bacteriology* 188 (23): 8307–12. https://doi.org/10.1128/JB.00850-06.

Köhl, Jürgen, Rogier Kolnaar, and Willem J. Ravensberg. 2019. "Mode of Action of Microbial Biological Control Agents against Plant Diseases: Relevance beyond Efficacy." *Frontiers in Plant Science* 10 (July): 1–19. https://doi.org/10.3389/fpls.2019.00845.

Kralik, Petr, and Matteo Ricchi. 2017. "A Basic Guide to Real Time PCR in Microbial Diagnostics: Definitions, Parameters, and Everything." *Frontiers in Microbiology* 8 (February): 1–9. https://doi.org/10.3389/fmicb.2017.00108.

Krishna, B. Meera, Munawwar Ali Khan, and Shams Tabrez Khan. 2019. "Next-Generation Sequencing (NGS) Platforms: An Exciting Era of Genome Sequence Analysis." In: Tripathi, V., Kumar, P., Tripathi, P., Kishore, A., and Kamle, M. (eds.), *Microbial Genomics in Sustainable Agroecosystems*, 89–119. Singapore: Springer. https://doi.org/10.1007/978-981-32-9860-6_6.

Kumar, Ashwani, and Anamika Dubey. 2020. "Rhizosphere Microbiome: Engineering Bacterial Competitiveness for Enhancing Crop Production." *Journal of Advanced Research* 24: 337–52. https://doi.org/10.1016/j.jare.2020.04.014.

Lagier, Jean Christophe, Sophie Edouard, Isabelle Pagnier, Oleg Mediannikov, Michel Drancourt, and Didier Raoult. 2015. "Current and Past Strategies for Bacterial Culture in Clinical Microbiology." *Clinical Microbiology Reviews* 28 (1): 208–36. https://doi.org/10.1128/CMR.00110-14.

Land, Miriam, Loren Hauser, Se Ran Jun, Intawat Nookaew, Michael R. Leuze, Tae Hyuk Ahn, Tatiana Karpinets et al. 2015. "Insights from 20 Years of Bacterial Genome Sequencing." *Functional and Integrative Genomics* 15 (2): 141–61. https://doi.org/10.1007/s10142-015-0433-4.

Lane, D. J., B. Pace, G. J. Olsen, D. A. Stahl, M. L. Sogin, and N. R. Pace. 1985. "Rapid Determination of 16S Ribosomal RNA Sequences for Phylogenetic Analyses." *Proceedings of the National Academy of Sciences of the United States of America* 82 (20): 6955–59. https://doi.org/10.1073/pnas.82.20.6955.

Lange, Markus, Nico Eisenhauer, Carlos A. Sierra, Holger Bessler, Christoph Engels, Robert I. Griffiths, Perla G. Mellado-Vázquez, Ashish A. Malik, Jacques Roy, and Stefan Scheu. 2015. "Plant Diversity Increases Soil Microbial Activity and Soil Carbon Storage." *Nature Communications* 6 (1): 1–8.

Langille, Morgan G. I., Jesse Zaneveld, J Gregory Caporaso, Daniel McDonald, Dan Knights, Joshua A. Reyes, Jose C. Clemente et al. 2013. "Predictive Functional Profiling of Microbial Communities Using 16S RRNA Marker Gene Sequences." *Nature Biotechnology* 31 (9): 814–21. https://doi.org/10.1038/nbt.2676.

Lei, Xia, En Tao Wang, Wen Feng Chen, Xin Hua Sui, and Wen Xin Chen. 2008. "Diverse Bacteria Isolated from Root Nodules of Wild Vicia Species Grown in Temperate Region of China." *Archives of Microbiology* 190 (6): 657–71. https://doi.org/10.1007/s00203-008-0418-y.

Levy, Asaf, Jonathan M. Conway, Jeffery L. Dangl, and Tanja Woyke. 2018. "Elucidating Bacterial Gene Functions in the Plant Microbiome." *Cell Host and Microbe* 24 (4): 475–85. https://doi.org/10.1016/j.chom.2018.09.005.

Levy, Asaf, Isai Salas Gonzalez, Maximilian Mittelviefhaus, Scott Clingenpeel, Sur Herrera Paredes, Jiamin Miao, Kunru Wang et al. 2018. "Genomic Features of Bacterial Adaptation to Plants." *Nature Genetics* 50 (1): 138–50. https://doi.org/10.1038/s41588-017-0012-9.

Lewis, Kim, Slava Epstein, Anthony D'Onofrio, and Losee L. Ling. 2010. "Uncultured Microorganisms as a Source of Secondary Metabolites." *Journal of Antibiotics* 63 (8): 468–76. https://doi.org/10.1038/ja.2010.87.

Li, Xia, Xiaoyan Geng, Rongrong Xie, Lei Fu, Jianxiong Jiang, Lu Gao, and Jianzhong Sun. 2016. "The Endophytic Bacteria Isolated from Elephant Grass (Pennisetum Purpureum Schumach) Promote Plant Growth and Enhance Salt Tolerance of Hybrid Pennisetum." *Biotechnology for Biofuels* 9 (1): 190. https://doi.org/10.1186/s13068-016-0592-0.

Li, Ying, Warren A. Dick, and Olli H. Tuovinen. 2004. "Fluorescence Microscopy for Visualization of Soil Microorganisms—A Review." *Biology and Fertility of Soils* 39 (5): 301–11. https://doi.org/10.1007/s00374-004-0722-x.

Liu, Weilong, Lv Li, Md Asaduzzaman Khan, and Feizhou Zhu. 2012. "Popular Molecular Markers in Bacteria." *Molecular Genetics, Microbiology and Virology* 27 (3): 103–7. https://doi.org/10.3103/S0891416812030056.

Liu, Wen et al. 2020. "Correlative Surface Imaging Reveals Chemical Signatures for Bacterial Hotspots on Plant Roots." *Analyst* 145 (2): 393–401. doi: 10.1039/C9AN01954E.

Loevinsohn, Michael, Jim Sumberg, Aliou Diagne, and Stephen Whitfield. 2013. "Under What Circumstances and Conditions Does Adoption of Technology Result in Increased Agricultural Productivity? A Systematic Review Prepared for the Department for

International Development," no. July. https://opendocs.ids.ac.uk/opendocs/bitstream/handle/123456789/3208/Productivity systematic review report 3.pdf;jsessionid=7DD2 717D91EF930A407AD3D81FBCDF43?sequence=1.

Louca, Stilianos, Martin F. Polz, Florent Mazel, Michaeline B. N. Albright, Julie A. Huber, Mary I. O'Connor, Martin Ackermann et al. 2018. "Function and Functional Redundancy in Microbial Systems." *Nature Ecology and Evolution* 2 (6): 936–43. https://doi.org/10.1038/s41559-018-0519-1.

Lozada, Mariana, and Hebe M. Dionisi. 2015. "Microbial Bioprospecting in Marine Environments." In: Kim, S. K. (ed.), *Handbook of Marine Biotechnology*. Springer Handbooks. Berlin, Heidelberg: Springer, 307–26. Springer. https://doi.org/10.1007/978-3-642-53971-8_11.

Mackelprang, Rachel, Alyssa M. Grube, Regina Lamendella, Ederson da C. Jesus, Alex Copeland, Chao Liang, Randall D. Jackson et al. 2018. "Microbial Community Structure and Functional Potential in Cultivated and Native Tallgrass Prairie Soils of the Midwestern United States." *Frontiers in Microbiology* 9 (August): 1–15. https://doi.org/10.3389/fmicb.2018.01775.

Macrae, Andrew. 2000. "The Use of 16S RDNA Methods in Soil Microbial Ecology." *Brazilian Journal of Microbiology* 31 (2): 77–82. https://doi.org/10.1590/S1517-83822000000200002.

Massol-Deya, A. A., D. A. Odelson, R. F. Hickey, and J. M. Tiedje. 1995. "Bacterial Community Fingerprinting of Amplified 16S and 16–23S Ribosomal DNA Gene Sequences and Restriction Endonuclease Analysis (ARDRA) BT." In: Akkermans, A. D. L., Van Elsas, J. D., and De Bruijn, F. J. (eds.), *Molecular Microbial Ecology Manual*, 289–96. Netherlands: Springer. https://doi.org/10.1007/978-94-011-0351-0_20.

Mathesius, Ulrike, Susan Mulders, Mengsheng Gao, Max Teplitski, Gustavo Caetano-Anollés, Barry G. Rolfe, and Wolfgang D. Bauer. 2003. "Extensive and Specific Responses of a Eukaryote to Bacterial Quorum-Sensing Signals." *Proceedings of the National Academy of Sciences of the United States of America* 100 (3): 1444–49. https://doi.org/10.1073/pnas.262672599.

Matteoli, Filipe P., Hemanoel Passarelli-Araujo, Régis Josué A. Reis, Letícia O. Da Rocha, Emanuel M. De Souza, L. Aravind, Fabio L. Olivares, and Thiago M. Venancio. 2018. "Genome Sequencing and Assessment of Plant Growth-Promoting Properties of a Serratia Marcescens Strain Isolated from Vermicompost." *BMC Genomics* 19 (1): 1–19. https://doi.org/10.1186/s12864-018-5130-y.

Meij, Anne van der, Sarah F. Worsley, Matthew I. Hutchings, and Gilles P. van Wezel. 2017. "Chemical Ecology of Antibiotic Production by Actinomycetes." *FEMS Microbiology Reviews* 41 (3): 392–416. https://doi.org/10.1093/femsre/fux005.

Mendes, Lucas William, and Siu Mui Tsai. 2018. "Distinct Taxonomic and Functional Composition of Soil Microbiomes along the Gradient Forest-Restinga-Mangrove in Southeastern Brazil." *Antonie van Leeuwenhoek, International Journal of General and Molecular Microbiology* 111 (1): 101–14. https://doi.org/10.1007/s10482-017-0931-6.

Metzker, Michael L. 2010. "Sequencing Technologies the next Generation." *Nature Reviews Genetics* 11 (1): 31–46. https://doi.org/10.1038/nrg2626.

Montagu, Marc Van. 2020. "The Future of Plant Biotechnology in a Globalized and Environmentally Endangered World." Genetics and *Molecular Biology* 43 (1 suppl 2): e20190040. https://doi.org/10.1590/1678-4685-GMB-2019-0040.

Montanari-Coelho, Katiúscia Kelli, Alessandra Tenório Costa, Julio Cesar Polonio, João Lúcio Azevedo, Silvana Regina Rockenbach Marin, Renata Fuganti-Pagliarini, Yasunari Fujita et al. 2018. "Endophytic Bacterial Microbiome Associated with Leaves

of Genetically Modified (AtAREB1) and Conventional (BR 16) Soybean Plants."
World Journal of Microbiology and Biotechnology 34 (4): 56. https://doi.org/10.1007/
s11274-018-2439-2.

Monteiro, Douglas Alfradique, Eduardo da Silva Fonseca, Renato de Aragão Ribeiro
Rodrigues, Jacqueline Jesus Nogueira da Silva, Elderson Pereira da Silva, Fabiano de
Carvalho Balieiro, Bruno José Rodrigues Alves, and Caio Tavora Coelho da Costa
Rachid. 2020. "Structural and Functional Shifts of Soil Prokaryotic Community Due
to Eucalyptus Plantation and Rotation Phase." *Scientific Reports* 10 (1): 9075. https://
doi.org/10.1038/s41598-020-66004-x.

Montella, Salvatore, Valeria Ventorino, Vincent Lombard, Bernard Henrissat, Olimpia Pepe,
and Vincenza Faraco. 2017. "Discovery of Genes Coding for Carbohydrate-Active
Enzyme by Metagenomic Analysis of Lignocellulosic Biomasses." *Scientific Reports*
7 (February): 1–15. https://doi.org/10.1038/srep42623.

Moustafa, Khaled, and Joanna M. Cross. 2016. "Genetic Approaches to Study Plant Responses
to Environmental Stresses: An Overview." *Biology* 5 (2): 1–18. https://doi.org/10.3390/
biology5020020.

Mühling, Martin et al. 2016. "Reconstruction of the Metabolic Potential of Acidophilic
Sideroxydans Strains from the Metagenome of an Microaerophilic Enrichment Culture
of Acidophilic Iron-Oxidizing Bacteria from a Pilot Plant for the Treatment of Acid
Mine Drainage Reveals Metabolic Versatility and Adaptation to Life at Low pH."
Frontiers in Microbiology 7: 2082. https://doi.org/10.3389/fmicb.2016.02082.

Muir, Paul, Shantao Li, Shaoke Lou, Daifeng Wang, Daniel J. Spakowicz, Leonidas Salichos,
Jing Zhang et al. 2016. "The Real Cost of Sequencing: Scaling Computation to Keep
Pace with Data Generation." *Genome Biology* 17 (1): 1–9. https://doi.org/10.1186/
s13059-016-0917-0.

Musumeci, Matías A., Mariana Lozada, Daniela V. Rial, Walter P. Mac Cormack, Janet
K. Jansson, Sara Sjöling, Jo Lynn Carroll, and Hebe M. Dionisi. 2017. "Prospecting
Biotechnologically-Relevant Monooxygenases from Cold Sediment Metagenomes: An
in Silico Approach." *Marine Drugs* 15 (4): 114. https://doi.org/10.3390/md15040114.

Muyzer, G., E. C. De Waal, and A. G. Uitterlinden. 1993. "Profiling of Complex Microbial
Populations by Denaturing Gradient Gel Electrophoresis Analysis of Polymerase
Chain Reaction-Amplified Genes Coding for 16S RRNA." *Applied and Environmental
Microbiology* 59 (3): 695–700. https://doi.org/10.1128/aem.59.3.695-700.1993.

Nadeem, Muhammad Azhar, Muhammad Amjad Nawaz, Muhammad Qasim Shahid, Yıldız
Doğan, Gonul Comertpay, Mehtap Yıldız, Rüştü Hatipoğlu et al. 2018. "DNA Molecular
Markers in Plant Breeding: Current Status and Recent Advancements in Genomic
Selection and Genome Editing." *Biotechnology and Biotechnological Equipment*
32 (2): 261–85. https://doi.org/10.1080/13102818.2017.1400401.

Nannipieri, P., J. Ascher, M. T. Ceccherini, L. Landi, G. Pietramellara, and G. Renella. 2017.
"Microbial Diversity and Soil Functions." *European Journal of Soil Science* 68 (1):
12–26. https://doi.org/10.1111/ejss.4_12398.

Nannipieri, P., L. Giagnoni, G. Renella, E. Puglisi, B. Ceccanti, G. Masciandaro, F. Fornasier,
M. C. Moscatelli, and S. Marinari. 2012. "Soil Enzymology: Classical and Molecular
Approaches." *Biology and Fertility of Soils* 48 (7): 743–62. https://doi.org/10.1007/
s00374-012-0723-0.

Navarrete, Acacio Aparecido, Rita de Cássia Bonassi, Juliana Heloisa Pinê Américo-Pinheiro,
Gisele Herbst Vazquez, Lucas William Mendes, Elisângela de Souza Loureiro,
Eiko Eurya Kuramae, and Siu Mui Tsai. 2021. "Methods to Identify Soil Microbial
Bioindicators of Sustainable Management of Bioenergy Crops." In: L. C. Carvalhais
and P. G. Dennis (eds.), *The Plant Microbiome. Methods in Molecular Biology*, 251–63.
New York: Humana Press.

Neufeld, Josh D., Jyotsna Vohra, Marc G. Dumont, Tillmann Lueders, Mike Manefield, Michael W. Friedrich, and Colin J. Murrell. 2007. "DNA Stable-Isotope Probing." *Nature Protocols* 2 (4): 860–66. https://doi.org/10.1038/nprot.2007.109.

Nichols, D., N. Cahoon, E. M. Trakhtenberg, L. Pham, A. Mehta, A. Belanger, T. Kanigan, K. Lewis, and S. S. Epstein. 2010. "Use of Ichip for High-Throughput in Situ Cultivation of 'Uncultivable Microbial Species.' " *Applied and Environmental Microbiology* 76 (8): 2445–50. https://doi.org/10.1128/AEM.01754-09.

O'Connell, Elise M., and Thomas B. Nutman. 2016. "Review Article: Molecular Diagnostics for Soil-Transmitted Helminths." *American Journal of Tropical Medicine and Hygiene* 95 (3): 508–14. https://doi.org/10.4269/ajtmh.16-0266.

Ong, Quang, Phuc Nguyen, Nguyen Phuong Thao, and Ly Le. 2016. "Bioinformatics Approach in Plant Genomic Research." *Current Genomics* 17 (4): 368–78. https://doi.org/10.2174/1389202917666160331202956.

Oraiopoulou, M. E., E. Tzamali, G. Tzedakis, E. Liapis, G. Zacharakis, A. Vakis, J. Papamatheakis, and V. Sakkalis. 2018. "Integrating in Vitro Experiments with in Silico Approaches for Glioblastoma Invasion: The Role of Cell-to-Cell Adhesion Heterogeneity." *Scientific Reports* 8 (1): 1–13. https://doi.org/10.1038/s41598-018-34521-5.

Orlofsky, Ezra, Maya Benami, Amit Gross, Michelle Dutt, and Osnat Gillor. 2015. "Rapid MPN-Qpcr Screening for Pathogens in Air, Soil, Water, and Agricultural Produce." *Water, Air, and Soil Pollution* 226 (9) 303. https://doi.org/10.1007/s11270-015-2560-x.

Paczia, Nicole, Anke Nilgen, Tobias Lehmann, Jochem Gätgens, Wolfgang Wiechert, and Stephan Noack. 2012. "Extensive Exometabolome Analysis Reveals Extended Overflow Metabolism in Various Microorganisms." *Microbial Cell Factories* 11: 1–14. https://doi.org/10.1186/1475-2859-11-122.

Paiva-Cavalcanti, M., C. G. Regis-da-Silva, and Y. M. Gomes. 2010. "Comparison of Real-Time PCR and Conventional PCR for Detection of Leishmania (Leishmania) Infantum Infection: A Mini-Review." *Journal of Venomous Animals and Toxins Including Tropical Diseases* 16 (4): 537–42. https://doi.org/10.1590/S1678-91992010000400004.

Park, Myoungsu, Chungwoo Kim, Jinchul Yang, Hyoungseok Lee, Wansik Shin, Seunghwan Kim, and Tongmin Sa. 2005. "Isolation and Characterization of Diazotrophic Growth Promoting Bacteria from Rhizosphere of Agricultural Crops of Korea." *Microbiological Research* 160 (2): 127–33. https://doi.org/10.1016/j.micres.2004.10.003.

Partida-Martínez, Laila P., and Martin Heil. 2011. "The Microbe-Free Plant: Fact or Artifact?" *Frontiers in Plant Science* 2 (December): 1–16. https://doi.org/10.3389/fpls.2011.00100.

Pereira e Silva, Michele C., Alexander V. Semenov, Heike Schmitt, Jan Dirk van Elsas, and Joana Falcão Salles. 2013. "Microbe-Mediated Processes as Indicators to Establish the Normal Operating Range of Soil Functioning." *Soil Biology and Biochemistry* 57: 995–1002. https://doi.org/10.1016/j.soilbio.2012.10.002.

Pham, Van H. T., and Jaisoo Kim. 2012. "Cultivation of Unculturable Soil Bacteria." *Trends in Biotechnology* 30 (9): 475–84. https://doi.org/10.1016/j.tibtech.2012.05.007.

Pinu, Farhana R., David J. Beale, Amy M. Paten, Konstantinos Kouremenos, Sanjay Swarup, Horst J. Schirra, and David Wishart. 2019. "Systems Biology and Multi-Omics Integration: Viewpoints from the Metabolomics Research Community." *Metabolites* 9 (4): 1–31. https://doi.org/10.3390/metabo9040076.

Pratama, Akbar Adjie, and Jan Dirk van Elsas. 2018. "The 'Neglected' Soil Virome – Potential Role and Impact." *Trends in Microbiology* 26 (8): 649–62. https://doi.org/10.1016/j.tim.2017.12.004.

Porras-Alfaro, Andrea, Liu, Kuan-Liang, Kruske, Cheryl R., Xie, Gary. 2014. "From Genus to Phylum: Large-Subunit and Internal Transcribed Spacer rRNA Operon Regions Show Similar Classification Accuracies Influenced by Database Composition."

Applied and Environmental Microbiology 80 (3): 829–40. https://doi.org/10.1128/AEM.02894-13.

Rachid, Caio T. C. C., Fabiano C. Balieiro, Eduardo S. Fonseca, Raquel Silva Peixoto, Guilherme M. Chaer, James M. Tiedje, and Alexandre S. Rosado. 2015. "Intercropped Silviculture Systems, a Key to Achieving Soil Fungal Community Management in Eucalyptus Plantations." *PLoS ONE* 10 (2): e0118515. https://doi.org/10.1371/journal.pone.0118515.

Rachid, Caio T. C. C., Adriana L. Santos, Marisa C. Piccolo, Fabiano C. Balieiro, Heitor L. C. Coutinho, Raquel S. Peixoto, James M. Tiedje, and Alexandre S. Rosado. 2013. "Effect of Sugarcane Burning or Green Harvest Methods on the Brazilian Cerrado Soil Bacterial Community Structure." *PLoS ONE* 8 (3): e59342. https://doi.org/10.1371/journal.pone.0059342.

Rachman, L. M. 2019. "Development of Technique to Determine Soil Quality Index for Assessing Soil Condition." *Journal of Physics: Conference Series* 1375 (1): 1–6. https://doi.org/10.1088/1742-6596/1375/1/012046.

Raes, Jeroen, Ivica Letunic, Takuji Yamada, Lars Juhl Jensen, and Peer Bork. 2011. "Toward Molecular Trait-Based Ecology through Integration of Biogeochemical, Geographical and Metagenomic Data." *Molecular Systems Biology* 7 (473): 1–9. https://doi.org/10.1038/msb.2011.6.

Rao, Zhiguo et al. 2017. "Relationship Between the Stable Carbon Isotopic Composition of Modern Plants and Surface Soils and Climate: A Global Review." Earth-Science Reviews 165, 110–19. https://doi.org/10.1016/j.earscirev.2016.12.007.

Rasheed, M. A., Syed Zaheer Hasan, P. L.S. Srinivasa Rao, Annapurna Boruah, V. Sudarshan, B. Kumar, and T. Harinarayana. 2014. "Application of Geo-Microbial Prospecting Method for Finding Oil and Gas Reservoirs." *Frontiers of Earth Science* 9 (1): 40–50. https://doi.org/10.1007/s11707-014-0448-5.

Rastogi, G., and R. K. Sani. 2011. "Molecular Techniques to Assess Microbial Community Structure, Function, and Dynamics in the Environment." In: Ahmad, I., Ahmad, F., and Pichtel, J. (eds.), *Microbes and Microbial Technology*, 28–57. New York, NY: Springer.

Rezene, Yayis, Kassahun Tesfaye, Mukankusi Clare, Allan Male, and Paul Gepts. 2018. "Rep-PCR Genomic Fingerprinting Revealed Genetic Diversity and Population Structure among Ethiopian Isolates of Pseudocercospora Griseola Pathogen of the Common Bean (Phaseolus Vulgaris L.)." *Journal of Plant Pathology & Microbiology* 9 (11). https://doi.org/10.4172/2157-7471.1000463.

Romano, Ida, Valeria Ventorino, and Olimpia Pepe. 2020. "Effectiveness of Plant Beneficial Microbes: Overview of the Methodological Approaches for the Assessment of Root Colonization and Persistence." *Frontiers in Plant Science* 11 (January): 1–16. https://doi.org/10.3389/fpls.2020.00006.

Ruppert, Krista M., Richard J. Kline, and Md Saydur Rahman. 2019. "Past, Present, and Future Perspectives of Environmental DNA (EDNA) Metabarcoding: A Systematic Review in Methods, Monitoring, and Applications of Global EDNA." *Global Ecology and Conservation* 17: e00547. https://doi.org/10.1016/j.gecco.2019.e00547.

Rychlik, Tomasz, Artur Szwengiel, Marta Bednarek, Edna Arcuri, Didier Montet, Baltasar Mayo, Jacek Nowak, and Zbigniew Czarnecki. 2017. "Application of the PCR-DGGE Technique to the Fungal Community of Traditional Wielkopolska Fried Ripened Curd Cheese to Determine Its PGI Authenticity." *Food Control* 73: 1074–81. https://doi.org/10.1016/j.foodcont.2016.10.024.

Ryder, Lauren S., Beverley D. Harris, Darren M. Soanes, Michael J. Kershaw, Nicholas J. Talbot, and Christopher R. Thornton. 2012. "Saprotrophic Competitiveness and Biocontrol Fitness of a Genetically Modified Strain of the Plant-Growth-Promoting Fungus Trichoderma Hamatum GD12." *Microbiology* 158 (1): 84–97. https://doi.org/10.1099/mic.0.051854-0.

Sagaram, Uma Shankar, Kristen M. Deangelis, Pankaj Trivedi, Gary L. Andersen, Shi En Lu, and Nian Wang. 2009. "Bacterial Diversity Analysis of Huanglongbing Pathogen-Infected Citrus, Using PhyloChip Arrays and 16S RRNA Gene Clone Library Sequencing." *Applied and Environmental Microbiology* 75 (6): 1566–74. https://doi.org/10.1128/AEM.02404-08.

Saiki, R. K., S. Scharf, F. Faloona, K. B. Mullis, G. T. Horn, H. A. Erlich, and N. Arnheim. 1992. "Enzymatic Amplification of Beta-Globin Genomic Sequences and Restriction Site Analysis for Diagnosis of Sickle Cell Anemia. 1985." *Biotechnology (Reading, Mass.)* 24: 476–80. https://doi.org/10.1007/BF00985904.

Saminathan, Thangasamy, Marleny García, Bandana Ghimire, Carlos Lopez, Abiodun Bodunrin, Padma Nimmakayala, Venkata L. Abburi, Amnon Levi, Nagamani Balagurusamy, and Umesh K. Reddy. 2018. "Metagenomic and Metatranscriptomic Analyses of Diverse Watermelon Cultivars Reveal the Role of Fruit Associated Microbiome in Carbohydrate Metabolism and Ripening of Mature Fruits." *Frontiers in Plant Science* 9 (January): 1–13. https://doi.org/10.3389/fpls.2018.00004.

Sarhan, Mohamed S., Sascha Patz, Mervat A. Hamza, Hanan H. Youssef, Elhussein F. Mourad, Mohamed Fayez, Brian Murphy, Silke Ruppel, and Nabil A. Hegazi. 2018. "G3 Phylochip Analysis Confirms the Promise of Plant-Based Culture Media for Unlocking the Composition and Diversity of the Maize Root Microbiome and for Recovering Unculturable Candidate Divisions/Phyla." *Microbes and Environments* 33 (3): 317–25. https://doi.org/10.1264/jsme2.ME18023.

Schaefer, Amy L., Colin R. Lappala, Ryan P. Morlen, Dale A. Pelletier, Tse Yuan, S. Lu, Patricia K. Lankford, Caroline S. Harwood, and E. Peter Greenberg. 2013. "LuxR- and LuxI-Type Quorum-Sensing Circuits Are Prevalent in Members of the Populus Deltoides Microbiome." *Applied and Environmental Microbiology* 79 (18): 5745–52. https://doi.org/10.1128/AEM.01417-13.

Schloter, Michael, Paolo Nannipieri, Søren J. Sørensen, and Jan Dirk van Elsas. 2018. "Microbial Indicators for Soil Quality." *Biology and Fertility of Soils* 54 (1). https://doi.org/10.1007/s00374-017-1248-3.

Schoch, Conrad L., Keith A. Seifert, Sabine Huhndorf, Vincent Robert, John L. Spouge, C. André Levesque, Wen Chen et al. 2012. "Nuclear Ribosomal Internal Transcribed Spacer (ITS) Region as a Universal DNA Barcode Marker for Fungi." *Proceedings of the National Academy of Sciences of the United States of America* 109 (16): 6241–46. https://doi.org/10.1073/pnas.1117018109.

Sekurova, Olga N., Olha Schneider, and Sergey B. Zotchev. 2019. "Novel Bioactive Natural Products from Bacteria via Bioprospecting, Genome Mining and Metabolic Engineering." *Microbial Biotechnology* 12 (5): 828–44. https://doi.org/10.1111/1751-7915.13398.

Seshadri, Rekha, Wayne G. Reeve, Julie K. Ardley, Kristin Tennessen, Tanja Woyke, Nikos C. Kyrpides, and Natalia N. Ivanova. 2015. "Discovery of Novel Plant Interaction Determinants from the Genomes of 163 Root Nodule Bacteria." *Scientific Reports* 5 (October): 1–9. https://doi.org/10.1038/srep16825.

Setten, L., G. Soto, M. Mozzicafreddo, A. R. Fox, C. Lisi, M. Cuccioloni, M. Angeletti, E. Pagano, A. Díaz-Paleo, and N. D. Ayub. 2013. "Engineering Pseudomonas Protegens Pf-5 for Nitrogen Fixation and Its Application to Improve Plant Growth under Nitrogen-Deficient Conditions." *PLoS ONE* 8 (5): e63666. https://doi.org/10.1371/journal.pone.0063666.

Shade, A., and J. Handelsman. 2012. "Beyond the Venn Diagram: The Hunt for a Core Microbiome." *Environmental Microbiology* 14 (1): 4–12. https://doi.org/10.1111/j.1462-2920.2011.02585.x.

Shade, A., and N. Stopnisek. 2019. "Abundance-Occupancy Distributions to Prioritize Plant Core Microbiome Membership." *Current Opinion in Microbiology* 49: 50–58. https://doi.org/10.1016/j.mib.2019.09.008.

Shen, Zongzhuan, Chao Xue, C. Ryan Penton, Linda S. Thomashow, Na Zhang, Beibei Wang, Yunze Ruan, Rong Li, and Qirong Shen. 2019. "Suppression of Banana Panama Disease Induced by Soil Microbiome Reconstruction through an Integrated Agricultural Strategy." *Soil Biology and Biochemistry* 128: 164–74. https://doi.org/10.1016/j.soilbio.2018.10.016.

Shrestha, Abhishek, Maja Grimm, Ichie Ojiro, Johannes Krumwiede, and Adam Schikora. 2020. "Impact of Quorum Sensing Molecules on Plant Growth and Immune System." *Frontiers in Microbiology* 11 (July): 1–11. https://doi.org/10.3389/fmicb.2020.01545.

Silva Marques, Eric de Lima, Eduardo Gross, João Carlos Teixeira Dias, Carlos Priminho Pirovani, and Rachel Passos Rezende. 2018. "Ammonia Oxidation (AmoA) and Nitrogen Fixation (NifH) Genes along Metasandstone and Limestone Caves of Brazil." *Geomicrobiology Journal* 35 (10): 869–78. https://doi.org/10.1080/01490451.2018.1482386.

Simpson, A. J. G. 2000. "The Complete Genome Sequence of the Plant Pathogen Xylella Fastidiosa." *Biochemical Society Transactions* 28 (5): A102–A102. https://doi.org/10.1042/bst028a102b.

Singh, Brajesh Kumar. 2010. "Exploring Microbial Diversity for Biotechnology: The Way Forward." *Trends in Biotechnology* 28 (3): 111–16. https://doi.org/10.1016/j.tibtech.2009.11.006.

Singh, Joginder, Deepansh Sharma, Gaurav Kumar, and Neeta Raj Sharma. 2018. "Small at Size, Big at Impact: Microorganisms for Sustainable Development." In: Singh, J., Sharma, D., Kumar, G., and Sharma, N. (eds.), Microbial Bioprospecting for Sustainable Development. Microbial Bioprospecting for Sustainable Development. Singapore: Springer. https://doi.org/10.1007/978-981-13-0053-0.

Singh, Rajesh Kumar, Pratiksha Singh, Hai Bi Li, Qi Qi Song, Dao Jun Guo, Manoj K. Solanki, Krishan K. Verma et al. 2020. "Diversity of Nitrogen-Fixing Rhizobacteria Associated with Sugarcane: A Comprehensive Study of Plant-Microbe Interactions for Growth Enhancement in Saccharum Spp." *BMC Plant Biology* 20 (1): 1–21. https://doi.org/10.1186/s12870-020-02400-9.

Sklarz, Menachem Y., Roey Angel, Osnat Gillor, and M. Ines M. Soares. 2009. "Evaluating Amplified RDNA Restriction Analysis Assay for Identification of Bacterial Communities." *Antonie van Leeuwenhoek, International Journal of General and Molecular Microbiology* 96 (4): 659–64. https://doi.org/10.1007/s10482-009-9380-1.

Starke, Robert, Nico Jehmlich, and Felipe Bastida. 2019. "Using Proteins to Study How Microbes Contribute to Soil Ecosystem Services: The Current State and Future Perspectives of Soil Metaproteomics." *Journal of Proteomics* 198 (October 2018): 50–58. https://doi.org/10.1016/j.jprot.2018.11.011.

Stefani, Franck O. P., Terrence H. Bell, Charlotte Marchand, Ivan E. De La Providencia, Abdel El Yassimi, Marc St-Arnaud, and Mohamed Hijri. 2015. "Culture-Dependent and -Independent Methods Capture Different Microbial Community Fractions in Hydrocarbon-Contaminated Soils." *PLoS ONE* 10 (6): 1–16. https://doi.org/10.1371/journal.pone.0128272.

Stevenson, Bradley S., Stephanie A. Eichorst, John T. Wertz, Thomas M. Schmidt, and John A. Breznak. 2004. "New Strategies for Cultivation and Detection of Previously Uncultured Microbes." *Applied and Environmental Microbiology* 70 (8): 4748–55. https://doi.org/10.1128/AEM.70.8.4748-4755.2004.

Stewart, Eric J. 2012. "Growing Unculturable Bacteria." *Journal of Bacteriology* 194 (16): 4151–60. https://doi.org/10.1128/JB.00345-12.

Stolze, Y., A. Bremges, M. Rumming et al. 2016. "Identification and Genome Reconstruction of Abundant Distinct Taxa in Microbiomes from One Thermophilic and Three Mesophilic Production-Scale Biogas Plants." Biotechnol Biofuels 9: 156. https://doi.org/10.1186/s13068-016-0565-3.

Subramaniam, Gopalakrishnan, Vivek Thakur, Rachit K. Saxena, Srinivas Vadlamudi, Shilp Purohit, Vinay Kumar, Abhishek Rathore, Annapurna Chitikineni, and Rajeev K. Varshney. 2020. "Complete Genome Sequence of Sixteen Plant Growth Promoting Streptomyces Strains." *Scientific Reports* 10 (1): 1–13. https://doi.org/10.1038/s41598-020-67153-9.

Swenson, Tami L., Stefan Jenkins, Benjamin P. Bowen, and Trent R. Northen. 2015. "Untargeted Soil Metabolomics Methods for Analysis of Extractable Organic Matter." *Soil Biology and Biochemistry* 80: 189–98. https://doi.org/10.1016/j.soilbio.2014.10.007.

Swenson, Tami L., Ulas Karaoz, Joel M. Swenson, Benjamin P. Bowen, and Trent R. Northen. 2018. "Linking Soil Biology and Chemistry in Biological Soil Crust Using Isolate Exometabolomics." *Nature Communications* 9 (1): 1–10. https://doi.org/10.1038/s41467-017-02356-9.

Thompson, Luke R., Jon G. Sanders, Daniel McDonald, Amnon Amir, Joshua Ladau, Kenneth J. Locey, Robert J. Prill et al. 2017. "A Communal Catalogue Reveals Earth's Multiscale Microbial Diversity." *Nature* 551 (7681): 457–63. https://doi.org/10.1038/nature24621.

Tilman, David, Kenneth G. Cassman, Pamela A. Matson, Rosamond Naylor, and Stephen Polasky. 2002. "Agricultural Sustainability and Intensive Production Practices." *Nature* 418 (6898): 671–77. https://doi.org/10.1038/nature01014.

Tosi, Micaela, Eduardo Kovalski Mitter, Jonathan Gaiero, and Kari Dunfield. 2020. "It Takes Three to Tango: The Importance of Microbes, Host Plant, and Soil Management to Elucidate Manipulation Strategies for the Plant Microbiome." *Canadian Journal of Microbiology* 66 (7): 413–33. https://doi.org/10.1139/cjm-2020-0085.

Tringe, Susannah G., and Philip Hugenholtz. 2008. "A Renaissance for the Pioneering 16S RRNA Gene." *Current Opinion in Microbiology* 11 (5): 442–46. https://doi.org/10.1016/j.mib.2008.09.011.

Trivedi, Pankaj, Manuel Delgado-Baquerizo, Chanda Trivedi, Hangwei Hu, Ian C. Anderson, Thomas C. Jeffries, Jizhong Zhou, and Brajesh K. Singh. 2016. "Microbial Regulation of the Soil Carbon Cycle: Evidence from Gene-Enzyme Relationships." *ISME Journal* 10 (11): 2593–2604. https://doi.org/10.1038/ismej.2016.65.

Turk-Kubo, Kendra A., Katherine M. Achilles, Tracy R. C. Serros, Mari Ochiai, Joseph P. Montoya, and Jonathan P. Zehr. 2012. "Nitrogenase (NifH) Gene Expression in Diazotrophic Cyanobacteria in the Tropical North Atlantic in Response to Nutrient Amendments." *Frontiers in Microbiology* 3 (November): 1–17. https://doi.org/10.3389/fmicb.2012.00386.

Vandenkoornhuyse P., Quaiser A., Duhamel M., Le Van A., and Dufresne A. 2015. "The Importance of the Microbiome of the Plant Holobiont." *New Phytol* 206 (4): 1196–206. doi: 10.1111/nph.13312.

Vartoukian, Sonia R., Aleksandra Adamowska, Megan Lawlor, Rebecca Moazzez, Floyd E. Dewhirst, and William G. Wade. 2016. "In Vitro Cultivation of 'unculturable' Oral Bacteria, Facilitated by Community Culture and Media Supplementation with Siderophores." *PLoS ONE* 11 (1): 1–19. https://doi.org/10.1371/journal.pone.0146926.

Vollú, Renata Estebanez, Simone Raposo Cotta, Diogo Jurelevicius, Deborah Catharine de Assis Leite, Cláudio Ernesto Taveira Parente, Olaf Malm, Denize Carvalho Martins, Álvaro Vilela Resende, Ivanildo Evódio Marriel, and Lucy Seldin. 2018. "Response of the Bacterial Communities Associated With Maize Rhizosphere to Poultry Litter as an Organomineral Fertilizer." *Frontiers in Environmental Science*. https://www.frontiersin.org/article/10.3389/fenvs.2018.00118.

Wang, Daojing, and Steven Bodovitz. 2011. "Single Cell Analysis: The New Frontier in 'Omics' Single Cell Analysis: Needs and Applications." *Trends in Biotechnology* 28 (6): 281–90. https://doi.org/10.1016/j.tibtech.2010.03.002.Single.

Wang, Yong, and Pei Yuan Qian. 2009. "Conservative Fragments in Bacterial 16S RRNA Genes and Primer Design for 16S Ribosomal DNA Amplicons in Metagenomic Studies." *PLoS ONE* 4 (10). https://doi.org/10.1371/journal.pone.0007401.

Watanabe, Kanako, James Nelson, Shigeaki Harayama, and Hiroaki Kasai. 2001. "ICB Database: The GyrB Database for Identification and Classification of Bacteria." *Nucleic Acids Research* 29 (1): 344–45. https://doi.org/10.1093/nar/29.1.344.

Watanabe, Kazuya, and Paul W. Baker. 2000. "Environmentally Relevant Microorganisms." *Journal of Bioscience and Bioengineering* 89 (1): 1–11. https://doi.org/10.1016/S1389-1723(00)88043-3.

Webster, R. 2017. "Replicate and Randomize, or Lie." *Environmental Microbiology* 19 (1): 25–28. doi.10.1111/1462-2920.13533.

Wei Wei, Kazuo Isobe, Yutaka Shiratori, Tomoyasu Nishizawa, Nobuhito Ohte, Shigeto Otsuka, and Keishi Senoo. 2014. "N2O Emission from Cropland Field Soil through Fungal Denitrification after Surface Applications of Organic Fertilizer." *Soil Biology and Biochemistry* 69: 157–67. https://doi.org/10.1016/j.soilbio.2013.10.044.

Weinert, Nicole, Yvette Piceno, Guo Chun Ding, Remo Meincke, Holger Heuer, Gabriele Berg, Michael Schloter, Gary Andersen, and Kornelia Smalla. 2011. "PhyloChip Hybridization Uncovered an Enormous Bacterial Diversity in the Rhizosphere of Different Potato Cultivars: Many Common and Few Cultivar-Dependent Taxa." *FEMS Microbiology Ecology* 75 (3): 497–506. https://doi.org/10.1111/j.1574-6941.2010.01025.x.

Wienhausen, Gerrit, Beatriz E. Noriega-Ortega, Jutta Niggemann, Thorsten Dittmar, and Meinhard Simon. 2017. "The Exometabolome of Two Model Strains of the Roseobacter Group: A Marketplace of Microbial Metabolites." *Frontiers in Microbiology* 8 (October): 1–15. https://doi.org/10.3389/fmicb.2017.01985.

Williams, John G. K., Anne R. Kubelik, Kenneth J. Livak, J. Antoni Rafalski, and Scott V. Tingey. 1990. "DNA Polymorphisms Amplified by Arbitrary Primers Are Useful as Genetic Markers." *Nucleic Acids Research* 18 (22): 6531–35. https://doi.org/10.1093/nar/18.22.6531.

Wilmes, Paul, and Philip L. Bond. 2009. "Microbial Community Proteomics: Elucidating the Catalysts and Metabolic Mechanisms That Drive the Earth's Biogeochemical Cycles." *Current Opinion in Microbiology* 12 (3): 310–17. https://doi.org/10.1016/j.mib.2009.03.004.

Wood, Stephen A., Maya Almaraz, Mark A. Bradford, Krista L. McGuire, Shahid Naeem, Christopher Neill, Cheryl A. Palm, Katherine L. Tully, and Jizhong Zhou. 2015. "Farm Management, Not Soil Microbial Diversity, Controls Nutrient Loss from Smallholder Tropical Agriculture." *Frontiers in Microbiology* 6 (March): 1–10. https://doi.org/10.3389/fmicb.2015.00090.

Xie, Hualin, Yanwei Zhang, Yongrok Choi, and Fengqin Li. 2020. "A Scientometrics Review on Land Ecosystem Service Research." *Sustainability (Switzerland)* 12 (7): 1–23. https://doi.org/10.3390/su12072959.

Xin, Bingyue, Hualin Liu, Jinshui Zheng, Chuanshuai Xie, Ying Gao, Dadong Dai, Donghai Peng, Lifang Ruan, Huanchun Chen, and Ming Sun. 2020. "In Silico Analysis Highlights the Diversity and Novelty of Circular Bacteriocins in Sequenced Microbial Genomes." *MSystems* 5 (3): 1–14. https://doi.org/10.1128/msystems.00047-20.

Xu, Jin, and Nian Wang. 2019. "Where Are We Going with Genomics in Plant Pathogenic Bacteria?" *Genomics* 111 (4): 729–36. https://doi.org/10.1016/j.ygeno.2018.04.011.

Yamamoto, S., and S. Harayama. 1995. "PCR Amplification and Direct Sequencing of GyrB Genes with Universal Primers and Their Application to the Detection and Taxonomic Analysis of Pseudomonas Putida Strains." *Applied and Environmental Microbiology* 61 (3): 1104–9. https://doi.org/10.1128/aem.61.3.1104-1109.1995.

Yang, Qing Song, Jun De Dong, Manzoor Ahmad, Juan Ling, Wei Guo Zhou, Ye Hui Tan,

Yuan Zhou Zhang, Dan Dan Shen, and Yan Ying Zhang. 2019. "Analysis of NifH DNA and RNA Reveals a Disproportionate Contribution to Nitrogenase Activities by Rare Plankton-Associated Diazotrophs." *BMC Microbiology* 19 (1): 1–12. https://doi.org/10.1186/s12866-019-1565-9.

Yang, Wanjie, Xiaolong Kang, Qingfeng Yang, Yao Lin, and Meiying Fang. 2013. "Review on the Development of Genotyping Methods for Assessing Farm Animal Diversity." *Journal of Animal Science and Biotechnology* 4 (1): 2–7. https://doi.org/10.1186/2049-1891-4-2.

Yeoh, Yun Kit et al. 2016. "The Core Root Microbiome of Sugarcanes Cultivated Under Varying Nitrogen Fertilizer Application." *Environmental Microbiology* 18 (5): 1338–51. https://doi.org/10.1111/1462-2920.12925.

Yilmaz, Pelin et al. 2014. The SILVA and "All-Species Living Tree Project (LTP)" Taxonomic Frameworks. *Nucleic Acids Research* 42 (D1) D643-D648. https://doi.org/10.1093/nar/gkt1209.

Yu, Lingling, Shasha Luo, Xia Xu, Yonggang Gou, and Jianwu Wang. 2020. "The Soil Carbon Cycle Determined by GeoChip 5.0 in Sugarcane and Soybean Intercropping Systems with Reduced Nitrogen Input in South China." *Applied Soil Ecology* 155 (May): 103653. https://doi.org/10.1016/j.apsoil.2020.103653.

Zengler, Karsten. 2009. "Central Role of the Cell in Microbial Ecology." *Microbiology and Molecular Biology Reviews* 73 (4): 712–29. https://doi.org/10.1128/mmbr.00027-09.

Zhang, Jin-Lin, Mina Aziz, Yan Qiao, Qing-Qiang Han, Jing Li, Yin-Quan Wang, Xin Shen, Suo-Min Wang, and Paul W. Paré. 2014. "Soil Microbe *Bacillus Subtilis* (GB03) Induces Biomass Accumulation and Salt Tolerance with Lower Sodium Accumulation in Wheat." *Crop and Pasture Science* 65 (5): 423–27. https://doi.org/10.1071/CP13456.

Zhang, Xian, Zhenghua Liu, Guanyun Wei, Fei Yang, and Xueduan Liu. 2018. "In Silico Genome-Wide Analysis Reveals the Potential Links between Core Genome of Acidithiobacillus Thiooxidans and Its Autotrophic Lifestyle." *Frontiers in Microbiology* 9 (June): 1–14. https://doi.org/10.3389/fmicb.2018.01255.

Žifčáková, Lucia, Tomáš Větrovský, Vincent Lombard, Bernard Henrissat, Adina Howe, and Petr Baldrian. 2017. "Feed in Summer, Rest in Winter: Microbial Carbon Utilization in Forest Topsoil." *Microbiome* 5 (1): 122. https://doi.org/10.1186/s40168-017-0340-0.

Zuñiga, Cristal, Livia Zaramela, and Karsten Zengler. 2017. "Elucidation of Complexity and Prediction of Interactions in Microbial Communities." *Microbial Biotechnology* 10 (6): 1500–22. https://doi.org/10.1111/1751-7915.12855.

3 Novel Metabolites from Endophytes
Potential Applications

Jhumishree Meher, Raina Bajpai,
Md. Mahtab Rashid, Basavaraj Teli,
and Birinchi Kumar Sarma
Institute of Agricultural Sciences, Banaras
Hindu University, Varanasi, India

CONTENTS

3.1 INTRODUCTION

The plant microbiome refers to highly complex phenomena occurring in nature between plants and a diverse group of microbes (Badri et al., 2009; Evangelisti et al., 2014). Among the experiments conducted on this aspect, most of them are extensively based on the symbiotic relationship between plants and rhizobacteria involved in nitrogen fixation (Oldroyd et al., 2011) or as phosphorus solubilizing arbuscular mycorrhizal fungi (Parniske, 2008). Therefore, vast research is necessary for

DOI: 10.1201/9781003110477-3

exploring the behaviour of the endophytic micro-organism residing in plants. De Bary (1866) was the first person, who proposed the word "endophyte". The biology of an endophyte was first studied by Darnelin in 1904– while working in a field weed (*Agrostemma githago* L.). The word "endophyte" was obtained from two Greek words *endon* meaning "within" and *phyton* meaning "plant". These are microorganisms, that is, maybe a fungus or bacteria inhabiting inside the plants, colonizing the cell of roots, seeds, flowers, leaves, stems, or other parts intercellularly as well as intracellularly (Toubal et al., 2018; Pylak et al., 2019). There are several species of endophytic fungus and bacteria are present inside the plant but each of them behaves differently. Based on their behaviour, the relationship may exist in three ways: mutualism, commensalism, and parasitism (Min et al., 2014; Lata et al., 2018).

Endophytic microbes are very important as they affect the vital functioning of plants by producing several novel chemical compounds, which ultimately enhances the nutrient uptake, plant growth and development and also triggers the plant defence response against wide ranges of biotic and abiotic stress (Le Cocq et al., 2017; Toubal et al., 2018). Osmoprotectants such as proline, exopolysaccharides and other volatile organic molecules are secreted by the endophytes which are helpful to mitigate the osmotic and ionic imbalance caused due to increased salt concentration in plant cells (Park et al., 2017). For example, endophytic bacteria such as *Pseudomonas fluorescens* is reported to produce ACC deaminase (E.C.4.1.99.4) in the tomato plant and that helps in reducing the adverse effect due to salinity stress (Ali et al., 2014). In addition to this, several metabolites or bioactive compounds released by certain endophytes like fistupyrone, pterocidin, bacillomycin, fengycin, 6-isoprenylindole-3-carboxylic acid, cytonic acid, pentaketide, have been also noted to defend its host from the attack of diverse pathogenic microbes (Brady and Clardy 2000; Guo et al., 2000; Lu et al., 2000; Gond et al., 2015a; Gond et al., 2015b; Bernardi et al., 2019). Furthermore, some endophytes are also used to degrade hazardous compounds found in soil or rhizosphere (DalCorso et al., 2019). Hence, many fungal and bacterial endophytic microorganisms species such as *Pseudomonas, Aspergillus, Klebsiella, Bacillus, Burkholderia, Curvularia, Streptomyces*, are now being explored for bioaugmentation, phytoremediation, and biostimulation processes to reduce the level of contaminants in soil (Adams et al., 2015; Baoune et al., 2019; Wu et al., 2019; Pietro-Souza et al., 2020).

Some members of the endophytic microbiome are known for their ability to modify the expression pattern of the genes in the host (Berendsen et al., 2015). Thus, for an in-depth analysis of the mechanisms behind this, the modern molecular technique, that is, next-generation sequencing (NGS) is mostly used. Apart from this "omics"-based approaches such as technology, metagenomics, comparative genomics, and metatranscriptomics are useful too (Kaul et al., 2016). Among these, NGS technologies are the most crucial techniques, which led to the development of metagenomics, that is, without growing microorganisms on any artificial media, genome analysis of the microorganisms is done only by DNA extraction (Wolinska, 2019). NGS technique is also applicable to identify and characterize different fungal and bacterial endophytes, their diversity, niches, and to reveal the unexplored phenomenon occurring as a result of interaction between the host and its endophyte (König et al., 2018).

3.2 NICHES OF ENDOPHYTES FOUND IN DIFFERENT PLANTS

3.2.1 BACTERIAL ENDOPHYTES

Almost in every plant, bacterial endophytes are found to internally colonize the plant tissues ubiquitously. However, the composition and distribution pattern of bacterial endophytes vary from organ to organ. The diversity of endophytes inhabiting the root endosphere region is more as compared to the microbiome residing in the bulk of soil from the rhizospheric zone (Koonin et al., 2007; Liu et al., 2017). Whereas in terms of frequency, the bacteria in soil from the rhizospheric zone is more, that is, 10^6–10^9 bacterial cells g^{-1} soil, as compared to the number of endophytes living within the plant tissue, that is, 10^4–10^8 bacterial cells g^{-1} plant tissue (Bulgarelli et al., 2013). Therefore, vast research is necessary to check, whether rhizospheric endophytes are getting more advantages or the bacterial endophytes residing inside other parts of the host cell (Rosenblueth and Martínez-Romero, 2006). Mostly, the modern techniques based on 16S rRNA are found useful for molecular characterization of endophytic bacteria residing in various plant parts like roots, leaves, stems, tubers, and so on (Franks et al., 2006). This revealed the ultimate five important genera with the greater intensity of colonization, that is, *Cellulomonas*, *Clavibacter*, *Curtobacterium*, *Pseudomonas*, and *Microbacterium* (Elvira-Recuenco and van Vuurde, 2000; Zinniel et al., 2002; Franks et al., 2006).

To enter into the host cells or tissues, especially into the cells of the roots, several mechanisms are employed by the endophytic microbiome. The tissue wounds and the cracks develop on primary and lateral root during plant growth are found as the most feasible entry point for the endophytic bacteria, except for seed-endophytes (Truyens et al., 2014; Sorensen and Sessitsch, 2006). From wounds present in the root, some metabolites may leak out, which also attracts a diverse range of bacteria (Hallmann et al., 1997). Furthermore, natural opening such as stomata pore, lenticels, root hair cells, and radicles are few more modes of entry for endophytes to enter into the host plant (Gagné et al., 1987). Several endophytic bacteria produce cellulose-degrading enzymes to dissolve the host cell wall and the capability of a bacterium to dissolve cellulose in it is directly associated with its endophytic nature. It is noted that the ability of *Enterobacter asburiae* to hydrolyse cellulose aids its entry into the cotton plant cells (Hallmann et al., 1997).

3.2.2 FUNGAL ENDOPHYTES

Fungal endophytes can be defined as the fungus inhabiting inside a living plant host in mycelial form or spending a small part of its life cycle in it. Hence, the presence of fungal hyphae in the plant tissue is an essential criterion of a fungal endophyte (Kaul et al., 2012). It has been found in association with a diverse group of crop plants in the natural ecosystem (Suryanarayanan, 2013). Furthermore, in distinct plant components such as roots, leaves, shoots, branches, and stems, different endophytic fungi are reported with distinct distribution patterns (Unterseher and Schnittler, 2010; Fürnkranz et al., 2012). In the roots of banana plants, more endophytes are found, that is, about 67% in contrast to the cortex (23%) and central cylinder (10%) (Pocasangre et al., 2000). Apart from this, the population density is also influenced by environmental conditions. In tropical areas, plant species are abundant in endo-phyte diversity as compared to the diversity found in temperate regions (Azevedo et al., 2000). Contrarily, in temperate areas, the fungal endophytes found are more host-specific than the tropical endophytes. For example, the *Xylaria* spp., a domi-nated fungal endophyte in tropical woody trees which is very low host specificity. However, *Pleuroplaconema* spp. is found in temperate forests and is observed to be confined to Californian redwood forests (Lodge et al., 1996; Giménez et al., 2007).

This diversified microbiota can also be divided into two broad classes: (a) clavi-cipitaceous and (b) non-clavicipitaceous, according to the type of host plant, taxo-nomic position, and ecological functions. The clavicipitaceous group includes the endophytes predominant in the grass host whereas, non-clavicipitaceous group includes the endophytes mostly inhabiting inside the higher plant cell (Pavithra et al., 2020). However, based on the manner of transmission of fungal endophytes, it can be broadly categorized as vertically as well as horizontally transmitted endophytic microbes (Saikkonen et al., 2002). When the endophytic microorganisms are trans-mitted to the progenies from the host directly through the seeds, they are termed as vertically transmitted or seed-transmitted endophytes (Schardl et al., 2013), whereas, in horizontally transmitted endophytic microbes, the inoculums are carried from one host to another by air. This mode is common in spore-producing fungi and is also known as spore-transmitted endophytes (Faeth and Fagan, 2002; Hartley and Gange, 2009).

Moreover, the endophyte may behave as a biotroph or as a necrotroph according to the host on which it is harbouring. Biotrophic fungal endophytes grow and with-draw nutrition from living host cells only, whereas the necrotrophic fungal endo-phytes grow and withdraw nutrition from the host cell after killing it (Delaye et al., 2013). It might produce some typical symptoms on its host, while in other cases, it is just associated with the host asymptomatically (Niere, 2001; Sikora et al., 2008). However, the switching between these two lifestyles may result due to periodic evolutionary and ecological changes. Some biotrophic fungal endophytes may interact as necrotrophs in the later phase when living cell is not available (Delaye et al., 2013). In healthy *Arabidopsis thaliana* plants, the fungus *Leptosphaeria maculans* inhabit inside the living cells as a biotroph asymptomatically but when the plant was under stressed conditions, its behaviour was changed to a necrotrophic pathogen (Junker et al., 2012).

3.3   FACTORS REGULATING THE POPULATION OF ENDOPHYTES

The population or distribution pattern of endophytes is being affected by several types of factors that can be grouped into two broad classes: (a) host factor and (b) environmental factor.

3.3.1   HOST FACTORS INFLUENCING THE POPULATION DYNAMICS OF ENDOPHYTES

According to the reports given by many researchers, it is concluded that the endo-phytes colonizing a plant is significantly affected by the host plant species and some endophytes are found in a particular host species only (Jia et al., 2016); though, the state of a host plant also influences the fitness of the endophytic fungi, in which age of the plant plays a vital role (Saikkonen et al., 2004; Sieber, 2007). Chareprasert (2006) has reported that in the matured leaves of teak (*Tectonagrandis* L.) and rain tree (*Samaneasaman* Merr.), greater diversity of endophytes was found with a higher frequency of colonization as compared to the young leaves. In the distinct plant parts like leaves, petiole, and twigs, the frequency of endophytes is also different and it may change with the season. For example, in the plant *Gingko biloba* L., *Phyllosticta* spp. was first detected in the month of August in both leaves and petioles. The population level was at a peak in October and after this, the population gradually declined to zero in the month of May, whereas *Phomopsis* spp. can be detected in twigs of the same plant all over the growing season (Thongsandee et al., 2012; Nair and Padmavathy, 2014).

Plant root exudates acting as a chemo-attractant notably influences the successful colonization by endophytes. Root exudates are comprised of various types of biomolecules viz., organic acids, flavonoids, polysaccharides, and amino acids (Kawasaki et al., 2016; Pétriacq et al., 2017). Flavonoids are very essential compounds, which are well established to enhance the colonization of both rhizobia and non-rhizobial endophytes. Accordingly, it was found that the endophytic *Serratia* spp. EDA2 and *Azorhizobium caulinodans* ORS571 colonizes the roots of rice and wheat more effectively upon perceiving the flavonoids (Arora and Mishra, 2016). Also, the genetic background of the plant plays a vital role in altering the distribution pattern of endophytes. Thus, the plant genotype determines the nature of the relationship that develops between endophytic microbe and the plant host, whether it'll become a mutualistic or a parasitic relation (Jia et al., 2016). This could be considered as a flexible interaction, where the consequences might be beneficial or harmful for the host plant, or neutral for both the partner, that depends on the gene expression pattern in both endophyte and host plant (Moricca and Ragazzi, 2008).

3.3.2 ENVIRONMENTAL FACTORS INFLUENCING THE POPULATION DYNAMICS OF ENDOPHYTES

Environmental features like temperature, moisture, light intensity, soil nutrient status, geographical location, and the nature of vegetation, remarkably influence the distribution pattern of endophytic fungi and bacteria. These factors directly influence the spore germination, colonization, reproduction, and survival of fungal endophytes (Song, 2008). In cold environmental conditions, only some specific species of the host could grow on which certain types of endophytes could be established. As a consequence, those particular types of endophytes develop regional specificity in such a cold environment (Jiang et al., 2010). It was noticed by many researchers that the fungal endophytes isolated from similar regions show a greater resemblance in terms of species taxonomy. In contrast, fungal endophytes extracted from a particular plant species of distinct regions show a very low resemblance (Jia et al., 2016). Temporal changes in the environmental condition may affect the relative abundance of endophytes significantly (Chareprasert et al., 2006). Generally, the endophytes inhabit the cells of root, stem, and leaf of the plant making them face lesser competition from the soil-inhabiting other bacteria for space and nutrients (Gaiero et al., 2013).

3.4 METABOLITES

Metabolites are the low molecular weight bioactive compounds released by endophytic micro-organisms which are exploited as therapeutic agents due to their greater potential to cure plant, animal, and human diseases (Strobel et al., 2004; Godstime et al., 2014; Gouda et al., 2016). Metabolites are helpful to activate the resistance mechanisms in plants and humans to combat the invasion of pathogenic microorganisms (Tan and Zou, 2001; Table 3.1). New unique metabolites with an antimicrobial property can also be used as a great alternative for overpowering the drug resistance imposed by various pathogens of both humans and plants (Yu et al., 2010; Song 2008). Furthermore, their antimicrobial activity is not only useful for making medical drugs but also applied as preservatives in the food industry to check food quality deterioration and thereby control foodborne infections (Lui et al., 2008). Conversely, the impact of the endophytes on the host plant secondary metabolite synthesis is also found by few researchers, but the mechanisms involved are still not clear (Khare et al., 2018). Irmer et al. (2015) reported that biosynthesis of pyrrolizidine alkaloids (PAs) in Crotalaria, which is crucial for activating the defence system against herbivores rely on the presence of the endophytic bacteria *Bradyrhizobium* spp. (nodulating bacteria).

The metabolites synthesized by endophytic microbes are the array of natural drugs with a unique structure comprising of terpenoids, tannins, flavonoids, alkaloids, tetralones, chinones, phenolic acids, antioxidants, quinones, and others (Godstime et al., 2014; Gouda et al., 2016). In plants with pharmaceutical importance, the endophytic bacteria residing in them produce novel antibiotics like ecomycins, pseudomycins, munumbicins, and kakadumycins. Hence, the endophytes inhabiting these crops are believed as the best probable source of useful novel bioactive compounds (Yang et al., 2010). These bioactive compounds could be directly isolated from the natural source but it has many disadvantages including political and climatic constraints, seasonal dependency, and other ecological constraints (Bicas et al., 2008; Shukla et al., 2014). Therefore, other various methods like the productions of microbes using a fermentation process and by microbial transformation technique are more feasible as compared to the extraction from natural sources. The biotechnological techniques are also very cost-effective and an inexhaustible method for extracting high-value bioactive products for applications (Hussain et al., 2012; Gouda et al., 2016).

Several new bioactive compounds were already extracted from various endophytes possessing antimicrobial properties. The endophytic fungus *Xylaria* sp. inhabiting inside different host plants produce antifungal compounds like "sordaricin" which is effective against *Candida albicans* (Pongcharoen et al., 2008). Another bioactive compound, that is, "7-amino-4-methyl coumarin" derived from *Xylaria* spp. strain YX-28 is having inhibitory effects against various food-borne microbes causing food spoilage. Microorganisms, therefore, can be developed as a natural food preservative (Liu et al., 2008). Whereas, one more strain of this fungus isolated from another host *Abiesholophylla* produces "griseofulvin" that can be used to manage plant pathogenic fungi (Park et al., 2005). Apart from antimicrobial properties, other compounds such as chlorinated metabolites like mycorrhizin A and cryptosporiopsin

TABLE 3.1
List of Novel Metabolites Produced by Different Endophytes with Their Possible Functions

Endophyte	Host Plant	Metabolite/bioactive Compound	Function	References
Streptomyces spp.TP-A0569	Allium fistulosum,	Fistupyrone,	Antifungal effect,	Bernardi et al., 2019
Streptomyces spp.TP-A0556,	Aucuba japonica,	Coumarins TPU-0031-A and B	Antibiotic activity,	
Streptomyces hygroscopicus	Pteridium aquilinum	Pteridic acids, Pterocidin	Promotion of growth	
Bacillus spp. Bacillus amyloliquefaciens	Maize Zea mays L.	Iturin A, Bacillomycin, Fengycin Indian popcorn	Antifungal properties	Gond et al., 2015a; Gond et al., 2015b
Neotyphodium spp. Epichloë spp.	Grasses	Peramine, (pyrrolopyrazine alkaloid)	Protects plants against insects like Argentine stem weevil	Rowan and latch, 1994
Phomopsis spp.	Salix gracilistyla var. melanostachys	Phomopsichalasin, Cytochalasin	Inhibiting microbes (Bacterial and fungal pathogens) example: Bacillus sp, Salmonella gallinarum, Staphylococcus aureus.	Horn et al., 1996
Neotyphodium spp.		Agroclavine, Chanoclavine, and Elymoclavine	Lethal to insects & mammals	Schardl and Phillips, 1997
Colletotrichum spp.	Artemisia annua	6-isoprenylindole-3-carboxylic acid,	Antagonistic action against both pathogenic bacteria and fungus. Example: Bacillus sp, Pseudomonas sp., Sarcina lutea and Staphylococcus aureus, Phytophthora capsici and Rhizoctonia cerealis)	Lu et al., 2000
Pestalotiopsis spp.	T. brevifolia	Pestalotiopsins A–C and 2α-hydroxydimeninol	-	Pulici et al., 1996a; Pulici et al., 1996b
Phyllosticta spp.	Abies balsamea	Hydroheptelidic and heptelidic acid	Lethal against Choristoneura fumiferana (spruce budworm)	Calhoun et al., 1992

(Continued)

TABLE 3.1 (Continued)
List of Novel Metabolites Produced by Different Endophytes with Their Possible Functions

Endophyte	Host Plant	Metabolite/bioactive Compound	Function	References
Epichloë typhina	Gaultheria procumbens	Sesquiterpenes chokols A–G	Antifungal activity against leaf spot pathogen (Cladosporium phlei)	Findlay et al., 1997; Koshino et al., 1989
Colletotrichum spp.	A. annua	Steriods like 3β,5α,6β-trihydroxyergosta-7,22-diene Ergosterol, etc.	Antagonistic against phytopathogens	Lu et al., 2000
Cryptosporiopsis quercina	Redwood	Cryptocandin	Antifungal activity,	Strobel et al., 1999
Aspergillus spp., Pezicula spp., and Cryptosporiopsis sp.	Pinus sylvestris and Fagus sylvatica	Cyclopeptides Echinocandins A, B, D, and H	Antifungal activity	Traber et al. 1979
Acremonium spp.	Taxus baccata	Leucinostatin A	Antimycotic and Anticancer properties	Strobel et al., 1997
Epichloë typhina	Phleum pratense	Phenolic derivatives (Tyrosol, p-hydroxyphenyl acetic acid, p-hydroxybenzoic acid, and cis and trans-p-coumaric acids)	Antimycotic properties	Koshino et al., 1988
Colletotrichum spp.	Artemisia mongolica	Colletotric acid	Antimicrobial Properties	Zou et al., 2000
Phomopsis spp.	Salix and non-willow plants	Phomodiol and Homoproline B	-	Horn et al., 1996
Fusarium spp.	Selaginella pallescens	Pentaketide	Antibiotic activity	Brady and Clardy, 2000
Aspergillus parasiticus	Redwood	Sequoiatones A and B	Antitumor properties	Stierle et al., 1999
Rhinocladiella spp.	Tripterygium wilfordii	Cytochalasin E	-	Wagenaar et al., 2000
Cytonaema spp.	Quercus sp.	Cytonic acids A and B	Antifungal properties	Guo et al., 2000

extracted from *Pezicula* strains also possess herbicidal properties (Schulz et al., 1995). While chlorinated benzophenone derivatives are isolated from *Pestalotiopsis adusta*, that is, "Pestalachlorides A" (C21H21Cl2NO5) and "B" (C20H18Cl2O5) are also reported to inhibit important fungal pathogens of plant, *Fusarium culmorum*, *Gibberella zeae*, and *Verticillium albo-atrum* (Jiang et al., 2008).

3.5 INTERACTION BETWEEN THE HOST PLANT AND ENDOPHYTES

3.5.1 ENDOPHYTES AS PLANT GROWTH PROMOTER

Various plant growth regulating phytohormones are produced by the endophytic microbiome to promote plant growth. Endophytic bacteria like *Azospirillum* spp., *Bacillus* spp., *Acinetobacter* spp., *Pseudomonas* spp. are known to produce auxins, cytokinins, and gibberellins that enhance nutrients and water uptake, photosynthetic efficiency, nitrogen metabolism, regulate stomatal opening and closing, and also encourage morphological alteration of roots (Figure 3.1) (Beneduzi et al., 2012). Few of the endophytes are also noted to release an enzyme, that is, 1-aminocyclopropane-1-carboxylic acid (ACC) deaminase that lowers the ethylene level, as this enzyme degenerate the precursor of ethylene and foster germination of seed and root initiation (Oliveira et al., 2003). However, a higher level of ACC deaminase may lead to detrimental effects on host plant-like, cell division, and DNA synthesis inhibition thereby leading to a decline in growth (Gaiero et al., 2013).

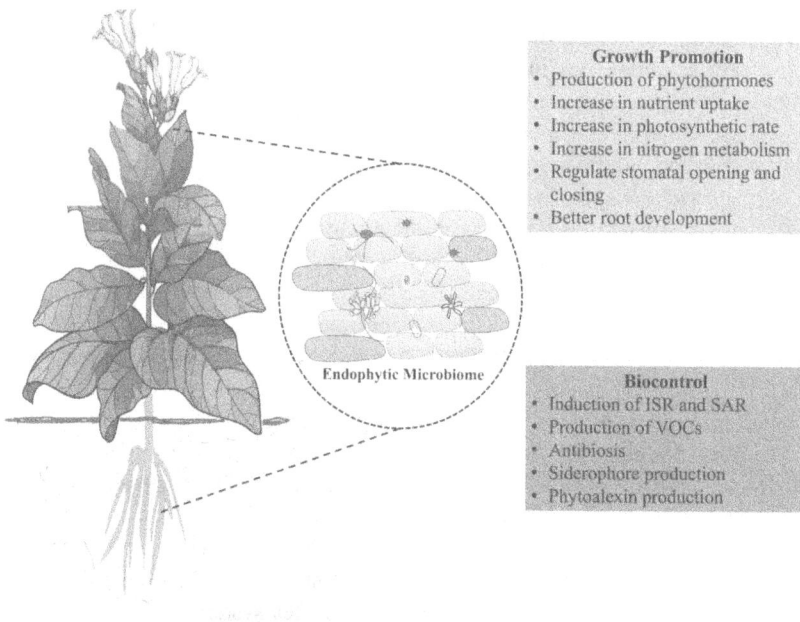

Growth Promotion
• Production of phytohormones
• Increase in nutrient uptake
• Increase in photosynthetic rate
• Increase in nitrogen metabolism
• Regulate stomatal opening and closing
• Better root development

Endophytic Microbiome

Biocontrol
• Induction of ISR and SAR
• Production of VOCs
• Antibiosis
• Siderophore production
• Phytoalexin production

FIGURE 3.1 Interaction between host plants and endophytes lead to plant growth promotion and suppression of plant diseases.

A significant increase in the chlorophyll content of leaves was also detected in beet on inoculation with *Bacillus pumilus*, *Chryseobacterium indologenes*, and *Acinetobacter rjohnsonii*, which in turn enhance the carbohydrates in the plant due to an increase in photosynthetic capacity (Shi et al., 2010). Some of the fungal endophytes are also reported to enhance height, biomass, and other growth parameters of a plant (Jaber and Enkerli, 2016; Jaber and Araj, 2017). In *Vicia faba*, an increase in plant height and weight was observed after the artificial inoculation of *Beauveria* spp. (Jaber and Enkerli, 2017). Lopez and Sword (2015) also reported enhancement in the growth and dry weight of cotton plants following the colonization of *Beauveria bassiana* and *Penicillium lilacinum*.

A remarkable increase in plant growth was resulted due to the bioavailability of nutrients in the presence of endophytic microorganisms. The barley plant inoculated with endophytic *Bacillus* sp. reflects a notable increase in the manganese (Mg), zinc (Zn), and copper (Cu) content (Canbolat et al., 2006). Nitrogen (N) availability to plant is increased due to the nitrogen-fixing ability of endophytic bacteria. In sugarcane crops, a 30% rise in the nitrogen content occurs after inoculation with an endophytic bacterial mixture like *Burkholderia* spp., *Azospirillum* spp., *Gluconacetobacter diazotrophicus*, and so on (Oliveira et al., 2002). Apart from this, the fixation of phosphorus is an important factor limiting phosphorus availability required by the plant for its growth (Khan et al., 2007; Dias et al., 2009). The plant endophytes are well reported to synthesize organic acids, such as oxalic, citric and tartaric acids, which helps in phosphorus solubilization (Busato et al., 2012).

3.5.2 ENDOPHYTES AS BIO-CONTROL AGENTS AGAINST PLANT DISEASES

Several studies suggested the existence of both bacterial and fungal endophytes is extremely advantageous for the host plant, as they activate the immune system in plants, thereby helping to reduce the pathogen invasion (Figure 3.1) (Miller et al., 2002; Gunatilaka, 2006). Potential bacterial endophytes viz., *Bacillus* spp., *Arthrobacter* spp., and *Nocardiopsis* spp. and fungal endophytes viz, *Gilmaniella* spp., *Colletotrichum* spp., *Serendipita* spp., *Trichoderma* spp., are reported to suppress phytopathogens in different crop plants, that is, maize, cocoa, barley, tomato (Gond et al., 2015; Dong et al., 2019; Sarkar et al., 2019; Yuan et al., 2019).

Various endophytes are also known to trigger both systemic acquired resistance (SAR) and induced systemic resistance (ISR) against a variety of fungal, bacterial, and other pathogens attacking different plants (Kloepper and Ryu, 2006). Salicylic acid (SA) and jasmonic acid (JA) are key diffusible molecules involved in activating the plant immune response. The gibberellin-producing endophytic microorganisms are found to modulate the SA and JA pathways to strengthen plant immunity to combat various pathogens and herbivores (Waqas et al., 2015; Khare et al., 2016). The antifungal lipopeptides (iturin A, bacillomycin, fengycin) are secreted by bacterial endophytes (*Bacillus amyloliquefaciens*) inhabiting several maize varieties that inhibits the disease development by up-regulating the synthesis of pathogenesis-related proteins (PR-proteins), which in turn activates systemic acquired resistance (Gond et al., 2015a). However, it was also reported that the pretreatment of endophytic bacterium (*Bacillus*) in plants enhance the JA-induced gene expression during plant-pathogen interaction as compared to that of untreated plants. Hu et al. (2018) revealed that the endophytic bacteria *Bacillus cereus* BCM2 activates the defense-related enzymes to impart resistance against the nematode *Meloidogyne incognita*. While in barley (*Hordeum vulgare*), an endophytic fungus *Serendipita vermifera* was observed to induce expression of the hydrolytic enzyme-coding genes to reduce the pathogen infection (Sarkar et al., 2019). According to the experiment conducted by Kavroulakis et al. (2007), the tomato foliar pathogen *Septoria lycopersici* was inhibited by *Fusarium solani* by eliciting ISR against it. Foliar endophytes are also found to influence plant-pathogen interactions. In *Theobroma cacao*, a phyllosphere fungal endophyte *Colletotrichum tropicale* is found to suppress the infection by *Phytophthora* spp. (Mejía et al., 2014).

Many volatile organic compounds (VOCs) are reported to be released by bacterial endophytes with antimicrobial property that could inhibit many pathogenic bacteria, fungi, and nematodes. *Pseudomonas putida* BP25, an endophytic bacterium inhabiting the black pepper plant, is reported to produce numerous volatile substances to inhibit the infection by *Phytophthora* spp., *Gibberella moniliformis*, *Pythium* spp., *Rhizoctonia solani*, and *Colletotrichum gloeosporioides*. These volatile organic compounds were also helpful to check the plant-parasitic nematodes such as *Radopholus similis* (Sheoran et al., 2015). Phytoalexins are antimicrobial compounds produced by healthy cell tissues to restrict phytopathogen infection. This phytoalexin is comprising of many bioactive metabolites viz, terpenoids, isoflavonoids, glycosteroids, and alkaloids (Gao et al., 2010). An increase in terpenoids content was observed in the plant *Euphorbia pekinensis* due to the existence of *Fusarium* spp. (Yong et al., 2009). The bacterial endophytes are also well reported to synthesize siderophores and antibiotics or bacteriocins to avoid the proliferation of harmful microorganisms (Gaiero et al., 2013). Therefore, the existence of fungal or bacterial endophytes in the plant tissues can hence enhance plant resistance towards phytopathogens, either by eliciting the host defence response or by secreting various antimicrobial compounds themselves.

3.6 GENETICS OF METABOLITE PRODUCTION IN ENDOPHYTES

It is considered that the secondary metabolites are generally the chemical signals for communication between the host and endophyte, which further helps to inhibit pathogenic microbes (Brakhage and Schroeckh, 2011). In the bacterial genomes, the clusters of functional genes are organized as operons, which are transcribed together as a single unit (Koonin, 2009). Similarly, in fungi also it is quite evident that genes in clusters regulate an array of biological functions in cells (Gutierrez et al., 1999; Rosewich and Kistler, 2000). These gene clusters also regulate the secondary metabolites production in endophytes (Trail et al., 1995; Lo et al., 2012). In fungi, usually these gene clusters code for two important classes of multimodular enzyme complexes, that is, non-ribosomal peptide synthetases [NRPS] and polyketide synthases [PKS], which consists of various domains with well-defined functions (Cane and Walsh, 1999; Brakhage and Schroeckh, 2011; Evans et al., 2011; Wang et al., 2014; Amoutzias et al., 2016). These form the backbone for the majority of secondary metabolites. A classical modular PKS consist of acyltransferase (AT), ketosynthase (KS), and acyl-carrier protein domains (Crawford et al. 2009). However, the NRPS system is composed of an adenylation (A) domain, a peptidyl carrier protein domain and a condensation (C) domain (Nikolouli and Mossialos, 2012). Examples of important derivatives of NRP used as antibiotics are penicillin and cephalosporin. The immune suppressant such as cyclosporine is also a derivative of NRP, while the lovastatin is a derivative of polyketide (Brakhage, 1998; Hoffmeister and Keller, 2007). Furthermore, mixed PKS-NRP hybrid origin compounds have also been isolated, for example, aspyridones (Bergmann et al., 2007).

The prediction and identification of putative genes responsible for secondary metabolites production in endophytes are increasing with the increased access to whole-genome sequences of fungi and bacteria. Several bioinformatics algorithms are developed by researchers to detect the gene clusters accountable for secondary metabolisms like SMURF [SM Unknown Regions Finder], antiSMASH and FungiFun (Priebe et al., 2011). These algorithms are useful for speculating the genes coding for core PKS and/or NRPS enzymes, as well as the role of neighbouring genes in the genome (Bergmann et al., 2007). Nearly about 50 and 27 gene clusters in *Aspergillus* spp. and *Arthroderma* spp. respectively are successfully identified using the above algorithms (Burmester et al., 2011). These clusters are commonly called "cryptic" or "orphan" clusters because the structural, as well as functional, characterization of the secondary metabolites synthesized from these gene clusters is still not explored (Bergmann et al., 2007; Hertweck, 2009b).

3.7   CONCLUSION

The plant endophytic bacteria together with the endophytic fungi are very important microorganisms inhabiting the cells of different plant components such as roots, leaves, stems, tubers, intercellularly as well as intracellularly. Several new molecular techniques, as well as bioinformatics algorithms, are available nowadays, which can be used to explore more novel useful metabolites produced by this endophytic microbiome and to search new gene clusters and pathways regulating their biosynthesis. In the natural ecosystem, the diversity and abundance of endophytes are affected by several environmental factors like temperature, moisture, light intensity, soil nutrient status, geographical location, nature of vegetation as well as by host factors such as plant genotype, root exudates, and so on. Therefore, a more extensive study of this section is required to develop the protocol for multiplying the endophytes *in vivo* conditions so that metabolites produced by these microorganisms could be easily isolated for further use as growth booster or as a protectant.

ACKNOWLEDGEMENT

JM and MMR are thankful to the Department of Science & Technology under the Ministry of Science & Technology, Government of India (DST/INSPIRE Fellowship/2017/IF170563) and University Grants Commission (NFO-2018-19-OBC-BIH-68765) for financial assistance.

REFERENCES

Adams, G.O., Fufeyin, P.T., Okoro, S.E. & Ehinomen, I. (2015). Bioremediation, biostimulation and bioagumentation: A review. *Journal of Bioremediation and Biodegradation*, 3, 28–39.

Ali, S., Charles, T.C. & Glick, B.R. (2014). Amelioration of high salinity stress damage by plant growth-promoting bacterial endophytes that contain ACC deaminase. *Plant Physiology and Biochemistry*, 80, 160–167.

Amoutzias, G., Chaliotis, A. & Mossialos, D. (2016). Discovery strategies of bioactive compounds synthesized by non-ribosomal peptide synthetases and type-I polyketide synthases derived from marine microbiomes. *Marine Drugs*, 14 (4), 80.

Arora, N.K. & Mishra, J. (2016). Prospecting the roles of metabolites and additives in future bioformulations for sustainable agriculture. *Applied Soil Ecology*, 107, 405–407.

Azevedo, J.L, Maccheroni, W., Pereira, J.O. & De Araujo, W.L. (2000). Endophytic microorganisms: a review on insect control and recent advances on tropical plants. *Electronic Journal of Biotechnology*, 3, 40.

Badri, D.V., Weir, T.L, van. Der. Lelie, D. & Vivanco, J.M. (2009). Rhizosphere chemical dialogues: Plant-microbe interactions. *Current Opinion in Biotechnology*, 20, 642–650.

Baoune, H., Aparicio, J.D., Pucci, G., ElHadj-Khelil, A.O. & Polti, M.A. (2019). Bioremediation of petroleum-contaminated soils using *Streptomyces* sp. Hlh1. *Journal of Soils and Sediments*, 19, 2222–2230.

Beneduzi, A., Ambrosini, A. & Passaglia, L.M. (2012). Plant growth-promoting rhizobacteria (PGPR): Their potential as antagonists and biocontrol agents. Genetics and Molecular Biology, 35 (4), 1044-1051.

Berendsen, R.L., van. Verk, M.C., Stringlis, I.A., Zamioudis, C., Tommassen, J., Pieterse, C.M. & Bakker, P.A.H.M. (2015). Unearthing the genomes of plant-beneficial Pseudomonas model strains WCS358, WCS374 and WCS417. *BMC Genomics*, 16, 539.

Bergmann, S., Schumann, J., Scherlach, K., Lange, C., Brakhage, A.A. & Hertweck, C. (2007). Genomics-driven discovery of PKS-NRPS hybrid metabolites from *Aspergillus nidulans*. *Nature Chemical Biology*, 3, 213–217.

Bernardi, D.I., das Chagas, F.O., Monteiro, A.F., dos Santos, G.F. & de Souza Berlinck, R.G. (2019). Secondary metabolites of endophytic Actinomycetes: Isolation, synthesis, biosynthesis, and biological activities. In *Progress in the Chemistry of Organic Natural Products*, Kinghorn, D.A., Falk, H., Gibbons, S., Kobayashi, J., Asakawa, Y. & Liu, J.K. (eds.). Switzerland: Springer Nature, pp. 219–265.

Bicas, J.L., Barros, F.F.C., Wagner, R., Godoy, H.T. & Pastore, G.M. (2008). Optimization of R-(+)-α-terpineol production by the biotransformation of R-(+)-limonene. Journal of Industrial Microbiology and Biotechnology, 35 (9), 1061–1070.

Brady, S.F. & Clardy, J. (2000). CR377, a new pentaketide antifungal agent isolated from an endophytic fungus. *Journal of Natural Products*, 63 (10), 1447–1448.

Brakhage, A.A. & Schroeckh, V. (2011). Fungal secondary metabolites–strategies to activate silent gene clusters. *Fungal Genetics and Biology*, 48, 15–22.

Brakhage, A.A. (1998). Molecular regulation of lactam biosynthesis in filamentous fungi. *Microbiology and Molecular Biology Reviews*, 62, 547–585.

Bulgarelli, D., Schlaeppi, K., Spaepen, S., Ver Loren van Themaat, E. & Schulze- Lefert, P. (2013). Structure and functions of the bacterial microbiota of plants. *Annual Review of Plant Biology*, 64, 807–838.

Burmester, A., Shelest, E., Glockner, G., Heddergott, C., Schindler, S., Staib, P., Heidel, A., Felder, M., Petzold, A., Szafranski, K. & Feuermann, M. (2011). Comparative and functional genomics provide insights into the pathogenicity of dermatophytic fungi. *Genome Biology*, 12 (1), 1–16.

Busato, J.G., Lima, L.S., Aguiar, N.O., Canellas, L.P. & Olivares, F.L. (2012). Changes in labile phosphorus forms during maturation of vermicompost enriched with phosphorus-solubilizing and diazotrophic bacteria. *Bioresource Technology*, 110, 390–395.

Calhoun, L.A., Findlay, J.A., Miller, J.D. & Whitney, N.J. (1992). Metabolites toxic to spruce budworm from balsam fir needle endophytes. *Mycological Research*, 96, 4, 281–286.

Canbolat, M.Y., Bilen, S., Çakmakç, R., Şahin, F. & Aydın, A. (2006). Effect of plant growth-promoting bacteria and soil compaction on barley seedling growth, nutrient uptake, soil properties and rhizosphere microflora. *Biology and Fertility of Soils*, 42 (4), 350–357.

Cane, D.E. & Walsh, C.T. (1999). The parallel and convergent universes of polyketide synthases and non-ribosomal peptide synthetases. *Chemistry and Biology*, 6, R319–R325.

Chareprasert, S., Piapukiew, J., Thienhirun, S., Whalley, A.J.S. & Sihanonth, P. (2006). Endophytic fungi of teak leaves *Tectona grandis* L. and rain tree leaves *Samanea saman* Merr. *World Journal of Microbiology and Biotechnology*, 22 (5), 481–486.

Crawford, J.M., Korman, T.P., Labonte, J.W., Vagstad, A.L., Hill, E.A., Kamari-Bidkorpeh, O., Tsai, S.C. & Townsend, C.A. (2009). Structural basis for biosynthetic programming of fungal aromatic polyketide cyclization. *Nature*, 461 (7267), 1139–1143.

DalCorso, G., Fasani, E., Manara, A., Visioli, G. & Furini, A. (2019). Heavy metal pollutions: State of the art and innovation in phytoremediation. *International Journal of Molecular Sciences*, 20, 3412.

De Bary, A. (1866). *Morphologie und Physiologie der Pilze, Flechten, und Myxomyceten.* Hofmeister's Handbook of Physiological Botany, Vol. 2, Leipzig : W. Engelmann.

Delaye, L., García-Guzmán, G. & Heil, M. (2013). Endophytes versus biotrophic and necrotrophic pathogens—-are endophytes fungal lifestyles evolutionarily stable traits? *Fungal Diversity*, 60, 125–135.

Dias, A.C.F., Costa, F.E.C., Andreote, F.D., Lacava, P.T., Teixeira, M.A., Assumpção, L.C., Araújo, W.L., Azevedo, J.L. & Melo, I.S. (2009). Isolation of micropropagated strawberry endophytic bacteria and assessment of their potential for plant growth promotion. *World Journal of Microbiology and Biotechnology*, 25(2), 189–195.

Dong, Z.Y., Rao, N.M.P., Wang, H.F., Fang, B.Z., Liu, Y.H., Li, L., Xiao, M. & Li, W.J. (2019). Transcriptomic analysis of two endophytes involved in enhancing salt stress ability of *Arabidopsis thaliana*. *Science of the Total Environment*, 10, 107–117.

Elvira-Recuenco, M. & van Vuurde, J.W.L. (2000). Natural incidence of endophytic bacteria in pea cultivars under field conditions. *Canadian Journal of Microbiology*, 46, 1036–1041.

Evangelisti, E., Rey, T. & Schornack, S. (2014). Cross-interference of plant development and plant–microbe interactions. *Current Opinion in Biotechnology*, 20, 118–126.

Evans, B.S., Robinson, S.J. & Kelleher, N.L. (2011). Surveys of non-ribosomal peptide and polyketide assembly lines in fungi and prospects for their analysis in vitro and in vivo. *Fungal Genetics and Biology*, 48 (1), 49–61.

Faeth, S.H. & Fagan, W.F. (2002). Fungal endophytes: Common host plant symbionts but uncommon mutualists. *Integrative and Comparative Biology*, 42, 360–368.

Findlay, J.A., Buthelezi, S., Li, G.Q., Seveck, M. & Miller, J.D. (1997). Insect toxins from an endophytic fungus from wintergreen. *Journal of Natural Products*, 60, 214–1215.

Franks, A., Ryan, P.R., Abbas, A., Mark, G.L. & O'Gara, F. (2006). *Molecular Tools for Studying Plant Growth-Promoting Rhizobacteria (PGPR): Molecular Techniques for Soil and Rhizosphere Microorganisms.* Wallingford, Oxfordshire: CABI Publishing.

Fürnkranz, M., Lukesch, B., Müller, H., Huss, H., Grube, M. & Berg, G. (2012). Microbial diversity inside pumpkins: microhabitat-specific communities display a high antagonistic potential against phytopathogens. *Microbial Ecology*, 63, 418–428.

Gagné, S., Richard, C., Rouseau, H. & Antoun, H. (1987). Xylem-residing bacteria in alfalfa roots. *Canadian Journal of Microbiology*, 33, 996–1000.

Gaiero, J.R., McCall, C.A., Thompson, K.A., Day, N.J., Best, A.S. & Dunfield, K.E. (2013). Inside the root microbiome: bacterial root endophytes and plant growth promotion. American Journal of Botany, 100 (9), 1738–1750.

Gao, F.K., Dai, C.C. & Liu, X.Z. (2010). Mechanisms of fungal endophytes in plant protection against pathogens. *African Journal of Microbiology Research*, 4, 1346–1351.

Giménez, C., Cabrera, R., Reina, M. & González-Coloma, A. (2007). Fungal endophytes and their role in plant protection. Current Organic Chemistry, 11 (8), 707–720.

Godstime, O.C., Enwa, F.O., Augustina, J.O. & Christopher, E.O. (2014). Mechanisms of anti-microbial actions of phytochemicals against enteric pathogens—A review. *Journal of Pharmaceutical, Chemical and Biological Sciences*, 2, 77–85.

Gond, S.K., Surendra, K., Bergen, M.S., Torres, M.S., White, J.F. & Kharwar, R.N. (2015). Effect of bacterial endophyte on expression of defense genes in Indian popcorn against *Fusarium moniliforme. Symbiosis*, 66, 133–140.

Gouda, S., Das, G., Sen, S.K., Shin, H.S. & Patra, J.K. (2016). Endophytes: A treasure house of bioactive compounds of medicinal importance. *Frontiers in Microbiology*, 7, 1538.

Gunatilaka, A.A.L. (2006). Natural products from plant-associated microorganisms: distribution, structural diversity, bioactivity, and implications of their occurrence. Journal of Natural Products, 69, 509–526.

Guo, L.D., Hyde, K.D. & Liew, E.C.Y. (2000). Identification of endophytic fungi from *Livistona chinensis* based on morphology and rDNA sequences. *New Phytologist*, 147 (3), 617–630.

Gutierrez, S., Fierro, F., Casqueiro, J. & Martin, J.F. (1999). Gene organization and plasticity of the lactam genes in different filamentous fungi. *Antonie Van Leeuwenhoek*, 75, 81–94.

Hallmann, J., Quadt-Hallmann, A., Mahaffee, W.F. & Kloepper, J.W. (1997). Bacterial endophytes in agricultural crops. *Canadian Journal of Microbiology*, 43, 895–914.

Hartley, S.E. & Gange, A.C. (2009). Impacts of plant symbiotic fungi on insect herbivores: mutualism in a multitrophic context. *Annual Review on Entomology*, 54, 323–342.

Hertweck, C. (2009). Hidden biosynthetic treasures brought to light. *Nature Chemical Biology*, 5, 450–452.

Hoffmeister, D. & Keller, N.P. (2007). Natural products of filamentous fungi: enzymes genes and their regulation. *Natural Product Reports*, 24, 393–416.

Horn, W.S., Simmonds, M.S., Schwartz, R.E. & Blaney, W.M. (1996). Variation in production of Phomodiol and Phomopsolide B by *Phomopsis* spp. *Mycologia*, 88, 588–595.

Hu, H., Wang, C., Li, X., Tang, Y., Wang, Y., Chen, S. & Yan, S. (2018). RNA-Seq identification of candidate defense genes targeted by endophytic *Bacillus cereus*-mediated induced systemic resistance against *Meloidogyne incognita* in tomato. *Pest Management Science*, 74, 2793–2805.

Hussain, M.S., Fareed, S., Ansari, S., Rahman, M.A., Ahmad, I.Z. & Saeed, M. (2012). Current approaches toward production of secondary plant metabolites. *Journal of Pharmacy and Bioallied Sciences*, 4, 10–20.

Irmer, S., Podzun, N., Langel, D., Heidemann, F., Kaltenegger, E., Schemmerling, B., Geilfus, C.M., Zörb, C. & Ober, D. (2015). New aspect of plant–rhizobia interaction: Alkaloid biosynthesis in crotalaria depends on nodulation. *Proceedings of the National Academy of Sciences*, 112 (13), 4164–4169.

Jaber, L.R. & Araj, S.E. (2017). Interactions among endophytic fungal entomopathogens (Ascomycota: Hypocreales), the green peach aphid *Myzus persicae* Sulzer (Homoptera: Aphididae), and the aphid endoparasitoid *Aphidius colemani* Viereck (Hymenoptera: Braconidae). *Biological Control*, 116, 53–61.

Jaber, L.R. & Enkerli, J. (2016). Effect of seed treatment duration on growth and colonization of *Vicia faba* by endophytic *Beauveria bassiana* and *Metarhizium brunneum*. *Biological Control*, 103, 187–195.

Jaber, L.R. & Enkerli, J. (2017). Fungal entomopathogens as endophytes: Can they promote plant growth? *Biocontrol Science and Technology*, 27, 28–41.

Jia, M., Chen, L., Xin, H.L., Zheng, C.J., Rahman, K, Han, T. & Qin, L.P. (2016). A friendly relationship between endophytic fungi and medicinal plants: A systematic review. *Frontiers in Microbiology*, 7, 906.

Jiang, S., Duan, J.A., Tao, J.H., Yan, H. & Zheng, J.B. (2010). Ecological distribution and elicitor activities of endophytic fungi in Changium smyrnioides. *Chinese Traditional and Herbal Drugs*, 1, 121–125.

Jiang, L., Li, E., Guo, L., Zhang, H. & Che, Y. (2008). Pestalachlorides A-C, antifungal metabolites from the plant endophytic fungus Pestalotiopsis adusta. Bioorganic and Medicinal Chemistry, 16 (17), 7894–7899.

Junker, C., Draeger, S. & Schulz, B. (2012). A fine line—endophytes or pathogens in *Arabidopsis thaliana. Fungal Ecology*, 5, 657–662.

Kaul, S., Gupta, S., Ahmed, M. & Dhar, M.K. (2012). Endophytic fungi from medicinal plants: A treasure hunt for bioactive metabolites. *Phytochemistry Reviews*, 11, 487–505.

Kaul. S., Sharma, T. & Dhar, M.K. (2016). "Omics" tools for better understanding the plant–endophyte interactions. *Frontiers in Plant Science*, 7, 955.

Kavroulakis, N.S., Zervakis, G.I., Ehaliotis, C., Haralampidis, K. & Papadopoulou, K.K. (2007). Role of ethylene in the protection of tomato plants against soil-borne fungal pathogens conferred by an endophytic *Fusarium solani* strain. *Journal of Experimental Botany*, 58, 3853–3864.

Kawasaki, A., Donn, S., Ryan, P.R., Mathesius, U., Devilla, R., Jones, A. & Watt, M. (2016). Microbiome and exudates of the root and rhizosphere of *Brachypodium distachyon*, a model for wheat. *PLoS One*, 11, e0164533.

Khan, M.S., Zaidi, A. & Wani, P.A. (2007). Role of phosphate-solubilizing microorganisms in sustainable agriculture: A review. *Agronomy for Sustainable Development*, 27 (1), 29–43.

Khare, E., Kim, K.M. & Lee, K.J. (2016). Rice OsPBL1 (ORYZA SATIVA ARABIDOPSIS PBS1-LIKE 1) enhanced defense of *Arabidopsis* against *Pseudomonas syringae* DC3000. *European Journal of Plant Pathology*, 146 (4), 901–910.

Khare, E., Mishra, J. & Arora, N.K. (2018). Multifaceted interactions between endophytes and plant: Developments and prospects. *Frontiers in Microbiology*, 9, 2732.

Kloepper, J.W. & Ryu, C.M. (2006). Bacterial endophytes as elicitors of induced systemic resistance. In *Microbial Root Endophytes*, Schulz, B.J.E., Boyle, C.J.C. & Sieber, T.N. (eds.). Berlin: Springer, pp. 33–52.

König, J., Guerreiro, M.A., Peršoh, D., Begerow, D. & Krauss, J. (2018). Knowing your neighbourhood—the effects of *Epichloë* endophytes on foliar fungal assemblages in perennial ryegrass in dependence of season and land-use intensity. *Peer J.*, 6, e4660.

Koonin, E.V. (2009). Evolution of genome architecture. *International Journal of Biochemistry & Cell Biology*, 41, 298–306.

Koonin, S., Otkur, M., Zhang, Z. & Tang, Q. (2007). Isolation and characterization of endophytic microorganisms in *Glaycyrrhiza inflat* Bat. From Xinjiang. *Microbiology*, 5, 867–870.

Koshino, H., Terada, S.I., Yoshihara, T., Sakamura, S., Shimanuki, T., Sato. T. &, Tajimi, A. (1988). Three phenolic acid derivatives from stromata of *Epichloë typhina* on *Phleum pretense. Phytochemistry*, 27 (5), 1333–1338.

Koshino, T., Yoshihara, S., Sakamura, T., Shimanuki, T., Sato, A. & Tajimi, A. (1989). A ring B aromatic sterol from stromata of *Epichloë typhina. Phytochemistry*, 28, 771–772.

Lata, R., Chowdhury, S., Gond, S.K. & White, J.F. (2018). Induction of abiotic stress tolerance in plants by endophytic microbes. *Letters in Applied Microbiology*, 66, 268–276.

Le Cocq, K., Gurr, S.J., Hirscg, P.R. & Mauchline, T.H. (2017). Exploitation of endophytes for sustainable agricultural intensification. *Molecular Plant Pathology*, 18, 469–473.

Liu, H., Carvalhais, L.C., Schenk, P.M. & Dennis, P.G. (2017). Effects of jasmonic acid signalling on the wheat microbiome differ between body sites. *Scientific Reports*, 7, 41766.

Liu, X., Dong, M., Chen, X., Jiang, M., Lv, X. & Zhou, J. (2008). Antimicrobial activity of an endophytic *Xylaria* spp. YX-28 and identification of its antimicrobial compound 7-amino- 4-methylcoumarin. *Applied Microbiology and Biotechnology*, 78 (2), 241–247.

Lo, H.C., Entwistle, R., Guo, C.J., Ahuja, M., Szewczyk, E., Hung, J.H., Chiang, Y.M., Oakley, B.R. & Wang, C.C. (2012). Two separate gene clusters encode the biosynthetic pathway for them eroterpenoids austinol and dehydroaustinol in *Aspergillus nidulans*. *Journal of the American Chemical Society*, 134, 4709–4720.

Lodge, D.J., Fisher, P.J. & Sutton, B.C. (1996). Endophytic fungi of *Manilkara bidentata* leaves in Puerto Rico. *Mycologia*, 1996 (88), 733.

Lopez, D.C. & Sword, G.A. (2015). The endophytic fungal entomopathogens *Beauveria bassiana* and *Purpureocillium lilacinum* enhance the growth of cultivated cotton (*Gossypium hirsutum*) and negatively affect survival of the cotton bollworm (*Helicoverpa zea*). *Biological Control*, 89, 53–60.

Lu, H., Zou, W.X., Meng, J.C., Hu, J. & Tan, R.X. (2000). New bioactive metabolites produced by *Colletotrichum* sp., an endophytic fungus in *Artemisia annua*. *Plant Science*, 151 (1), 67–73.

Mejía, L.C., Herre, E.A., Sparks, J.P., Winter, K., García, M.N., Van Bael, S.A., Stitt, J., Shi, Z., Zhang, Y., Guiltinan, M.J. & Maximova, S.N. (2014). Pervasive effects of a dominant foliar endophytic fungus on host genetic and phenotypic expression in a tropical tree. *Frontiers in Microbiology*, 5, 479.

Miller, J.D., Mackenzie, S., Foto, M., Adams, G.W. & Findlay, J.A. (2002). Needles of white spruce inoculated with rugulosin-producing endophytes contain rugulos in reducing spruce budworm growth rate. *Mycological Research*, 106, 471–479.

Min, Y.J., Park, M.S., Fong, J.J., Quan, Y., Jung, S. & Lim, Y.W. (2014). Diversity and saline resistance of endophytic fungi associated with *Pinus thunbergii* in coastal shelterbelts of Korea. *Journal of Microbiology and Biotechnology*, 24, 324–333.

Moricca, S. & Ragazzi, A. (2008). Fungal endophytes in Mediterranean oak forests: Alesson from *Discula quercina*. *Phytopathology*, 98, 380–386.

Nair, D.N. & Padmavathy, S. (2014). Impact of endophytic microorganisms on plants, environment and humans. *The Scientific World Journal*, Article ID 250693.

Niere, B.I. (2001). Significance of Non-pathogenic Isolates of *Fusarium oxysporum* Schlecht: Fries for the Biological Control of the Burrowing Nematode *Radopholus similis (Cobb) Thorne on Tissue Cultured Banana*. Ph.D. thesis, University of Bonn, Bonn.

Nikolouli, K. & Mossialos, D. (2012). Bioactive compounds synthesized by non-ribosomal peptide synthetases and type-I polyketide synthases discovered through genome-mining and metagenomics. *Biotechnology Letters*, 34 (8), 1393–1403.

Oldroyd, G.E., Murray, J.D., Poole, P.S. & Downie, J.A. (2011). The rules of engagement in the legume rhizobial symbiosis. *Annual Review of Genetics*, 45, 119–144.

Oliveira, A.L.M., Urquiagas S. & Baldani, J.I. (2003). Processos e mecanismos envolvidos na influência de microrganismos sobre o crescimento vegetal. *Embrapa Agrobiologia, Documentos*, 161. https://ainfo.cnptia.embrapa.br/digital/bitstream/CNPAB-2010/28064/1/doc161.pdf.

Oliveira, A.L.M., Urquiaga, S., Döbereiner, J. & Baldani, J.I. (2002). The effect of inoculating endophytic N2 fixing bacteria on micropropagated sugarcane plants. Plant and Soil, 242 (2) 205–215.

Park, J.H., Choi, G.J., Lee, H.B., Kim, K.M., Jung, H.S., Lee, S.W., Jang, K.S., Cho, K.Y. & Kim, J.C. (2005). Griseofulvin from *Xylaria* spp. strain F0010, an endophytic fungus of *Abies holophylla* and its antifungal activity against plant pathogenic fungi. *Journal of Microbiology and Biotechnology*, 15 (1), 112–117.

Park, Y.H., Kim, Y., Mishra, R.C. & Bae, H. (2017). Fungal endophytes inhabiting mountain-cultivated ginseng (*Panax ginseng* Meyer): Diversity and biocontrol activity against ginseng pathogens. *Scientific Reports*, 7, 16221.

Parniske, M. (2008). Arbuscular mycorrhiza: the mother of plant root endosymbioses. *Nature Reviews Microbiology*, 6 (10), 763–775.

Pavithra, G., Bindal, S., Rana, M. & Srivastava, S. (2020). Role of Endophytic Microbes Against Plant Pathogens: A Review. *Asian Journal of Plant Sciences*, 19(1), 54–62.

Pétriacq, P., Williams, A., Cotton, A., McFarlane, A.E., Rolfe, S.A. & Ton, J. (2017). Metabolite profiling of non-sterile rhizosphere soil. *The Plant Journal*, 92 (1), 147–162.

Pietro-Souza, W., de Campos Pereira, F., Mello, I.S., Stachack, F.F.F., Terezo, A.J., Nunes da Cunha, C., White, J.F., Li, H. & Soares, M.A. (2020). Mercury resistance and bioremediation mediated by endophytic fungi. *Chemosphere*, 240, 124874.

Pocasangre, L., Sikora, R.A., Vilich, V. & Schuster, R.P. (2000). Survey of banana endophytic fungi from Central America and screening for biological control of the burrowing nematode (*Radopholus similis*). *The International Journal on Banana and Plantain*, 9, 3.

Pongcharoen, W., Rukachaisirikul, V., Phongpaichit, S., Kühn, T., Pelzing, M., Sakayaroj, J. & Taylor, W.C. (2008). Metabolites from the endophytic fungus *Xylaria* spp. PSU-D14. *Phytochemistry*, 69 (9), 1900–1902. PMID: 18495187.

Priebe, S., Linde, J., Albrecht, D., Guthke, R. & Brakhage, A.A. (2011). Fungi Fun a web-based application for functional categorization of fungal genes and proteins. *Fungal Genetics and Biology*, 48, 353–358.

Pulici, M., Sugawara, F., Koshino, H., Uzawa, J., Yoshida, S., Lobkovsky, E. & Clardy, J. (1996b). A new isodrimeninol from *Pestalotiopsis* sp. *Journal of Natural Products*, 59, 1, 47–48.

Pulici, M., Sugawara, F., Koshino, H., Uzawa, J., Yoshida, S., Lobkovsky, E. & Clardy, J. (1996a). Pestalotiopsins A and B: new caryophyllenes from an endophytic fungus of *Taxus brevifolia*. *Journal of Organic Chemistry*, 61 (6), 2122–2124.

Pylak, M., Oszust, K. & Frac, M. (2019). Review report on the role of bio-products, biopreparations, biostimulants and microbial inoculants in organic production of fruit. *Reviews in Environmental Science and Biotechnology*, 18, 597–616.

Rosenblueth, M. & Martínez-Romero, E. (2006). Bacterial endophytes and their interactions with hosts. *Molecular Plant-Microbe Interactions*, 19, 827–837.

Rosewich, U.L. & Kistler, H.C., 2000. Role of horizontal gene transfer in the evolution of fungi. *Annual Review of Phytopathology*, 38, 325–363.

Rowan, D.D. & Latch, G.C.M. (1994). Biotechnology of endophytic fungi of grasses. In *Utilization of Endophyte-Infected Ryegrasses for Increased Insect Resistance*, Bacon C.W. & White J.F. Jr. (eds.). Boca Raton: CRC Press, pp 169–183.

Saikkonen, K., Ion, D. & Gyllenberg, M. (2002). The persistence of vertically transmitted fungi in grass metapopulations. *Proceedings of the Royal Society B: Biological Sciences*, 269, 1397–1403.

Saikkonen, K., Wäli, P., Helander, M. & Faeth, S.H. (2004). Evolution of endophyte-plant symbioses. *Trends in Plant Science*, 9, 275–280.

Sarkar, D., Rovenich, H., Jeena, G., Nizam, S., Tissier, A., Balcke, G.U., Mahdi, L.K., Bonkowski, M., Langen, G. & Zuccar, A. (2019). The inconspicuous gatekeeper: Endophytic *Serendipita vermifera* acts as extended plant protection barrier in the rhizosphere. *New Phytologist*, 224, 886–901.

Schardl, C.L. & Phillips, T.D. (1997). Protective grass endophytes: Where are they from and where are they going? *Plant Disease*, 81 (5), 430–438.

Schardl, C.L., Young, C.A., Pan, J., Florea, S., Takach, J.E., Panaccione, D.G., Farman, M.L., Webb, J.S., Jaromczyk, J., Charlton, N.D., Nagabhyru, P., Chen, L., Shi, C. & Leuchtmann, A. (2013). Currencies of mutualisms: sources of alkaloid genes in vertically transmitted epichloae. *Toxins*, 5, 1064–1088.

Schulz, B., Sucker, J., Aust, H.J., Krohn, K., Ludewig, K., Jones, P.G. & Döring, D. (1995). Biologically active secondary metabolites of endophytic *Pezicula* species. *Mycological Research*, 99 (8), 1007–1015.

Sheoran, N., Valiya Nadakkakath, A., Munjal, V., Kundu, A., Subaharan, K., Venugopal, V., Rajamma, S., Eapen, S.J. & Kumar, A. (2015). Genetic analysis of plant endophytic *Pseudomonas putida* BP25 and chemo-profiling of its antimicrobial volatile organic compounds. *Microbiological Research*, 173, 66–78.

Shi, Y., Lou, K. & Li, C. (2010). Growth and photosynthetic efficiency promotion of sugar beet (*Beta vulgaris* L.) by endophytic bacteria. *Photosynthesis Research*, 105 (1), 5–13.

Shukla, S.T., Habbu, P.V., Kulkarni, V.H., Jagadish, K.S., Pandey, A.R. & Sutariya, V.N. (2014). Endophytic microbes: A novel source for biologically/ pharmacologically active secondary metabolites. Asian Journal of Pharmacology and Toxicology, 2, 1–16.

Sieber, T.N. (2007). Endophytic fungi in forest trees: Are they mutualists? *Fungal Biology Reviews*, 21, 75–89.

Sikora, R.A., Pocasangre, L., Zum Felde, A., Niere, B., Vu, T.T. & Dababat, A. (2008). Mutualistic endophytic fungi and in-planta suppressiveness to plant parasitic nematodes. *Biological Control*, 46, 15–23.

Song, J.H. (2008). What's new on the antimicrobial horizon?. *International Journal of Antimicrobial Agents*, 32 (4), S207–S213.

Sorensen, J. & Sessitsch, A. (2006). Plant-associated bacteria lifestyle and molecular interactions. In *Modern Soil Microbiology*, 2nd ed., van Elsas, J.D., Janet., K. & Trevors, J.T. (eds.). Boca Raton: CRC Press, pp. 211–236.

Stierle, A.A., Stierle, D.B. & Bugni, T. (1999). Sequoiatones A and B: Novel antitumor metabolites isolated from a redwood endophyte. *Journal of Organic Chemistry*, 64 (15), 5479–5484.

Strobel, G., Daisy, B., Castillo, U. & Harper, J. (2004). Natural products from endophytic microorganisms. *Journal of Natural Products*, 67 (2), 257–268.

Strobel, G.A., Miller, R.V., Martinez-Miller, C., Condron, M.M., Teplow, D.B. & Hess, W.M. (1999). Cryptocandin, a potent antimycotic from the endophytic fungus *Cryptosporiopsis cf. quercina*. *Microbiology*, 145 (8), 1919–1926.

Strobel, G.A., Torczynski, R. & Bollon, A. (1997). 4 Acremonium sp.—a leucinostatin A producing endophyte of European yew (*Taxus baccata*). *Plant Science*, 128 (1), 97–108.

Suryanarayanan, T.S. (2013). Endophyte research: Going beyond isolation and metabolite documentation. *Fungal Ecology*, 6, 561–568.

Tan, R. X. & Zou, W. X. (2001). Endophytes: A rich source of functional metabolites. *Natural Product Reports*, 18(4), 448–459.

Thongsandee, W., Matsuda, Y. & Ito, S. (2012). Temporal variations in endophytic fungal assemblages of *Ginkgo biloba* L. *Journal of Forest Research*, 17 (2), 213–218.

Toubal, S., Bouchenak, O., Elhaddad, D., Yahiaoui, K., Boumaza, S. & Arab, K. (2018). MALDI-TOF MS detection of endophytic bacteria associated with great nettle (*Urtica dioica* L.), grown in Algeria. *Polish Journal of Microbiology*, 67, 67–72.

Traber, R., Keller-Juslén, C., Loosli, H.R., Kuhn, M. & Von Wartburg, A. (1979). Cyclopeptide antibiotics from *Aspergillus* species. Structure of echinocandins C and D. *Helvetica Chimica Acta*, 62 (4), 1252–1267.

Trail, F., Mahanti, N., Rarick, M., Mehigh, R., Liang, S.H., Zhou, R. & Linz, J.E. (1995). Physical and transcriptional map of an aflatoxin gene cluster in *Aspergillus parasiticus* and functional disruption of a gene involved early in the aflatoxin pathway. *Applied and Environmental Microbiology*, 61, 2665–2673.

Truyens, S., Weyens, N., Cuypers, A. & Vangronsveld, J. (2014). Bacterial seed endophytes: genera, vertical transmission and interaction with plants. *Environmental Microbiology Reports*, 7, 40–50.

Unterseher, M. & Schnittler, M. (2010). Species richness analysis and ITS rDNA phylogeny revealed the majority of cultivable foliar endophytes from beech (*Fagus sylvatica*). *Fungal Ecology*, 3, 366–378.

Wagenaar, M.J., Corwin, G., Strobel, G.A. & Clardy, J. (2000). Three new chytochalasins produced by an endophytic fungus in the genus Rhinocladiella. *Journal of Natural Products*, 63, 1692–1695.

Wang, X., Wang, H., Liu, T. & Xin, Z. (2014). A PKS I gene-based screening approach for the discovery of a new polyketide from *Penicillium citrinum* Salicorn. *Applied Microbiology and Biotechnology*, 98 (11), 4875–4885.

Waqas, M., Khan, A.L., Hamayun, M., Shahzad, R., Kang, S.M., Kim, J.G. & Lee, I.J. (2015). Endophytic fungi promote plant growth and mitigate the adverse effects of stem rot: An example of *Penicillium citrinum* and *Aspergillus terreus*. *Journal of Plant Interactions*, 10, 280–287.

Wolinska, A. (2019). Metagenomic achievements in microbial diversity determination in croplands: A review. In *Microbial Diversity in Genomic Era*, Das, S. & Dash, H.R. (eds.). Cambridge, MA: Academic Press, pp. 15–35.

Wu, T., Xu, J., Liu, J., Guo, W.H., Li, X.B., Xia, J.B., Xie, W.J., Yao, Z.G., Zhang, Y.M. & Wang, R.Q. (2019). Characterization and initial application of endophytic *Bacillus safensis* strain ZY16 for improving phytoremediation of oil-contaminated saline soils. *Frontiers in Microbiology*, 10, 991.

Yang, J., Xu, F., Huang, C., Li, J., She, Z., Pei, Z. & Lin, Y. (2010). Metabolites from the mangrove endophytic fungus Phomopsis sp. (#zsu-H76). European Journal of Organic Chemistry, 5, 3692–3695.

Yong, Y.H., Dai, C.C., Gao, F.K., Yang, Q.Y. & Zhao, M. (2009). Effects of endophytic fungi on growth and two kinds of terpenoids for Euphorbia pekinensis. *Chinese Traditional and Herbal Drugs*, 40, 18–22.

Yu, H., Zhang, L., Zheng, C., Li, L., Guo, L., Li. W., Sun, P., & Qin, L. (2010). Recent developments and future prospects of antimicrobial metabolites produced by endophytes. Microbiological Research, 165(6), 437–449.

Yuan, J., Zhang, W., Sun, K., Tang, M.J., Chen, P.X., Li, X. & Dai, C.C. (2019). Comparative transcriptomics and proteomics of *Atractylodes lancea* in response to endophytic fungus *Gilmaniella* spp. AL12 reveals regulation in plant metabolism. *Frontiers in Microbiology*, 10, 1208.

Zinniel, D.K., Lambrecht, P., Harris, N.B., Feng, Z., Kuczmarski, D., Higley, P., Ishimaru, C.A., Arunakumari, A., Barletta, R.G. & Vidaver, A.K. (2002). Isolation and characterization of endophytic colonizing bacteria from agronomic crops and prairie plants. *Applied and Environmental Microbiology*, 68 (5), 2198–2208.

Zou, W.X., Meng, J.C., Lu, H., Chen, G.X., Shi, G.X., Zhang, T.Y. & Tan, R.X. (2000). Metabolites of *Colletotrichum gloeosporioides,* an endophytic fungus in *Artemisia mongolica. Journal of Natural Products*, 63 (11), 1529–1530.

4 Bioactive-Molecules Biosynthesis for Agroforestry Applications

Cleverson Busso and Priscila Vaz de Arruda
Federal University of Technology—Paraná (UTFPR),
Toledo, Brazil

CONTENTS

4.1 INTRODUCTION

The microbial diversity and chemical composition of the soil are essential factors for healthy plant growth. It is believed that there are approximately 4×10^6 taxa of microorganisms per ton of soil, with an average of 10^9 cells per gram of soil sample. Although soil presents a great diversity, there are limited studies on the characterization and potential uses of these microorganisms in agroforestry systems.

Prokaryotes are the largest taxon present in the soil environment, but approximately 99% are not cultivable in the laboratory, which limits more in-depth studies of their potential applications. Fungi also constitute another abundant and important group in the soil, and together with other microorganisms are able to synthesize useful biomolecules for plant development and protection.

The development of agroforestry depends on the knowledge of new techniques of soil management, especially in studies and applications of different microorganisms that produce biomolecules. The microbial identification by molecular biology techniques and the characterization of the potential uses of cellular metabolites produced by different groups of microorganisms, currently constitute an excellent opportunity for the expansion of agroforestry systems.

DOI: 10.1201/9781003110477-4

4.2 MYCORRHIZA BIODIVERSITY AND IMPORTANCE IN THE PRODUCTION OF BIOMOLECULES

Plant roots and fungi are capable of establishing an extraordinarily versatile mutualistic relationship called mycorrhiza. The plants are the hosts and receive mineral nutrients obtained by the fungi hyphae from the soil, the latter benefit from carbon sources from the photosynthetic process of the plants. Morphological, paleobotanical and phylogenetic analysis indicate that this plant-fungus interaction co-evolved more than 400 million years ago, promoting morphological changes and constituting different types of mycorrhizae (Brundrett 2002).

Approximately 90% of vascular plant families have associations with fungi. This association is characterized according to the fungal growth on the plant: ectomycorrhiza, when the fungus hyphae develop on the root surface; endomycorrhiza, where the presence of hyphae is observed inside the root tissues; and ecto-endomycorrhiza, when there is the presence of hyphae on the surface and interior of the plant tissues (Winagraski et al. 2019). Arbuscular mycorrhizal fungi (AMF) are endomycorrhizal fungi, so named because of the formation of an intraradial mycelium present within the root cells and can be found in angiosperms, gymnosperms, pteridophytes and some bryophytes (Wang and Qiu 2006). The AMF belong to the phylum Glomeromycota and are necessarily symbiotic with hyphae that facilitate the access of low mobility minerals present in the soil, stabilize the soil through the formation of aggregates and improve the absorption mainly of Ca, S, K, Cu and Mg (Gildon and Tinker 1983). Ectomycorrhizal fungi (EMF) are part of the phylum Ascomycota and Basidiomycota and are characterized by forming a network of intercellular hyphae on root cells called Hartig's network. This mycelial network extends to the soil and is involved in processes of mobilization, absorption and translocation of water and nutrients from the soil to the roots of the plants. EMF are found mainly in some families of gymnosperms and dicotyledonies.

Phylogenetic data indicate that AMF are predominant, coinciding evolutionarily with the origin of terrestrial plants. However, EMF evolved in parallel and independently, contributing to the diversification of host plants and symbiotic fungi. The advantage of this symbiosis for the plant depends on the type of mycorrhiza formed, that is, for adult trees formed predominantly by EMF, there is a favor for the development of new seedlings over those where the predominance is colonization by AMF. This occurs particularly due to the presence of biomolecules that favor greater transfer of nitrogen from EMF, especially when considering environments limited in nitrogen such as forests. Another factor is the ability of EMF to colonize root surfaces providing extra protection against antagonistic microorganisms (Wang and Qiu 2006; Bennett et al. 2017). The combination of EMF and AMF in the same root is favored by a synergistic effect with the plant-soil dynamics, where extreme environmental fluctuations, such as soil moisture, temperature and nutrient availability can help in this interaction. In addition, these extreme factors also encourage root colonization by EMF, while AMF are favored by normal environmental conditions.

The interaction between tree and type of mycorrhiza has a fundamental role in soil integrity, studies carried out in the United Kingdom suggest that trees containing EMF, synthesize substances that contribute to soil acidification. This interaction favors the decomposition of rock grains containing calcite much faster than plants with AMF; thus, fast-growing angiosperms containing EMF would be a good option for this type of soil (Thorley et al. 2015). Trees with EMF present a better yield in infertile soils, mainly due to the greater capacity of these microorganisms to absorb N and P directly from organic sources. However, trees containing AMF in fertile soils have greater specificity and stability inside the roots, which drastically reduces maintenance costs of mycorrhizal symbiosis (Mao et al. 2019). An analysis of 164 species of AMF present in forest ecosystems in Brazil allowed us to conclude that these microorganisms are not specific to the host, but selected from the environmental variations of where the host is located. Thus, species such as *Acaulospora scrobiculata, Acaulospora foveata, Clareoideglomus etunicatum* and *Glomus macrocarpum* seem to be adapted to different stages and forest management, which may be good candidates for use as inoculants of forest species (Winagraski et al. 2019).

The use of inoculants in tree-seedling nurseries has been shown to be an efficient method for obtaining plants with desirable EMF colonization. The identification and physiological knowledge of the species are essential for the success of the technique, especially to avoid cross-contamination with other undesirable species. In addition, it is essential that the fungus has a rapid growth due to the short incubation period in the nursery (Marx 1980). Processes such as the soil itself containing EMF, spore solution or suspension and also the fungus mycelium have been used as an inoculation method in plant seedlings. The use of hyphae is the most advantageous, these strains are already selected and adapted, have uniformity and have a low rate of bacterial contamination when obtained in solid medium (Cannel and Moo-Young 1980). The use of AMF inoculants in culture techniques has been a great challenge, since the biggest problem is in obtaining it on a large scale with desirable quality and low cost (Ijdo, Cranenbrouck, and Declerck 2011).

Some experiments using inoculants in the greenhouse with the plant *Morus alba* showed that the spores of AMF *Glomus mosseae* and *Glomus inraradices* when inoculated individually resulted in a significant improvement in the physiological performance and increase in plant size. However, when both fungi were inoculated together, the effect did not show significant results, suggesting that competition between these microorganisms may influence the AMF community, interfering in the symbiosis process (Lu et al. 2015). Another study evaluated whether the combination of three AMF species (*G. mosseae, G. claroideum* and *G. intraradices*) provided more phosphorus and supported greater growth in *Allium porrum* than a single AMF species. Using real-time PCR, it was possible to observe that the plant colonized by a mixture of *G. claroideum* and *G. intraradices* acquired more phosphorus than with either of the two AMF separately. Competition between species represents a major challenge in interpreting experiments with mixed inoculations, but this is greatly facilitated by the modern use of molecular biology techniques (Jansa, Smith, and Smith 2008).

A better understanding of the physiology of mycorrhizae in the plant–soil interaction with the characterization of the biomolecules involved in this process can provide new beneficial strategies for plant growth, as well as applications in the restoration of the soil's natural fertility and in the control of plant pathogens. In addition to the environmental impact of nutrient cycling, mycorrhizae can also benefit in socioeconomic aspects by increasing productivity in horticulture and contributing to higher yields in the production and development of trees of commercial interest.

4.3 PLANT GROWTH PROMOTING MICROORGANISMS

The rhizosphere is a thin layer where the soil and the roots of plants establish an intense metabolic activity. In addition to the physical-chemical interaction, this region has considerable microbiological diversity, with fungi, protozoa, algae and, in greater quantities, bacteria with approximately 10^{10} cells per cm^2. Many of these microorganisms can promote the direct or indirect growth of plants, thus playing an important role in the cycling of nutrients with a consequent increase in soil fertility. The main direct effects consist of biofertilization, that is, mechanisms of nitrogen fixation, ion solubilization and production of phytohormones such as gibberiline, auxin and ethylene. Indirect mechanisms are associated with biocontrol processes such as the production of lytic enzymes amylases, proteases and esterases as well as the production of siderophores, which are small molecules involved in the sequestration of ions, such as Fe^{3+}, thus competing with other phytopathogenic microorganisms (Gupta et al. 2015).

Considering that bacteria are the most abundant microorganisms in the rhizosphere, their greatest influence on plant physiology is notable, especially regarding the competitiveness in root colonization. These bacteria are known as plant growth-promoting rhizobacteria (PGPR). The PGPR can be located outside the plant in the rhizosphere or in the cellular spaces of the root cortex. However, there are PGPRs located in internal nodular root structures, the latter belonging to the Rhizobiaceae family and include the genera *Allorhizobium*, *Bradyrhizobium*, *Mesorhizobium* and *Rhizobium*, in addition to endophytes and some species of *Frankia*, all associated with the process of atmospheric nitrogen fixation. Bacteria belonging to the genera *Agrobacterium*, *Arthrobacter*, *Azotobacter*, *Azospirillum*, *Bacillus*, *Burkholderia*, *Caulobacter*, *Chromobacterium*, *Erwinia*, *Flavobacterium*, *Micrococcous*, *Pseudomonas* and *Serratia* are the most common found externally on the surface of plant roots (Bhattacharyya and Jha 2012). In general, the bacterial diversity in the rhizosphere is represented mainly by members of the phyla Proteobacteria, Firmicutess and Actinobacteria (Zuluaga et al. 2020).

Many studies using Rhizobacteria as growth promoters have been applied mainly to gymnosperm trees, such as the genera *Picea*, *Pinus*, *Tsuga* and *Pseudotsuga* (Mafia et al. 2009) although their use in angiosperms has also been practiced. The inoculation of bacterial strains in pine and fir seedlings destined for reforestation was able to promote a significant gain between 32 and 49% after one year of cultivation (Chanway 1997). *Pinus taeda* seedlings had an increase in root and shoot biomass by 67 and 33% respectively, when inoculated with *Bacillus subtilis* after 180 days of emergency (Santos et al. 2018). Many of these bacteria are capable of producing auxins stimulating plant growth; in addition, they can synthesize enzymes that are associated with phosphate solubilization, which stimulates the growth of some ectomycorrhizal fungi (Heredia-Acuña et al. 2019). Fruit angiosperms also show excellent results when inoculated with rhizobacteria. In an experiment carried out for two years in the Anatolia region in Turkey, apple trees (*Malus domestica* Borckh) newly planted and inoculated with rhizobacteria had an average length increase of 60% greater when compared to the non-inoculated group. In addition, fruit yield reached close to 120%. Analysis indicated that these bacteria were able to produce indole acetic acid (AIA) as well as dissolve phosphate, thus stimulating the plant's growth and fruit production (Aslantaş, Çakmakçi, and Şahin 2007).

The antagonistic effect against phytopathogens exerted by rhizobacteria can indirectly contribute to plant growth. Bacterial isolates obtained from avocado trees that survived root rot infestations showed antagonistic activity against the pathogenic oomycete *Phytophthora cinnamomi*. One of the isolates was identified as *Bacillus acidiceler* and was able to inhibit the growth of oomycete by up to 76%. Mass spectrometry analyzes (GC-MS) the volatile compounds produced by this bacterium found the presence of 2,3,5-trimethylpyrazine, 6,10-dimethyl-5-9-undecadien-2-one and 3-amino-1, 3-oxozolidin-2-one, all compounds known to have antifungal activity (Méndez-Bravo et al. 2018).

In addition to producing compounds with recognized antagonistic activity against phytopathogens, the use of bioinoculants from plant growth promoting rhizobacteria can contribute ecologically by reducing or even replacing inorganic fertilizers that are responsible for depositing heavy metals in the soil and plants, resulting in serious problems to environmental and human health when bioaccumulated. Studies of the molecular and physiological mechanisms of the effects of this bacterium-plant interaction have contributed especially to bioprospecting and the development of new bioinoculants that contribute to the strategy in the management of plants of environmental and economic interest.

4.4 SYNTHESIS OF NANOPARTICLES AND PRODUCTION OF BIOINOCULANTS OF AGROFORESTRY INTEREST

Nanotechnology is a science that makes it possible to restructure matter on a nanometric scale (between 1 and 100 nm) and consequently allows the development of unusual chemical, physical and biological properties in materials converted to the nanoscale. Nanoparticles (NPs) are structures resulting from this new technology, where beneficial properties of organic, inorganic or even hybrid materials can result in structures capable of being used by man in areas such as industry, medicine and agriculture. NPs can be produced not only artificially in laboratories but also naturally from processes of photochemical reactions, volcanic eruptions, and also from the metabolism of plants and microorganisms (Shang et al. 2019).

Several organisms can be used for the biosynthesis of nanoparticles, acting as reducing agents and/or stabilizing agents (Zewde et al. 2016). Many of these reactions are promoted by enzymes such as nitrate reductase. The bacteria *Bacillus subtilis, B. licheniformis, Pseudomonas proteolytica* and *P. meridiana* have successfully synthesized nanoparticles containing Ag^+ maintaining the stability of these compounds for 5–8 months. Some studies indicate that plant extracts are also considered fast and efficient methods for obtaining nanoparticles, the use of *Pinus eldarica* extract made it possible to obtain spherical particles with size ranging between 10 and 40 nm in diameter (Iravani and Zolfaghari 2013). Fungi are one of the most versatile systems, mainly due to their high tolerance to metals and easy handling. The synthesis in these microorganisms can occur in an intracellular and extracellular way; in the first case, the metal is added in the culture medium and then internalized by the fungus, in the second case, the precursor metal is added in the aqueous filtrate containing only the biomolecules of the fungus that results in the formation of nanoparticles in the dispersion. Fungi like *Aspergillus fumigatus*, *A. oryzae* and *Trichoderma viride* are well known in the production of silver nanoparticles (Guilger-Casagrande and de Lima 2019).

Metallic NPs are being used more and more in agriculture and forestry as alternatives to growth-stimulating compounds, pesticides and even fertilizers. The unique properties obtained with this technology have favored the development of high quality seeds for agronomic crops and trees with a significant increase in productivity (Polischuk et al. 2019). Studies suggest that leaf treatments using AgNPs in non-toxic concentrations significantly influence biotic interactions, increasing the production of phytohormones as well as controlling colonization by ectomycorrhizae. The use of concentrations of 5, 25 and 50 ppm of AgNPs is able to stimulate the formation of mycorrhiza in pedunculated oak seedlings, obtaining the best effect on the intermediate concentration (Aleksandrowicz-Trzcińska et al. 2019).

Although the use of nanoparticles has aroused interest in the production of several plants, it still needs more conclusive studies, as is the case with AgNPs. Some studies have pointed out the interference of these nanoparticles on the development of white fungi on *Trifolium repens*. While low concentrations of AgNPs (0.01 mg.kg^{-1}) are able to reduce fungal development, high concentrations (0.1 and mg.kg^{-1} soil) significantly increase the extent of plant root colonization. This dubiousness is believed to be due to the agglomeration of nanoparticles in high concentrations, which, therefore, would reduce dissolution and consequently less capture by the plant (Feng 2013).

Several other studies involving nanoparticles with inorganic components have been used to manage plant diseases. Compounds such as ZnO, MgO, TiO$_2$, MnO, Cu, SiO$_2$, and CaO were evaluated in the fight against bacterial and fungal diseases, as well as in the control of weeds in different cultures (Servin et al. 2015). ZnO and MgO NPs used in concentrations below 100 mg.L^{-1} had a high rate of inhibition in the spore germination of the fungi *Alternaria alternata*, *Fusarium oxysporum*, *Rhizopus stolonifer* and *Mucor plumbeus*. In vitro studies have shown that ZnO nanoparticles in concentrations of 25 μg.mL^{-1} showed significant responses in inhibiting growth against the bacterium *Pseudomonas aeruginosa* (Jayaseelan et al. 2012).

Some compounds have, in addition to their antimicrobial capacity, a plant growth-stimulating effect. TiO$_2$ is capable of inhibiting between 69 and 91% the infection caused by *P. syringae* and *P. cubensis* in cucumbers, in addition to increasing the photosynthetic rate up to 30% when compared to the group without nanoparticles (Cui 2009). There are several mechanisms of action of NPs on microorganisms, from interaction with the cell wall and plasma membrane-altering osmolarity, permeability and electron transport to interactions with ribosomal DNA and phosphorus and sulfur-binding sites in proteins (Składanowski 2017).

Numerous studies have suggested the NPs absorption processes in plant cells. In general, these particles are thought to bind to carrier proteins on the cell surface through aquaporins, ion channels or endocytosis through the creation of new pores. Once inside the cell, the displacement of the NPs can occur through spaces external to the cell membranes (apoplast), or through the cytoplasm of the cells of the epidermis and cortex, which constitute the symplast (Rico et al. 2011). Although, there are still conflicting reports on toxicity in humans and the environment promoted by nanoparticles (Pramanik et al. 2020), further studies are needed on the application of these compounds in plant management, especially because it is a new technology but with great prospects for its use in forestry.

The exponential increase in the world population is one of the most important concerns in the next years due to environmental harm caused by the fertilizers and pesticides on agriculture to ensure a global food security. In this sense, the use of beneficial microbes in agriculture, as microbial inoculants—denominated as biofertilizers in some countries—is an attractive alternative for sustainable way to enhance the agriculture production in the world (Barea 2015; Arora et al. 2020; Dilnashin et al. 2020). The bioprospecting of microorganisms capable of controlling phytopathogens and also those that provide nutrients to plants through symbiotic association in roots, constitute an important biotechnological area for plant protection and also as an adjuvant in soil health (Raaijmakers and Lugtenberg 2013).

The first inoculation technology was developed over 130 years ago, by U.S. Department of Agriculture employing *Rhizobium* for legume crops during experimental researches of nitrogen fixation (Schneider 1892). Today, according to Market Data Forecast, the global microbial soil inoculants market was estimated to be approximately US$452.07 million in 2022 and it is expected to generate more than US$751.51 million by 2025. The microbial inoculant market in Latin America is expected to be the fastest growing due to corn and soybean crops' commercialized cultivation. The leading producers dominating this market are Monsanto BioAg, Kiwa Bio-Tech Products Group Corporation, Lallemand Inc., Camson Biotechnologies Limited and Agrinos AS, Basf S.E., E.I. Dupont De Nemours Company, Bayer Crop Science, Novozymes A/S, Verdesian Life Science LLC, Advanced Biological Marketing Inc, Brettyoung, Precision Laboratories LLC, Queensland Agricultural Seeds Pty Limited and Xitebio Technologies Incorporated (Mordor Intelligence 2020; Market Data Forecast 2020).

Inoculants are commercial biological agricultural inputs of liquid or solid nature, depending on the physical state of the medium in which microbial cells are arranged and contain a high concentration of bacterial or fungal or both, pure culture or mixtures of different microorganisms varying according to the target culture that was designed (Arora et al. 2020).

An important aspect of microbial inoculant technology is the choice of carrier for the microorganisms, which could determine the cell viability and facilities of application in the crop. O'Callaghan (2016) reported that microbial seed inoculation in an industrial scale involves techniques and processes that have "trade secrets" and "in-house knowledge", although the literature describes high numbers of methods for this propose, but in laboratory scale.

In general, there are two main technologies that have been used in a commercial scale, a solid inoculant, which is also known as peat, and liquid formulations (Santos et al. 2019; Arora et al. 2020). The peaty matrix is a solid material rich in organic matter growing from the soil naturally, occurring in specific environments and formed after a long geological period (Santos et al. 2019). According to these authors, this kind of carrier provides physical protection for the microorganisms, allows better cell survival especially during critical culture conditions as water restriction and elevated temperatures. However, this carrier has limitation related to its obtaining, handling and the sterilization of peat, which suggest new formulations need to be tested, developed and adequately assessed in the field (Araujo et al. 2012). In this sense, liquid formulations began to gain space, especially because they are easily manipulated, no cause mechanical failures in sowing machines, guarantee high cell levels in soil, is the easiness of sterilization, and facilitating the absence of contaminants (O'Callaghan 2016).

Liquid formulations is composed by (%): specific broth (10–40), suspension ingredient (1–3), surfactant (3–8), dispersant (1–5) and carrier, which could be oil or water with broth combination (Sharma, Sharma and Salwan 2020). However, these types of microbial inoculants show some difficulties related to storage and transportation, since they require refrigeration and homogeneous mixtures, and coating in the seeds is limited (Araujo et al. 2012).

Other carriers for the microorganisms, such by-products of agricultural residues as sugarcane bagasse, industrial waste as wastewater sludge and polymer blends have been studied in order to increase the inoculant quality and efficiency and to reduce costs and environmental impacts (Rebah et al. 2007; Araujo et al. 2012; Unnikrishnan and Vijayaraghavan 2019).

There are two main origins of microbial-based bioinoculants formulation: bacterial and fungal, which are constituted mainly by the microorganism's genera *Bacillus, Pseudomonas, Rhizobia, Trichoderma, Mycorrhiza, Endophytes,* and other types could be found in blue-green algae, microbial metabolites or other additives along with cell-based bioinoculants (Arora et al. 2020).

The diazotrophic bacteria *Rhyzobium* ssp. is the most frequently used microbe as inoculants in Brazilian and in the worldwide crops, mainly in the soybean cultivation. This fact could be explained since the bacteria improve the biological nitrogen fixation (BNF) by rhizosphere colonization and nodules establishment in the roots of the host plants (Russo et al. 2012; Souza et al. 2015; Santos et al. 2019). Among the *Rhizobium* species, *Bradyrhyzobium japonicum* and *B. elkanii* are the most widely used species as inoculant in leguminous plant (Souza et al. 2015). The BNF by the bacteria save millions of dollars which otherwise would be spent on chemical fertilizers; for example, in Brazil, according to the same authors, 70% of nitrogenous fertilizers are imported and the use of microbial inoculant could be saving of US$7.0 billion/year.

Arora et al. (2020) describe in their review that BNF is a complex metabolic interaction between plants and microbes that result in the nodule formation in leguminous plants mainly by the signals of flavonoids and isoflavonoids. These are produced by the rhizospheric microbes, which are attracted toward plants and inducing the cascade of releasing nod factors which take up these signals and bind to the transcriptional regulator NodD of conserved *nod* genes.

Different studies have been reported that the application of a mixture of strains, which could have stimulated the synergy between them, resulted in the environmental adaptation and increase the efficiency of the product (Araujo et al. 2012). Table 4.1 shows some microbe-based inoculations and target crops.

TABLE 4.1
Some Microbe-Based Commercial Inoculants and Target Crops

Micro-Based Formulation	Claimed Strains	Inoculant Benefits	Commercial Name of Manufactured Products	Reference
Bacterial	*Bradyrhizobium japonicum*	Enhancement the selectivity and the protection of seed and soil to pathogens.	Vault® HP	Herrmann et al. (2015), Basf-Corporation (2020a)
Bacterial	*Bradyrhizobium elkanii*	Contributes to the increase in grain yield and plant efficiency in the first year of cultivation. Increases soil organic matter levels.	Adhere® 60	Basf-Corporation (2020b)
Mixed bacterial	*Azospirillum lipoferum* and *Azospirillum brasilense*	*Improving plant biomass yields and quality*, reduce soil erosion, improvement of root zone and phosphate release for horticulture crops.	Bio-N®	Cruz et al. (2012)
Fungal	*Penicillium bilaiae*	Significant impact on maize yield (enhanced until 80% in large plots for field trials)	Jumpstart®	Parnell et al. (2016)
Mixed fungal	*Arthrobotrys conoides*, *Purpureocillium lilacinus* and *Pochonia chlamydosporium.*	Improving soil food web ratios. Applicable at all stages of crop production. Biodegradable and target specific. No with holding period. Contains three species of bio-balancing fungi. A powerful probiotic and plant growth promoter.	Nutri-Life Root-Guard™	NTS (2020)
Mixed bacterial-fungal	*Bacillus amyloliquefaciens* and *Trichoderma virens*	Liberate bound phosphate making this nutrient more available to plant roots.	QuickRoots®	Parnell et al. (2016)
Mixed bacterial-fungal	*Bradyrhizobium* and *Trichoderma*	Improve nutrient and water efficiency, yield advantages for soybeans crop	Excalibre-SA™	Parnell et al. (2016), ABM (2020)

4.5 MICROBIAL SYSTEMS APPLIED TO BIOREMEDIATION

This is a biotechnological process that has been intensively studied and recommended by the current scientific community as a viable alternative for the treatment of contaminated environments, such as surface water and groundwater; soil, industrial and effluent waste in landfill areas. It is an alternative for conventional treatment such as physicochemical or electrochemical since several microorganisms such as bacteria, fungi, and microalgae could be used with highly efficient, low cost technology, hence high technology is not necessary (Chen et al. 2015; Dangi et al. 2018; Kazemalilou et al. 2020).

According to Dangi et al. (2018), there are two types of pollutants, the old and the new. The first one could be characterized by chemical solvents, petroleum, pesticides, herbicides, heavy metals, paints, PAHs (polycyclic aromatic hydrocarbons), industrial solvents, among others. On the other hand, according to the same authors, the new prominent pollutants releases from carbon-nanomaterials (carbon nanotubes, graphene, metal and metal oxide nanoparticles).

However, the bioremediation process using single/isolate microbes has a limitation that could be related to an incomplete or ineffective degradation of the pollutant in mixed wastes and there are little-known facts about enzymatic aspects in contaminated soils remediated by composting or compost (Chen et al. 2015; Dangi et al. 2018). In this sense, a microbe's consortia associated with plants have been used as techniques for remediation of polluted soils. Biological remediation is the process, which uses green plants and/or their association with microorganisms to assimilate, transform, metabolize or detoxify environmental pollutants, such as metals (Cu, Fe, Mo, Zn and others). This behavior is possible because plants uptake metals and they are essential for plant health (Ye et al. 2017; Kazemalilou et al. 2020).

Das and Chandran (2011) reported there are two main routes of microbial remediation technology in degradation of petroleum hydrocarbon contaminants in soil: the biostimulation and the bioaugmentation. Biostimulation is characterized by enhancing the quality of soil by nutrients or other growth-limiting cosubstrates addition for the native microbes. Bioaugmentation is a pool of specific microbes (tolerate multiple heavy metals) that are added to native microflora to remove target pollutants (Das and Chandran 2011; Gao et al. 2014).

Many environmental factors such as physical, chemical and biological nature influence the ability of a microbial system to biodegrade a molecule. In this sense, the following can be highlighted as the main physical factors: the nature of the matrix (soil, water or sediment), temperature and light. For example, in the regions with low average annual temperatures, the metabolic activity of microorganisms can be reduced which decreases the degradation rate of the pollutants in those areas (Cooney et al. 1985; Das and Chandran, 2011). As for chemical factors, it should be noted that the chemical composition of the matrix, such as pH, humidity, dissolved oxygen content, redox potential of the medium and the pollutant chemical structure, are also important factors. In this case, for example, the presence of heavy metals can interact with enzymes produced by microorganisms, inhibiting their activity and, consequently, their degradative capacity (Cooney et al. 1985; Delangiz et al. 2020).

TABLE 4.2

Some Microbes Used in Bioremediation Application in Different Contaminated Environments

	Contaminant	Contaminating Resource	Microbes	References
Heavy metals	Mercury (Hg)	Hospital waste, fluorescent lamps, electrical industry	*Pseudomonas putida, Geobacter sulfurreducens*	Zhang et al. (2012); Schaefer et al. (2011)
	Cadmium (Cd) and Copper (Cu)	Soil	*Clitocybe maxima*	Liu et al. (2015)
	Copper (Cu), Zinc (Zn) and Lead (Pb)	Soil	Microbial consortium	Xu et al. (2013)
Organic pollutants	Carbon nanotubes (CNTs), graphene (GRA) and their derivatives	Drug carriers, electronics, biosensors, sorbents, and fuel cells	*Acidithiobacillus ferrooxidans* CFMI-1, *Burkholderia kururiensis, Delftia acidovorans, and Stenotrophomonas maltophilia*	Chen et al. (2017)
	2,4,5-trichlorophenol (TCP)	Soil	*Clitocybe maxima*	Liu et al. (2015)
	Naphthalene, phenanthrene, pyrene and crude oil	Soil	Microbial consortium	Xu et al. (2013)

There are a large number of microorganisms that biodegraded the pollutants present in the environment. According to Das and Chandran (2011), hydrocarbons are primarily taken up by bacteria, yeast and fungi. On the other hand, some researchers reported that mixed population could be required to degrade a crude oil in soil, for example, because of its complex mixtures of hydrocarbons (Cooney et al. 1985; Gao et al. 2014; Chen et al. 2015). Table 4.2 provides some microbes used in bioremediation examples with toxic chemicals and their contaminating resources.

4.6 IDENTIFICATION OF MICROORGANISMS THAT PRODUCE BIOMOLECULES BY MOLECULAR TECHNIQUES

Factors such as temperature, availability of nitrogen and organic carbon in the soil are closely correlated with the microbial community of the habitat, both in taxonomic levels and in functional terms of the gene. Temperature is one of the abiotic factors that interferes most with genetic functionality, significantly altering the microbial community resulting in the succession of forests (Cong et al. 2015). In order to understand this dynamic, it is essential to identify and characterize the microbial diversity present in the soil, a task that is not easy, above all, due to factors such as the isolation and limited in-vitro growth of most microorganisms. It is believed that 1g of soil has approximately 10^9 cells and only 1% of these isolates can be cultivated by classical laboratory techniques (Van Elsas and Rutgers 2006; Davis, Joseph and Jansen 2005).

The greatest difficulty in identifying and classifying microorganisms in the soil lies in the fact that there is little knowledge about the species' identity, which confers limitations in classifying them in taxonomic standards (Stefanis 2013). Modern techniques for sequencing the microbial genome brought about a great revolution in the study of the taxonomy of soil microorganism populations (Lozupone and Knight 2008). Amplicons from the ITS region (Internal Transcribed Spacer), 16s rDNA region in prokaryotes and the large subunit (LSU) of ribosomal RNA genes in eukaryotes, have been used frequently in the identification and characterization of microbial diversity in soil in forests.

The most promising technique for analyzing soil microbial diversity is metagenomics. In this technique, genetic material can be recovered from microbial cells separated from the soil with subsequent cell lysis and DNA isolation for analysis. Another way is the direct lysis of the cells contained in the soil with the recovery of the genetic material. In both cases, the DNA is fragmented, linked into a cloning vector, and finally introduced into recombinant vectors that will amplify the genetic material used in the identification of new useful sequences (Daniel 2005). Although it is a promising technique, metagenomics exhibits difficulties regarding the coextraction of interfering substances, as is the case with humic extracts that interfere with enzymes used in the digestion and amplification of genetic material.

Microarray (or microchip) technology has been widely investigated as a tool in the study of gene expression and regulation on a genomic scale (Wu et al. 2001). In this technique, DNA samples (or probes) are immobilized on the surface of a slide in which they hybridize with RNA extracted from a biological sample marked with fluorophores. The greater the amount of RNA-DNA binding, the greater the expression of the gene under study. This methodology is useful because it indirectly assesses the level of gene expression; however, studies indicate a sensitivity of 100–10,000 times lower when comparing conventional PCR techniques (Daniel 2005).

Currently, the biggest challenge of modern molecular biology techniques is to characterize the diversity of dynamics of soil microbial communities that vary widely in space and time. The use of a genomic library of isolated microorganisms and mainly of a metagenomic library associated with new techniques that monitor gene expression in communities can bring new understandings about the diversity and functional attributes of the soil microbial community, especially in forest management areas.

4.7 CONCLUDING REMARKS

Although the high diversity of microbial is found in the terrestrial rhizosphere, knowledge about the potential benefits of microorganism–plant interactions is limited, especially in the agroforestry area. The biomolecules synthesized by these microorganisms can influence plant development through the production of phytohormones, nutrient cycling, and even competitive inhibition against phytopathogens. In addition, some studies suggest the use of microbial nanoparticles as compounds that influence biotic interactions, such as the formation of mycorrhizae. Modern molecular studies have made it possible to identify and characterize metabolic pathways of different biomolecule-producing microorganisms and thus use them as commercial bioinoculants. Understanding the dynamics and physiology of the microorganism–plant interaction is fundamental for the development of a healthy, productive, and commercially profitable agroforestry system.

REFERENCES

ABM—Advanced Biological Market. 2020. Excalibre-SA for Soybenans. https://www.abm1st.com/products/soybeans/excalibre-sa-for-soybeans/ (accessed October 1, 2020).

Aleksandrowicz-Trzcińska, M., J. Olchowik, M. Studnicki, and A. Urban. 2019. Do silver nanoparticles stimulate the formation of ectomycorrhizae in seedlings of pedunculate oak (*Quercus robur L.*)? *Symbiosis* 79: 89–97.

Araujo, A. S. F., L. F. C. Leite, B. F. Iwata, M. A. Lira Jr., G. R. Xavier, and M. V. B. Figueiredo. 2012. Microbiological process in agroforestry systems. A review. *Agron Sustain Dev* 32(1): 215–26.

Arora, N. K., T. Fatima, I. Mishra, and S. Verma. 2020. Microbe-based Inoculants: role in next green revolution. In *Environmental Concerns and Sustainable Development*, eds. V. Shukla and N. Kumar, 191–246. Singapore: Springer.

Aslantaş, R., R. Çakmakçi, and F. Şahin. 2007. Effect of plant growth promoting rhizobacteria on young apple tree growth and fruit yield under orchard conditions. *Sci Hortic* 111(4): 371–377.

Barea, J. M. 2015. Future challenges and perspectives for applying microbial biotechnology in sustainable agriculture based on a better understanding of plant-microbiome interactions. *J Soil Sci Plant Nutr* 15(2): 261–82.

Basf-Corporation. 2020a. Vault® HP, inoculante para soja. https://agriculture.basf.com/ar/es/proteccion-de-cultivos-y-semillas/productos/vault-hp.html (accessed October 1, 2020).

Basf-Corporation. 2020b. Adhere® 60, inoculante biológico para a cultura da soja. https://agriculture.basf.com/br/pt/protecao-de-cultivos-e-sementes/produtos/adhere-60.html (accessed October 1, 2020).

Bennett, J. A., H. Maherali, K. O. Reinhart, Y. Lekberg, M. M. Hart, and J. Klironomos. 2017. Plant-soil feedbacks and mycorrhizal type influence temperate forest population dynamics. *Science* 355: 181–84.

Bhattacharyya, P. N., and D. K. Jha. 2012. Plant growth-promoting rhizobacteria (PGPR): emergence in agriculture. *World J Microbiol Biotechnol* 28: 1327–50.

Brundrett, M. C. 2002. Coevolution of roots and mycorrhizas of land plants. *New Phytol* 154: 275–304.

Cannel, E., and M. Moo-Young. 1980. Solidstate fermentation systems. *Process Biochem* 15(4): 24–29.

Chanway, C. P. 1997. Inoculation of tree roots with plant growth promoting rhizobacteria: An emerging technology for reforestation. *For Sci* 43: 99–112.

Chen, M., P. Xu, G. Zeng, C. Yang, D. Huang, and J. Zhang. 2015. Bioremediation of soils contaminated with polycyclic aromatic hydrocarbons, petroleum, pesticides, chlorophenols and heavy metals by composting: Applications, microbes and future research needs. *Biotechnol Adv.* 33: 745–55.

Chen, M., X. Qin, and G. Zeng. 2017. Biodegradation of carbon nanotubes, graphene, and their derivatives. *Trends Biotechnol.* 35: 836–46.

Cong, J., Y. Yang, X. Liu, H. Lu, X. Liu, J. Zhou, D. Li, H. Yin, J. Ding, and Y. Zhang. 2015. Analyses of soil microbial community compositions and functional genes reveal potential consequences of natural forest succession. *Sci Rep* 5: 10007. https://doi.org/10.1038/srep10007.

Cooney, J.J., S. A. Silver, and E. A. Beck. 1985. Factors influencing hydrocarbon degradation in three freshwater lakes. *Microbial Ecology* 11(2): 127–37.

Cruz, P. C., N. P. Banayo, S. R. Marundan, A. M. Magnaye, and D. J. Lalican. 2012. Bio-inoculant and foliar fertilizer in combination with soil applied fertilizer on the yield of lowland rice. *Philipp J Crop Sci* 37: 85–94.

Cui, H., P. Zhang, W. Gu, and J. Jiang. 2009. Application of anatasa TiO_2 sol derived from peroxotitannic acid in crop diseases control and growth regulation. *NSTI-Nanotech* 2: 286–289.

Dangi, A. K., B. Sharma, R. T. Hill, and P. Shukla. 2018. Bioremediation through microbes: Systems biology and metabolic engineering approach. *Crit Rev Biotechnol* 39(1): 79–98.

Daniel, R. 2005. The Metagenomics of Soil. *Nat Rev Microbiol* 3(6): 470–8.

Das, N., and P. Chandran, 2011. Microbial degradation of petroleum hydrocarbon contaminants: an overview. *Biotechnol Res Int* 11: 1–13 http://doi:10.4061/2011/941810. Article ID 941810.

Davis, K. E. R., S. J. Joseph, and P. H. Jansen. 2005. Effects of growth medium, inoculum size and incubation on culturability and isolation of soil bacteria. *Appl Environ Microbiol* 71: 826–34.

Delangiz, N., M. B. Varjovi, B. A. Lajayer, and M. Ghorbanpour. 2020. Beneficial microorganisms in the remediation of heavy metals. In *Molecular Aspects of Plant Beneficial Microbes in Agriculture*, eds. V. Sharma, R. Salwan, L. K. T. Al-Ani, 417–23. London: Academic Press.

Dilnashin, H., H. Birla, T. X. Hoat, H. B. Singh, S. P. Singh, and C. Keswani. 2020. Applications of agriculturally important microorganisms for sustainable crop production. In *Molecular Aspects of Plant Beneficial Microbes in Agriculture*, eds. V. Sharma, R. Salwan, L. K. T. Al-Ani, 403–15. London: Academic Press.

Feng, Y., X. Cui, S. He, G. Dong, M. Chen, J. Wang, and X. Lin. 2013. The role of metal nanoparticles in influencing arbuscular mycorrhizal fungi effects on plant growth. *Environ Sci Technol* 47(16): 9496–9504.

Gao, Y. C., S. H. Guo, J. N. Wang, D. Li, H. Wang, and D. H. Zeng. 2014. Effects of different remediation treatments on crude oil contaminated saline soil. *Chemosphere*. 117: 486–493.

Gildon, A., and P. B. Tinker. 1983. Interactions of vesicular arbuscular mycorrhizal infections and heavy metals in plants. I. The effect of heavy metals on the development of vesicular arbuscular mycorrhizas. *New Phytol* 95: 247–61.

Guilger-Casagrande, M., and R. de Lima. 2019. Synthesis of silver nanoparticles mediated by fungi: A review. *Front Bioeng Biotechnol* 7: 287. https://doi.org/10.3389/fbioe.2019.00287.

Gupta, G., S. S. Parihar, N. K. Ahirwar, S. K. Snehi, and V. Singh. 2015. Plant growth promoting rhizobacteria (PGPR): Current and future prospects for development of sustainable agriculture. *J Microbial Biochem Technol* 7: 96–102.

Heredia-Acuña, C., J. J. Almaraz-Suarez, R. Arteaga-Garibay, R. Ferrera-Cerrato, and D. Pineda-Mendoza. 2019. Isolation, characterization and effect of plant-growth-promoting rhizobacteria on pine seedlings (*Pinus pseudostrobus* Lindl.). *J For Res* 30: 1727–34. https://doi.org/10.1007/s11676-018-0723-5.

Herrmann, L., M. Atieno, L. Brau, and D. Lesueur. 2015. Microbial quality of commercial inoculants to increase BNF and nutrient use efciency. In *Biological Nitrogen Fixation*, ed. F. J. de Bruijn, 1031–40. Blackwell, UK; Hoboken, NJ: Wiley. https://doi.org/10.1002/9781119053095.ch101.

Ijdo, M., S. Cranenbrouck, and S. Declerck. 2011. Methods for large-scale production of AM fungi: past, present, and future. *Mycorrhiza* 21: 1–16. https://doi.org/10.1007/s00572-010-0337-z.

Iravani, S., and B. Zolfaghari. 2013. Green synthesis of silver nanoparticles using pinus eldarica bark extract. *Biomed Res Int*, Article ID 639725. https://doi.org/10.1155/2013/639725.

Jansa, J., F. A. Smith, and S. E. Smith. 2008. Are there benefits of simultaneous root colonization by different arbuscular mycorrhizal fungi?. *New Phytol* 177(3):779–89.

Jayaseelan, C., A. A. Rahuman, A. V. Kirthi, S. Marimuthu, T. Santhoshkumar, A. Bagavan, K. Gaurav, L. Karthik, and K. V. Rao. 2012. Novel microbial route to synthesize ZnO nanoparticles using *Aeromonas hydrophila* and their activity against pathogenic bacteria and fungi. *Spectrochim Acta A Mol Biomol Spectrosc* 90: 78–84.

Kazemalilou, S., N. Delangiz, B. A. Lajayer, and M. Ghorbanpour. 2020. Insight into plant-bacteria-fungi interactions to improve plant performance via remediation of heavy metals: An overview. In *Molecular Aspects of Plant Beneficial Microbes in Agriculture*, eds. V. Sharma, R. Salwan, L. K. T. Al-Ani, 123–32. London: Academic Press.

Liu, H., S. Guo, K. Jiao, J. Hou, H. Xie, and H. Xu. 2015. Bioremediation of soils co-contaminated with heavy metals and 2,4,5-trichlorophenol by fruiting body of *Clitocybe maxima*. *J Hazard Mater* 294: 121–27.

Lozupone, C. A., and R. Knight. 2008. Species divergence and the measurement of microbial diversity. *FEMS Microbiol Rev* 32(4): 557–78.

Lu, N., X. Zhou, M. Cui, M. Yu, J. Zhou, Y. Qin, and Y. Li. 2015. Colonization with arbuscular mycorrhizal fungi promotes the growth of *Morus alba* L. Seedlings under greenhouse conditions. *Forests* 6: 734–47.

Mafia, R. G., A. C. Alfenas, E. M. Ferreira, D. H. B. Binoti, G. M. Ventura, and A. H. Mounteer. 2009. Root colonization and interaction among growth promoting rhizobacteria isolates and eucalypts species. *Rev Árvore 33*(1): 1–9. https://doi.org/10.1590/S0100-67622009000100001.

Mao, Z., A. Corrales, K. Zhu, Z. Yuan, F. Lin, J. Ye, Z. Hao, and X. Wang. 2019. Tree mycorrhizal associations mediate soil fertility effects on forest community structure in a temperate forest. *New Phytol* 223: 475–86.

Market Data Forecast. 2020. Microbial Soil Inoculants Market. https://www.marketdata-forecast.com/market-reports/microbial-soil-inoculants-market (accessed October 25, 2020).

Marx, D.H. 1980. Ectomycorrhizal fungus inoculations: A tool for improving forestation practices. In *Tropical Mycorrhiza Research*, ed. P. Mikola, 13–71. London: Oxford University Press.

Méndez-Bravo, A., E. M. Cortazar-Murillo, E. Guevara-Avendaño, O. Ceballos-Luna, B. Rodríguez-Haas, A. L. Kiel-Martínez, O. Hernández-Cristóbal, J. A. Guerrero-Analco, and F. Reverchon. 2018. Plant growth-promoting rhizobacteria associated with avocado display antagonistic activity against *Phytophthora cinnamomi* through volatile emissions. *PLoS ONE* 13(3): e0194665. https://doi.org/10.1371/journal.pone.0194665.

Mordor Intelligence. 2020. Global Biofertilizers Market-Growth, Trends and Forecast. (2020–2025). https://www.mordorintelligence.com/industry-reports/global-biofertilizers-market-industry (accessed October 13, 2020).

NTS—Nutri Tech Solutions. 2020. Nutri-Life Root-Guard™. https://www.nutri-tech.com.au/factsheets/root-guard.pdf (accessed October 1, 2020).

O'Callaghan, M. 2016. Microbial inoculation of seed for improved crop performance: issues and opportunities. *Appl Microbiol Biotechnol* 100: 5729–5746.

Parnell, J.J., R. Berka, H. A. Young, J. M. Sturino, Y. Kang, D. M. Barnhart, and M. V. Di Leo. 2016. From the lab to the farm: An industrial perspective of plant beneficial microorganisms. *Front Plant Sci* 7: 1110. https://doi.org/10.3389/fpls.2016.01110.

Polischuk, S., G. Fadkin, D. Churilov, V. Churilova, and G. Churilov. 2019. The stimulating effect of nanoparticle suspensions on seeds and seedlings of Scotch pine (*Pínus sylvéstris*). *IOP Conf Ser: Earth Environ Sci* 226: 012020.

Pramanik, P., P. Krishnan, A. Maity, N. Mridha, A. Mukherjee, and V. Rai. (2020) Application of nanotechnology in agriculture. In *Environmental Nanotechnology*, eds. N. Dasgupta, S. Ranjan, E. Lichtfouse, v. 4: 317–348. Cham: Springer.

Raaijmakers, J. M., and B. J. J. Lugtenberg, 2013. Perspectives for rhizosphere research. In *Molecular Microbial Ecology of the Rhizosphere*, ed. F. J. de Bruijn, 1227–32. Hoboken, NJ: Wiley Blackwell. https://doi.org/10.1002/9781118297674.ch118.

Rebah, F. B., D. Prévost, A. Yezza, and R. D. Tyagi. 2007. Agro-industrial waste materials and wastewater sludge for rhizobial inoculant production: A review. *Bioresour Technol* 98(18): 3535–46.

Rico, C. M., S. Majumdar, M. Duarte-Gardea, J. R. Peralta-Videa, and J. L. Gardea-Torresdey. 2011. Interaction of nanoparticles with edible plants and their possible implications in the food chain. *J Agric Food Chem* 59(8): 3485–98. https://doi: 10.1021/jf104517j.

Russo, A., G. P. Carrozza, L. Vettori, C. Felici, F. Cinelli, and A. Toffanin. 2012. Plant beneficial microbes and their application in plant biotechnology. In *Innovations in Biotechnology*, ed. E. C. Agbo, 57–72. InTechOpen, https://www.intechopen.com/books/innovations-in-biotechnology/plant-beneficial-microbes-and-their-application-in-plant-biotechnology.

Santos, M. S., M. A. Nogueira, and M. Hungria. 2019. Microbial inoculants: Reviewing the past, discussing the present and previewing an outstanding future for the use of beneficial bacteria in agriculture. *AMB Express*, 9(1): 205. https://doi.org/10.1186/s13568-019-0932-0.

Santos, R. F., S. P. Cruz, G. R. Botelho and A. V. Flores. 2018. Inoculation of *Pinus taeda* seedlings with plant growth-promoting rhizobacteria. *Floresta Amb* 25(1): e20160056. http://dx.doi.org/10.1590/2179-8087.005616.

Schaefer, J. K., S. S. Rocks, W. Zheng, L. Liang, B. Gu, and F. M. Morel. 2011. Active transport, substrate specificity, and methylation of Hg (II) in anaerobic bacteria. *Proc. Natl. Acad. Sci.* 108(21): 8714–19.

Schneider, A. 1892. Observations on some American Rhizobia *Torrey Bot Club* 19(7): 203–18.

Servin, A., W. Elmer, A. Mukherjee, R. D. Torre-Roche, H. Hamdi, J. C. White, P. Bindraban, and C. Dimkpa. 2015. A review of the use of engineered nanomaterials to suppress plant disease and enhance crop yield. *J Nanopart Res* 17: 92. https://doi.org/10.1007/s11051-015-2907-7.

Shang, Y., M. K. Hasan, G. J. Ahammed, M. Li, H. Yin, and J. Zhou. 2019. Applications of nanotechnology in plant growth and crop protection: A review. *Molecules* 24(14): 2558. https://doi.org/10.3390/molecules24142558.

Sharma, V., A. Sharma, and R. Salwan. 2020. Overview and challenges in the implementation of plant beneficial microbes. In *Molecular Aspects of Plant Beneficial Microbes in Agriculture*, eds. V. Sharma, R. Salwan, L. K. T. Al-Ani, 1–18. London: Academic Press. https://doi.org/10.1016/B978-0-12-818469-1.00001-8.

Składanowski, M., M. Wypij, D. Laskowski, P. Golińska, H. Dahm, and M. Rai. 2017. Silver and gold nanoparticles synthesized from *Streptomyces* sp. isolated from acid forest soil with special reference to its antibacterial activity against pathogens. *J Clust Sci* 28: 59–79.

Souza, R., A. Ambrosini, and L. P. M. Passaglia, 2015. Plant growth-promoting bacteria as inoculants in agricultural soils. *Genet Mol Biol* 38(4): 401–19. https://doi.org/10.1590/S1415-475738420150053>.

Stefanis, C., A. Alexopoulos, C. Voidarou, S. Vavias, and E. Bezirtzoglou. 2013. Principal methods for isolation and identification of soil microbial communities. *Folia Microbiol* 58(1): 61–8.

Thorley, R. M. S., L. L. Taylor, S. A. Banwart, J. R. Leake, and D. J. Beerling. 2015. The role of forest trees and their mycorrhizal fungi in carbonate rock weathering and its significance for global carbon cycling. *Plant Cell Environ* 38(9): 1947–61.

Unnikrishnan, G., and R. Vijayaraghavan. 2019. Utilization of liquid fertilizers for Agro-Industrial waste management and reducing challenges through Nano-encapsulation-A review. *Indian J Agric Res* 53(6): 641–45.

Van Elsas, J. D., and M. Rutgers. 2006. Estimating soil microbial diversity and community composition. In *Microbiological methods for assessing soil quality*, eds. J. Bloem, D. W. Hopkins, A. Benedetti. Cambridge: Cabi.

Wang, B., and Y.-L. Qiu. 2006. Phylogenetic distribution and evolution of mycorrhizas in land plants. *Mycorrhiza* 16(5): 299–363.

Winagraski, E., G. Kaschuk, P. H. R. Monteiro, G. Auer, and A. R. Higa. 2019. Diversity of arbuscular mycorrhizal fungi in forest ecosystems of brazil: A review. *Cerne* 25(1): 25–35.

Wu, L., D. K. Thompson, G. Li, R. A. Hurt, J. M. Tiedje, and J. Zhou. 2001. Development and evaluation of functional gene arrays for detection of selected genes in the environment. *Appl Environ Microbiol* 67(12): 5780–90.

Xu, N., M. Bao, P. Sun, and Y. Li. 2013. Study on bioadsorption and biodegradation of petroleum hydrocarbons by a microbial consortium. *Bioresour Technol* 149: 22–30.

Ye, S., G. Zeng, H. Wu, C. Zhang, J. Dai, J. Liang, J. Yu, et al. 2017. Biological technologies for the remediation of co-contaminated soil. *Crit Rev Biotechnol.* 37(8): 1062–76. https://doi: 10.1080/07388551.2017.1304357.

Zewde, B., A. Ambaye, J. III Stubbs, R. Dharmara. 2016. A review of stabilized silver nanoparticles—synthesis, biological properties, Characterization, and potential areas of applications. *JSM Nanotechnol Nanomed* 4(2): 1043.

Zhang, W., L. Chen, and D. Liu. 2012. Characterization of a marine-isolated mercury-resistant *Pseudomonas putida* strain SP1 and its potential application in marine mercury reduction. *Appl Microbiol Biotechnol* 93(3): 1305–14. https://doi.org/10.1007/s00253-011-3454-5.

Zuluaga, M. Y. A., K. M. L. Milani, L. S. A. Gonçalves, and A. L. M. de Oliveira. 2020. Diversity and plant growth-promoting functions of diazotrophic/N-scavenging bacteria isolated from the soils and rhizospheres of two species of Solanum. *PLoS One* 15(1): e0227422. https://doi.org/10.1371/journal.pone.0227422.

5 Potential Applications of Plant Growth-Promoting Rhizobacteria in Soil

Debosmita Chakraborty[†], Puneet Kumar Singh[†],
Bhagyashree Puhan, Pratikhya Mohanty,
and Snehasish Mishra
Kalinga Institute of Industrial Technology
Deemed University, Bhubaneswar, India

CONTENTS

5.1 INTRODUCTION

Every plant has an association with a well-managed and well-structured community of microbes (Tewari and Arora 2014). In fact, most of the multi-cellular organisms have a relationship with the microbiome. Indeed, these presumably originated before plants colonized the land (Compant et al. 2019). Since the dawn of evolution,

---

[†] Equal first author contribution

DOI: 10.1201/9781003110477-5

this community of microbes has been related to land plants so that they can aid these preliminary terrestrial plants for problems like nutrient accessibility, disease-causing organisms, and novel conditions that were generally stressful. All the significant parts of a plant, including the root, stem, fruit, flower, and leaves, have an association with microbes like bacteria, fungi, and so on (Compant et al. 2019). One of the most crucial components of soil is the different genera of bacteria. Among all the community of microbes that have an association with plants, the microbial community inhabiting the roots have the highest population and complexity. This network of microbes associated to plant roots is referred to as the rhizomicrobiome. The involvement of these rhizomicrobes with the ecosystem of the soil and the biotic activities in it helps to make the soil active for the turnover of nutrients and sustainable to produce crops. These rhizomicrobes help in promoting the growth of plants via mobilization of soil nutrients, protection of plants from phytopathogens via their control or inhibition, production of many regulators for plant growth, sequestration of toxic heavy metals, and degradation of xenobiotics compounds to improve the structure of soil and for the bioremediation of polluted soil (Ahemad and Kibret 2014). Bacteria dwelling in and around plant roots (*Rhizobacteria*) have more versatility in transformation, mobilization, and solubilization of nutrients mechanism in contrast to the nutrients in bulk soil (Hayat et al. 2010). As a result, the deriving forces that prevail in the process of nutrient recycle in the soil are the rhizobacteria. Thus, these *Rhizobacteria* are vital for the fertility of the soil (Glick 2012). There has been a consistent symbiosis between plants and various types of soil microorganisms so that plants can grow and develop properly and get other such requirements. Microbial revitalization by PGPR, which includes different groups of bacteria that form colonies on plant roots and enhance their differentiation, has been an area currently utilized for enhanced agricultural productivity in several geographical regions. It is accomplished via two approaches, indirect and direct. In direct approach, the plant is provided with vital materials by bacteria in a direct way, which helps in plant growth promotion. Rhizo-remediation, biofertilization, and control of plant stress are certain techniques that help directly. The ability of absorbing water, temperature and other nutrients are some of the environmental factors that strongly restrict terrestrial plant's growth. By introducing PGPR, there is an improvement in the growth of plants due to increased access to water and nutrient uptake from a restricted pool of nutrients in soil via biofertilization. It also involves rhizo-remediation, which is the interaction of soil and microbes with the plant to enhance plant growth. Plant stress neutralization is another important factor of PGPR where it neutralizes both the stress types, biotic and abiotic. In this chapter, we discuss about the potential applications of PGPR in soil from both aspects.

5.2 RHIZOSPHERE

The term 'rhizosphere', proposed by a German plant physiologist and agronomist Lorenz Hiltner, portrays the collaboration between plants and roots. It is the zone around the roots of plants where a distinctive population of microbes dwell, and their striving is affected by the chemicals which the roots release. From the following year onwards, the definition of rhizosphere has gone through various modifications

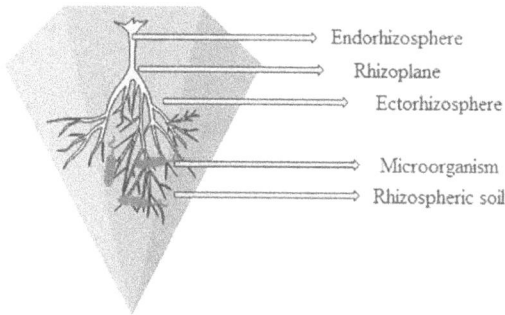

Endorhizosphere
Rhizoplane
Ectorhizosphere

Microorganism
Rhizospheric soil

FIGURE 5.1 Schematic presentation of the rhizosphere.

so that it can incorporate three zones which are characterized dependent on their relative vicinity to, and consequently impact from, the root. A schematic diagram of the different parts of rhizosphere is shown in Figure 5.1. As indicated by the intrinsic intricacy and the diversification in the root system of plants, the rhizosphere is anything but an area of determinable shape and size. In fact, for the rhizosphere, the physical, chemical, and biological properties vary along the root in both radial and longitudinal directions. Characterization of this zone in based on intensified biological activity on account of the root releasing chemicals for stimulation or inhibition of rhizomicrobiome. The dynamics and the complex nature of the rhizosphere region can be portrayed by the association between the plant, the soil, and the rhizomicrobiome. The plant interacts with microorganisms and fauna of various kinds around the area of the rhizosphere. Nitrogen fixation, mycorrhizal interaction, and pathogenic association are the various forms of interactions that take place between plants and other life forms in the rhizosphere area. Considering the ecology of the rhizosphere, success might be obtained by carrying out managerial approaches like biological control and bioremediation (De Luna et al. 2011).

5.3 PGPR IN PLANT GROWTH AND WELL-BEING

The rhizosphere is the dwell for various useful microorganisms that help in plant growth-promoting factors. Plant growth-promoting rhizobacteria or PGPR are a particular type of bacteria that form colonies in roots or any other part of plants and promote the growth of plants while protecting it from abiotic stresses and diseases by using a broad range of mechanisms (Table 5.1). The three main attributes that define a plant growth-promoting rhizobacteria are (a) ability to form colony in the roots or other parts of plants, (b) used as probiotic or bioinoculants for plant, and (c) ability to promote growth of plants (Gosal et al. 2017). On the basis of their functional activities, PGPR can be categorized as biofertilizers to increase the nutrient availability to plants, as phytostimulators that promote plant growth, via phytohormones, as phytoremediators that can degrade organic pollutants, and as biopesticides that can control disease by producing antibiotics and metabolites that are antifungal (Ahemad and Kibret 2014). Various physiological activities that may profoundly affect the development of plants, as well as plant health, are associated with the capacity to

TABLE 5.1
The Role of PGPR on Plant Well-Being

PGPR	Plant Growth-Promoting Characteristic	References
Pseudomonas aeruginosa	IAA, siderophores, HCN, ammonia,	(Ahemad and Khan 2012)
Mesorhizobium sp.	IAA, HCN and exopolysaccharides	(Ahemad and Khan 2010)
Acinetobacter sp.	IAA, phosphate solubilization, siderophores	(Rokhbakhsh-Zamin et al. 2011)
Pseudomonas sp.	Phosphate solubilization, IAA, Siderophore, HCN, biocontrol ppotentials	(Kumar et al. 2019)
Rhizobium sp. (lentil)	IAA, siderophores, HCN, ammonia, exo-polysaccharides	(Kumar et al. 2017; Kumar et al. 2019)
Stenotrophomonas maltophilia	Nitrogenase activity, phosphate solubilization, IAA, ACC deaminase	Ahemad and Kibert 2014
Paenibacillus polymyxa	IAA, siderophores	(Phi et al. 2010)
Pseudomonas fluorescens	ACC deaminase, phosphate solubilization	(Shaharoona et al. 2008)
Bradyrhizobium sp.	IAA, siderophores, HCN, ammonia, exo-polysaccharides	(Ahemad and Khan 2010)
Rhizobium phaseoli	IAA	(Zahir et al. 2010)
Burkholderia	ACC deaminase, IAA, siderophore, heavy metal solubilization, phosphate solubilization	(Jiang et al. 2008)
Pseudomonas aeruginosa 4E	Siderophores	(Naik and Dubey 2011)
Klebsiella oxytoca	IAA, phosphate solubilization, nitrogenase activity	(Jha and Kumar 2007)

promote plant growth. A significant inhibition in almost all agricultural ecosystems while producing saleable yields are the soil-borne plant pathogens (Dardanelli 2010). Various negative impacts are caused due to the expanded application of chemicals on plants. These negative impacts include developing resistance towards pathogens when particular chemical agents are applied and also their non-target natural effects (Dardanelli 2010). Nevertheless, there has been a need to find substitutes for such chemical agents due to the developing expense of pesticides, especially in less prosperous locales of the world, and the increasing customer interest in food without pesticide. For quite a while, there have been descriptions, characterization, and tests on a large variety of *Rhizobacteria* to understand whether they can function as an agent for bio-controlling diseases brought about by soil-borne plant microbes (Dardanelli 2010). Different chemicals like antibiotics, lytic enzymes, iron-chelating siderophores, detoxication enzymes, and biocidal volatiles are synthesized to mediate various biocontrol activities of the PGPR (Dardanelli 2010). The strains of PGPR that belong to genera *Rhizobium, Arthrobacter, Bacillus, Azotobacter, Azospirillum, Flavobacterium, Acinetobacter, Enterobacter, Erwinia, Serratia, Alcaligenes, Beijerinckia, Burkholderia* are presently being utilized worldwide to upgrade agricultural profitability (Bharti et al. 2013; Kumar et al. 2019).

5.4 PGPR AND PLANT GROWTH PROMOTION MECHANISMS

Production of different substances leads to change in the entire microbial network in the niche of the rhizosphere, which causes the promotion of plant growth mediated by PGPR (Ahemad and Kibret 2014). In general, there are two methods in which PGPR directly promote the growth of plants. In the first method, PGPR facilitates the acquisition of various resources like phosphorus, nitrogen, and essential minerals. In the second method, PGPR modulates the levels of hormones in plants. PGPR can also promote plant growth in an indirect method by functioning as biocontrol agents and lead to a decrease in the inhibitory effects of several pathogens on the growth of plants (Figure 5.2) (Ahemad and Kibret 2014; Glick 2012).

5.4.1 NITROGEN FIXATION

For the growth and productivity of plants, the most essential nutrient is nitrogen. Even though almost 78% of the atmosphere is nitrogen, it is not directly available for plants during their growth phase. The conversion of atmospheric nitrogen into a particular form that can be utilized by plants is done by the process of biological nitrogen fixation. In this process, there is a transformation of nitrogen into ammonia with the help of nitrogen-fixing microbes that utilize nitrogenase, a complex enzyme (Ahemad and Kibret 2014). As a matter of fact, the process of biological nitrogen fixation represents roughly 66% of the N_2 fixed around the world, while the remainder of the nitrogen is synthesized in the industry by the Haber–Bosch technique

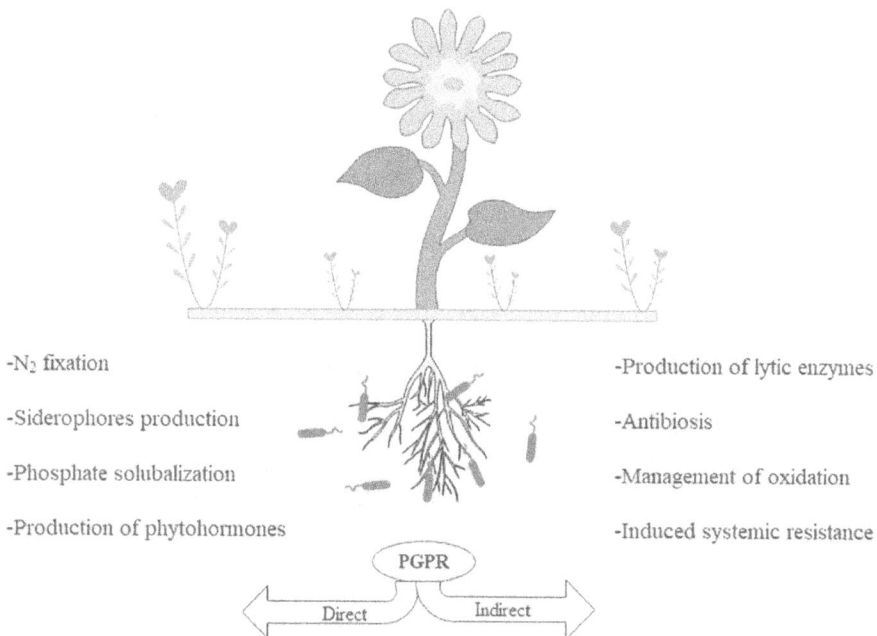

FIGURE 5.2 Role of PGPR in direct and indirect mechanisms for plant growth promotion.

(Ahemad and Kibret 2014). In general, the process of biological fixation of nitrogen takes place in moderate temperature with the help of nitrogen-fixing microbes, which have a wide distribution in nature (Ahemad and Kibret 2014). Besides, the process of biological nitrogen fixation accounts for a more monetarily advantageous and naturally stable option in contrast to chemical fertilizers. Commonly, nitrogen-fixing microbes are classified into two main forms:

1. Symbiotic form of N_2 fixers, which have a symbiotic relationship with leguminous plants, such as rhizobia, as well as plants that are non-leguminous, such as Frankia.
2. Non-symbiotic form of N_2 fixers, which are either free-living, associative, or endophytes-like cyanobacteria (*Nostoc and Anabaena*), *Azoarcus, Azotobacter, Azospirillum*, and *Gluconoacetobacter diazotrophicus* (Ahemad and Kibret 2014).

Involvement of a complex interplay must be present within symbiont and host to establish the symbiotic relationship bringing about the development of nodules within which the rhizobia form colonies as intracellular symbionts (Ahemad and Kibret 2014). Diazotrophs is another name for nitrogen-fixing PGPR, which can interact with the host plant in a non-obligate manner. The nitrogenase complex, a complex enzyme, helps in the process of nitrogen-fixing.

The nitrogenase enzyme structure suggests that it is of two components, dinitrogenase reductase (Fe-protein) and dinitogenase (Mo-Fe protein). Electrons having a high reducing capacity are provided by the enzyme dinitrogenase reductase. On the other hand, these electrons are utilized by the enzyme dinitrogenase for the reduction of nitrogen to ammonia. There are three different systems of nitrogen-fixing: Mo-nitrogenase, V-nitrogenase, and Fe-nitrogenase. Based on the structure, the system of nitrogen-fixing is different for different genera of bacteria. Most of the nitrogen-fixing process is performed by the action of the molybdenum nitrogenase enzyme, which is present in all diazotrophs. Both the symbiotic system and the free-living system contain the nitrogen-fixing genes known as *nif* genes. The *nif* or nitrogenase genes consist of several genes, including structural genes, a gene required for the biosynthesis of iron-molybdenum cofactor, genes for donating electrons, genes which can regulate the synthesis of enzyme and its function, genes needed for activating the iron protein (Ahemad and Kibret 2014). The nif genes are generally arranged in the form of clusters of about 20–24 kb consisting of 7 operon coding for 20 distinct proteins (Ahemad and Kibret 2014; Glick 2012). *nifDK* and *nifH* are the two genes that code for the two-component proteins of the complex enzyme molybdenum nitrogenase. The protein *NifDK* is heteromeric, made up of two diameters, 'a' and 'b'. These two diameters 'a' and 'b' are connected via a twofold symmetry to form 'a2b2' (Ahemad and Kibret 2014). Each of the 'a' subunit of the active site of the *NifDK* protein contains a single iron-molybdenum cofactor (FeMo-co) (Rubio and Ludden 2008). In case of *Rhizobium*, a low concentration of oxygen has a direct effect on the symbiotic activation of the nif genes, which, therefore, is controlled by a different set of genes known as fix-genes. For both the system of nitrogen fixation, free-living and symbiotic, these fix-genes are quite common (Ahemad and Kibret 2014).

5.4.2 Phosphate Solubilization

The second most significant nutrient for limiting the growth of plants is phosphorus, which has an abundant availability in both inorganic and organic soils (Ahemad and Khan 2012; Khan et al. 2009). Regardless of huge store of P, the measure of forms accessible for plants is commonly low (Ahemad and Khan 2012). Most of the phosphorus present in the soil is insoluble that includes organic forms like phospho-monoesters, phosphotriesters, inositol phosphate, and inorganic minerals like apatite (Ahemad and Khan 2012; Glick 2012). Fertilizers containing phosphorus are regularly applied to agricultural fields for overcoming the lack of phosphorus in soils. Out of the total phosphatic fertilizer applied, only a lesser portion is absorbed by plants, and there is a conversion of the remaining portion into complexes that are not soluble in the soil. However, if the phosphatic fertilizers are regularly applied to the agricultural fields, it will be unwanted both economically and environmentally. Due to this, there has been a requirement to look for an alternative that is safe for the environment and financially feasible for enhancing agricultural production in soils having lower phosphorus (Ahemad and Khan 2012). In this regard, organisms that can solubilize phosphate, generally referred to as PSM or phosphate-solubilizing microorganisms, might be able to supply the absorbable forms of phosphorus for plants and henceforth be a feasible alternative to synthetic phosphatic fertilizers (Ahemad and Khan 2012; Khan et al. 2009). Among the different phosphate-solubilizing microorganisms that occupy the rhizosphere, the phosphate-solubilizing bacteria, also known as PSB(s), are regarded as suitable biofertilizers. This is because PSB(s) will be able to provide phosphorus from sources, which, in general, are inadequately accessible by different mechanisms (Ahemad and Khan 2012; Zaidi et al. 2009). According to reports, the most important phosphate solubilizing bacterias fall under the genera *Bacillus, Enterobacter, Flavobacterium, Pseudomonas, Serratia, Burkholderia, Azotobacter, Beijerinckia, Rhizobium, Erwinia,* and *Microbacterium* (Ahemad and Kibret 2014; Bhattacharyya and Jha 2012). In general, the molecular weights of organic acids that the different soil bacteria synthesize are low, because of which the inorganic phosphorus can be solubilized (Ahemad and Khan 2012; Zaidi et al. 2009).

Although the occurrence of PSB(s) are common in almost all soils, environmental factors like stress have a severe influence on their establishment and also their performances (Ahemad and Khan 2012; Ahemad and Kibret 2014). In any case, there have been reports suggesting that the utilization of PBS inoculation alone (Ahemad and Khan 2012; Ahemad and Kibret 2014) or in association with other rhizospheric microorganisms have advantageous impacts. Other than helping in phosphorus supply to plants, the growth of plants is also enhanced by the phosphate-solubilizing bacteria. To enhance plant growth, PSB(s) stimulate the capability of the BNF via the synthesis of significant substances for plant growth, which in turn enhances the availableness of other trace elements.

5.4.3 Production of Siderophores

Iron is an important nutrient for most organisms. Apart from some specific lactobacilli, iron is required by all the microbes known until now (Ahemad and Kibret 2014). In an oxygenated environment, iron is mainly present in the form of Fe^{3+} and may probably produce hydroxides and oxyhydroxides that are insoluble, which in turn will make it unavailable to microbes as well as plants (Rajkumar et al. 2010). In general, iron is acquired by bacteria by secreting iron chelators with low molecular mass, known as siderophores, having high association constants to form iron complexes. Almost all siderophores are soluble in water and are categorized into two types, intracellular and extracellular siderophores. In general, rhizobacteria varies concerning the capacity to cross-utilize siderophores. One type of siderophores can utilize siderophores produced by rhizobacteria belonging to the same genus, known as homologous siderophores. The other type is capable of utilizing siderophores that are produced by rhizobacteria belonging to different genera, known as heterologous siderophores (Khan et al. 2009). For both Gram-positive and Gram-negative rhizobacteria, the Fe3+-siderophore complex has iron in Fe^{3+} form in it, which undergoes reduction to Fe^{2+}, and this Fe^{2+} is additionally released into the bacterial cells using a gating system through which the outer membranes and inner membranes of the bacterial cells are connected (Ahemad and Kibret 2014). In course of this process of reduction, the siderophore either undergoes destruction or recycling (Rajkumar et al. 2010). As a result, when the iron content is low, the siderophores can solubilize iron from various sources like organic compounds and minerals (Ahemad and Kibret 2014; Indiragandhi et al. 2008). Apart from iron, several other heavy metals like Zn, Cd, Al, Ga, In, Cu, Pb, and also radioactive elements like U and Np that pose a threat to the environment form stable complex with siderophores (Ahemad and Kibret 2014). There is an increase in the concentration of soluble metal when metal binds to siderophores (Ahemad and Khan 2012; Rajkumar et al. 2010). As a result, bacterial siderophores can aid in the reduction of stress that a high level of heavy metals in soil imposes on plants. Chelating and releasing iron, uptake of siderophore-Fe complexes directly, the reaction of ligand exchange are the various mechanisms utilizing which plant incorporate iron from bacteria.

5.4.4 PHYTOHORMONE

5.4.4.1 1-Aminocyclopropane-1-Carboxylate (ACC) Deaminase

In general, ethylene is a vital nutrient to help plants grow and develop normally (Ahemad and Khan 2012; Khalid et al. 2006). The production of this plant growth hormone occurs roughly within all plants. Ethylene production also occurs by various processes in the soil (biotic and abiotic) and is significant in prompting diverse physiological changes in plants (Ahemad and Kibret 2014). Besides its function as a plant growth regulator, ethylene also functions as a stress hormone. There is a significant increase in endogenous ethylene level under various stress conditions created by drought, salinity, water logging, pathogenicity, and heavy metals that has a negative impact on the overall growth of plant (Ahemad and Kibret 2014). For example, defoliation and various other processes of the cell are caused due to increased levels of ethylene that may prompt decreased performance of the crops (Bhattacharyya and Jha 2012). The level of ethylene is decreased by particular plant growth-promoting rhizobacteria that contains the enzyme 1-aminocyclopropane-1-carboxylate deaminase (ACC deaminase). This decrease in ethylene levels induces tolerance towards salt and reduces drought stress in plants, which further enhances the growth and development of plants (Ahemad and Khan 2012). Recently, there has been the identification of a wide variety of genera of bacterial strains containing the ACC deaminase enzyme. These genera include *Rhizobium, Ralstonia, Enterobacter, Bacillus, Alcaligenes, Achromobacter, Serratia, Burkholderia, Agrobacterium, Pseudomonas,* and *Azospirillum*(Ahemad and Kibret 2014; Kang et al. 2010). ACC, which is the precursor of ethylene, is absorbed and converted into 2-oxobutanoate and NH_3 by such rhizobacteria (Ahemad and Kibret 2014; Arshad, Saleem, and Hussain 2007). Impacts of phytopathogenic microbes like bacteria, fungi, virus, and resistance towards stress caused by heavy metals, insect predation, wounding, polyaromatic hydrocarbons, draft, radiation, high salt concentration, flooding, high intensity of light, high temperature are the various form of stresses from which the ACC deaminase-producing rhizobacteria protect a plant form (Ahemad and Khan 2010; Glick 2012). Consequently, elongation of the root, shoot growth promotion, improvement in uptake of N, P, K, and nodulation of rhizosphere are significant effects that can be noticed when seeds or roots of plants are inoculated with ACC deaminase producers (Ahemad and Khan 2010; Glick 2012).

5.4.4.2 Indole-3-Acetic Acid
Being the precursor of auxin, the formation of Indole-3-acetic acid (IAA) takes place in leaves that are either in the stage of development or are matured. Indole-3-acetic acid helps in the formation of lateral primordia in apical meristems for shoots as well as root (Kumar et al. 2019). Regulation of basipetal and acropetal transport of auxin was done for the coordination of the responses of root and shoot to stimuli of the environment, based upon the organ that detects the stimulus (Kumar et al. 2019). For maintenance of growth of root under saline conditions, there is a need for increased activity of auxin, which is regarded as an adaptive response for stress conditions of salinity and drought (Albacete et al. 2008; Kumar et al. 2019). For the prevention of the stress due to saline conditions, production of IAA is a comparatively common characteristics of PGPR (Dodd 2004; Kumar et al. 2019).

5.4.4.3 Cytokinin
The phytohormone cytokinin enhances cell division in roots and shoots of plants as well as has a role to play in the growth and differentiation of plants. It also has an effect of senescence of leaves, growth of axillary bud, and apical dominance (Kumar et al. 2019). Considering that cytokinin is a comparatively usual characteristic of PGPR, the selection of bacterial inoculants that produce high levels of cytokinin must be given more consideration for the possible enhancement of the salt stress (Dodd 2004; Kumar et al. 2019). Cytokinin is produced by *Bacillus subtilis*, which is capable of stimulating biomass of shoot of lettuce plant which grows in soil that is drying. This suggests substantial root-to-shoot signaling of cytokinin (Kumar et al. 2019). For the most part, there is the declination in the root-and-shoot growth and level of cytokinin in tomato plants.

5.5 ABIOTIC STRESS

The growth and development of plants are negatively affected by various abiotic stresses like drought, waterlogging, mineral toxicity, extreme temperatures, and salinity, which in turn have a negative influence on the quality of seed and all over the yield of plants. It has been anticipated that there will be a rise in the shortage of fresh water in the future, which will eventually lead to a rise in abiotic stresses. As a result, there is an immediate need for the development of particular varieties of crops that can adapt to abiotic stresses to guarantee the security of food and safety in the future. Roots act as the first line of defense for plants of abiotic stress. The possibility of survival for plants under stressful conditions will rise if the soil that holds the plant has good health and is diverse biologically. The disarrangement in the ratio of sodium and potassium in the plant cell cytoplasm is one of the essential reactions to abiotic stress like highly saline conditions. Certain phytohormones (like abscisic acid) assume a significant function concerning the resistance of plants to various abiotic stresses like drought, highly saline conditions, physical injury, and extreme temperatures (Table 5.2) (Ahemad and Khan 2012; Kumar et al. 2019).

TABLE 5.2
Role of PGPR in Managing Various Crop Stress Factors

PGPRs	Stress Factor	Effect on Indicator Parameters	Crop	References
Azospirillum	Salt	Plant height, shoot dry weight and grain yield, photosynthetic pigments and solute accumulation	Wheat	(Nia et al. 2012)
Azotobacter	Salt	Plant growth and potassium and phosphorus intake	Maize	(Rojas-Tapias et al. 2012)
Pseudomonas Aeruginosa	Salt	Plant growth	Sunflower	(Tewari and Arora 2014)
Piriformospora indica and *Azospirillum*	Salt	Seedling growth, solute accumulation (proline and sugars), and photosynthetic pigments	Wheat	(Zarea et al. 2012)
Xanthobacter autotrophicus BM13, Enterobacter aerogenes BM10, and *Bacillus brevis FK2*	Salt	Plant growth, yield, and mineral content	Eggplant	(Abd El-Azeem et al. 2012)
Burkholderia phytofirmans strain PsJN and *Enterobacter spp. FD17*	Drought	Growth, water status and photosynthetic activity	Maize	(Zahir et al. 2010)
Glomus deserticola	Salt and Heavy metal	Growth	Egg plant and Sudan grass	(Miransari 2011)
P. fluorescens, P. aeruginosa and *P. stutzeri*	Salt	Branching and leaf count	Tomato	(Tank and Saraf 2010)

5.5.1 Drought and Flood

The time frame during which a particular area receives precipitation less than the general level is referred to as drought. The decrease in groundwater or content of soil moisture, the reduction in the flow of the stream, the destruction of crops, and overall water deficiency are some outcomes caused by the absence of sufficient precipitation. Drought happens to be one of the most damaging abiotic stresses that expanded in power over the past decades and have an influence on the food security of the world. The duration of drought stress may go from short and medium to very extreme and delayed length, which in turn restricts the yield of crops. It has been anticipated that by the year 2050, growth of crops for over half of the Earth's cultivable land will face critical issues (Vurukonda et al. 2016). The turgor pressure and water potential of plants get influenced by drought to such an extent that the normal functions of plants get disrupted. As a result, there is an alteration in the morphology and physiology of plants (Rahdari and Hoseini 2012). Several studies have been done on the mechanism behind the reduction of growth in crops (wheat, rice, maize) due to drought conditions.

Some of the common growth factors for crops that get influenced by drought situation is the content of water and the fresh weight (Jaleel et al. 2009). Moreover, access to nutrients in the soil and their transportation is affected under drought conditions because water is the carrier of soil nutrients to plants via root. Consequently, the mass flow of nutrients soluble in water (like calcium, nitrate, magnesium, sulfate, silicon) and diffusion of nutrients, in general, undergoes a reduction under drought conditions (Farooq et al. 2012). During drought situations, ROS or reactive oxygen species (hydrogen peroxide, superoxide radicals, hydroxyl radicals) and defense by antioxidants get affected by free radicals, which leads to oxidative stress. In plants, peroxidation of internal lipid and degradation of membranes, lipids, proteins, and nucleic acid, can be caused due to increased ROS concentration. Nonetheless, a reduction in the chlorophyll content during conditions of drought was a side effect of photooxidation (Vurukonda et al. 2016).

5.5.2 Heavy Metals

The last few decades have experienced an increase in various anthropogenic activities like industrialization and use of modern technologies for agriculture. As a result of such anthropogenic activities, the environment has been contaminated by heavy metals which is toxic for living organisms (Miransari 2011; Naik and Dubey 2011). The use of fertilizers pesticides, compost wastes, release of heavy metals from mines and smelting industries has led to contamination of large areas by heavy metals. Heavy metals do exist naturally in the earth. However, it is a serious threat to the environment, if an excess amount of heavy metals is released into the Earth's environment as a result of anthropogenic activities (Singh et al. 2015). This is the scenario where the *Rhizobacteria* has a crucial role to play. Various strains of plant growth-promoting *Rhizobacteria* that have the capacity to reduce metals have been recognized. Bioremediation, which is the process of breakdown of environmental pollutants by microbes, relies upon the functionality of the rhizospheric microbes and the environmental conditions that supports their development. An eco-friendly, economical, and sustainable substitute to physicochemical remediation is the utilization of PGPR that have the capability to absorb heavy metallic ions. During the process of bioremediation of heavy metal contamination in soil, a crucial role is played by the plant growth-promoting rhizobacteria dwelling in the roots of plants. These bacteria break down the heavy metals by chelation of the metallic ions (Singh et al. 2015). Dual benefit can be obtained by applying PGPR to soil-bioremediation of heavy metals and promotion of plant growth. A variety of metabolites are secreted by PGPR for the purpose of soil remediation. Siderophores is one such metabolite, which is a small extracellular organic compound. It is produced when the iron is low to solubilize iron and help in its transportation. Although siderophores can specifically carry out chelation of iron, the high stability constant between metal and siderophore allow the binding of siderophores to other metals like actinides and heavy metals which are divalent. Siderophores have the capability to bind with several heavy metals like aluminium, copper, lead, arsenic, cadmium, magnesium, strontium, zinc, nickel, and cobalt. This characteristics of siderophores can be utilized to remove heavy metals from soil that are toxic in nature (Kumar et al. 2017).

5.5.3 SALINITY

Highly saline soil effects the physiological processes of plants because salt in any form is harmful for the health of plants. High salinity has a negative effect on the decomposition of residue, respiration and biodiversity of soil, activity of microbes in soil and nitrification-denitrification of soil and other related soil processes (Schirawski and Perlin 2018). When high amount of salt containing fertilizers is applied to soil, the productivity of the crop decreases (Rütting, Aronsson, and Delin 2018). An osmotic effect is imposed when fertilizers with high salt index is applied to the soil because of which water extraction from soil by the plants becomes difficult. The production of crops in soil containing high concentration of salt is also affected by certain farming practices. An increase in the water evaporation from surface of soil occurs due to ploughing and tilling of soil which increases the deposition of salts. Moreover, when irrigation water contains salt, it might lead to an increase in the salinity of the soil consequently decreasing the productivity of the soil (Mishra 2018). Various strategies and mechanisms have been developed by the ST-PGPR (Salt-tolerant plant growth-promoting *Rhizobacteria*) so that they can tolerate salinity stress. The various mechanisms include efflux systems, compatible solutes being formed and accumulated to balance external osmotic pressure, relative oxygen species (ROS) formation, and formation of secondary metabolites. In the presence of salinity stress, a number of metabolites and genes function to maintain plant–microbe interactions and cell integrity. However, not much is known about the mechanism by which ST-PGPR provide support to their symbiotic and also to themselves under salinity stress. Current research on ST-PGPR suggest that there is a huge scope of remediation and an improvement in the agricultural productivity that are under salinity stress (Egamberdieva et al. 2019).

5.6 CONCLUSION

The PGPR form colonies in roots or any other part of plants and promote the growth of plants while protecting it from abiotic stresses and diseases. Production of different substances leads to a change in the entire microbial network in the niche of the rhizosphere, which causes the promotion of plant growth mediated by PGPR. In general, nitrogen fixation, phosphate solubilization, siderophore production, and phytohormones are the mechanisms through which PGPR maintains plant well-being. Abiotic stress such as drought, salinity, mineral toxicity, excessive watering, and extreme temperatures (cold, frost, and heat) negatively impact the growth, development, yield, and seed quality of crop and other plants. The ACC deaminase, IAA and cytokinin are the examples of enzymes produced by PGPR for the protection and development of plant. The phyto-microbes and plant relationship are age old and result of coevolution on the Earth for their growth and development symbiotically.

REFERENCES

Abd El-Azeem, S.A., Elwan, M.W., Sung, J.K. and Ok, Y.S. 2012. Alleviation of salt stress in eggplant (*Solanum melongena* L.) by plant-growth-promoting *Rhizobacteria*. *Communications in Soil Science and Plant Analysis* 43(9): 1303–1315.

Ahemad, M. and Khan, M.S. 2010a. Influence of selective herbicides on plant growth promoting traits of phosphate solubilizing *Enterobacter asburiae* strain PS 2. *Research Journal of Microbiology* 5(9): 849–857.

Ahemad, M. and Khan, M.S. 2012. Alleviation of fungicide-induced phytotoxicity in greengram [*Vigna radiata* (L.) Wilczek] using fungicide-tolerant and plant growth promoting *Pseudomonas* strain. *Saudi Journal of Biological Sciences* 19(4): 451–459.

Ahemad, M. and Kibret, M. 2014. Mechanisms and applications of plant growth promoting *Rhizobacteria*: current perspective. *Journal of King Saud University-Science* 26(1): 1–20.

Albacete, A., Ghanem, M.E., Martínez-Andújar, C., Acosta, M., Sánchez-Bravo, J., Martínez, V., Lutts, S., Dodd, I.C. and Pérez-Alfocea, F. 2008. Hormonal changes in relation to biomass partitioning and shoot growth impairment in salinized tomato (*Solanum lycopersicum* L.) plants. *Journal of Experimental Botany* 59(15): 4119–4131.

Arshad, M., Saleem, M. and Hussain, S. 2007. Perspectives of bacterial ACC deaminase in phytoremediation. *TRENDS in Biotechnology* 25(8): 356–362.

Bharti, N., Yadav, D., Barnawal, D., Maji, D. and Kalra, A. 2013. Exiguobacterium oxidotolerans, a halotolerant plant growth promoting *Rhizobacteria*, improves yield and content of secondary metabolites in *Bacopa monnieri* (L.) Pennell under primary and secondary salt stress. *World Journal of Microbiology and Biotechnology* 29(2): 379–387.

Bhattacharyya, P.N. and Jha, D.K. 2012. Plant growth-promoting *Rhizobacteria* (PGPR): Emergence in agriculture. *World Journal of Microbiology and Biotechnology* 28(4): 1327–1350.

Compant, S., Samad, A., Faist, H. and Sessitsch, A. 2019. A review on the plant microbiome: ecology, functions, and emerging trends in microbial application. *Journal of Advanced Research* 19: 29–37.

Dardanelli, M.S. 2010. Benefits of plant growth-promoting *Rhizobacteria* and rhizobia in agriculture. In *Plant Growth and Health Promoting Bacteria*, eds. S.M. Carletti, N.S. Paulucci, D.B. Medeot, E.R. Caceres, F.A. Vita, M. Bueno, M.V. Fumero and M.B. Garcia, 1–20. Berlin: Springer.

De Luna, L.Z., Kennedy, A.C., Hansen, J.C., Paulitz, T.C., Gallagher, R.S. and Fuerst, E.P. 2011. Mycobiota on wild oat (*Avena fatua* L.) seed and their caryopsis decay potential. *Plant Health Progress* 12(1): 20.

Dodd, I.C. 2004. Will modifying plant ethylene status improve plant productivity in water-limited environments. In *Handbook and Abstracts for the 4th International Science Congress*, eds. A.A. Belimov, W.Y. Sobeih, V.I. Safronova, D. Grierson and W.J. Davies, 134. Brisbane: Australian Society of Agronomy Inc.

Egamberdieva, D., Wirth, S., Bellingrath-Kimura, S.D., Mishra, J. and Arora, N.K. 2019. Salt-tolerant plant growth promoting *Rhizobacteria* for enhancing crop productivity of saline soils. *Frontiers in Microbiology* 10: 2791.

Farooq, M., Hussain, M., Wahid, A. and Siddique, K.H.M. 2012. Drought stress in plants: an overview. *Plant Responses to Drought Stress* 1–33. https://doi.org/10.1007/978-3-642-32653-0_1.

Glick, B.R. 2012. Plant growth-promoting bacteria: mechanisms and applications. *Scientifica* 2012: 15.

Gosal, S.K., Kaur, J. and Kaur, J. 2017. Plant growth-promoting rhizobacteria: a probiotic for plant health and productivity. In *Probiotics and Plant Health*, eds. V. Kumar, M. Kumar, S. Sharma and R. Prasad, 589–600, Singapore: Springer. doi.org/10.1007/978-981-10-3473-2_27.

Hayat, R., Ali, S., Amara, U., Khalid, R. and Ahmed, I. 2010. Soil beneficial bacteria and their role in plant growth promotion: a review. *Annals of Microbiology* 60(4): 579–598.

Indiragandhi, P., Anandham, R., Madhaiyan, M. and Sa, T.M. 2008. Characterization of plant growth–promoting traits of bacteria isolated from larval guts of diamondback moth *Plutella xylostella* (Lepidoptera: Plutellidae). *Current Microbiology* 56(4): 327–333.

Jaleel, C.A., Manivannan, P.A.R.A.M.A.S.I.V.A.M., Wahid, A., Farooq, M., Al-Juburi, H.J., Somasundaram, R.A.M.A.M.U.R.T.H.Y. and Panneerselvam, R. 2009. Drought stress in plants: a review on morphological characteristics and pigments composition. *International Journal of Agriculture Biology* 11(1): 100–105.

Jha, P.N. and Kumar, A. 2007. Endophytic colonization of Typha australis by a plant growth-promoting bacterium *Klebsiella oxytoca* strain GR3. *Journal of Applied Microbiology* 103(4): 1311–1320.

Jiang, C.Y., Sheng, X.F., Qian, M. and Wang, Q.Y. 2008. Isolation and characterization of a heavy metal-resistant *Burkholderia* sp. from heavy metal-contaminated paddy field soil and its potential in promoting plant growth and heavy metal accumulation in metal-polluted soil. *Chemosphere* 72(2): 157–164.

Kamran, S., Shahid, I., Baig, D.N., Rizwan, M., Malik, K.A. and Mehnaz, S. 2017. Contribution of zinc solubilizing bacteria in growth promotion and zinc content of wheat. *Frontiers in Microbiology* 8: 2593.

Kang, B.G., Kim, W.T., Yun, H.S. and Chang, S.C. 2010. Use of plant growth-promoting rhizobacteria to control stress responses of plant roots. *Plant Biotechnology Reports* 4(3): 179–183.

Khalid, A., Akhtar, M.J., Mahmood, M.H. and Arshad, M. 2006. Effect of substrate-dependent microbial ethylene production on plant growth. *Microbiology* 75(2): 231–236.

Khan, M.S., Zaidi, A., Wani, P.A. and Oves, M. 2009. Role of plant growth promoting *Rhizobacteria* in the remediation of metal contaminated soils. *Environmental Chemistry Letters* 7(1): 1–19.

Kumar, A., Patel, J.S., Meena, V.S. and Ramteke, P.W. 2019. Plant growth-promoting *Rhizobacteria*: strategies to improve abiotic stresses under sustainable agriculture. *Journal of Plant Nutrition* 42(11–12): 1402–1415.

Kumar, A., Maurya, B.R., Raghuwanshi, R., Meena, V.S. and Islam, M.T. 2017. Co-inoculation with Enterobacter and Rhizobacteria on yield and nutrient uptake by wheat (*Triticum aestivum* L.) in the alluvial soil under indo-gangetic plain of India. *Journal of Plant Growth Regulation* 36(3): 608–617.

Miransari, M. 2011. Hyperaccumulators, arbuscular mycorrhizal fungi and stress of heavy metals. *Biotechnology Advances* 29(6): 645–653.

Mishra, J. 2018. Role of secondary metabolites from plant growth-promoting *Rhizobacteria* in combating salinity stress. In *Plant microbiome: Stress response*, eds. T. Fatima and N.K. Arora, 127–163. Singapore: Springer.

Naik, M.M. and Dubey, S.K. 2011. Lead-enhanced siderophore production and alteration in cell morphology in a Pb-resistant *Pseudomonas aeruginosa* strain 4EA. *Current Microbiology* 62(2): 409–414.

Nia, S.H., Zarea, M.J., Rejali, F. and Varma, A. 2012. Yield and yield components of wheat as affected by salinity and inoculation with *Azospirillum* strains from saline or non-saline soil. *Journal of the Saudi Society of Agricultural Sciences* 11(2): 113–121.

Phi, Q.T., Park, Y.M., Seul, K.J., Ryu, C.M., Park, S.H., Kim, J.G. and Ghim, S.Y. 2010. Assessment of root-associated *Paenibacillus polymyxa* groups on growth promotion and induced systemic resistance in pepper. *J Microbiology Biotechnology* 20(12): 1605–1613.

Rahdari, P. and Hoseini, S.M. 2012. Drought stress: a review. *International Journal of Agronomy and Plant Production* 3(10): 443–446.

Rajkumar, M., Ae, N., Prasad, M.N.V. and Freitas, H. 2010. Potential of siderophore-producing bacteria for improving heavy metal phytoextraction. *Trends in Biotechnology* 28(3): 142–149.

Rojas-Tapias, D., Moreno-Galván, A., Pardo-Díaz, S., Obando, M., Rivera, D. and Bonilla, R. 2012. Effect of inoculation with plant growth-promoting bacteria (PGPB) on amelioration of saline stress in maize (*Zea mays*). *Applied Soil Ecology* 61: 264–272.

Rokhbakhsh-Zamin, F., Sachdev, D., Kazemi-Pour, N., Engineer, A., Pardesi, K.R., Zinjarde, S., Dhakephalkar, P.K. and Chopade, B.A. 2011. Characterization of plant-growth-promoting traits of *Acinetobacter* sp. isolated from rhizosphere of *Pennisetum glaucum*. *Journal of Microbiology Biotechnology* 21(6): 556–566.

Rubio, L.M. and Ludden, P.W. 2008. Biosynthesis of the iron-molybdenum cofactor of nitrogenase. *Annual Review of Microbiology* 62: 93–111.

Rütting, T., Aronsson, H. and Delin, S. 2018. Efficient use of nitrogen in agriculture. *Nutrient Cycling in Agroecosystems* 110: 1–5.

Schirawski, J. and Perlin, M.H. 2018. Plant–microbe interaction 2017—the good, the bad and the diverse. *International Journal of Molecular Sciences* 19(5): 1374.

Shaharoona, B., Naveed, M., Arshad, M. and Zahir, Z.A. 2008. Fertilizer-dependent efficiency of Pseudomonads for improving growth, yield, and nutrient use efficiency of wheat (*Triticum aestivum* L.). *Applied Microbiology and Biotechnology* 79(1): 147–155.

Singh, R., Singh, S., Parihar, P., Singh, V.P. and Prasad, S.M. 2015. Retrograde signaling between plastid and nucleus: a review. *Journal of Plant Physiology* 181: 55–66.

Tank, N. and Saraf, M. 2010. Salinity-resistant plant growth promoting *Rhizobacteria* ameliorates sodium chloride stress on tomato plants. *Journal of Plant Interactions* 5(1): 51–58.

Tewari, S. and Arora, N.K. 2014. Multifunctional exopolysaccharides from *Pseudomonas aeruginosa* PF23 involved in plant growth stimulation, biocontrol and stress amelioration in sunflower under saline conditions. *Current Microbiology* 69(4): 484–494.

Vurukonda, S.S.K.P., Vardharajula, S., Shrivastava, M. and SkZ, A. 2016. Enhancement of drought stress tolerance in crops by plant growth promoting *Rhizobacteria*. *Microbiological Research* 184: 13–24.

Zahir, Z.A., Shah, M.K., Naveed, M. and Akhter, M.J. 2010. Substrate-dependent auxin production by *Rhizobium phaseoli* improves the growth and yield of Vigna radiata L. under salt stress conditions. *Journal of Microbiology Biotechnology* 20(9): 1288–1294.

Zaidi, A., Khan, M., Ahemad, M. and Oves, M. 2009. Plant growth promotion by phosphate solubilizing bacteria. *Acta Microbiologica et Immunologica Hungarica* 56(3): 263–284.

Zarea, M.J., Hajinia, S., Karimi, N., Goltapeh, E.M., Rejali, F. and Varma, A. 2012. Effect of Piriformospora indica and Azospirillum strains from saline or non-saline soil on mitigation of the effects of NaCl. *Soil Biology and Biochemistry* 45: 139–146.

6 Roles of Novel Bacterial and Fungal Metabolites in Agriculture

Richa Salwan[1], Randhir Kaur[2], and Vivek Sharma[2]
[1]Dr. Yashwant Singh Parmar University of Horticulture and Forestry, Himachal Pradesh, India
[2]Chandigarh University, Punjab, India

CONTENTS

6.1 INTRODUCTION

The demand for agricultural commodities is anticipated to surge at least 70% by 2050. Therefore, modern agriculture practices are being explored at global scale keeping in view the environmental as well as economical sustainability concerns of food products at world level (Altieri, 2004). The microorganisms and their secondary metabolites have played vital and significant roles in our life and surrounding ecosystem (Bisen et al., 2012). The exploration of soil microbiology-based practices offers a sustainable, efficient and effective practice of enhanced crop production without disturbing the biosphere. Plant beneficial microorganisms play fundamental role in agriculture, due to their ability to improve plant health, without spoiling the

DOI: 10.1201/9781003110477-6

soil quality (Barea et al., 2013; Lugtenberg, 2015). These microorganisms are one of the major resources of secondary metabolites. Both bacteria and fungi associated to plants from diverse trophic/living habitats have been reported for their various roles as saprophytes or in symbiotic relationships with the host plant. Mostly, these microbes exist either in the rhizospheric soil or as endophytes or rhizoplane (Porras-Alfaro and Bayman, 2011; Hardoim and van Elsas, 2013; Brader et al., 2014; Barea, 2015: Mercado-Blanco, 2015). Plant beneficial microbes, such as "rhizobia", fix N_2 in association with legume host plants (Olivares et al., 2013; de Bruijn, 2015), whereas bacteria belonging to actinomycetes, genus *Frankia*, are capable of fixing N_2 freely in non legume plants (Normand et al., 2007). The other group of microorganisms belonging arbuscular mycorrhiza (AM) fungi phylum Glomeromycota establish associations with plant species in rhizosphere (Schüßler et al., 2001; Smith and Read, 2008; van der Heijden et al., 2015) ().

The secondary metabolites from different microbes have been explored for their role in agriculture, nutrition, healthcare and even in basic research. In agriculture sector, microbial secondary metabolites have been explored for their role as plant growth promoters, acquisition of minerals, antimicrobial, herbicidal and insecticidal properties (Kieslich, 1986). In addition, approximately, 80% of anti-cancerous molecules and around 50% of FDA-approved drugs are directly or indirectly derived from natural products which are products of secondary metabolism and tried for ailments with restricted treatments (Milshteyn et al., 2014). The microbial secondary metabolites are low molecular organic compounds produced during or at the end of the stationary growth phase (Bérdy, 2005). The role of these metabolites in the development, growth and reproduction of microorganism is still not explored (Singh et al., 2017; Katoch et al., 2020). However, the secondary metabolites have been well explored for a wide range of biological beneficial activities such as antimicrobial activities, enzymes inhibitors, antitumor, immunosuppressive (Demain, 1999), plant growth promotion, herbicidal, insecticidal, and others activities (Salwan et al., 2019).

In general, the microbial secondary metabolites are secreted in late growth phase. The production is modulated through specific regulatory mechanisms which is generally suppressed during logarithmic phase and activated in stationary phase of life cycle. The molecular skeleton of secondary metabolite is distinctive and approximately 40% of the microbial secondary metabolites cannot be synthesized chemically (Feher and Schmidt, 2003; Thirumurugan et al., 2018). Since the discovery of penicillin by Fleming in 1929 referred to as golden age of antibiotics in 1960 (Singh et al., 2017), continuous research led to the discovery of several antibiotics including anthracyclins, ansamacrolids, aminoglycosides, cephalosporin, chloramphenicol, macrolids, tetracycline, peptide inhibitors, glycopeptides and antifungal antibiotics (Kieslich, 1986; Hassan et al. 2012; Singh et al., 2017). In agriculture, biocontrol species of *Bacillus* (Salwan and Sharma, 2020a), *Pseudomonas fluorescens*, *P. aureofaciens* (Shanmugam et al., 2008), *Streptomyces* spp. (Salwan and Sharma, 2020b; Salwan et al., 2020) and *Trichoderma* (Sharma et al., 2018) have been explored to counter wide range of plant pathogens such as *Fusarium*, *Phytopthora infestans*, *Alternaria solani* and *Colletotrichum* (Canan, 2013; Chatzipavlidis et al., 2013; Singh et al., 2017; Sharma et al., 2016a; Sharma et al., 2017). The volatile secondary metabolites have been explored as repellents or attractants, whereas non-volatile compounds have been used to boost immune response in plants, suppress plant pathogens and even kill the insect, or alleviate other stresses of host plants (Salwan et al., 2019; Zhang et al., 2020).

The demarcation of primary and secondary metabolites is not clear since several intermediates of both overlap with each other. The ambiguity in intermediates indicate common biosynthetic pathways and precursors being shared in primary and secondary metabolic pathways. It is also reported that secondary metabolites act as buffer zone where excess amount of carbon and nitrogen may be stored from primary metabolism as inactive part (Thirumurugan et al., 2018). The worldwide market of beneficial microorganisms and their products in 2014 was estimated approximately $143.5 billion and expected to attain $306 billion at approximately 14.6% of annual growth rate between 2015 and 2020 (Singh et al., 2017). The products derived from microbial sources include organic acids, enzymes, coloring agents, flavoring agents, pharmaceutical products and several metabolites of agricultural importance (Demain, 2007).

6.2 MICROBIAL SECONDARY METABOLITES

Microbial secondary metabolites are potential source of bioactive compounds (Barkal et al., 2017). The major compounds such as peptides, isoprenes and polyketides (PKs) are produced through secondary metabolic biosynthetic pathways (Ruiz et al., 2010). The two major secondary metabolite groups have been categorized according to their polarity, evaporation and diffusion abilities into (i) volatile organic compounds (VOCs), (ii) soluble ones (Tyc et al., 2017). The former can evaporate and diffuse whereas latter are water soluble mainly due to their polarity. Secondary metabolites such as indole, pyrazines and certain sulphur-containing compounds are of VOC type whereas lipopeptides, bacteriocins and PKs are of soluble type. The dimethyl disulphide, a volatile compound of microbial origin, has been commercially approved under the name palandin by the Environmental Protection Agency, United States, as fumigants to manage nematodes and other soil-borne pathogens (Tyc et al., 2017).

In general, the secondary metabolites are low molecular weight compounds of less than 1000 Da (Pinu et al., 2017). The entire metabolites of an organism are known as metabolome (Aldridge and Rhee 2014). First proposed in 1997, the metabolome is divided into three domains; intracellular, extracellular (Vaidyanathan, 2005) and the ones present in the headspace (Beale et al., 2016). These metabolites can be either polar and water soluble, nonpolar insoluble in water and even volatile (Villas-Boas et al., 2007). Moreover, the categories of metabolites are not clearly defined and hence open to clarification (Vaidyanathan, 2005). During metabolism, metabolites are biosynthesized as well as transformed into various compounds. The metabolome reveals the biological capabilities and function of an organism and is closely linked to the phenotype (Macintyre et al., 2014). Secondary metabolites although are not vital for the regular growth and development of an organism but their role in providing survival advantage to the producing organism against environmental stresses or uptake of nutrients has been studied (Sansinenea and Ortiz, 2011; Breitling et al., 2013). Secondary metabolites are generally produced under nutrient depletion during stationary growth phase (Ruiz et al., 2010). The metabolites produced during stationary phase have higher biological activity. The maximum production of antimicrobial bacteriocins by *B. amyloliquefaciens* An6 in the stationary phase also support same hypothesis (Ayed et al., 2015). Different secondary metabolites are either biosynthesized using ribosomal or nonribosomal unit usually located on chromosome (Ruiz et al., 2010). The accumulation of secondary metabolites either into extracellular medium or inside cell is observed to be higher compared to primary metabolites (Covington et al., 2017; Pinu et al., 2017).

The secondary metabolites are produced during the idiophase under nutrient-limited conditions. These molecules are not vital for the routine growth and reproduction of microbes. The biological activities of these compounds ranges from antimicrobial, herbicidal, insecticidal, antiparasitic, plant growth promotion to nutrient acquisition activities in agriculture, veterinary and medical sciences (Barrios-Gonzalez et al., 2005). Diverse array of bacteria such as *Pseudomonas, Bacillus* (Salwan and Sharma 2020a) and *Streptomyces* (Salwan et al., 2020) have been explored for the production of secondary metabolites such as anthracyclines, aminoglycosides, bithiazoles, cyclic lipopeptides, isoquinoline, quinones, polyketides, peptides, polypeptides, phloroglucinols, pyrroles, phenazines, penes macrolides, isocoumarins, bacteriocins, aminosugars, siderophores with plethora of biological activities due to their antibacterial and antifungal activities (Pathma et al., 2021).

Pseudomonas have been reported for the production of VOCs like hydrogen sulfide, 2-phenylethanol that can suppress both fungal and bacterial pathogens of plants (Piechulla et al., 2017). Similarly, the VOCs produced by *Bacillus amyloliquefaciens* are reported to change the community structure of microorganisms (Yuan et al., 2017). Besides bacteria, fungi such as *Trichoderma* are potential alternate to synthetic pesticides against different phytopathogenic fungi such as oomycetes and nematodes. The detailed mechanisms of *Trichoderma* are well documented and beyond the scope of this chapter. The culture filtrate of *T. harzianum* and *T. virens* containing proteins, metabolites and VOCs suppresses the bacterial pathogens. The production of antimicrobial molecules by *T. virens* and *T. harzianum* has been affected by VOCs of bacteria indicating the effect of both on each other (Sharma et al. 2016b; Li et al., 2019). The effect of VOCs of *Trichoderma* during interaction with plants, suppression of plant pathogenic fungi pathogens, bacteria and interaction with other microbes in rhizosphere have been explored in several studies (Wheatley et al., 1997; Lee et al., 2016; Martı́nez-Medina et al., 2017; Meena et al., 2017; Li et al., 2018; Sharma et al., 2018b). All these finding revealed the importance of VOCs of *Trichoderma* as a fumigant, pathogen suppressor as well modulator of microbial community in the rhizosphere. The role of VOCs in *F. oxysporum* and *Trichoderma* interaction is also studied (Li et al., 2018; Li et al., 2019).

6.3   CLASSIFICATION OF SECONDARY METABOLITES

Plant beneficial microorganisms such as *Bacillus, Pseudomonas* and *Streptomyces* as well as fungal genera such as *Trichoderma* and several others produced diverse array of secondary metabolites (Shanmugam et al., 2008; Sharma et al., 2013; Sharma et al., 2018b; Sharma et al., 2018a; Sharma et al., 2017; Sharma et al., 2017; Salwan and Sharma, 2020a; Salwan and Sharma, 2020b; Salwan et al., 2020). Among these, *Trichoderma*-based bio formulations contribute to over 60% of the registered biopesticides worldwide (Mukherjee et al., 2013; Singh et al., 2018; Li et al., 2019). The secondary metabolites of these microorganisms are responsible for various plant-beneficial attributes like plant growth promotion, nutrient acquisition and pathogen suppression (Ruiz et al., 2010), quorum-sensing and antimicrobial activities (Katoch et al., 2020). The basic "scaffold" of secondary metabolites is biosynthesized using polyketides (PK), nonribosomal peptides (NRPs), isoprenoids and shikimate derivatives (Lange et al., 2000; Schwarzer et al., 2003), which are subsquently converted into final product through post modification like acylation, macrocyclization, glycosylation, oxidation/reduction, methylation halogenation, phosphorylation, sulfation, into final metabolite (Savile et al., 2010). Over several thousands of secondary metabolites including alkaloids, nonribosomal polypeptides, terpenoids, fatty acid-derived substances and polyketides have been used for agricultural applications (Lange et al., 2000).

Alkaloids are biologically synthesized through a multifunctional complex unit without involving any transcription. These are produced from acyl precursors including acetyl CoA, propionyl CoAand methylmalonyl CoA (Lange et al., 2000). The non-ribosomal polypeptides, terpenoids and other steroids are biosynthesized using isopentenyl diphosphate units. Presently, more than 35,000 of terpenoid and steroid compounds have been characterized. Terpenoids have unrelated structures, whereas steroids are composed of a tetracyclic carbon moiety and are modified terpenoids obtained from the triterpene lanosterol processing.

The plant beneficial bacterial are prolific producers of cyclic lipopeptides (Rangarajan and Clarke 2016). Microbial genomes of plant beneficial microorganisms such as *Bacillus*, *Pseudomonas*, *Streptomyces* and *Trichoderma* harbor diversity of biosynthetic cluster of genes (BCGs) which encode polyketide synthetases and nonribosomal peptide synthetases for the production of a variety of polykeitdes (PKs) and peptides (Aleti et al., 2015). These cyclic lipopeptides are categorized into iturin, fengycin, surfactinand kurstakin (Hathout et al., 2000; Frikha-Gargouri et al., 2017). These lipopeptides are composed of a tail of lipid and oligopeptide unit of linear or cyclic type (Tyc et al., 2017) which are responsible for lipophilic and hydrophilic nature and contribute to amphiphilic behavior (Perez-Garcia et al., 2011; Rangarajan and Clarke, 2016). These peptides exhibit diverse biological actions mainly due to amino acid residues, peptide cyclization and nature of the fatty acid units, that is, nature, length and branching (Frikha-Gargouri et al., 2017). Due to tolerance and stability, the lipopeptides can be targeted for agricultural applications such as biocontrol agents or biofertilizers. The synergistic action of lipopeptides is reported to enhance each other's activities (Mongkolthanaruk, 2012; Aleti et al., 2015).

The surfactins based on the lipid homology and the amino acid residues may be categorized into A, B and C types (Wang et al., 2015). These peptides differ based on the amino acid location near to lactone ring. Surfactin A possess L-leucine, B type contain L-valine whereas type C is composed of L-isoleucine type (Wang et al., 2015). The different strains of *B. subtilis* vary in their efficiency to produce surfactin (Paraszkiewicz et al., 2017). Another cyclic lipopeptides iturin obtained from *Bacillus* has been explored in agriculture. It differs from surfactin, due to cyclic lipopeptide structure (Aleti et al., 2015). The bacillomycin D, bacillomycin DC and mycosubtilin are also produced by *Bacillus* (Deravel et al., 2014; Gong et al., 2014;

Gu et al., 2017; Jin et al., 2017). Fengycin is composed of a lactone ring which comprises a fatty acid chain attached to decapeptide. The two variants of fengycin, that is, A and B, differs in the presence of D-alanine in former while latter contain D-valine at 6-position (Rangarajan and Clarke 2016; Zhao and Kuipers, 2016). Other cyclic lipopeptide linked to bacterial cells and spores from *Bacillus thuringiensis* subsp. *kurstaki* has been reported for its biocontrol activity (Hathout et al., 2000; Bechet et al., 2012; Gelis-Jeanvoine et al., 2017).

6.3.1 POLYKETIDES

Polyketides (PKs), synthesized nonribosomally, represent one of the major classes of secondary metabolites in microorganisms using known PKs synthetases (Mojid Mondol et al., 2013; Aleti et al., 2015; Palazzini et al., 2016; Frikha-Gargouri et al., 2017). Structurally diverse with various biological activities in *Bacillus*, PKs include bacillaene, macrolactins and difficidin (Hamdache et al., 2011; Aleti et al., 2015; Palazzini et al., 2016). Due to the antimicrobial activity and boosting plant defense systems, PKs are of vital importance in plant disease management (Aleti et al., 2015). Bacillaene antibiotic property with a linear structural organization of two amide bonds is specific to prokaryotes (Hamdache et al., 2011). Macrolactins contain a macrolide-like structure, whereas difficidin inhibit the proteins biosynthesis (Hamdache et al., 2011). The production of bacillaene, macrolactin, and difficidin along with hydrolytic enzymes is responsible for the biocontrol efficacy of *B. amyloliquefaciens* subsp. *plantarum* strain 32a (Ben Abdallah et al., 2018).

6.3.2 BACTERIOCINS

The antimicrobial peptide bacteriocins are produced ribosomally (Ayed et al., 2015; Berini et al., 2018). So far, bacteriocins have been reported from each major bacterial class and around 99% of bacteria are reported to produce these peptides (Tyc et al., 2017). Among bacteria, *Bacillus* has been reported for the production of diversity of bacteriocins like coagulin, cerein, subtilin, subtilosin A, sublancin, ericin, mersacidin, haloduracin, thurincin, megacin, and lichenin (Abriouel et al., 2011). Nisin, most studied bacteriocin obtained from *Lactococcus lactis*, is used in food industry as a preservative (McAuliffe et al., 2001; Lee and Kim, 2011). However, compared to lactic acid bacteria, bacteriocins of *Bacillus* sp. have been used in agricultural fields (Abriouel et al., 2011). The bacteriocin obtained from *B. amyloliquefaciens* is used as nematicidal agent (Liu et al., 2013; Chowdhury et al., 2015). The bacteriocins of *Bacillus* are found to exhibit inhibitory activities (Ayed et al., 2015), differs in their chemical structural composition, mode of action, thermal stability, enzymatic sensitivity and other parameters (Sumi et al., 2015). These molecules are classified as I, II and III bacteriocins (Balciunas et al., 2013).

The first class of bacteriocins categorized as class I lantibiotics are composed of a short chain of amino acids (Arguelles Arias et al., 2013; Balciunas et al., 2013). The category I bacteriocins are thermostable (Arguelles Arias et al., 2013) can be categorized as lanthionine-containing (Fuchs et al., 2011) reported from diverse bacteria (Abriouel et al., 2011). Based on their structural composition and net charge at neutral pH, these are further classified into Type A and Type B lantibiotics (Arguelles Arias et al., 2013; Wang et al., 2015). Lantibiotics after post-translational modification forms lanthionine or methyllanthionine thioether bridges (Abriouel et al., 2011). Subtilin lantibiotic is characterized by the presence of unusual amino acids such as lanthionine, dehydroalanine, D-alanine, methyllanthionine, and dehydrobutyrine (Abriouel et al., 2011). The lantibiotics are produced by *Bacillus* sp in response to environmental stimuli like nutrient deficiency as a part of their adaptive response under nutrient-defiant conditions (Abriouel et al., 2011). Another peptide amylolysin inhibit the cell wall biosynthesis after its interaction with lipid and the peptidoglycan (Sumi et al., 2015).

The bacteriocins of class II lack lanthionine are further divided into five subclasses (McAuliffe et al., 2001). The class III bacteriocins represents high molecular weight peptides (>30 kDa) and are labile (McAuliffe et al., 2001; Lee and Kim, 2011). The class I peptides with molecular weight of less than 5 kDa consist of a linear secondary structure and are positively charged whereas class II are globular and either negatively charged or neutral like mersacidin (Arguelles Arias et al., 2013; Wang et al., 2015).

6.4 MINING MICROBIAL GENOME FOR SECONDARY METABOLITES ENCODING CLUSTER OF GENES

Microorganisms are one of the biggest sources of various secondary metabolites with potential applications in biomedical, biotechnological and agricultural sectors. Recent advancements in technology led to mining of microbial genomes, rewire/ engineer novel biosynthetic gene clusters through genetic engineering for enhanced production of targeted secondary metabolites in native producers as well as in the heterologous host systems (Figure 6.1). The application of genetic engineering tools

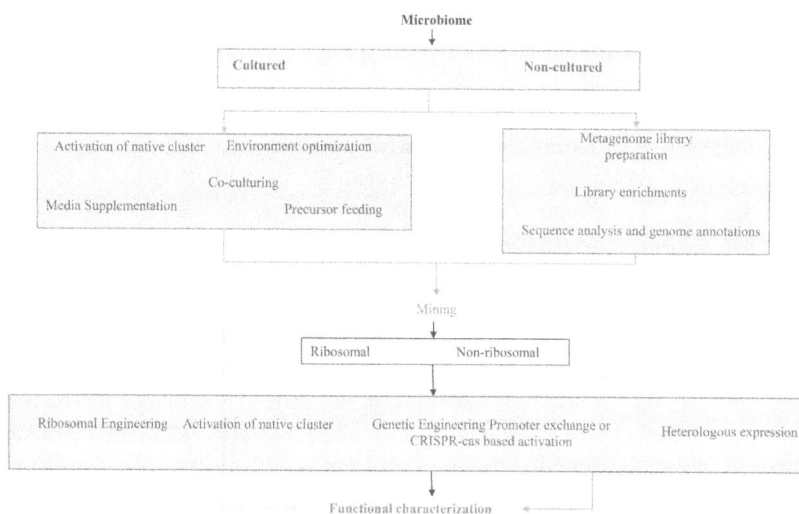

FIGURE 6.1 Overview of approach of mining secondary metabolites and their encoding pathways of microorganisms.

like gene editing for the biosynthesis of microbial secondary metabolites, engineering of synthetic promoters and expression systems has been explored (de Frias et al., 2018). The mining of genomes led to identification and characterization of several putative and silent BCGs in the genomes of microorganisms which can be targeted for the production of novel compounds that were undetected under lab conditions (Abdelmohsen et al., 2015).

The prediction of these BCGs using bioinformatics tools is of vital importance to identify and characterize novel natural products. The extensive genome sequencing and bioinformatics tools such as antiSMASH, np.searcher and ClustScan can now be easily applied to mine the microbial biosynthetic clusters of genes in unassembled genomes, then prioritize the activation

of silent gene clusters at greater pace for novel molecule discovery (Starcevic et al., 2008; Li et al., 2009; Medema et al., 2011; Milshteyn et al., 2014). For example, the bioinformatic analysis of the draft genome of plant beneficial *Bacillus* and *Streptomyces* species revealed diversity for various secondary metabolites encoding biosynthetic cluster of genes (BCG) (Udwary et al., 2007; Salwan and Sharma, 2020a; Salwan and Sharma 2020b). Additionally, exploration of unculturable microorganisms using metagenome-based mining is nowadays explored at rapid pace for the identification, characterization and expression of several cryptic clusters for novel biosynthetic clusters (Figure 6.1). Although slow but functional metagenomics approach have been used to enhance the identification of bioactive small molecules using direct screening of clones generated from environmental DNA for the activities of our interest. Further, high throughput screening is suitable for providing deep mining of these samples (Torsvik et al., 1998; Rappe and Giovannoni, 2003).

Moreover, optimization of process parameters can be used in single strain as "one strain many compounds" approach (OSMAC) as an efficient method for silent pathway activation or even expression of poorly expressed metabolites (Bode et al., 2002, Paranagama et al., 2007, Wei et al., 2010; Abdelmohsen et al., 2014). Alternate methods like co-cultivation are also deployed for the production of enzymes, food additives and other products. However, its usage for secondary metabolites is still underexplored (Bader et al., 2010) (Figure 6.1). Another approach, involving challenging microbes with externa stimuli referred to as "elicitors", has been used for the production of biologically active novel metabolites (Chiang et al., 2011; Müller and Wink, 2014). The perception of signals by producer strains can lead to change in metabolites profile (Abdelmohsen et al., 2015). Since fermentation-based methods of activation rely on trial and error to activate cryptic pathways, availability of full genomic informations can play vital role in the activation of cluster induction studies. Effect of environmental conditions on secondary metabolite production has been reported in some studies. Studies on role of pressure and pH on secondary metabolites production are limited only (Schmidt et al., 2019). The higher production of secondary metabolites has been recorded for actinobacteria isolated from alkaline soils (Gonzalez et al., 1999; Sagova-Mareckova et al., 2015) whereas higher number of low-molecular weight compounds were reported later on from acidic soil (Sagova-Mareckova et al., 2015). In another study, confrontation of *Pseudomonas fluorescens* Pf0–1 with different bacterial species triggered the biosynthesis of broad-spectrum antibiotics active against phtyopathogenic fungi (Garbeva et al., 2011).

6.5 APPLICATIONS OF MICROBIAL SECONDARY METABOLITES IN AGRICULTURE

Microbial secondary metabolites are also extensively used in agriculture fields. The usage of biopesticides like application of *Phytopthora* as bioherbicides, *Trichoderma* as biofungicides and *Bacillus* spp as bioinsecticides have gained attentions (Canan, 2013). The secondary metabolites have been explored as competitive weapons against microorganisms, plants, insects, animals and metal-chelating agents, for their role in interaction with other organisms, morphogenesis like sporulation and germination and as well as signaling molecules during communication (Demain, 2007; Demain and Fang, 2000). Further, the secondary metabolites are also known for diverse biological activities including antimicrobial, antiparasitic, antitumor agents, inhibitors of enzyme, immunosuppressive agents and others (Demain, 1999). Chemically synthesized pesticides like halogenated, carbamate and organophosphorus that are extensively used in agriculture have already caused severe problems to our ecosystem (Lacey and Siegel, 2000; Canan, 2013; Nawaz et al., 2016). The global biopesticides business was $4.40 billion in 2019 and expected to achieve $10.63 billion by 2027 (https://www.fortune-businessinsights.com/industry-reports/biopesticides-market-100073). The biochemicals such as kasugamycin and polyoxins have been used as biopesticides whereas *Bacillus thuringiensis* crystals, spinosyns and nikkomycin are recommended as bioinsecticides and bioherbicides (bialaphos). On the other hand, plant hormones like gibberellins, indole acetic acid (IAA) as growth regulators and siderophores for metal chelation from microbial sources are also explored (Demain, 1999).

6.5.1 MICROBIAL SECONDARY METABOLITES AS ION CHELATORS

Microbial secondary metabolites such as siderophores known for their ferric iron or other ions-chelating activity can play vital role under iron-deficient conditions (Crosa, 1989). These siderophores have been classified as catecholate and hydroxamates types. The catecholate types of siderophore consists of two oxygen units and form a hexagonal octahedric complex the Fe^{3+} ions. Microorganisms like *Pseudomonas* are found to produce a mixture of catecholates and hydroxamates siderophores (Gulledge et al., 2002). The protonation of catecholate sierophores or conversion of Fe^{3+} to Fe^{2+} in a cell can help in achieving iron depletion (Stintzi et al., 2000). The hydroxamate types of siderophore are one of the most widely distributed siderophores reported from bacteria, fungi and cyanobacteria. Structurally, the hydroxamate type of siderophores consist of a $O = CR-N(OH)R'$ group, where R or R' moiety can either be an amino acid residue or their derivative. The oxygen atoms in the hydroxamate siderophore help in forming an octahedral complex with Fe^{3+}. The carboxylate type of siderophores ligate to iron through carboxyl and hydroxyl moieties (Rao et al., 2002; Tomá et al., 2018). Secondary metabolites as herbicidal agents have been found effective and eco-friendly alternate in agriculture fields for managing weeds. The herbicidal agents obtained from microbial sources are categorized into microbes as herbicidal agents and microbes-derived products. Products obtained from plant pathogenic microorganisms like *Colletotrichum* and *Phytophthora* are also used to control weeds (Demain and Sanchez 2009; Chandler et al. 2011).

6.5.2 MICROBIAL SECONDARY METABOLITES AS BIOPESTICIDES

Biological pesticides are target-specific and their application in agriculture and public health is considerable. The worldwide bioherbicides market was about $973 million in 2015 and is anticipated to grow at a CAGR of 10.87% between 2015 and 2020. Currently, hundreds of bacteria have been explored as pathogen of insects. Among these, *Bacillus thuringiensis* and genes have been used due to its insecticidal activity against caterpillar, mosquito larvae, fly and beetles (Argolo-Filho and Loguercio 2014; Nawaz et al., 2016). The endotoxin protein known as Bt and produced during spore formation after binding is capable of destroying digestive tract of insects which can lead to their death (Schunemann et al., 2014). This protein specifically targets caterpillars including butterflies and moths, mosquito larvae, blackflies and simuliid, and hence recommended for fruit and vegetable crops (Meadows, 1993).

Secondary metabolites as biofungicides have been deployed against plant pathogenic fungi. *Trichoderma* species, *Metarhizium anisopliae* and *Beauveria bassiana* antagonist to *Fusarium*, *Rhizoctionia*, *Pythium*, *Colletotrichum*, *Sclerotinia sclerotiorum* and other soil-borne pathogens are one of the most common fungal biopesticides used in agriculture. *Beauveria bassiana* and *Metarhizium anisopliae* are insect parasitic fungi and are found effective against pests such as thrips, aphids and even pesticide-resistant whitefly strains. *Metarhizium anisopliae* is also used to control spittlebugs on sugarcane, grassland and even against grasshopper and locust in Africa and Australia. Similarly, mycoparasitic fungi *Coniothyrium minitans* is used to manage the *Sclerotinia sclerotiorum* (Chandler et al., 2011; Chatzipavlidis et al., 2013).

6.5.3 BIONEMATICIDES

Plant pathogenic nematodes globally causes significant damage in various agricultural crops. The biocontrol agents like *Bacillus*, *Pseduomonas* (Aleti et al., 2015; Berini et al., 2018) and *Trichoderma* are deployed as nematicidal agents and their role can be attained by directly using the inoculum of the antagonistic strain (Horak et al., 2019). The biocontrol species of *Bacillus* are recommended as safe microorganisms by United States Food and Drug Administration (Cawoy et al., 2011; Frikha-Gargouri et al., 2017). The ability of these biocontrol agents to target nematode is ascribed to the secondary metabolites production along with hydrolytic enzymes. Presently, over 350 biopesticidal agents have been registered for the control of insects, mites and nematodes in the United States and among these 57 are derived from microbes (Arthurs and Dara, 2019).

6.5.4 PLANT GROWTH REGULATORS

Plant growth regulators adjust the growth and other development of cells and tissues of plants (Spence and Bais, 2015). The plant growth regulators market was evaluated at USD million in 2020 and it is projected to register a CAGR of 8.5% between 2022 and 2027 (https://www.mordorintelligence.com/industry-reports/global-plant-growth-regulators-market-industry). Among different plant growth regulators, gibberellins alone account for the major market share then followed by cytokinins and auxins. At global level, largely used for fruit and vegetable production at large scale, seed germination, elongation of stem, flowering and embryo growth, the annual demand of gibberellic acid (GA) is approximately 60 tons. The FMC Corporation, Syngenta AG (Switzerland), the Dow Chemical Company from the United States, Nufarm Limited (Australia) and BASF SE (Germany) are leading companies. Beside this, a large number of microorganisms have been reported for the production of plant growth regulators so far (Yurekli et al., 1999; MacMillan, 2001; Karadeniz et al., 2006; Tsavkelova et al., 2006).

6.6 CHALLENGES AND FUTURE SCOPE

The intensive agricultural practices led to enhanced emission of greenhouse gases resulting in rise of Earth's temperature which severely affected the biosphere stability and also impacted productivity of agricultural fields and even natural ecosystems. Additionally, the application of chemicals in fields to counter biotic and abiotic stresses have created hazardous conditions in our surrounding environment and even threatened human health. Researchers are continuously exploring different alternate to provide a sustainable solution to mitigate the challenges. Microorganisms are one of the intrinsic parts of ecological system on earth. Microorganisms of terrestrial and marine are known to produce diverse array of secondary metabolites which are well known for their role as antibiotics, antitumor, immuno-suppressor, enzyme inhibitors and various other plant beneficial attributes that can be used for agricultural applications. For timely production of these secondary metabolites, the microbial cultures should be harvested from the stationary growth phase and processed because early or delayed sampling can hamper secondary production or their degradation. Beside this, the major challenge is identification of new metabolites and then linking corresponding biosynthetic pathways. Recent developments in molecular biology have played big role in elucidating the microbial genomes or even mine metagenomes for natural products discovery and development of technologies to enhance productivity, reassembly of novel biosynthetic pathways for overexpression into native or heterologous hosts. The genetic engineering tools such as synthetic biology have been developed to synthetically engineer the pathways for secondary metabolites biosynthesis. Even though the microbial genomes can be explored now with a greater accuracy, still a number of questions needs to be addressed. For example, a major share of microbial gene clusters has been reported silent under routine laboratory conditions, role of these metabolites on host-microbiome interactions, ecology and others are largely underexplored area.

In addition, tools such as functional metagenomics are capable of identifying bioactive natural products even from non-culturable microorganisms. To fully utilize the potential, advancement in methods such as capabilities to clone large environmental DNA samples, can help in cloning and subsequent characterization of gene cluster on individual clone. Exploration of other alternate host for heterologous expression like *Streptomyces*, preparation of enriched metagenomic library, efficient screening processes and activation of cryptic types of secondary metabolic pathways can be useful for the efficient expression of plethora of genes clusters for the discovery of novel metabolites. The availability of genomic data and ease of comprehensive bioinformatic tools for genome mining for cryptic genes followed by designing rational approaches for activation through strategic incorporation of robust constitutive pro-moters coupled with synthetic biology can be really useful strategy for the activation of biosynthetic pathways. The concept of "ribosome engineering" through mutations in the native RNA polymerase and ribosomal proteins could lead to activation of otherwise cryptic clusters.

The limitation linked to identification of promoters and ribosomal binding sites in heterologous host often led to poor or even no expression. Now, the attentions have shifted to fine-tune, increase the production of industrially relevant metabolites, even of cryptic origin. To overcome such problems, replacement of the native transcrip-tional regulatory elements with already characterized ones within the biosynthetic gene cluster can be helpful in the activation of cryptic genes in the heterologous host. The success stories of refactoring entire biosynthetic gene clusters for heter-ologous expression have motivated researchers to develop new tools. Methods such as use of Gibson and Golden Gate led to rapid assembly of even multiple PCR frag-ments in a user-designed way. All these methods have been successfully deployed for developing codon-optimized synthetic constructs for functional purposes. The other methods like use of seamless ligation cell extract (SLiCE) eliminate multiple transformation and cell growth steps and facilitate cloning as well as genetic manip-ulation of comparatively large DNA fragments with cell extracts from *E. coli* and lambda prophage optimized Red/ET system.

In a nut shell, advancements in molecular biology have significantly altered and increased the pace of natural products discovery. To further increase the pace of natural product discovery, availability of sequenced genomes, potential to extend limitations of whole genome sequencing as well as metagenomes, long-read nano-pore and get complete genomic information from a single cell sequencing will help in alleviating and providing access to cryptic clusters of environmental microbiomes and will also help us in alleviating the conventional barriers to discovery of sec-ondary metabolites for diverse applications. Therefore, the challenge is to develop universal heterologous expression systems, which can even activate the plethora of cryptic gene clusters present in the Earth's microbiome.

6.7 CONCLUDING REMARKS

The microbial secondary metabolites have immense potential, largely due to their applications in diverse sectors. Further, the advancement in genomic tools have already played vital role in decoding the future possibilities from both culturable and non-culturable sources. However, the challenges are to explore the major share of these microbial gene clusters normally silent under routine laboratory conditions for novel compounds and establish their biological role which can be used for several industrial applications.

ACKNOWLEDGMENT

Dr. Vivek Sharma and Dr. Richa Salwan are thankful to SEED Division, Department of Science and Technology, GOI for providing financial benefits (SP/YO/125/2017) and (SEED-TIASN-023-2018).

REFERENCES

Abdelmohsen UR, Bayer K, and Hentschel U. (2014). Diversity, abundance and natural products of marine sponge-associated actinomycetes. *Natural Product Reports* 31: 381–399.

Abdelmohsen UR, Grkovic T, Balasubramanian S, Kamel MS, Quinn RJ, and Hentschel U. (2015). Elicitation of secondary metabolism in actinomycetes. *Biotechnology Advances* 33(6): 798–811. https://doi.org/10.1016/j.biotechadv.2015.06.003

Abriouel H, Franz CMAP, Omar NB, and Gálvez A. (2011). Diversity and applications of *Bacillus* bacteriocins. *FEMS Microbiology Reviews* 35(1): 201–232. https://doi.org/10.1111/j.1574-6976.2010.00244.x

Aldridge BB, and Rhee KY. (2014) Microbial metabolomics: innovation, application, insight. *Current Opinion in Microbiology* 19: 90–96.

Aleti G, Sessitsch A, and Brader G. (2015). Genome mining: prediction of lipopeptides and polyketides from *Bacillus* and related Firmicutes. *Computational and Structural Biotechnology Journal* 13: 192–203. doi: 10.1016/j.csbj.2015.03.003

Altieri MA. (2004). Linking ecologists and traditional farmers in the search for sustainable agriculture. *Frontiers in Ecology and the Environment* 2: 35–42.

Argolo-Filho RC, and Loguercio LL. (2014). *Bacillus thuringiensis* is an environmental pathogen and host-specificity has developed as an adaptation to human-generated ecological niches. *Insects* 5(1): 62–91. doi:10.3390/insects5010062

Arguelles Arias A, Ongena M, Devreese B, Terrak M, Joris B, and Fickers P. (2013) Characterization of amylolysin, a novel lantibiotic from *Bacillus amyloliquefaciens* GA1. *PLoS ONE* 8: e83037.

Arthurs S, and Dara SK. (2019). Microbial biopesticides for invertebrate pests and their markets in the United States. *Journal of Invertebrate Pathology* 165, 13–21.

Ayed HB, Maalej H, Hmidet N, and Nasri M. (2015) Isolation and biochemical characterisation of a bacteriocin-like substance produced by *Bacillus amyloliquefaciens* An6. *Journal of Global Antimicrobial Resistance* 3: 255–261.

Bechet M, Caradec T, Hussein W, Abderrahmani A, Chollet M, Leclere V, Dubois T, Lereclus D., et al. (2012). Structure, biosynthesis, and properties of kurstakins, nonribosomal lipopeptides from Bacillus spp. *Applied Microbiology and Biotechnology* 95: 593–600.

Bader J, Mast-Gerlach E, Popovic MK, Bajpai R, and Stahl U. (2010). Relevance of microbial coculture fermentations in biotechnology. *Journal of Applied Microbiology* 109: 371–387.

Balciunas EM, Castillo Martinez FA, Todorov SD, Franco *Balciunas EM, Castillo Martinez FA, Todorov SD et al.*, Converti A, and Oliveira RPDS. (2013). Novel biotechnological applications of bacteriocins: a review. *Food Control* 32: 134–142.

Barea JM. (2015). Future challenges and perspectives for applying microbial biotechnology in sustainable agriculture based on a better understanding of plant-microbiome interactions. *Journal of Soil Science and Plant Nutrition* 15(2): 261–282.

Barea JM, Pozo MJ, Azcón R, and Azcón Aguilar C. (2013). Microbial interactions in the rhizosphere. In: FJ de Bruijn (ed.), *Molecular Microbial Ecology of the Rhizosphere*, vol 1, pp: 29–44. Hoboken, NJ: Wiley Blackwell.

Barkal LJ, Procknow CL, Álvarez-García YR, Niu M, Jiménez-Torres JA, Brockman-Schneider RA, Gern JE, Denlinger LC, Theberge AB, Keller NP, Berthier E, and Beebe DJ. (2017). Microbial volatile communication in human organotypic lung models. *Nature Communications* 8(1): 1770. doi:10.1038/s41467-017-01985-4

Barrios-Gonzalez J, Fernández FJ, Tomasini A, and Mejía A. 2005. Production of secondary metabolites by solid- state fermentation. *Malaysian Journal of Microbiology* 1: 1–6.

Beale D, Kouremenos K, and Palombo, E. (2016). *Microbial Metabolomics: Applications in Clinical, Environmental and Industrial Microbiology.* Cham, Switzerland: Springer.

Ben Abdallah D, Frikha-Gargouri O, and Tounsi S. (2018) Rizhospheric competence, plant growth promotion and biocontrol efficacy of *Bacillus amyloliquefaciens* subsp. *plantarum* strain 32a. *Biological Control* 124: 61–67.

Bérdy J. (2005). Bioactive microbial metabolites. *Journal of Antibiotics* 58(1): 1–26.

Berini F, Katz C, Gruzdev N, Casartelli M, Tettamanti G, and Marinelli F. (2018) Microbial and viral chitinases: attractive biopesticides for integrated pest management. *Biotechnology Advances* 36: 818–838.

Bisen PS, Debnath M, and Prasad GB. (2012). *Microbes: Concepts and Applications.* Wiley-Blackwell.

Bode HB, Bethe B, Hofs R, and Zeeck A. (2002). Big effects from small changes: possible ways to explore nature's chemical diversity. *Chembiochem* 3: 619–627.

Brader G, Compant S, Mitter B, Trognitz F, and Sessitsch A. (2014). Metabolic potential of endophytic bacteria. *Current Opinion in Biotechnology* 27: 30–37.

Breitling R, Jankevics A, Takano E, and Ana C. (2013). Metabolomics for secondary metabolite research. *Metabolites* 3: 1076–1083.

Canan U. (2013). Microorganisms in biological pest control—a review (bacterial toxin application and effect of environmental factors). In: M Silva-Opps (ed.), *Current Progress in Biological Research*, InTech. doi:10.5772/55786

Cawoy H, Wagner B, Fickers P, and Ongena M. (2011). Bacillus-based biological control of plant diseases. In: M Stoytcheva (ed.), *Pesticides in the Modern World – Pesticides Use and Management*, pp. 272–302. Rijeka: InTech.

Chandler D, Bailey AS, Tatchell GM, et al. (2011). The development, regulation and use of biopesticides for integrated pest management. *Philosophical Transactions of the Royal Society B* 366: 1987–1998.

Chatzipavlidis I, Kefalogianni L, Venieraki A, and Holzapfel W. (2013). Status and trends of the conservation and sustainable use of microorganisms in agroindustrial processes. http://www.fao.org/ docrep/meeting/028/mg339e.pdf

Chiang YM, Chang SL, Oakley BR, and Wang CC. (2011). Recent advances in awakening silent biosynthetic gene clusters and linking orphan clusters to natural products in microorganisms. *Current Opinion in Chemical Biology* 15: 137–143.

Chowdhury SP, Hartmann A, Gao X, and Borriss R. (2015). Biocontrol mechanism by root-associated *Bacillus amyloliquefaciens* FZB42 – a review. *Frontiers in Microbiology* 6: 1–11.

Covington BC, McLean JA, and Bachmann BO. (2017). Comparative mass spectrometry-based metabolomics strategies for the investigation of microbial secondary metabolites. *Natural Product Reports* 34: 6–24.

Crosa JH. (1989). Genetics and molecular biology of siderophore-mediated iron transport in bacteria. *Microbiological Reviews* 53(4): 517–530.

de Bruijn FJ. (2015). Biological nitrogen fixation. In: B Lugtenberg (ed.), *Principles of Plant-Microbe Interactions*, pp. 215–224. Heidelberg: Springer.

Demain AL. (2007). The business of biotechnology. *Industrial Biotechnology* 3: 269–283.

Demain AL, and Fang A. (2000). The natural functions of secondary metabolites. *Advances in Biochemical Engineering/Biotechnology* 69: 1–39.

Demain AL, and Sanchez S. (2009). Microbial drug discovery: 80 years of progress. *Journal of Antibiotics* 62: 5–16.

Demain AL. (1999). Pharmaceutically active secondary metabolites of microorganisms. *Applied Microbiology and Biotechnology* 52(4): 455–463.

Deravel J, Lemiere S, Coutte F, Krier F, Hese NV, Bechet M, Sourdeau N, Hofte M., et al. (2014). Mycosubtilin and surfactin are efficient, low ecotoxicity molecules for the biocontrol of lettuce downy mildew. *Applied Microbiology and Biotechnology* 98: 6255–6264.

Feher M, and Schmidt JM. (2003). Property distribution: difference between drugs, natural products and molecules from combinatorial chemistry. *Journal of Chemical Information and Computer Sciences* 43(1): 218–227.

Frias UA, De Kelly G, Pereira B, Guazzaroni M, and Silva-rocha R. (2018). Boosting secondary metabolite production and discovery through the engineering of novel microbial biosensors. *Biomed Research International*. doi:10.1155/2018/7021826

Frikha-Gargouri O, Ben Abdallah D, Ghorbel I, Charfeddine I, Jlaiel L, Triki MA, and Tounsi S. (2017). Lipopeptides from a novel *Bacillus methylotrophicus* 39b strain suppress agrobacterium crown gall tumours on tomato plants. *Pest Management Science* 73: 568–574.

Fuchs SW, Jaskolla TW, Bochmann S, Kotter P, Wichelhaus T, Karas M, Stein T, and Entian K-D. (2011). Entianin, a novel subtilin-like lantibiotic from *Bacillus subtilis* subsp. *spizizenii* DSM [15029.sup.T] with high antimicrobial activity. *Applied and Environmental Microbiology* 77: 1698–1707.

Garbeva P, Tyc O, Remus-Emsermann MN, van der Wal A, Vos M, Silby M, et al. (2011). No apparent costs for facultative antibiotic production by the soil bacterium *Pseudomonas fluorescens* Pf0-1. *PLoS One* 6: e27266.

Gelis-Jeanvoine S, Canette A, Gohar M, Caradec T, Lemy C, Gominet M, Jacques P, Lereclus D, et al. (2017). Genetic and functional analyses of krs, a locus encoding kurstakin, a lipopeptide produced by *Bacillus thuringiensis*. *Research in Microbiology* 168: 356–368.

Gong Q, Zhang C, Lu F, Zhao H, Bie X, and Lu Z. (2014) Identification of bacillomycin D from *Bacillus subtilis* fmbJ and its inhibition effects against *Aspergillus flavus*. *Food Control* 36: 8–14.

Gonzalez I, Niebla A, Lemus M, Gonzalez L, Iznaga IO, Perez ME, et al. (1999). Ecological approach of macrolide-lincosamides- streptogramin producing actinomyces from Cuban soils. *Letters in Applied Microbiology* 29: 147–150.

Gu Q, Yang Y, Yuan Q, Shi G, Wu L, Lou Z, Huo R, Wu H., et al. (2017). Bacillomycin D produced by *Bacillus amyloliquefaciens* is involved in the antagonistic interaction with the plant-pathogenic fungus *Fusarium graminearum*. *Applied and Environmental Microbiology* 83: e01075–e0117.

Gulledge BM, Aggen JB, Huang HB, Nairn AC, and Chamberlin AR (2002). The microcystins and nodularins: Cyclic polypeptide inhibitors of PP1 and PP2A. *Current Medicinal Chemistry* 9(22): 1991–2003.

Hamdache A, Lamarti A, Aleu J, and Collado IG. (2011). Non-peptide metabolites from the genus *Bacillus*. *Journal of Natural Products* 74: 893–899.

Hardoim PR, and van Elsas JD. (2013). Properties of bacterial endophytes leading to maximized host fitness. In: FJ de Bruijn (ed.), *Molecular Microbial Ecology of the Rhizosphere*, vol 1, pp. 405–411. Hoboken, NJ: Wiley Blackwell.

Hassan M, Kjos M, Nes IF, et al. (2012). Natural antimicrobial peptides from bacteria: characteristics and potential applications to fight against antibiotic resistance. *Journal of Applied Microbiology* 113(4): 723–736. doi:10.1111/j.1365-2672.2012.05338.x

Hathout Y, Ho Y-P, Ryzhov V, Demirev P, and Fenselau C. (2000). Kurstakin: a new class of lipopeptides isolated from Bacillus thuringiensis. *Journal of Natural Products* 63: 1492–1496.

Horak I, Engelbrecht G, Rensburg PJJ, and Van, Claassens S. (2019). Microbial metabolomics: Essential definitions and the importance of cultivation conditions for utilizing Bacillus species as bionematicides. *Journal of Applied Microbiology* 127(2): 326–343. https://doi.org/10.1111/jam.14218

Jin P, Wang H, Liu W, Fan Y, and Miao W. (2017). A new cyclic lipopeptide isolated from *Bacillus amyloliquefaciens* HAB-2 and safety evaluation. *Pesticide Biochemistry and Physiology* 147: 40–45.

Karadeniz A, Topcuog˘lu SF, and I˙nan S. (2006). Auxin, gibberellin, cytokinin and abscisic acid production in some bacteria. *World Journal of Microbiology and Biotechnology* 22(10): 1061–1064.

Katoch S, Kumari N, Salwan R, Sharma V, and Sharma PN. (2020). Recent developments in social network disruption approaches to manage bacterial plant diseases. *Biological Control* 150: 104376. https://doi.org/10.1016/j.biocontrol.2020.104376

Kieslich K. (1986). Production of drugs by microbial biosynthesis and biotransformation: possibilities, limits and future developments. 1st communication. *Arzneimittelforschung* 36(4): 774–778.

Lacey LA, and Siegel JP. (2000). Safety and ecotoxicology of entomopathogenic bacteria, in entomopatgenic bacteria: From laboratory to field application. In: JF Charles, A Delécluse, and CNL Roux. (eds.), *Entomopathogenic Bacteria: From Laboratory to Field Application*. Dordrecht: Springer. doi:10.1007/978-94-017-1429-7_14

Lange BM, Rujan T, Martin W, and Croteau R. (2000). Isoprenoid biosynthesis: The evolution of two ancient and distinct pathways across genomes. *Proceedings of the National Academy of Sciences of the United States of America* 97(24): 13172–13177. doi:10.1073/pnas.240454797

Lee S, Yap M, Behringer G, Hung R, and Bennett JW. (2016). Volatile organic compounds emitted by *Trichoderma* species mediate plant growth. *Fungal Biology and Biotechnology* 3: 7. https://doi.org/10.1186/s40694-016- 0025-7

Lee H, and Kim HY. (2011). Lantibiotics, class I bacteriocins from the genus *Bacillus*. *Journal of Microbiology and Biotechnology* 21: 229–235.

Li N, Alfiky A, Wang W, Islam M, Nourollahi K, Liu X, et al. (2018). Volatile compound-mediated recognition and inhibition between *Trichoderma* biocontrol agents and *Fusarium oxysporum*. *Frontiers in Microbiology* 9: 1–16.

Li MH, Ung PM, Zajkowski J, Garneau-Tsodikova S, and Sherman DH. (2009). Automated genome mining for natural products. *BMC Bioinformatics* 10: 185.

Li N, Islam MT, Kang S. (2019). Secreted metabolite-mediated interactions between rhizosphere bacteria and Trichoderma biocontrol agents. PloS one, 14(12), e0227228.

Liu Z, Budiharjo A, Wang P, Shi H, Fang J, Borriss R, Zhang K, and Huang X. (2013). The highly modified microcin peptide plantazolicin is associated with nematicidal activity of *Bacillus amyloliquefaciens* FZB42. *Applied Microbiology and Biotechnology* 97: 10081–10090.

Lugtenberg, B. (2015). Life of microbes in the rhizosphere. In: B Lugtenberg (ed.), *Principles of Plant-Microbe Interactions*, pp. 7–15. Switzerland, Heidelberg: Springer.

Macintyre L, Zhang T, Viegelmann C, Martinez IJ, Cheng C, Dowdells C, Abdelmohsen UR, Gernert C., et al. (2014). Metabolomic tools for secondary metabolite discovery from marine microbial symbionts. *Marine Drugs* 12: 3416–3448.

MacMillan J. (2001). Occurrence of gibberellins in vascular plants, fungi, and bacteria. *Journal of Plant Growth Regulation* 20: 387–442.

Martı´nez-Medina A, Van Wees SCM, and Pieterse CMJ. (2017). Airborne signals from *Trichoderma* fungi stimulate iron uptake responses in roots resulting in priming of jasmonic acid-dependent defences in shoots of *Arabidopsis thaliana* and *Solanum lycopersicum*. *Plant, Cell & Environment* 40: 2691–2705. https://doi.org/10.1111/pce.13016

McAuliffe O, Ross RP, and Hill C. (2001) Lantibiotics: Structure, biosynthesis and mode of action. *FEMS Microbiology Reviews* 25: 285–308.

Meadows MP (1993) *Bacillus thuringiensis* in the environment—ecology and risk assessment. In: PF Entwistle, JS Cory, MJ Bailey, and S Higgs (eds.), *Bacillus Thuringiensis: An Environmental Biopesticide; Theory and Practice*, pp. 193–220. Chichester: Wiley.

Medema MH, Blin K, Cimermancic P, de Jager V, Zakrzewski P, Fischbach MA, Weber T, Takano E, and Breitling R. (2011). AntiSMASH: Rapid identification, annotation and analysis of secondary metabolite biosynthesis gene clusters in bacterial and fungal genome sequences. *Nucleic Acids Research* 39: W339–W346.

Meena M, Swapnil P, Zehra A, Dubey MK, and Upadhyay RS. (2017). Antagonistic assessment of *Trichoderma* spp. by producing volatile and non-volatile compounds against different fungal pathogens. *Archives of Phytopathology and Plant Protection* 50: 629–648. https://doi.org/10.1080/03235408.2017.1357360

Mercado-Blanco J. (2015). Life of microbes inside the plant. In: B Lugtenberg (ed.), *Principles of Plant-Microbe Interactions*, pp. 25–32. Heidelberg: Springer.

Milshteyn A, Schneider JS, and Brady SF. (2014). Mining the metabiome: identifying novel natural products from microbial communities. *Chemistry & Biology* 21(9): 1211–1223. doi:10.1016/j.chembiol.2014.08.006

Mojid Mondol MA, Shin HJ, and Islam MT. (2013). Diversity of secondary metabolites from marine Bacillus species: chemistry and biological activity. *Mar Drugs* 11: 2846–2872.

Mongkolthanaruk, W. (2012). Classification of *Bacillus* beneficial substances related to plants, humans and animals. *World Journal of Microbiology and Biotechnology* 22: 1597–1604.

Mukherjee PK, Horwitz BA, Herrera-Estrella A, Schmoll M, and Kenerley CM. (2013). *Trichoderma* research in the genome era. *Annual Review of Phytopathology* 51: 105–129. https://doi.org/10.1146/annurev-phyto-082712-102353

Müller R, and Wink J. (2014). Future potential for anti-infectives from bacteria – how to exploit biodiversity and genomic potential. *International Journal of Medical Microbiology* 304: 3–13.

Nawaz M, Mabubu JI, and Hua H (2016). Current status and advancement of biopesticides: microbial and botanical pesticides. *Journal of Entomology and Zoology Studies* 4(2): 241–246.

Normand P, Queiroux C, Tisa LS, Benson DR, Rouy Z, Cruveiller S, and Medigue C. (2007). Exploring the genomes of *Frankia*. *Physiologia Plantarum* 130: 331–343.

Olivares J, Bedmar EJ, and Sanjuan J. (2013). Biological nitrogen fixation in the context of global change. *Molecular Plant-Microbe Interactions* 26: 486–494.

Palazzini JM, Dunlap CA, Bowman MJ, and Chulze SN. (2016). *Bacillus velezensis* RC 218 as a biocontrol agent to reduce Fusarium head blight and deoxynivalenol accumulation: genome sequencing and secondary metabolite cluster profiles. *Microbiological Research* 192: 30–36.

Paranagama PA, Wijeratne EMK, and Gunatilaka AAL. (2007). Uncovering biosynthetic potential of plant-associated fungi: Effect of culture conditions on metabolite production by *Paraphaeosphaeria quadriseptata* and *Chaetomium chiversii*. *Journal of Natural Products* 70: 1939–1945.

Paraszkiewicz K, Bernat P, Siewiera P, Moryl M, Paszt LS, Trzcinski P, Jałowiecki Ł, and Płaza G. (2017). Agricultural potential of rhizospheric *Bacillus subtilis* strains exhibiting varied efficiency of surfactin production. *Scientia Horticulturae* 225: 802–809.

Pathma J, Rahul GR, Kennedy KR, Subashri R, and Sakthivel N. (2021). Secondary metabolite production by bacterial antagonists Secondary metabolite production by bacterial antagonists. *Journal of Biological Control* 25(3): 165–181.

Perez-García A, Romero D, and de Vicente A. (2011). Plant protection and growth stimulation by microorganisms: biotechnological applications of Bacilli in agriculture. *Current Opinion in Biotechnology* 22: 187–193.

Piechulla B, Lemfack MC, and Kai M. (2017). Effects of discrete bioactive microbial volatiles on plants and fungi. *Plant, Cell & Environment* 40: 2042–2067. https://doi.org/10.1111/pce.13011

Pinu FR, Villas-Boas SG, and Aggio R. (2017). Analysis of intracellular metabolites from microorganisms: quenching and extraction protocols. *Metabolites* 7: 53.

Porras-Alfaro A, and Bayman P. (2011). Hidden fungi, emergent proper-ties: endophytes and microbiomes. *Annual Review of Phytopathology* 49: 291–315.

Rangarajan V, and Clarke KG. (2016). Towards bacterial lipopeptide products for specific applications – a review of appropriate downstream processing schemes. *Process Biochemistry* 51: 2176–2185.

Rao PV, Gupta N, Bhaskar AS, and Jayaraj R. (2002). Toxins and bioactive compounds from cyanobacteria and their implications on human health. *Journal of Environmental Biology* 23(3): 215–224.

Rappe MS, and Giovannoni SJ (2003). The uncultured microbial majority. *Annual Review of Microbiology* 57: 369–394. [PubMed: 14527284]

Ruiz B, Chavez A, Forero A, Garcia-Huante Y, Romero A, Sanchez M, Rocha D, Sanchez B, Rodriguez-Sanoja R, Sanchez S, and Langley E. (2010). Production of microbial secondary metabolites: Regulation by the carbon source. *Critical Reviews in Microbiology* 36:146–167.

Sagova-Mareckova M, Ulanova D, Sanderova P, Omelka M, Kamenik Z, Olsovska J, et al. (2015). Phylogenetic relatedness determined between antibiotic resistance and 16S rRNA genes in actinobacteria. *BMC Microbiology* 15: 81.

Salwan, Richa, Nidhi Rialch, and Vivek Sharma. (2019) "Bioactive volatile metabolites of Trichoderma: An overview." Secondary Metabolites of Plant Growth Promoting Rhizomicroorganisms 87–111.

Salwan R, and Sharma V. (2020a). Genomics Genome wide underpinning of antagonistic and plant beneficial attributes of Bacillus sp. SBA12 Fusarium oxysporum Phytopthora infestans Sclerotinia sclerotiorum. *Genomics* 112(4): 2894–2902. https://doi.org/10.1016/j.ygeno.2020.03.029

Salwan R, and Sharma V. (2020b). Molecular and biotechnological aspects of secondary metabolites in actinobacteria. *Microbiological Research* 231(2020): 126374. https://doi.org/10.1016/j.micres.2019.126374

Salwan R, Sharma V, Sharma A, and Singh A. (2020). Molecular imprints of plant beneficial *Streptomyces* sp. AC30 and AC40 reveal differential capabilities and strategies to counter environmental stresses. *Microbiological Research* 235: 126449. https://doi.org/10.1016/j.micres.2020.126449

Sansinenea E, and Ortiz A. (2011). Secondary metabolites of soil *Bacillus* spp. *Biotechnology Letters* 33: 1523–1538.

Savile CK, Janey JM, Mundorff EC, et al. (2010). Biocatalytic asymmetric synthesis of chiral amines from ketones applied to sitagliptin manufacture. *Science* 329(5989): 305–309. doi:10.1126/science.1188934

Schmidt R, Ulanova D, Wick LY, and Bode HB. (2019). Microbe-driven chemical ecology: past, present and future. *ISME Journal* 11: 2656–2663. doi:10.1038/s41396-019-0469-x

Schunemann R, Knaak N, and Fiuza LM. (2014). Mode of action and specificity of *Bacillus thuringiensis* toxins in the control of caterpillars and stink bugs in soybean culture. *ISRN Microbiol* 135675: 12. doi:10.1155/2014/135675

Schüßler A, Schwarzott D, and Walker C. (2001). A new fungal phylum, the Glomeromycota, phylogeny and evolution. *Mycological Research* 105: 1413–1421.

Schwarzer D, Finking R, and Marahiel MA. (2003). Nonribosomal peptides: from genes to products. *Natural Product Reports* 20: 275–287.

Shanmugam V, Singh Ajit N, Verma R, and Sharma V. (2008). Diversity and differentiation among fluorescent pseudomonads in crop rhizospheres with whole-cell protein profiles. *Microbiological Research* 163(5): 571–578. https://doi.org/10.1016/j.micres.2006.08.006

Sharma V, Bhandari P, Singh B, Bhatacharya A, and Shanmugam V. (2013). Chitinase expression due to reduction in fusaric acid level in an antagonistic *Trichoderma harzianum*

S17TH. *Indian Journal of Microbiology* 53(2): 214–220. https://doi.org/10.1007/s12088-012-0335-2

Sharma V, Salwan R, and Shanmugam V. (2018a). Molecular characterization of β-endoglucanase from antagonistic *Trichoderma saturnisporum* isolate GITX-Panog (C) induced under mycoparasitic conditions. *Pesticide Biochemistry and Physiology* 149: 73–80. https://doi.org/10.1016/j.pestbp.2018.06.001

Sharma V, Salwan R, and Shanmugam V. (2018b). Unraveling the multilevel aspects of least explored plant beneficial *Trichoderma saturnisporum* isolate GITX-Panog (C). *European Journal of Plant Pathology* 152(1): 169–183. https://doi.org/10.1007/s10658-018-1461-4

Sharma V, Salwan R, and Sharma PN. (2016a). Differential Response of extracellular proteases of *Trichoderma harzianum* against fungal phytopathogens. *Current Microbiology* 73(3): 419–425. https://doi.org/10.1007/s00284-016-1072-2

Sharma V, Salwan R, Sharma PN, and Kanwar SS. (2016b). Molecular cloning and characterization of ech46 endochitinase from *Trichoderma harzianum*. *International Journal of Biological Macromolecules* 92: 615–624. https://doi.org/10.1016/j.ijbiomac.2016.07.067

Sharma V, Salwan R, Sharma PN, and Kanwar SS. (2017). Elucidation of biocontrol mechanisms of *Trichoderma harzianum* against different plant fungal pathogens: universal yet host specific response. *International Journal of Biological Macromolecules* 95: 72–79. https://doi.org/10.1016/j.ijbiomac.2016.11.042

Singh A, Shukla N, Kabadwal BC, Tewari AK, and Kumar J. (2018). Review on plant-*Trichoderma*-pathogen inter-action. *International Journal of Current Microbiology and Applied Science* 7: 2382–2397. https://doi.org/10.20546/ijcmas.2018.702.291

Singh R, Kumar M, Mittal A, and Mehta PK. (2017). Microbial metabolites in nutrition, healthcare and agriculture. *3 Biotech* 7(1): 15. https://doi.org/10.1007/s13205-016-0586-4

Smith SE, and Read DJ. (2008). *Mycorrhizal Symbiosis*, 3rd ed. San Diego, CA: Academic Press.

Spence C, and Bais H. (2015). Role of plant growth regulators as chemical signals in plant–microbe interactions: a double edged sword. *Current Opinion in Plant Biology* 27: 52–58. doi:10.1016/j.pbi.2015.05.028

Starcevic, A, Zucko J, Simunkovic J, Long PF, Cullum J, and Hranueli D. (2008). ClustScan: an integrated program package for the semi-automatic annotation of modular biosynthetic gene clusters and in silico prediction of novel chemical structures. *Nucleic Acids Research* 36: 6882–6892.

Stintzi A, Barnes C, Xu J, and Raymond KN. (2000) Microbial iron transport via a siderophore shuttle: a membrane ion transport paradigm. *Proceedings of the National Academy of Sciences of the United States of America USA* 97(20): 10691–10696.

Sumi CD, Yang BW, Yeo I-C, and Hahm YT. (2015) Antimicrobial peptides of the genus *Bacillus*: a new era for antibiotics. *Canadian Journal of Microbiology* 61: 93–103.

Thirumurugan D, Cholarajan A, Raja Suresh SS, and Vijayakumar R. (September 5, 2018). An introductory chapter: secondary metabolites. In: R Vijayakumar and Suresh SS Raja (ed.), *Secondary Metabolites - Sources and Applications*, IntechOpen. doi:10.5772/intechopen.79766. Available from: https://www.intechopen.com/books/secondary-metabolites-sources-and-applications/an-introductory-chapter-secondary-metabolites

Tomá Ř, Palyzová A, Sigler K, and Sigler K. (2018). Isolation and identification of siderophores produced by cyanobacteria. *Folia Microbiologica* 63: 569–579.

Torsvik V, Daae FL, Sandaa RA, and Ovreas L, (1998). Novel techniques for analysing microbial diversity in natural and perturbed environment. *Journal of Biotechnology* 64: 53–62.

Tsavkelova EA, Klimova S-Yu, Cherdyntseva TA, and Netrusov AI. (2006). Hormones and hormone-like substances of microorgan- isms: a review. *Applied Biochemistry and Microbiology* 42(3): 229–235.

Tyc O, Song C, Dickschat JS, Vos M, and Garbeva P. (2017) The ecological role of volatile and soluble secondary metabolites produced by soil bacteria. *Trends in Microbiology* 25: 280–292.

Udwary D, et al. (2007). Genome sequencing reveals complex secondary metabolome in the marine actinomycete Salinispora tropica. *Proceedings of the National Academy of Sciences of the United States of America* 104: 10376–10381.

Vaidyanathan S. (2005). Profiling microbial metabolomes: what do we stand to gain? *Metabolomics* 1: 17–28.

van der Heijden MGA, Martin FM, Selosse MA, and Sanders IR. (2015). Mycorrhizal ecology and evolution: the past, the present, and the future. *New Phytologist* 205: 1406–1423.

Villas-Boas SG, Roessner U, Hansen MAE, Smedsgaard J, and Nielsen J. (2007). *Metabolome Analysis: An Introduction*. Hoboken, NJ: John Wiley & Sons.

Wang T, Liang Y, Wu M, Chen Z, Lin J, and Yang L. (2015). Review: natural products from Bacillus subtilis with antimicrobial properties. *Chinese Journal of Chemical Engineering* 23: 744–754.

Wei H, Lin Z, Li D, Gu Q, and Zhu T. (2010). OSMAC (one strain many compounds) approach in the research of microbial metabolites. *Wei Sheng Wu Xue Bao* 50: 701–709.

Wheatley R, Hackett C, Bruce A, and Kundzewicz A. (1997). Effect of substrate composition on production of volatile organic compounds from *Trichoderma* spp. inhibitory to wood decay fungi. *International Biodeterioration & Biodegradation* 39: 199–205. https://doi.org/10.1016/S0964-8305(97)00015-2

Yuan J, Zhao M, Li R, Huang Q, Raza W, Rensing C, et al. (2017). Microbial volatile compounds alter the soil microbial community. *Environmental Science and Pollution Research* 24: 22485–22493. https://doi.org/10.1007/s11356- 017-9839-y

Yurekli F, Yesilada O, Yurekli M, and Topcuoglu SF. (1999). Plant growth hormone production from olive oil mill and alcohol factory wastewaters by white rot fungi. *World Journal of Microbiology and Biotechnology* 15: 503–505.

Zhang L, Fasoyin OE, Molnár I, and Xu Y. (2020). Secondary metabolites from hypocrealean entomopathogenic fungi: novel bioactive compounds. *Natural Product Reports* 37(9): 1181–1206. https://doi.org/10.1039/c9np00065h

Zhao X, and Kuipers OP. (2016). Identification and classification of known and putative antimicrobial compounds produced by a wide variety of Bacillales species. *Genomics* 17: 882.

7 Antimicrobial Resistance in Agriculture
An Emerging Threat to the Ecosystem

Vijay Laxmi Shrivas[1], P. Hariprasad[1],
Anil K. Choudhary[2], and Shilpi Sharma[3]

[1]Indian Institute of Technology Delhi, New Delhi, India
[2]Central Potato Research Institute, Himachal Pradesh, India
[3]Indian Institute of Technology Delhi, New Delhi, India

CONTENTS

7.1 INTRODUCTION

Since the 19th century, antimicrobial compounds produced by microorganisms to kill other microbes have been termed "antibiotics" (Waksman 1947). Antibiotics include a wide variety of natural, synthetic, and semi-synthetic chemicals. Based on the mode of action against pathogens, antibiotics are either cidal (killing of microbes) or static (inhibiting the growth of microbes). Their effectiveness against pathogens categorizes them as broad or narrow range antibiotics (Martínez 2012; Milić et al. 2013). A narrow range antibiotic refers to one with efficiency against selective microbial species, whereas broad range antibiotics are effective against a range of microbial infections. Further, antibiotics can be classified based on their mechanisms of action, such as inhibition of synthesis of cell wall, proteins, and nucleic acids, alteration of cell membrane structure and permeability, and so on (Kümmerer 2009).

Antibiotics are widely used in different sectors such as healthcare, veterinary, and agriculture. The trend of antibiotic consumption, expressed in terms of defined daily dose (DDD), was analyzed over the period of 2000–2015 in 76 countries (Klein et al. 2018). An increase of 65% from 21.1 to 34.8 billion DDD in consumption was reported during the period. In the year 2010, major consumption was reported in China (23%), United States (13%), Brazil (9%), and 3% each in Germany and India (Amábile-Cuevas 2016; van Boeckel et al. 2015).

Indiscriminate use of antibiotics over time leads to the evolution of bacteria that resist the antimicrobials, resulting in antimicrobial resistance (AMR). Resistance occurs due to mutations in microbes or transfer of resistance due to horizontal gene transfer under the selection pressure of antimicrobials (Fair and Tor 2014). Various mechanisms involved in the development of resistance in microbes include modification in cell membrane permeability, degradation, and protein alteration. Resistance can either be developed within the bacterial chromosome or in the plasmid (Dever and Dermody 1991). As per the World Health Organization, AMR has surfaced as one of the ten major health threats globally in 2019 (WHO 2019).

7.2 APPLICATION OF ANTIBIOTICS IN AGRICULTURE AND ANIMAL FARMING

Antibiotics are popularly used in agriculture and veterinary sectors including animal husbandry and aquaculture, to treat various microbial infections and diseases. Administration of antibiotics in veterinary can be further categorized into three groups based on their applications, *viz.* therapeutic agent, prophylactic, and growth-promoting agent. Therapeutic agents are a group of antimicrobials that act against the pathogens either by reducing the metabolic activity or by killing the pathogen. A prophylactic medication works as a preventive measure for any disease to occur, generally used in the case of asymptomatic diseases. To increase the natural growth and productivity of animals, feeding antibiotics in low dosage works as a growth promoter (Wegener et al. 1999). According to the available data, the annual consumption of antibiotics in the agricultural sector varies from 63,000 to 240,000 tons globally (United Nations 2015). However, the data is probably an underrepresentation due to limited data collection and weak surveillance systems in several countries (van Boeckel et al. 2015). Though half of the antibiotics used in agriculture are of medical importance for humans, they are commonly used in the veterinary sector as well (O'Neill 2015). In 2015, as per sales statistics, three major groups of antibiotics belonging to penicillin, macrolides, and tetracyclines, with a cost of ~ US$500 million each, were being sold worldwide (Laxminarayan et al. 2015). The usual annual intake of antibiotics in animal farming has been estimated to be around 14.8 g per kg for chicken, 17.2 g per kg for pig, and 4.5 g per kg for cattle (van Boeckel et al. 2015).

7.3 ROUTE OF ENTRY OF ANTIMICROBIAL-RESISTANT BACTERIA AND GENES (ARB AND ARGs) IN AGRICULTURE

In the agro-systems, antibiotics enter as contaminants through various sources including the application of antibiotics for crop protection. Animal manure, loaded with a variety of antibiotics as residues, is also applied as fertilizer (Heuer et al. 2009). Further, antibiotics have been used as livestock feed for growth promotion in animal farming (Wegener et al. 1999). Irrigation with water contaminated with antibiotic residues leads to contamination of groundwater and surface water (Du and Liu 2012). Contamination also occurs due to a huge percentage of antibiotics being released in agriculture through treated sewage sludge biosolid or manure as composts (Thiele-Bruhn 2003). Aerial transport through particulate matter also acts as a source of contamination in agriculture (McEachran et al. 2015). The possible sources and fates of antibiotics in the agricultural system have been illustrated in Figure 7.1.

7.4 IMPACT OF MANURE APPLICATION IN AGROECOSYSTEMS ON THE SPREAD OF ANTIBIOTIC RESISTANCE

Excessive use of antimicrobials in animal farming leads to the accumulation of antibiotics in animal-based products, and their release into the environment through animal waste (manure and urine) (Manyi-Loh et al. 2018; Quaik et al. 2020). Over the last two decades, many studies have reported the spread of antibiotic resistance

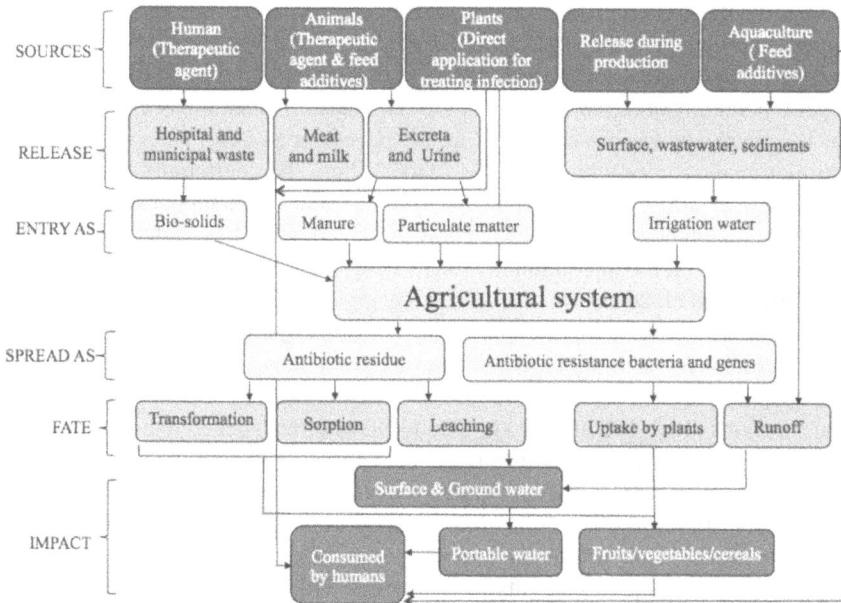

FIGURE 7.1 Entry and fate of antibiotics in the agricultural system.

through manure application (Kumar et al. 2005; Udikovic-Kolic et al. 2014; Grenni et al. 2018). It has been observed that manure fertilization increases the abundance of several ARGs, integrons, and plasmids in the soil as well as biosolids (Munir and Xagoraraki 2011; Marti et al. 2013; McKinney et al. 2018). The efficiency of horizontal gene transfer is enhanced in the soil upon application of manure loaded with antibiotics and ARB (Heuer et al. 2011). The amount of antibiotics released by livestock in their excreta depends on the type of drugs administered to these animals (Sarmah et al. 2006). For instance, of the total antibiotic fed to the animals approximately 90%, 65%, and 50–100%, for sulphonamides, tetracyclines, and macrolides, respectively, were detected in their excreta (Halling-Sørensen et al. 2001; Winckler and Grafe 2001; Arikan et al. 2009). Winckler and Grafe (2001) checked the stability of tetracycline in pig manure under different storage conditions, and found that about 72% of the tetracycline could be recovered at the end of the experiment. Similarly, about 95% of the ^{14}C-labeled difloxacin administered was excreted in the pig manure (Sukul et al. 2009). When the residual antibiotic levels were compared in animals' excreta (pig, cattle, and poultry), the levels were found to be maximum in pig manure (Zhang et al. 2008). Studies have reported that the metabolite's retransformation to the original compound further enhances the concentration of antibiotics in animal excreta (Kuppusamy et al. 2018).

7.5   POTENTIAL EFFECTS OF ANTIBIOTICS ON THE SOIL MICROBIOME

Microbes play key roles in numerous biochemical processes to ensure soil's health by maintaining the nutrient balance and organic matter content, which enhances soil fertility by stabilizing the soil structure (Bian et al. 2014; Mendes et al. 2015). Microbes also exhibit defensive roles against various pathogens within the soil matrix. For the proper functioning of the soil microbiome, it is crucial to maintain its diversity, which may be affected by various ecological factors (Cycoń et al. 2019; Shrivas et al. 2019). Antimicrobial activity within the soil matrix alters the community compositions (Kotzerke et al. 2008; Molaei et al. 2017). Various studies have supported the fact that the effect of antibiotics can be seen not only within an individual bacterial population but also on the microbial community as a whole (Hammesfahr et al. 2008; Reichel et al. 2015; Xu et al. 2016).

Hammesfahr et al. (Hammesfahr et al. 2008) analyzed the microbial community structure using techniques like denaturing gradient gel electrophoresis (DGGE) and phospholipid fatty acids (PLFAs) from agricultural soils containing sulfadiazine concentrations of 10–100 mg per kg. The results showed a decrement in the proportion of bacteria and fungi as estimated by PLFA analysis. The negative impact of sulfadiazine was also seen on β-*Proteobacteria* and pseudomonads. When the effect of different concentrations of sulfonamide group of antibiotics like sulfamethoxazole, sulfadimethoxine, and sulfamethazine was assessed on the bacterial communities using PLFA profiles, similar results were obtained with bacterial versus fungal ratios being dependent on the antibiotic dosage (Gutiérrez et al. 2010).

In recent times, the change in total bacterial diversity has been assessed in different environments by employing techniques like quantitative polymerase chain reaction (qPCR) and amplicon sequencing of 16S rRNA gene (Xu et al. 2016; Lopatto et al. 2019). Ding et al. (2014) concluded that the application of animal manure may result in the alteration of selective taxa that could further suppress the other bacterial taxa. In the study, the effect of sulphadiazine on the bacterial community was analyzed by pyrosequencing of the 16S rRNA gene after 193 days of application of the antibiotic. A decrement in the stability of the community was seen in the soil amended with sulphadiazine antibiotic as compared to the initial bulk composition. Additionally, several human pathogens including *Shinella*, *Clostridium*, *Leifsonia*, *Devosia*, *Gemmatimonas*, *Peptostreptococcus*, and *Stenotrophomonas* could be detected in the soil. In a recent study, the impact of antibiotic-amended manure on 15-year-old agricultural fields was assessed using high throughput qPCR (Wang et al. 2020). The result revealed an increased abundance of antibiotic resistance genes in manure-laden treatment as compared to the chemical-treated field. Moreover, the presence of mobile genetic elements (MGEs) in manure-laden treatment indicated the spread of ARGs via horizontal gene transfer, which is responsible for shaping the bacterial community. Recently, it has been reported that the presence of native soil microbes may reduce the spread of antibiotic resistance in soil (Pérez-Valera et al. 2019). Antibiotics have been shown to affect even the endophytic microbial population in crops. In a recent study, endophytic bacteria isolated from the leafy vegetables showed resistance to various antibiotics and heavy metals (Karmakar et al. 2019).

Many reports have targeted specific microbial functions like soil enzyme activities and biogeochemical cycles of nitrogen, carbon, phosphorus, iron, and sulphur (Cui et al. 2014; Liu et al. 2015; Ma et al. 2016). Activities of soil enzymes have been reported to exhibit a direct correlation with the properties of antibiotics and their dosage, and soil properties (Thiele-Bruhn and Beck 2005; Fang et al. 2014; Cao et al. 2015). Enzyme activity has been reported to be reduced at higher concentrations of antibiotics, while lower concentration of antibiotics has been observed to have a positive impact on enzymes (Boleas et al. 2005; Thiele-Bruhn and Beck 2005; Cao et al. 2015). Chen et al. (2013) assessed the effect of oxytetracycline on various enzymatic activities such as alkaline phosphatase, arylsulfatase, dehydrogenase, and urease. Results suggested that small and moderate amounts of oxytetracycline (~15 mg per kg of soil) enhanced the activity of urease, while oxytetracycline at 1 mg per kg of soil increased the activity of arylsulfatase. Increasing the concentration of oxytetracycline up to 200 mg per kg of soil was reported to adversely affect the activities of all the enzymes tested. Similarly, the combined effect of two antibiotics (chlortetracycline and sulfapyridine) was studied on activities of alkaline phosphatase, dehydrogenase, and urease on the 1st, 4th, and 21st day of application (Molaei et al. 2017). Results concluded that both antibiotics had an immediate negative impact on the enzyme activities, *viz.* sulfapyridine negatively affected the enzyme alkaline phosphatase on all the sampling points, while the effect of tetracycline on alkaline phosphatase was reduced after the 4th day. Chlortetracycline and sulfapyridine were found to be effective on the very first day in impacting urease activity.

However, the efficacy of sulfapyridine reduced progressively with time with respect to its effect on dehydrogenase activity.

The adverse effect of various narrow (olaquindox, oxolinic acids, and tylosin) and broad (aminoglycosides, sulfonamides, tetracyclines) range of antibiotics has been reported on nitrification (Halling-Sørensen 2001; Kumar et al. 2005). Kumar et al. (2005) observed that the application of manure with antibiotics hampered the process of mineralization and decomposition in soil. Oxytetracycline was found to reduce the nitrification process because of elevated ammonium ions in the soil system (Cao et al. 2015). Similarly, the effect of sulphadiazine was seen on nitrification at a higher dose of antibiotic (around 100 mg per kg), while the lower dose (10 mg per kg) did not exert any negative impact (Kotzerke et al. 2008). A higher concentration of sulphonamide group of antibiotics resulted in the blockage of the iron reduction process in soil up to 50 days (Toth et al. 2011).

7.6   THE SPREAD OF ANTIBIOTIC RESISTANCE IN SOIL AND ITS INFLUENCE ON CROP PLANTS AND HUMAN HEALTH

Upon entering into the soil system, antibiotics do not degrade easily, hence are retained in the system for a while (Aust et al. 2010; Ostermann et al. 2013). Antibiotic residues on the soil surface get absorbed by the plants and subsequently result in bioaccumulation and phytotoxicity. There are multiple studies where accumulation, absorption, and transport of antibiotics have been reported in vegetable crops such as cabbage, onion, potato, and leek (Kumar et al. 2005; Dolliver et al. 2007; Wang and Han 2008; Zhang et al. 2019). A study conducted to check the migration patterns of veterinary antibiotics revealed that water transport and passive absorption were the main sources of migration of antibiotics to vegetables (Hu et al. 2010). The order of distribution was leaf > stem > root. The concentration of humus has been reported to play a role in the bioaccumulation of antibiotics in plants with an increasing concentration of humus negatively affecting the absorption of antibiotics by plants (Jjemba 2002). The studies related to accumulation of antibiotics in plants and vegetables have been compiled in Table 7.1.

Consumption of antibiotic-accumulated plants pose a high risk to human health as well (Chang et al. 2015). Forsberg et al. (2012) investigated the nucleotide identity link between soil bacteria (multi-drug resistant with aminoglycosides, β-lactams sulfonamides, and tetracyclines) and a pathogen of human origin. Antibiotic resistance cassettes found in soil bacteria against a variety of antibiotics (aminoglycosides, β-lactams, sulfonamides, and tetracyclines) had comparable nucleotide similarity in mobile elements and non-coding sections, indicating the occurrence of lateral exchange. Human interactions with the resistant bacteria in the agricultural system, and/or the consumption of raw vegetables loaded with resistant bacteria, enhances the chances of the transfer of resistance factor between the soil microbiome and humans (Kuppusamy et al. 2018). This may result in the development of incurable diseases in humans. The process of development of resistance in human pathogens from agriculture is complex and not completely understood. Smith et al. (2005) suggested the risk of transmission of resistance from agri-based systems to humans to be comparable to hospital-transmitted infection.

TABLE 7.1

Selected Studies Reporting Accumulation of Antibiotics in Plants and Vegetables

Type/Class of Antibiotic	Antibiotic	Crop/Plant	Plant/Vegetable Parts	Accumulation Range (in μg per kg)	Source of Antibiotic	Reference
Amphenicols	Florfenicol	Lettuce	Lettuce-leaves	15	Soil spiking	Boxall et al. 2006
		Carrot	Whole	5		
Cephalosporin	Cephalexin	Pakchoi	Edible part	26.4–48.1	-	Zhang, H. et al. 2017
Chloramphenicol	Chloramphenicol	Lettuce	Leaf	3.20	Irrigation with wastewater and fertilization with animal manure	Pan and Chu 2017
			Root	1.18		
		Carrot	Leaf	1.07		
			Root	1.62		
		Tomato	Fruit	4.34		
			Leaf	2.33		
			Root	1.21		
Fluoroquinolones	Norfloxacin	Lettuce	Leaf	2.88	Irrigation with wastewater and fertilization with animal manure	Pan and Chu 2017
			Root	2.64		
		Carrot	Leaf	3.03		
			Root	2.52		
		Tomato	Fruit	3.42		
			Leaf	2.70		
			Root	2.40		
	Enrofloxacin	Lettuce	Leaves	6–170	Soil spiking	Boxall et al. 2006
		Carrot	Whole			

(Continued)

TABLE 7.1 *(Continued)*
Selected Studies Reporting Accumulation of Antibiotics in Plants and Vegetables

Type/Class of Antibiotic	Antibiotic	Crop/Plant	Plant/Vegetable Parts	Accumulation Range (in µg per kg)	Source of Antibiotic	Reference
Ionophoric	Monensin	Spinach	–	3.0	Turkey and hog manure	Kang et al. 2013
		Lettuce				
		Cabbage				
		Carrot				
		Radish				
		Onion				
		Garlic				
		Tomato				
		Green Bell Pepper				
		Sweet Corn				
		Potato				
Imidazothiazoles	Levamisole	Lettuce	Leaves	170	Soil spiking	Boxall et al. 2006
		Carrot	Whole	<11		
Macrolides	Erythromycin	Lettuce	Leaf	Not detectable	Irrigation with wastewater and fertilization with animal manure	Pan and Chu 2017
			Root			
		Carrot	Leaf			
			Root			
		Tomato	Fruit			
			Leaf			
			Root			

(Continued)

TABLE 7.1 (Continued)
Selected Studies Reporting Accumulation of Antibiotics in Plants and Vegetables

Type/Class of Antibiotic	Antibiotic	Crop/Plant	Plant/Vegetable Parts	Accumulation Range (in µg per kg)	Source of Antibiotic	Reference
Streptogramin	Virginiamycin	Spinach, Lettuce, Cabbage, Carrot, Radish, Onion, Garlic, Tomato, Green Bell Pepper, Sweet Corn, Potato		0.4	Turkey and hog manure	Kang et al. 2013
Sulphonamides	Sulfadiazine	Lettuce	Leaves	≤17	Soil spiking	Boxall et al. 2006
		Carrot	Whole	≤6.1		Dolliver et al. 2007
		Lettuce		100–1200		
		Potato				
		Spinach, Lettuce, Cabbage, Carrot, Radish, Onion, Garlic, Tomato, Green Bell Pepper, Sweet Corn, Potato		0.8	Turkey and hog manure	Kang et al. 2013
	Sulfamethoxazole	Pakchoi	Edible part	18.1 to 35.3	–	Zhang, H. et al. 2017

(Continued)

TABLE 7.1 *(Continued)*
Selected Studies Reporting Accumulation of Antibiotics in Plants and Vegetables

Type/Class of Antibiotic	Antibiotic	Crop/Plant	Plant/Vegetable Parts	Accumulation Range (in µg per kg)	Source of Antibiotic	Reference
Tetracycline	Tetracycline	Lettuce	Leaf	~1.32	Irrigation with wastewater and fertilization with animal manure	Pan and Chu 2017
			Root			
		Carrot	Leaf			
			Root			
			Fruit			
		Tomato	Leaf			
			Root			
		Pakchoi	Edible part	6.9–11.8	-	Zhang, H. et al. 2017
	Chlortetracycline	Spinach	-	0.4	Turkey and hog manure	Kang et al. 2013
		Lettuce				
		Cabbage				
		Carrot				
		Radish				
		Onion				
		Garlic				
		Tomato				
		Green Bell Pepper				
		Sweet Corn				
		Potato				
	Oxytetracycline	Peanut	Stem	15.33–14.22	-	Zhao et al. 2019
			Leaves	21.84–24.95		
			Kernel	20.06–13.42		
			Shells	14.92–21.35		

7.7 TRANSFER OF ANTIBIOTIC RESISTANCE TO GROUNDWATER

Antibiotics from the surface water and soil reach the groundwater by the process of leaching and run-off. Sulphonamide has been reported to be leached into the groundwater from manure-treated fields (Hamscher et al. 2005). Blackwell et al. (2007) estimated the concentrations of oxytetracycline (0.9 μg L^{-1}) and sulfachloropyridazine (26 μg L^{-1}) in surface run-off samples. In a recent study, the manure composition shaped the resistome of an agricultural drainage system and led to the elevation of ARGs in agricultural water samples (Smith et al. 2019). Similarly, the transportation of lincomycin (~0.005 μg L^{-1}) was recorded in the groundwater from the manured field (Kuchta et al. 2009). Various studies have described the leaching of numerous antibiotics (sulfadiazine, sulfamethoxazole, sulfapyridine, sulfadimidine, dehydratoerythromycin, tylosin, tetracycline, and trimethoprim) into the surface and groundwater (Wehrhan et al. 2007; Burke et al. 2016). Burke et al. (2016) estimated the sorption affinity of antibiotic trimethoprim towards sandy aquifer material to be 5.7, which was responsible for greater leaching ability in water. Recently, a comparison between resistance profiles of antibiotics in surface water and groundwater was done by analyzing the abundance of ARGs using qPCR (Wu et al. 2020). The results revealed the dominance of sulphonamide, aminoglycosides, and multidrug-resistant genes in all the samples. Groundwater, river water, and sediment had a total of 116 ARGs in common, which accounted for 67.1% of all the genes detected.

7.8 DEGRADATION PATTERNS OF ANTIMICROBIALS IN THE ENVIRONMENT

The amount of antibiotic residues present in the environment may be affected by many factors (biotic and abiotic) and processes such as transformation (Duan et al. 2017; Reichel et al. 2013), sorption (Leal et al. 2013; Martínez-Hernández et al. 2016), leaching, runoff (Kuchta et al. 2009; Park and Huwe 2016; Pan and Chu 2017), plant uptake, and so on (Kumar et al. 2005; Dolliver et al. 2007; Kuchta et al. 2009; Carter et al. 2014). Factors like soil texture, moisture content, pH, temperature, organic matter content also affect the degradation rate (Koba et al. 2017). Koba et al. (2017) recorded the transformation of 3 different antibiotics (clindamycin, sulfamethoxazole, and trimethoprim) in 12 different soils up to 61 days. While antibiotics degraded over time, their metabolites remained persistent in the soil. When the degradation rate of oxytetracycline in two different agricultural soils was compared, the DT50 values were found to be 30.2 (for low organic carbon content) and 39.4 (for high organic carbon) days (Ling-Ling et al. 2010). Similarly, degradation of sulfadiazine (Sittig et al. 2014), sulfamethoxazole (Srinivasan and Sarmah 2014), sulfachloropyridazine (Accinelli et al. 2007), and chlortetracycline (Halling-Sørensen et al. 2005) was found to be dependent on the soil types. Srinivasan and Sarmah (2014) reported that an increased temperature led to a faster loss of sulfamethoxazole. In another study, the half-life of virginiamycin was found to be inversely linked with soil pH (Weerasinghe and Towner 1997). DT50 of virginiamycin was adversely affected by soil pH for different agricultural soils, ranging from 87 to 173 days. Also, the antibiotic concentration present in the environment plays a major role in its degradation. An increasing dose of ciprofloxacin was found to reduce the rate of degradation of the antibiotic (Cui et al. 2014). Similar trends were found for other antibiotics, for example, sulfadimethoxine (Wang et al. 2006), ofloxacin, norfloxacin, azithromycin, doxycycline, tetracycline, ciprofloxacin, gemfibrozil, triclosan (Walters et al. 2010), chlortetracycline (Fang et al. 2016), and sulphadiazine (Zhang, Y. et al. 2017).

7.9   STRATEGIES FOR MITIGATION OF THE EFFECTS OF ANTIBIOTIC RESISTANCE

To overcome the problem associated with the spread of antimicrobial resistance, targeted approaches need to be adopted, *viz.* organized use of antimicrobials in agriculture and animal farming to minimize the spread from animal waste (Kuppusamy et al. 2018), orientation towards sustainable processes, proper recycling of livestock manure before application, use of natural products as feed additives, technology development for better pretreatment of livestock manure (Tullo et al. 2019; Han et al. 2020), and identification of novel antimicrobial peptides for crop protection (Amso and Hayouka 2019). In this context, WHO (2020) has suggested regularizing the use of veterinary antibiotics for a sustainable ecosystem. It also recommends that improved hygiene and animal care on farms will be helpful to increase biosecurity and prevent diseases.

7.10   CONCLUSIONS

In the current scenario, antibiotics play an important role in the prevention of numerous diseases, hence holds medical and agricultural significance. Despite their beneficial roles, the adverse impact, in terms of the development of antimicrobial resistance, cannot be ignored. Several studies have reported the prevalence of antibiotic resistance in manure samples coming from animal feeding operations. However, there has been limited focus on understanding the physical and chemical parameters that govern the mobility of antibiotics in soil and water systems. Also, few reports are available in context to the tracking of antibiotic consumption in livestock and the uptake of antibiotics by plants. Given the growing concerns regarding the use of antibiotics, the spread of AMR, and its negative impact on the agricultural ecosystem, extensive research is required towards gaining a comprehensive understanding of its application in agriculture, the impact of various factors on its degradation, as well as tracking its entire course from source to sink. Surveillance studies need to be undertaken aggressively for conclusive determination of the antibiotic footprints in the agricultural systems, and to strategically curb its application and/or adverse effects.

ACKNOWLEDGMENT

The authors wish to acknowledge the funding received from Department of Biotechnology, Government of India (BT/PR27680/BCE/8/1434/2018).

REFERENCES

Accinelli, C., Koskinen, W.C., Becker, J.M. and Sadowsky, M.J. 2007. Environmental fate of two sulfonamide antimicrobial agents in soil. *Journal of Agricultural and Food Chemistry* 55: 2677–2682.

Amábile-Cuevas, C.F. 2016. *Antibiotics and Antibiotic Resistance in the Environment.* Leiden, The Netherlands: CRC Press.

Amso, Z. and Hayouka, Z. 2019. Antimicrobial random peptide cocktails: A new approach to fight pathogenic bacteria. *Chemical Communications* 55: 2007–2014.

Arikan, O.A., Mulbry, W. and Rice, C. 2009. Management of antibiotic residues from agricultural sources: Use of composting to reduce chlortetracycline residues in beef manure from treated animals. *Journal of Hazardous Materials* 164: 483–489.

Aust, M.O., Thiele-Bruhn, S., Seeger, J., Godlinski, F., Meissner, R. and Leinweber, P. 2010. Sulfonamides leach from sandy loam soils under common agricultural practice. *Water, Air & Soil Pollution* 211: 143–156.

Bian, R., Joseph, S., Cui, L., Pan, G., Li, L., Liu, X., Zhang, A., Rutlidge, H., Wong, S., Chia, C. and Marjo, C. 2014. A three-year experiment confirms continuous immobilization of cadmium and lead in contaminated paddy field with biochar amendment. *Journal of Hazardous Materials* 272: 121–128.

Blackwell, P.A., Kay, P. and Boxall, A.B. 2007. The dissipation and transport of veterinary antibiotics in a sandy loam soil. *Chemosphere* 67: 292–299.

Boleas, S., Alonso, C., Pro, J., Fernández, C., Carbonell, G. and Tarazona, J.V. 2005. Toxicity of the antimicrobial oxytetracycline to soil organisms in a multi-species-soil system (MS·3) and influence of manure co-addition. *Journal of Hazardous Materials* 122: 233–241.

Boxall, A.B., Johnson, P., Smith, E.J., Sinclair, C.J., Stutt, E. and Levy, L.S. 2006. Uptake of veterinary medicines from soils into plants. *Journal of Agricultural and Food Chemistry* 54: 2288–2297.

Burke, V., Richter, D., Greskowiak, J., Mehrtens, A., Schulz, L. and Massmann, G. 2016. Occurrence of antibiotics in surface and groundwater of a drinking water catchment area in Germany. *Water Environment Research* 88: 652–659.

Cao, J., Ji, D. and Wang, C. 2015. Interaction between earthworms and arbuscular mycorrhizal fungi on the degradation of oxytetracycline in soils. *Soil Biology and Biochemistry* 90: 283–292.

Carter, L.J., Harris, E., Williams, M., Ryan, J.J., Kookana, R.S. and Boxall, A.B. 2014. Fate and uptake of pharmaceuticals in soil–plant systems. *Journal of Agricultural and Food Chemistry* 62: 816–825.

Chang, Q., Wang, W., Regev-Yochay, G., Lipsitch, M. and Hanage, W.P. 2015. Antibiotics in agriculture and the risk to human health: How worried should we be? *Evolutionary Applications* 8: 240–247.

Chen, W., Liu, W., Pan, N., Jiao, W. and Wang, M. 2013. Oxytetracycline on functions and structure of soil microbial community. *Journal of Soil Science and Plant Nutrition* 13: 967–975.

Cui, H., Wang, S.P., Fu, J., Zhou, Z.Q., Zhang, N. and Guo, L. 2014. Influence of ciprofloxacin on microbial community structure and function in soils. *Biology and Fertility of Soils* 50: 939–947.

Cycoń, M., Mrozik, A. and Piotrowska-Seget, Z. 2019. Antibiotics in the soil environment—degradation and their impact on microbial activity and diversity. *Frontiers in Microbiology* 10: 338.

Dever, L.A. and Dermody, T.S. 1991. Mechanisms of bacterial resistance to antibiotics. *Archives of Internal Medicine* 151: 886–895.

Ding, G.C., Radl, V., Schloter-Hai, B., Jechalke, S., Heuer, H., Smalla, K. and Schloter, M. 2014. Dynamics of soil bacterial communities in response to repeated application of manure containing sulfadiazine. *PLoS ONE* 9: e92958.

Dolliver, H., Kumar, K. and Gupta, S. 2007. Sulfamethazine uptake by plants from manure-amended soil. *Journal of Environmental Quality* 36: 1224–1230.

Du, L. and Liu, W. 2012. Occurrence, fate, and ecotoxicity of antibiotics in agro-ecosystems. A review. *Agronomy for Sustainable Development* 32: 309–327.

Duan, M., Li, H., Gu, J., Tuo, X., Sun, W., Qian, X. and Wang, X. 2017. Effects of biochar on reducing the abundance of oxytetracycline, antibiotic resistance genes, and human pathogenic bacteria in soil and lettuce. *Environmental Pollution* 224: 787–795.

Fair, R.J. and Tor, Y. 2014. Antibiotics and bacterial resistance in the 21st century. *Perspectives in Medicinal Chemistry* 6: 14459.

Fang, H., Han, L., Cui, Y., Xue, Y., Cai, L. and Yu, Y. 2016. Changes in soil microbial community structure and function associated with degradation and resistance of carbendazim and chlortetracycline during repeated treatments. *Science of the Total Environment* 572: 1203–1212.

Fang, H., Han, Y., Yin, Y., Pan, X. and Yu, Y. 2014. Variations in dissipation rate, microbial function and antibiotic resistance due to repeated introductions of manure containing sulfadiazine and chlortetracycline to soil. *Chemosphere* 96: 51–56.

Forsberg, K.J., Reyes, A., Wang, B., Selleck, E.M., Sommer, M.O. and Dantas, G. 2012. The shared antibiotic resistome of soil bacteria and human pathogens. *Science* 337: 1107–1111.

Grenni, P., Ancona, V. and Caracciolo, A.B. 2018. Ecological effects of antibiotics on natural ecosystems: A review. *Microchemical Journal* 136: 25–39.

Gutiérrez, I.R., Watanabe, N., Harter, T., Glaser, B. and Radke, M. 2010. Effect of sulfonamide antibiotics on microbial diversity and activity in a Californian Mollic Haploxeralf. *Journal of Soils and Sediments* 10: 537–544.

Halling-Sørensen, B., Jacobsen, A.M., Jensen, J., Sengeløv, G., Vaclavik, E. and Ingerslev, F. 2005. Dissipation and effects of chlortetracycline and tylosin in two agricultural soils: A field-scale study in Southern Denmark. *Environmental Toxicology and Chemistry: An International Journal* 24: 802–810.

Halling-Sørensen, B., Jensen, J., Tjørnelund, J. and Montforts, M.H.M.M. 2001. Worst-case estimations of predicted environmental soil concentrations (PEC) of selected veterinary antibiotics and residues used in Danish agriculture. In Kümmerer, K. *(ed.), Pharmaceuticals in the Environment*, 143–157. Heidelberg: Springer.

Hammesfahr, U., Heuer, H., Manzke, B., Smalla, K. and Thiele-Bruhn, S. 2008. Impact of the antibiotic sulfadiazine and pig manure on the microbial community structure in agricultural soils. *Soil Biology and Biochemistry* 40: 1583–1591.

Hamscher, G., Pawelzick, H.T., Höper, H. and Nau, H. 2005. Different behavior of tetracyclines and sulfonamides in sandy soils after repeated fertilization with liquid manure. *Environmental Toxicology and Chemistry: An International Journal* 24: 861–868.

Han, Y., Yang, L., Chen, X., Cai, Y., Zhang, X., Qian, M., Chen, X., Zhao, H., Sheng, M., Cao, G. and Shen, G. 2020. Removal of veterinary antibiotics from swine wastewater using anaerobic and aerobic biodegradation. *Science of the Total Environment* 709: 136094.

Heuer, H., Kopmann, C., Binh, C.T., Top, E.M. and Smalla, K, 2009. Spreading antibiotic resistance through spread manure: Characteristics of a novel plasmid type with low% G+ C content. *Environmental Microbiology* 11: 937–949.

Heuer, H., Schmitt, H. and Smalla, K. 2011. Antibiotic resistance gene spread due to manure application on agricultural fields. *Current Opinion in Microbiology* 14: 236–243.

Hu, X., Zhou, Q. and Luo, Y. 2010. Occurrence and source analysis of typical veterinary antibiotics in manure, soil, vegetables and groundwater from organic vegetable bases, northern China. *Environmental Pollution* 158: 2992–2998.

Jjemba, P.K. 2002. The potential impact of veterinary and human therapeutic agents in manure and biosolids on plants grown on arable land: A review. *Agriculture, Ecosystems & Environment* 93: 267–278.

Kang, D.H., Gupta, S., Rosen, C., Fritz, V., Singh, A., Chander, Y., Murray, H. and Rohwer, C. 2013. Antibiotic uptake by vegetable crops from manure-applied soils. *Journal of Agricultural and Food Chemistry* 61: 9992–10001.

Karmakar, R., Bindiya, S. and Hariprasad, P. 2019. Convergent evolution in bacteria from multiple origins under antibiotic and heavy metal stress, and endophytic conditions of host plant. *Science of the Total Environment* 650: 858–867.

Klein, E.Y., Van Boeckel, T.P., Martinez, E.M., Pant, S., Gandra, S., Levin, S.A., Goossens, H. and Laxminarayan, R. 2018. Global increase and geographic convergence in antibiotic consumption between 2000 and 2015. *Proceedings of the National Academy of Sciences* 115: E3463–E3470.

Koba, O., Golovko, O., Kodešová, R., Fér, M. and Grabic, R. 2017. Antibiotics degradation in soil: A case of clindamycin, trimethoprim, sulfamethoxazole and their transformation products. *Environmental Pollution* 220: 1251–1263.

Kotzerke, A., Sharma, S., Schauss, K., Heuer, H., Thiele-Bruhn, S., Smalla, K., Wilke, B.M. and Schloter, M. 2008. Alterations in soil microbial activity and N-transformation processes due to sulfadiazine loads in pig-manure. *Environmental Pollution* 153: 315–322.

Kuchta, S.L., Cessna, A.J., Elliott, J.A., Peru, K.M. and Headley, J.V. 2009. Transport of lincomycin to surface and ground water from manure-amended cropland. *Journal of Environmental Quality* 38: 1719–1727.

Kumar, K., Gupta, S.C., Baidoo, S.K., Chander, Y. and Rosen, C.J. 2005. Antibiotic uptake by plants from soil fertilized with animal manure. *Journal of Environmental Quality* 34: 2082–2085.

Kümmerer, K. 2009. The presence of pharmaceuticals in the environment due to human use–present knowledge and future challenges. *Journal of Environmental Management* 90: 2354–2366.

Kuppusamy, S., Kakarla, D., Venkateswarlu, K., Megharaj, M., Yoon, Y.E. and Lee, Y.B. 2018. Veterinary antibiotics (VAs) contamination as a global agro-ecological issue: a critical view. *Agriculture, Ecosystems & Environment* 257: 47–59.

Laxminarayan, R., Van Boeckel, T. and Teillant, A. 2015. The economic costs of withdrawing antimicrobial growth promoters from the livestock sector. OECD Food, Agriculture and Fisheries Papers, No. 78, Paris: OECD Publishing. https://doi.org/10.1787/5js64kst5wvl-en.

Leal, R.M.P., Alleoni, L.R.F., Tornisielo, V.L. and Regitano, J.B. 2013. Sorption of fluoroquinolones and sulfonamides in 13 Brazilian soils. *Chemosphere* 92: 979–985.

Ling-Ling, L.I., Huang, L.D., Chung, R.S., Ka-Hang, F.O.K. and Zhang, Y.S. 2010. Sorption and dissipation of tetracyclines in soils and compost. *Pedosphere* 20: 807–816.

Liu, B., Li, Y., Zhang, X., Wang, J. and Gao, M. 2015. Effects of chlortetracycline on soil microbial communities: Comparisons of enzyme activities to the functional diversity via Biolog EcoPlates™. *European Journal of Soil Biology* 68: 69–76.

Lopatto, E., Choi, J., Colina, A., Ma, L., Howe, A. and Hinsa-Leasure, S. 2019. Characterizing the soil microbiome and quantifying antibiotic resistance gene dynamics in agricultural soil following swine CAFO manure application. *PLoS ONE* 14: e0220770.

Ma, W., Jiang, S., Assemien, F., Qin, M., Ma, B., Xie, Z., Liu, Y., Feng, H., Du, G., Ma, X. and Le Roux, X. 2016. Response of microbial functional groups involved in soil N cycle to N, P and NP fertilization in Tibetan alpine meadows. *Soil Biology and Biochemistry* 101: 195–206.

Manyi-Loh, C., Mamphweli, S., Meyer, E. and Okoh, A. 2018. Antibiotic use in agriculture and its consequential resistance in environmental sources: potential public health implications. *Molecules* 23: 795.

Marti, R., Scott, A., Tien, Y.C., Murray, R., Sabourin, L., Zhang, Y. and Topp, E. 2013. Impact of manure fertilization on the abundance of antibiotic-resistant bacteria and frequency of detection of antibiotic resistance genes in soil and on vegetables at harvest. *Applied and Environmental Microbiology* 79: 5701–5709.

Martínez, J.L, 2012. Natural antibiotic resistance and contamination by antibiotic resistance determinants: the two ages in the evolution of resistance to antimicrobials. *Frontiers in Microbiology* 3: 1.

Martínez-Hernández, V., Meffe, R., López, S.H. and de Bustamante, I. 2016. The role of sorption and biodegradation in the removal of acetaminophen, carbamazepine, caffeine, naproxen and sulfamethoxazole during soil contact: a kinetics study. *Science of the Total Environment* 559: 232–241.

McEachran, A.D., Blackwell, B.R., Hanson, J.D., Wooten, K.J., Mayer, G.D., Cox, S.B. and Smith, P.N. 2015. Antibiotics, bacteria, and antibiotic resistance genes: Aerial transport from cattle feed yards via particulate matter. *Environmental Health Perspectives* 123: 337–343.

McKinney, C.W., Dungan, R.S., Moore, A. and Leytem, A.B. 2018. Occurrence and abundance of antibiotic resistance genes in agricultural soil receiving dairy manure. *FEMS Microbiology Ecology* 94: fiy010.

Mendes, L.W., Tsai, S.M., Navarrete, A.A., De Hollander, M., van Veen, J.A. and Kuramae, E.E. 2015. Soil-borne microbiome: Linking diversity to function. *Microbial Ecology* 70: 255–265.

Milić, N., Milanović, M., Letić, N.G., Sekulić, M.T., Radonić, J., Mihajlović, I. and Miloradov, M.V. 2013. Occurrence of antibiotics as emerging contaminant substances in aquatic environment. *International Journal of Environmental Health Research* 23: 296–310.

Molaei, A., Lakzian, A., Datta, R., Haghnia, G., Astaraei, A., Rasouli-Sadaghiani, M. and Ceccherini, M.T. 2017. Impact of chlortetracycline and sulfapyridine antibiotics on soil enzyme activities. *International Agrophysics* 31: 499.

Munir, M. and Xagoraraki, I. 2011. Levels of antibiotic resistance genes in manure, biosolids, and fertilized soil. *Journal of Environmental Quality* 40: 248–255.

O'Neill, J. 2015. Antimicrobials in Agriculture and the Environment: Reducing Unnecessary Use and Waste. *Review on Antimicrobial Resistance*, UK. https://wellcomecollection. org/works/x88ast2u (Accessed 2 August 2021).

Ostermann, A., Siemens, J., Welp, G., Xue, Q., Lin, X., Liu, X. and Amelung, W. 2013. Leaching of veterinary antibiotics in calcareous Chinese croplands. *Chemosphere* 91: 928–934.

Pan, M. and Chu, L.M. 2017. Transfer of antibiotics from wastewater or animal manure to soil and edible crops. *Environmental Pollution* 231: 829–836.

Park, J.Y. and Huwe, B. 2016. Effect of pH and soil structure on transport of sulfonamide antibiotics in agricultural soils. *Environmental Pollution* 213: 561–570.

Pérez-Valera, E., Kyselková, M., Ahmed, E., Sladecek, F.X.J., Goberna, M. and Elhottová, D. 2019. Native soil microorganisms hinder the soil enrichment with antibiotic resistance genes following manure applications. *Scientific Reports* 9: 1–10.

Quaik, S., Embrandiri, A., Ravindran, B., Hossain, K., Al-Dhabi, N.A., Arasu, M.V., Ignacimuthu, S. and Ismail, N. 2020. Veterinary antibiotics in animal manure and manure laden soil: scenario and challenges in Asian countries. *Journal of King Saud University-Science* 32: 1300–1305.

Reichel, R., Michelini, L., Ghisi, R. and Thiele-Bruhn, S. 2015. Soil bacterial community response to sulfadiazine in the soil–root zone. *Journal of Plant Nutrition and Soil Science* 178: 499–506.

Reichel, R., Rosendahl, I., Peeters, E.T., Focks, A., Groeneweg, J., Bierl, R., Schlichting, A., Amelung, W. and Thiele-Bruhn, S. 2013. Effects of slurry from sulfadiazine-(SDZ) and difloxacin-(DIF) medicated pigs on the structural diversity of microorganisms in bulk and rhizosphere soil. *Soil Biology and Biochemistry* 62: 82–91.

Sarmah, A.K., Meyer, M.T. and Boxall, A.B. 2006. A global perspective on the use, sales, exposure pathways, occurrence, fate and effects of veterinary antibiotics (VAs) in the environment. *Chemosphere* 65: 725–759.

Shrivas, V.L., Singh, U., Weisskopf, L., Hariprasad, P. and Sharma, S. 2019. Effect of organic farming on structural and functional diversity of soil microbiome: Benefits and risks. In Varma, A., Tripathi, S. and Prasad, R. (eds.), *Plant Biotic Interactions*, 129–146. Cham: Springer.

Sittig, S., Kasteel, R., Groeneweg, J., Hofmann, D., Thiele, B., Köppchen, S. and Vereecken, H. 2014. Dynamics of transformation of the veterinary antibiotic sulfadiazine in two soils. *Chemosphere* 95: 470–477.

Smith, D.L., Dushoff, J. and Morris Jr., J.G. 2005. Agricultural antibiotics and human health. *PLoS Medicine* 2: 232.

Smith, S.D., Colgan, P., Yang, F., Rieke, E.L., Soupir, M.L., Moorman, T.B., Allen, H.K. and Howe, A. 2019. Investigating the dispersal of antibiotic resistance associated genes from manure application to soil and drainage waters in simulated agricultural farmland systems. *PloS ONE* 14: e0222470.

Srinivasan, P. and Sarmah, A.K. 2014. Dissipation of sulfamethoxazole in pasture soils as affected by soil and environmental factors. *Science of the Total Environment* 479: 284–291.

Sukul, P., Lamshöft, M., Kusari, S., Zühlke, S. and Spiteller, M. 2009. Metabolism and excretion kinetics of 14C-labeled and non-labeled difloxacin in pigs after oral administration, and antimicrobial activity of manure containing difloxacin and its metabolites. *Environmental Research* 109: 225–231.

Thiele-Bruhn, S. and Beck, I.C. 2005. Effects of sulfonamide and tetracycline antibiotics on soil microbial activity and microbial biomass. *Chemosphere* 59: 457–465.

Thiele-Bruhn, S., 2003. Pharmaceutical antibiotic compounds in soils–A review. *Journal of Plant Nutrition and Soil Science* 166: 145–167.

Toth, J.D., Feng, Y. and Dou, Z. 2011. Veterinary antibiotics at environmentally relevant concentrations inhibit soil iron reduction and nitrification. *Soil Biology and Biochemistry* 43: 2470–2472.

Tullo, E., Finzi, A. and Guarino, M. 2019. Environmental impact of livestock farming and precision livestock farming as a mitigation strategy. *Science of the Total Environment* 650: 2751–2760.

Udikovic-Kolic, N., Wichmann, F., Broderick, N.A. and Handelsman, J. 2014. Bloom of resident antibiotic-resistant bacteria in soil following manure fertilization. *Proceedings of the National Academy of Sciences* 111: 15202–15207.

United Nations 2015. World population prospects: The 2015 revision, key findings and advance tables. United Nations, Department of Economic and Social Affairs. Population division working paper no. ESA/P/WP, 241.

van Boeckel, T.P., Brower, C., Gilbert, M., Grenfell, B.T., Levin, S.A., Robinson, T.P., Teillant, A. and Laxminarayan, R. 2015. Global trends in antimicrobial use in food animals. *Proceedings of the National Academy of Sciences* 112: 5649–5654.

Waksman, S.A. 1947. What is an antibiotic or an antibiotic substance? *Mycologia* 39: 565–569.

Walters, E., McClellan, K. and Halden, R.U. 2010. Occurrence and loss over three years of 72 pharmaceuticals and personal care products from biosolids–soil mixtures in outdoor mesocosms. *Water Research* 44: 6011–6020.

Wang, F., Han, W., Chen, S., Dong, W., Qiao, M., Hu, C. and Liu, B. 2020. Fifteen-year application of manure and chemical fertilizers differently impacts soil ARGs and microbial community structure. *Frontiers in Microbiology* 11: 62.

Wang, J. and Han, J.Z. 2008. Effects of heavy metals and antibiotics on soil and vegetables. *Journal of Ecology and Rural Environment* 4: 90–93.

Wang, Q.Q., Bradford, S.A., Zheng, W. and Yates, S.R. 2006. Sulfadimethoxine degradation kinetics in manure as affected by initial concentration, moisture, and temperature. *Journal of Environmental Quality* 35: 2162–2169.

Weerasinghe, C.A. and Towner, D. 1997. Aerobic biodegradation of virginiamycin in soil. *Environmental Toxicology and Chemistry: An International Journal* 16: 1873–1876.

Wegener, H.C., Aarestrup, F.M., Jensen, L.B., Hammerum, A.M. and Bager, F. 1999. Use of antimicrobial growth promoters in food animals and *Enterococcus faecium* resistance to therapeutic antimicrobial drugs in Europe. *Emerging Infectious Diseases* 5: 329.

Wehrhan, A., Kasteel, R., Simunek, J., Groeneweg, J. and Vereecken, H. 2007. Transport of sulfadiazine in soil columns—Experiments and modelling approaches. *Journal of Contaminant Hydrology* 89: 107–135.

Winckler, C. and Grafe, A. 2001. Use of veterinary drugs in intensive animal production. *Journal of Soils and Sediments* 1: 66.

World Health Organization (WHO). 2019. Ten threats to global health. https://www.who.int/news-room/spotlight/ten-threats-to-global-health-in-2019 (Accessed on 2 August 2021).

World Health Organization (WHO). 2020. Antibiotic resistance. https://www.who.int/news-room/fact-sheets/detail/antibiotic-resistance (Accessed on 2 August 2021).

Wu, D.L., Zhang, M., He, L.X., Zou, H.Y., Liu, Y.S., Li, B.B., Yang, Y.Y., Liu, C., He, L.Y. and Ying, G.G. 2020. Contamination profile of antibiotic resistance genes in ground water in comparison with surface water. *Science of the Total Environment* 715: 136975.

Xu, Y., Yu, W., Ma, Q., Wang, J., Zhou, H. and Jiang, C. 2016. The combined effect of sulfadiazine and copper on soil microbial activity and community structure. *Ecotoxicology and Environmental Safety* 134: 43–52.

Zhang, H., Li, X., Yang, Q., Sun, L., Yang, X., Zhou, M., Deng, R. and Bi, L. 2017. Plant growth, antibiotic uptake, and prevalence of antibiotic resistance in an endophytic system of pakchoi under antibiotic exposure. *International Journal of Environmental Research and Public Health* 14: 1336.

Zhang, H.M., Zhang, M.K. and Gu, G.P., 2008. Residues of tetracyclines in livestock and poultry manures and agricultural soils from North Zhejiang Province. *Journal of Ecology and Rural Environment* 24: 69–73.

Zhang, Y., Hu, S., Zhang, H., Shen, G., Yuan, Z. and Zhang, W. 2017. Degradation kinetics and mechanism of sulfadiazine and sulfamethoxazole in an agricultural soil system with manure application. *Science of the Total Environment* 607: 1348–1356.

Zhang, Y.J., Hu, H.W., Chen, Q.L., Singh, B.K., Yan, H., Chen, D. and He, J.Z. 2019. Transfer of antibiotic resistance from manure-amended soils to vegetable microbiomes. *Environment International* 130: 104912.

Zhao, F., Yang, L., Chen, L., Li, S. and Sun, L. 2019. Bioaccumulation of antibiotics in crops under long-term manure application: Occurrence, biomass response and human exposure. *Chemosphere* 219: 882–895.

8 Microalgae Based Biofertilizers and Biostimulants for Agricultural Crops

Alex Consani Cham Junior[1], Ana Claudia Zanata[1], Sofia de Souza Oliveira[1], Eduardo Bittencourt Sydney[2], and Andréia Anschau[1]
[1]Universidade Tecnológica Federal do Paraná – Campus Dois Vizinhos, Paraná, Brazil
[2]Universidade Tecnológica Federal do Paraná – Campus Ponta Grossa, Paraná, Brazil

CONTENTS

8.1 INTRODUCTION

Microalgae have attracted interest not only from the agrochemical industry but also from agricultural producers, mainly for their biofertilizer and biostimulant characteristics. A scientific review summarizes the bioactive compounds present in the microalgae biofertilizers and biostimulants, and examines the research that supports the use of substances to manage productivity and abiotic stress in crops (Ronga et al., 2019). Considering the growth of the world population, among the challenges that agriculture faces, is that of satisfying the demand for food and improving the sustainability of agriculture.

Microalgae biomass has applications in different industrial sectors, such as food, cosmetics, pharmaceuticals, biofuels, animal feed and agricultural production. For agricultural production, microalgae present a high content of essential macro- and micronutrients for crop development with potential use as biofertilizers and biostimulants. Microalgae have been currently used in the agrochemical industry and have recently aroused the interest of producers in order to increase the sustainability of agricultural production.

Biofertilizers and biostimulants are presented as economic and ecological alternatives in comparison with the respective synthetic products (chemical fertilizers and plant growth regulators). Biostimulants applied in low doses to seeds crops, and soils, can regulate and increase the physiological processes of crops. Biofertilizers contain natural substances or live microorganisms being able to improve both the biological and chemical soil characteristics, stimulating the plant growth and restoring the soil fertility.

At present, the interest in biostimulants and biofertilizers in agricultural crops, and especially in organic agricultural crops, is increasing strongly. Organic crops are mostly characterized by lower yields when comparing to conventional methods of production. This difference in production is mainly due to the fertilizers and crop protection product used. Several studies on microalgae biostimulants have shown that these products can affect photosynthesis, cell respiration, nucleic acid synthesis, and ion assimilation in plants. Furthermore, biostimulants can improve: the nutrients availability in the soil, the antioxidant content in plants, cellular metabolism, and the chlorophyll content in leaves.

Several agricultural products based on microalgae have been commercialized as innovative products and to increase crop yields. Usually, the labels of microalgae biofertilizers and biostimulants inform their composition and respective concentrations of amino acids, elemental minerals and phytohormones.

Microalgal biomass contains macro- and micronutrients such as nitrogen (N), potassium (K) and phosphorus (P), and can also be described as a slow-release fertilizer. Microalgae present compounds that are growth promoters (amino acids, cytokinins, auxins, vitamins, and polyamines) in plants. In the next sections, microalgae biofertilizers and biostimulants are presented as alternative techniques or applied together with synthetic fertilizers, plant growth regulators and crop protection products.

8.2   MICROALGAL BIOFERTILIZERS (MBF)

The use of microalgae-based biofertilizer is attractive for several reasons. The biofertilizer derived from microalgae which fixed CO_2 from the atmosphere helped to reduce the greenhouse effect. Furthermore, N, K and P, required for algae growth, can be provided by wastewaters.

Since 2012 researchers from the University of Kentucky (UK, USA) have been evaluating the potential for CO_2 bio-mitigation using microalgae at power plants

(Crofcheck et al., 2012; Santillan-Jimenez et al., 2016; Wilson et al., 2014) and also have licensed its technology to a commercial partner in China (Melanson & Wells, 2016). Currently, academic studies characterizing the use of MBF are largely lacking and several reports suggest that microalgae could represent a promising alternative to commercial organic fertilizers (Hastings et al., 2014; Khan et al., 2019; Uysal et al., 2015).

The effects of the irrational use of pesticides and fertilizers have caused environmental problems, especially with regard to soil and water pollution, toxic effects to humans and non-human biota (Carvalho, 2017; Vasconcelos, 2018). Research around the world focuses on developing sustainable alternatives, one of which is the use and applicability of microalgae in the agriculture, especially in the biofertilizers production (Garcia-Gonzalez & Sommerfeld, 2016).

Biofertilizers produced from microalgae are ecological, sustainable, economical and easily available (Singh et al., 2019). In addition, microalgae have high growth rate and simple structure and cellular components such as carbohydrates, proteins, and lipids (Nesamma et al., 2015).

The use of MBF allows to increase agricultural production and to decrease the environmental pollution (Kawalekar, 2013). MBF are living microorganisms or natural compounds obtained from microorganisms such as fungi, bacteria and algae that chemically and biologically aid the soil properties, in addition to stimulating plant growth (Abdel-Raouf et al., 2012).

For the ideal development of crops, it is essential that nutrients are favorable to growth. For this reason, microalgae are important in agricultural production, where they have in their macro and micronutrient compositions that act as potentiators in the application as biostimulants and biofertilizers (Garcia-Gonzalez & Sommerfeld, 2016; Khan et al., 2009).

MBF potentially prevent the loss of nutrients by releasing N, K and P (Coppens et al., 2016; Mulbry et al., 2007). Microalgae can also assimilate N and P present in wastewaters and concentrate these elements inside the cells (Coppens et al., 2016; Schreiber et al., 2018). Some varieties of microalgae are prominent in the production of essential biofertilizers such as *Spirulina platensis* and *Chlorella vulgaris*. These are the main species studied, given their commercial relevance, as a source of high concentration of amino acids, proteins, fatty acids, and vitamins (Ahmed et al., 2011).

Studies using MBF have pointed out that content of minerals, proteins, amino acids, pigments, carotenes and chlorophyll present in microalgal biomass can promote development and plant growth (Oliveira et al., 2013). The microalgal treatment can increase the growth performance at the early phases of growth, in addition to increased seed germination of cucumber (Lv et al., 2020), onion (Dineshkumar et al., 2018), maize (Dineshkumar et al., 2017), green gram (Dineshkumar et al., 2020) and grass pasture (Lorentz et al., 2020) for example. Table 8.1 presents some MBF applications in plants and their respective effects.

Extracts of *C. vulgaris* and *Scenedesmus quadricauda* were evaluated in beet production. Treated seedlings present a longer root length and a higher number of root tips than the control. Microalgal extracts also regulated genes associated with biological pathways and processes, such as metabolisms and transport between cells (Barone, Baglieri et al., 2018).

TABLE 8.1
Microalgae Applied as Biofertilizer in Several Cultivations

Microalgae	Plant	Effects	Reference
Chlorella sp.	Willow (*Salix viminalis* L.)	Improved physiological performance and plant growth.	(Grzesik, et al., 2017)
Chlorella spp.	Maize (*Zea mays* L.)	Increase in soil organic carbon.	(Yilmaz & Sönmez 2017)
Chlorella vulgaris	Maize (*Zea mays* L.) and wheat (*Triticum* spp.)	The amount of soil organic matter and the water-holding capacity were improved.	(Uysal et al., 2015)
Chlorella vulgaris and *Scenedesmus dimorphus*	Rice (*Oryza sativa*)	The plant heights treated with polyculture of microalgal biomass were like or better than the urea treatment.	(Jochum et al., 2018)
Chlorella vulgaris and *Scenedesmus* sp.	*Urachloa brizantha* cv. Marandu	The biological treatment had similar plant productivity compared to the treatment with chemical conventional fertilizer.	(Lorentz et al., 2020)
Chlorella vulgaris and *Spirulina platensis*	Green gram (*Vigna radiata* L.)	*C. vulgaris* and *S. platensis* have very wide prolific positive effects on the growth parameters of green gram.	(Dineshkumar et al., 2020)
Chlorella vulgaris and *Spirulina platensis*	Maize (*Zea mays* L.)	Increase in plant height growth, yield characters, biochemical and mineral components as well as the germinability of the seeds produced.	(Dineshkumar et al., 2017)
Chlorella vulgaris and *Spirulina platensis*	Onion (*Allium cepa* L.)	The growth parameters, yield attributes, bio-chemical composition and minerals were higher than the control onion plants.	(Dineshkumar et al., 2018)
Microalgal consortia	Wheat (*Triticum aestivum*)	Increase in plant growth in terms of chlorophyll. Improving the plant nutritional status, and increase in grain yield and grain micronutrient content.	(Renuka et al., 2018)
Nannochloropsis oculate	Tomato (*Solanum lycopersicum* cv. "Maxifort" and *Solanum lycopersicon* cv. "Merlice")	Increase in carotenoid and sugar content of fruits.	(Coppens et al., 2016)
Sargassum johnstonii	Tomato (*Lycopersicon esculentum* Mill.)	Increase in vegetative growth, as well as reproductive and biochemical indicators.	(Kumari et al., 2011)
Scenedesmus quadricauda	Cucumber (*Cucumis sativus*)	Significant impact on the growth of cucumber and microbial abundance in the cucumber rhizosphere.	(Lv et al., 2020)
Spirulina sp.	Red beet (*Beta vulgaris*)	Increase in productivity, dry and wet mass.	(Oliveira et al., 2013)

Microalgae are viable in the production of biofertilizers because they present low cultivation and production costs, are environmentally correct and sustainable, besides increasing the production when compared to chemical fertilizers (Kawalekar, 2013). To meet the commercial agriculture requirements, a large amount of nutrients is needed, it means that high quantities of microalgal biomass will be needed to provide these nutrients as biofertilizer. In chemical fertilizers, the anhydrous ammonia contains over 80% of nitrogen. Microalgal biomass contains up to 10% of N (Cabanelas et al., 2013). Thus, it is needed fifteen times more microalgal biomass to reach a similar level of nitrogen fertilization than chemical fertilizers.

Live MBF present benefits, since it continuously sequestered nutrients during the plant's growth phases, prevent soil erosion and nutrients leaching, and assist to maintain the fertility and structure from soil. Moreover, microalgae biomass contains C (40–60%), P (1–4%), and other elements (Cabanelas et al., 2013). Compared to chemical fertilizers, the use of MBF has the great advantage of increasing soil organic carbon (SOC). Another challenge related to MBF is the inadequate information on the unified dose and the application method. Algae strains must be identified to be used successfully on a field scale in different soil types and geographical environments. However, in last years, great progress has been made in the use of MBF on a commercial scale. Algae-based biofertilizers are already available on the market, ensuring efficiency in increasing the soil fertility and plant productivity. Del Monte Fresh Produce Inc. field applied MBF at Arizona's raw desert (Renuka et al., 2018) and the algae fertilization restored an abandoned land and increased crop productivity.

The use of MBF tends to grow even more with the advances in biotechnology, preserving the environment and improving the quality of human life (Poyo, 2013). In addition, MBF can be used in agriculture in several applications, the most common being in soil or leaves (Ronga et al., 2019).

8.2.1 Soil Applications

Evaluating microalgal biofertilizers in tomato, an increase in sugar and carotenoids concentration was observed, in addition to higher quality and higher economic value of the fruits (Coppens et al., 2016). MBF contains carotenoids which can highlighted the quality of flowers and stimulate the development of yellow and orange colorations on the petals (Lachman et al., 2001). Lettuce seeds (*Lactuca sativa* L.) cultivated in a few concentrations of MBF of *C. vulgaris* showed significant increase in seed growth, soluble carbohydrates and proteins and total free amino acids in comparison to the control. Soil application of MBF also optimized the pigment concentration and the weight of the seedlings (Faheed & Fattah, 2008).

The microalgae application on the soil for rice cultivation also showed important rule. The use of blue-green microalgae (*Aulosira* spp., *Anabaena* spp., *Nostoc* spp., *Scytonema* spp., and *Westiellopsis* spp.) can be alternative sources of N increasing the rice productivity (Paudel & Pradhan, 2012). Studies also report improvements in seed germination, root and aerial growth, grain weight and protein concentration using MBF from *Aulosira fertilissima*, *Hapalosiphon* spp. and *Nostoc* spp. (Misra & Kaushik, 1989; Singh & Trehan, 1973). Cyanobacteria act as growth promoters and thus stimulate the development of microbial populations in the soil and culture (Singh & Trehan, 1973).

8.2.2 FOLIAR APPLICATIONS

In the soil application, some micronutrients may not meet cultivation requirements due to the chemical and physical difficulties in the soil. Thus, foliar fertilization by spraying allows to satisfy the demand for nutrients of the crop, arriving in productions with high yield (Arif et al., 2006).

Foliar application is a new practice that provides macro- and micronutrients in an environmentally safe way and satisfactorily assists in sustainable agricultural performance (Shaaban, 2001a, 2001b). The microalgae extracts applied in a foliar way increased the content of N in the root and tissues of the aerial part (Hellal et al., 2009), which is justified by the better absorption of nutrients and regulation of physiological mechanisms of the plant (Shaaban, 2001a).

The nutrients present in the microalgae biofertilizers applied to the leaves are quickly absorbed by the pores and stomata of the cuticle. In the morning, the absorption is even greater, since the pores are open (Battacharyya et al., 2015).

The plants correspond and provide better responses when they receive foliar nutrients comparing when applying soil correction methods (Broadley, 2012). Microalgae extracts may contain enzymes, vitamins, and phytohormones, which optimize the assimilation of nutrients, increasing significantly the yield (Shaaban, 2001a, 2001b).

8.3 MICROALGAL BIOSTIMULANTS

Plant biostimulants have attracted interest in modern agriculture as a system to increase the crop performance and the efficiency of nutrients assimilation (Chiaiese et al., 2018). Although it is widely known that microalgae produce numerous complex macromolecules which are active in higher plants, their applications in agricultural sciences are still incipient.

Several recent experimental studies have evaluated the influence of microalgal extracts on tomato, pepper, lettuce and red amaranth cultivations under greenhouse conditions and open fields. The microalgae extracts stimulated the germination, seedling growth, and root and aerial biomass in crops (Arroussi et al., 2018; Barone, Baglieri et al., 2018; Barone, Puglisi et al., 2018).

Extracts of *C. vulgaris* promoted the growth of lettuce in the early phases of development (Faheed & Fattah, 2008). The authors linked this factor to the stimulation of the biosynthesis of chlorophyll and carotenoids, which may have increased the photosynthetic activity. Application of *Spirulina*-based fertilizer enhanced the growth of several leafy vegetables such as arugula, pak choi and red spinach (Wuang et al., 2016). Pepper and tomato cultivars presented also positively influenced by microalgae extracts (Arroussi et al., 2018; Garcia-Gonzalez & Sommerfeld, 2016).

In some studies, seed treatment was more efficient compared to foliar spray of microalgae extracts (Supraja et al., 2020). Positive response on crop performance was also verified in hydroponic tomato cultivations with *S. quadricauda*, *C. vulgaris* or *C. infusionum* (Barone, Puglisi et al., 2018; Zhang et al., 2017). In tomato cultivations, extracts of *Acutodesmus dimorphus* were applied as seed primer, foliar spray, and also as biofertilizer, and both strategies stimulated the seed germination, plant growth and fruit production (Garcia-Gonzalez & Sommerfeld, 2016).

Barone, Baglieri et al. (2018) reported the biostimulant effect of *C. vulgaris* and *S. quadricauda* extracts on sugar beet growth. The treated seedlings present changes in the root architecture. The differences induced by the microalgal extracts on the root morphology have also been reflected at molecular level with an upregulation of genes involved in primary and secondary metabolic pathways.

A possible mechanism involved in the biostimulating action of microalgae extracts is the generation and excretion of hormones in the growing substrate/soil and in the surrounding (Jäger et al., 2010; Renuka et al., 2018). MBF can also attenuate the harmful effects over crops resulting from abiotic stressors, salinity and drought (Arroussi et al., 2018; El-Baky et al., 2010).

Studies reported that the plant biostimulation effect by using microalgae extracts is associated with the modulation of the microbiomes in the phyllosphere and rhizosphere (Manjunath et al., 2016; Priya et al., 2015; Ranjan et al., 2016; Renuka et al., 2018). The soil microbial community is positively affected by the exopolysaccharides produced and excreted by some microalgae species, supplying organic carbon for the development of beneficial microorganisms (Xiao & Zheng, 2016). The association between exopolysaccharides and soil components optimizes the bioavailability, mineralization, and solubilization of micro and macronutrients, improving the development of the crop (Drever & Stillings, 1997).

Mutale-joan et al. (2020) analyzed the biostimulant effects of eighteen extracts from cyanobacteria and microalgae in tomato plant growth, nutrient absorption, chlorophyll content, and metabolic profile. The author's suggested that the biostimulant properties of microalgae extracts are species-specific and related the effects of the bioextracts on biochemical pathways and plant physiology.

8.4 MICROALGAE AS BIOLOGICAL CONTROL AGENTS

8.4.1 MICROALGAL BIOPESTICIDES

In the last years, with environmental studies and the growing concern with food, new outlooks for the agriculture development have emerged with the purpose of maintaining the environmental balance and the good yield of cultivars. Among the common methods of disease control in agriculture, the application of agrochemicals for control of several phytopathogens control sticks out, generating great discussion regarding its implications to human health (Gomiero, 2018; van Lenteren et al., 2018).

The process of controlling insects and microorganisms in agriculture has increasingly encouraged the use of integrated pest management tools, combining different biological and chemical control tools to prevent and treat infestation episodes in crops. Due to the growing search for new chemical molecules and the often uncontrolled use to inhibit the growth of pests, reports of resistance to agrochemicals have become more frequent and have generated concern among farmers and researchers (Heimpel & Mills, 2017; Huang et al., 2005).

As an advantage to chemical control, biological control is based on the natural competition of agents and the induction of the plant response system, thus generating a rapidly growing market for biological control, which seeks new products through microorganisms. Among the possibilities of biological control, there are studies of interactions of microorganism cells with plants obtaining bioactive substances extracted from the fermentation of microorganisms (Glare et al., 2012).

The main justification of the agricultural rush to create alternatives to the agrochemicals is the expanding concern of consumers with the quality of the products they consume, with a great rejection of foods and commercialized products that have undergone processes involving agrochemicals. The consumer's main perception of food is related to environmental impacts and damage to health. Thus, considering the search for new technologies, biological control has corroborated with the launch of more sustainable and effective alternatives (Gupta & Dikshit, 2010; Gyarmati, 2017).

A few years ago, it was believed that only a few microorganisms would have the potential for use in biological control, as an example is some species of *Bacillus*, which arrived in the sector promising a great advance in the prevention and treatment of diseases in the crop. However, with the course of the research, other microorganisms began to be discovered and thus used as soil conditioners and as biopesticides (Bueno et al., 2012; Gazzoni, 2012).

In relation to the production of biopesticides, microalgae have gained prominence, as they are single-celled microorganisms with simpler nutritional needs for growth, presenting greater biomass generation by area and not being affected by climatic conditions. Among the possibilities of using microalgae is related to its application as a soil conditioner, or as a biopesticide. The wide application of microalgae allows the use of all products generated from biomass to post-fermentation filtrate (Castro et al., 2017; Gross et al., 2013), as presented in Figure 8.1.

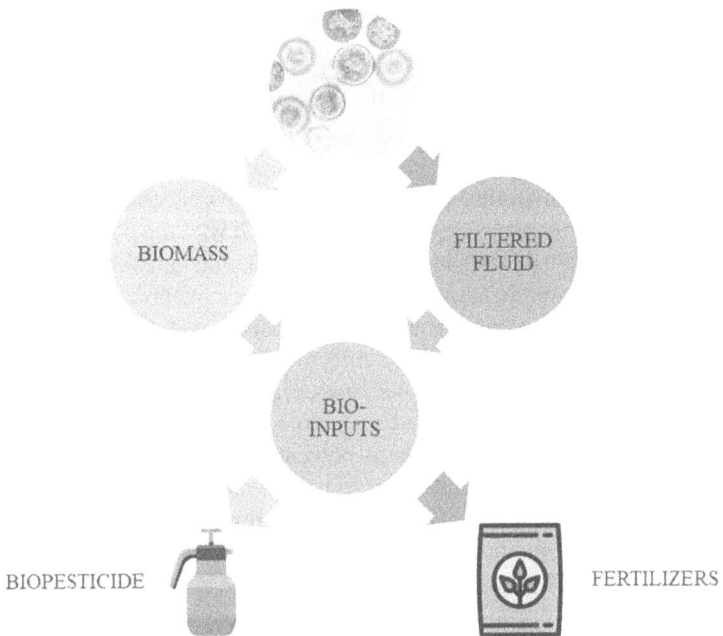

FIGURE 8.1 Use of biomass and metabolites for the production of biofertilizer and biopesticide.

For soil conditioning, the usual methodology is to obtain biomass by photobio-reactors and formulation using a combination of minerals, since in its composition the percentages of carbon and nitrogen sources are sufficient for good soil nutrition. Zhou et al. (2017) highlights the soil improvement and the reduction of the erosion process with *Spirulina platensis* strains, corroborating with the studies carried out by Priya et al. (2014), in which microalgae were used to produce biofertilizers, acting on soil removal and recovering heavy metals.

There are still few products formulated with microalgae and registered by regulatory agencies. However, the perspective is that this will change in the coming years with the growing market for biologicals and the already advanced researches on the microalgae production.

Biopesticides are characterized by their natural production from microorganisms such as microalgae, plant and bioactive extracts. The classification of products obtained from living organisms is given by their form of production and their origin. Biochemical biopesticides are obtained from obtaining metabolites from living organisms, while those obtained from microorganisms' own biomass are classified as microbial biopesticides.

However, microalgae have shown excellent results in both forms, both the metabolites obtained from fermented permeate and biomass. The inhibitory effect caused by microalgae is justified by the production of terpenes, phenolic compounds, growth regulators, and the induction of the plant's defense system (Aldrich & Cantelo, 1999).

The bioactive compounds produced by microorganisms, from the process point of view, are considered more promising than those obtained from plant extracts, bearing in mind that obtaining the extracts depends on factors that are difficult to control, such as the climate, cultivation, extraction conditions and the most optimized methods. The biopesticidal production systems, in turn, are simpler from the control point of view, generally obtained from solid state fermentation (SSF) and submerged fermentation (SmF), with variations between the two forms of cultivation.

Microalgae, especially cyanobacteria, are the main biological organisms used in the control of phytopathogenic microorganisms, as they produce secondary metabolites, which are indirectly developed for the defense of microorganisms, generating competitive advantages in the environment. The secondary metabolites of microalgae are described by the biocidal effect, with action on fungi, bacteria and nematodes (El-Mougy & Abdel-Kader, 2013; Zulpa et al., 2003).

The bioactive compounds produced are attributed to polyphenols, tocopherols, proteins, oils, sesquiterpenes and saponins. The action mechanism depends directly on the type of bioactive, and may act by inhibiting the synthesis of proteins and restriction enzymes and disrupt the cytoplasmic membrane (Hernández-Carlos & Gamboa-Angulo, 2011).

Several studies have reported the potential of the bioactive compounds produced from microalgae. Among the applications of agricultural interest, the use as herbicide and pesticides, represents great potential, given the relevance in the discovery of new bioactive molecules and resistance to conventional pest protection systems.

The use of biological products represent ecologically and effective correct strategies to control pathogenic organism transmitted by the soil. Although bacteria and fungi are the most elucidated microorganisms for product development as biological control agents, microalgae, especially cyanobacteria, are gaining prominence due to the potential in the control of pathogens and pests (Bardin et al., 2015).

Cyanobacteria in particular are responsible for having great importance due to the accumulation of various nutrients and producing large amounts of metabolites. Green microalgae are reported for their antibiotic properties. These extracts have a significant effect on both bacteria and phytopathogenic fungi (Falaise et al., 2016).

Despite the few studies relating the utilization of cyanobacteria in agriculture, *in vitro* studies already suggest efficiency under the main phytopathogens that are difficult to control. The treatment used is based on seed treatment and leaf spray tests.

The studies carried out by Prasanna et al. (2008), evaluated the fungicidal potential of seventy strains of cyanobacteria, verifying that the extracts obtained had a high concentration of hydrolytic enzymes with effect on the growth of species of fungi. Gupta and Baruah (2020) in similar studies observed the potential of *Calothrix spp.* filtrates to generate an effective response in the control of plant diseases. Among the fungi studied in the studies using cyanobacteria, the main fungi studied were *Fusarium oxysporum* (withering and yellowing of *Fusarium*), *Pythium debaryanum* (blight, or root rot), *Pythium aphanidermatum* (tipping, result of root rot) and *Rhizoctonia solani* (root rot).

Species of *Anabaena*, *Calothrix*, *Nostoc*, *Nodularia* and *Oscillatoria*, revealed antifungal activity to *Alternaria alternate*, *Botrytis cinerea* and *Rhizopus stolonifera*. The effect on nematodes, although little elucidated, is due to the production of peptide toxins produced by cyanobacteria, which suggests another potential target agent.

Khan et al. (1999) observed that when the cyanobacteria *Microcoleus vaginatus* was inoculated before *Meloidogyne incognita* growth on tomato plants, the suppressive effect of this nematode was reduced and the galls were also reduced significantly. Khan et al. (2005) found that after dipping seedlings in *M. vaginatus* filtrates, a reduction of up to 97.55% in the nematode population was obtained.

Using *Oscillatoria chlorina* in powder formulation reduces the nematodes quantity, reducing the gall formation in tomato plants by up to 97.6%, in comparison with the untreated soil. Chandel (2009) reported that the filtrate of *Aulosira fetilissima* inhibited the outbreak of nematodes of the species *Meloidogyne triticoryzae*. The toxin isolated from the cyanobacterium of the genus *Scytonemin* present insecticidal activity against *Helicoverpa armígera* at a concentration of 0.01%.

The effects of cyanobacteria were also responsible for decreasing the mosquito population in the rice fields (Victor & Reuben, 2000). Antifungal effects were also noted in the extracts of cyanobacteria *Nostoc*, generating effect against nine target fungi organisms while its insecticidal action was noted against *Helicoverpa armigera*, nematicide (against *Caenorhabditis elegans*) and herbicide (Biondi et al., 2004).

In addition to the direct effect of bioactive compounds production, cyanobacteria are described for their ability to improve plant immunity in triggering a defense response through the production of antioxidants and enzymes. The elucidation of this theme will be approached with more emphasis in the next topic.

The regulation of biopesticides is not yet well established, in countries like Brazil they do not have their own legislation, however, due to the growing contribution of the sector in agriculture and economic development, new measures for regulation are already underway to ensure good quality and the efficiency of those bioproducts.

European countries already act in the regulation of these products, being necessary to comply with essential requirements for the liberation of its use in agriculture, among them the performance of toxicological tests, environmental impacts and effect on the resident microbiota (Medina-Pastor & Triacchini, 2020).

The major concern in relation to synthetic pesticides approved in agriculture refers to the negative effects on humans, animals and the environment. The problem found in synthetic pesticides is in the action mechanism, which may occur during the inhibition of enzymes, alteration of the signaling system, electrolyte disturbance, degradation of lipophilic membranes and generation of free radicals with direct effect on the tissues, DNA and proteins of the organisms. Due to this non-selective effect, reactions to animals and to humans are frequently reported, including with regard to the deleterious effect on the genetic material of animals, pollinating insects and humans (Alavanja et al., 2004).

Unlike synthetic pesticides, biopesticides are formulated to affect only the target pest, because they are produced with microorganisms and bioactive microorganisms that do not affect human health. Although biopesticides have considerable advantages over synthetic ones, caution is necessary, given that high concentrations of certain biopesticides such as those obtained from algae and microalgae may present toxic metabolites when used in high concentrations.

8.5 INDUCTION AND STIMULATION OF THE DEFENSE OF PHYTOPATHOGENS

Cyanobacteria, despite their natural independence in the environment and their historical evolution in the ability to convert light into sugars, is widely reported in terms of their symbiosis with various organisms, including plants, fungi, sponges, among others. Its ability to fix nitrogen is a great advantage to the host, given its high capacity to be a source of carbon and nitrogen to the organism.

The association of plants and cyanobacteria still needs studies, however, preliminary results already demonstrate the efficiency of these organisms in contributing to the fixation of oxygen in plant roots, this characteristic is seen mainly by their flexibility in modifying the metabolism and promoting greater generation of nitrogen sources than of carbon, since the environment causes less activity of the existing chloroplasts.

Although little elucidated, the use of cyanobacteria brought excellent perspectives using the leaves and stems to increase the plant's own defense system, as well as barriers creation by the exopolysaccharides production. The production of exopolysaccharides by cyanobacteria has showed a essential funciton in the defense of this organism protecting contra desiccation, protozoas and attack by antimicrobials, and forming biofilms that allow resistance to other abiotic factors.

The observed effect regarding abiotic factors is highlighted mainly by the interaction with the plants that host them, which generates a specific response to stress in their host. Singh et al. (2011) indicate that the use of species of *Oscillatoria acuta* and *Plectonema boryanum* in rice cultivars increase the activities of phenylalanine ammonia lyase, phenylpropanoid and peroxidase, important compounds of response to abiotic factors.

As for abiotic aspects, microalgae such as *Nostoc calcicola*, *N. linkia* and *Anabaena variabilis*, satisfactory results were observed in increasing the roots length and the compliance of the plants, resulting from the formation of exopolysaccharides. Manchanda et al. (2018) reported that the use of microalgae may give to the plants more resistance to salinity and promote growth through the production of phytohormones.

Research on cyanobacteria has associated its use in biocontrol, essentially from the plant defense induction mechanism. The response is reached through the production of antioxidants and enzymes, such as catalase, chitinase, endoglucanase, polyphenol oxidase and peroxidase. *Microcystis aeruginosa*, *Anabaena* sp. and *Chlorella* sp. inoculations may also promote the plant's defense response, by rising the RNA activity (Grzesik et al., 2017). MBF improved substantially the activity of nutrient-assimilating enzymes in *Salix viminalis* leaves.

The modulation of the defense mechanism is triggered by several factors, including the action of microorganisms on the leaf surface and in the rhizosphere. In response to the attack of phytopathogens, the defense system induces the activation of specific genes to produce enzymes and metabolites such as antioxidants, thus generating the inhibitory and biocidal effect on fungi and bacteria. The process of applying microalgae to crops is also described for those already known, that is, leaf or soil treatment. Both applications have particular characteristics. The use in the soil shows excellent results in the recovery of organic matter, production of metabolites which stimulate the production of antioxidants and enzymes. Figure 8.2 presents some of the metabolites produced and the main purposes.

The cyanobacteria *Calothrix elenkinii* modulate defense mechanisms in plants activating the immune system related to plant defense, producing enzymes such as

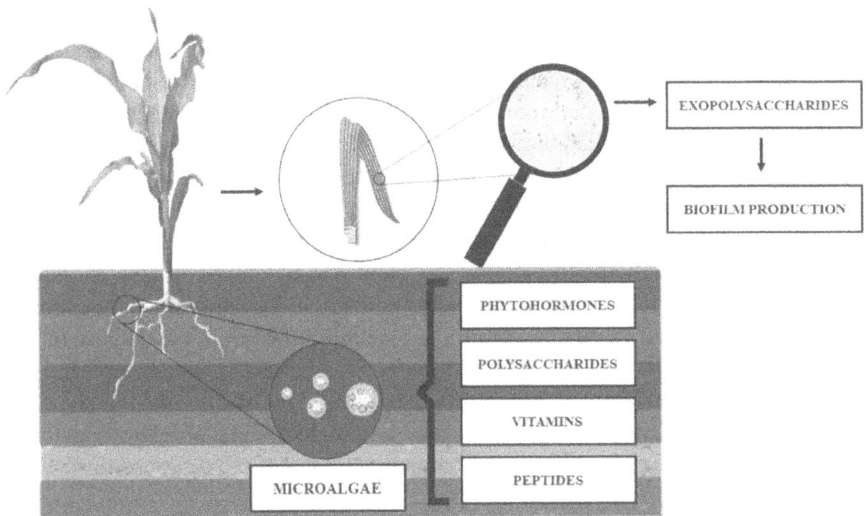

FIGURE 8.2 Stimulation of plant defense and application of microalgae to the roots for the production of several metabolites and foliar application for the generation of biofilm.

catalase, chitinase, β-1,3 endoglucanase, peroxidase, phenylalanine ammonia lyase, and polyphenol oxidase. For example, *C.* significantly elevated the activity of some defense enzymes in roots and aerial parts of rice cultivars (Priya et al., 2015) spice plants (Kumar et al., 2013). Babu et al. (2014) reported the effect of defense enzymes such as peroxidase (POD), phenylalanine ammonia lyase (PAL) and polyphenol oxidase (PPO) activities of wheat plants inoculated with different cyanobacteria, highlighting the results obtained using *Calothrix* sp.

Despite the excellent prospects for the use of microalgae in biological control, further studies have to elucidate the specific mechanisms generated by the symbiosis between plants and microalgae. Although, preliminary studies have already showed isolated efficiency in some cultivars, with great outlooks for the development of new microalgae bioproducts.

8.6 MICROALGAE TECHNOLOGY IN THE PRODUCTION OF BIOACTIVE COMPOUNDS

The production of solid, liquid and gas wastes are intrinsic to human activities. To support the population growth, men have manipulated land and consumed natural resources at a rate incompatible with the speed with which they recover. Due to the environmental impacts and the perception that they negatively affect our quality of life (and subsistence), concerns on developing a more sustainable model where humans and nature could live in harmony began to be discussed. What was considered utopian in the very beginning has shaped the technical and scientific development in such a way that we are living a transition moment in the history of human life where sustainable technologies will be imperative.

The agro-industry is responsible for the production and processing of food and non-food renewable resources. However, the necessary exploration of lands, transportation of raw materials and final products, high hydric and energy demand, and the great waste production rates have posed many challenges towards the sustainability (Barros et al., 2020). The valorization and reuse of solid, liquid and gas wastes from the agro-industry can contribute to the sustainability of this sector and promotes the circular economy, optimizing the utilization of natural resources, generating new bioproducts, reducing environmental impacts and the costs to avoid it (Figure 8.3).

Microalgae technology is related to the circular economy because they can easily reuse nutrients from wastewaters and reincorporate CO_2 from the atmosphere or waste gas streams (Banu et al., 2020). Like other microorganisms, microalgae nutrition depends largely on C, N, P, and micronutrients to produce biomass. Depending on the species the variety and proportion between these nutrients may vary, but the peculiar ability among the microscopic organisms to use inorganic sources of carbon, such as CO_2 and its salts (bicarbonates and carbonates), is probably among the most important mechanism to be explored in the construction of a sustainable economy. These are combined with a microalgal biodiversity estimated between 50,000 and 1 million species (Chiaiese et al., 2018) where very few (<10) are produced at commercial scales, suggesting that our knowledge is limited and the potential of microalgae to agriculture (and other markets) is enormous.

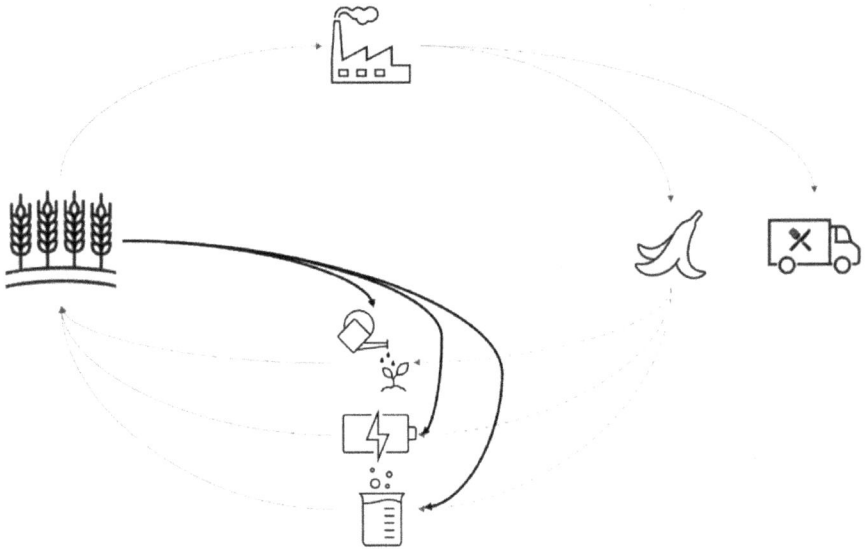

FIGURE 8.3   The valorization and reuse of the wastes produced in the agroindustry and in the field are essential to produce new products being reinserted in the economy and substitute no-renewable ones.

8.6.1   Overview on the Use of Wastewaters for Microalgae Production

Medium for microalgae cultivation mostly have more complex composition than those for fungi and bacteria, which came from the necessity of the addition of several salts as micronutrients. In fungi and bacteria cultivation, these nutrients are present in complex sources of carbon and nitrogen generally used, such as molasses, hydrolyzed biomass, yeast extract and peptone. As microalgae are photosynthetic organisms, their cultivation is independent of organic sources of carbon, but nitrogen and phosphorus continue to play significant role for biomass production. Magnesium is another essential element as it is the central molecule of chlorophyll, being part of the cellular system for the capture of solar (or artificial light) energy.

Wastewaters are generally produced at high rates in the agro-industry and need to be properly treated prior disposal (Siddeeg et al., 2019). The reuse of these waters for microalgae cultivation have environmental benefits because during biomass production, the available nutrients are used and consequently the pollutant potential is reduced. Besides the logical reduction in COD and BOD due to microalgae growth (Sydney et al., 2019), significant reductions in the nitrogen and phosphorous content of agro-industrial wastewaters are also achieved (Sydney et al., 2011; Wang et al., 2017) and is of great importance to avoid eutrophication in water. Economic benefits are also observed because wastewaters contain some or several nutrients (depending on their origin) required for microalgae growth, reducing production costs. On the other hand, wastewater poses some problems related to (i) the presence of toxic compounds which can inhibit or significantly reduce microalgae biomass production (Pacheco et al., 2020); (ii) turbidity, which prevents the incidence of light and, consequently, the energy availability for cell multiplication (Valev et al., 2020); and (iii) nutrient imbalance, which demands supplementation (Mohsenpour et al., 2021). The removal of toxic compounds can be achieved by different pretreatment methods (depending on the target compound) while the most common method to reduce turbidity are dilution with water, dilution with synthetic medium, and suspended solids removal by filtration, sedimentation or centrifugation.

The development of a microalgae technology for agro-industrial wastewater valorization should start from the residue physicochemical characterization. The importance of this step underlies in the understanding of the limitations and potential of the use of wastewater as nutrient source and to strategically evaluate the need of a pretreatment prior use. The choice of the microalgae strain to be used is probably the most complex stage because it depends on the desired products, on the adaptation of the strain to each wastewater, on the robustness related to the presence of possible microbial contaminants and cultivation conditions (which is associated to the type of reactor used – open ponds are more susceptible to contamination and temperature variations while closed reactor can be easily controlled, for example), biomass and/ or products productivity, and facilitated (and less costly) recovery of the biomass and bioproducts (strains with thick cell walls require more extreme cell lysis method, which can affect the bioproduct, for example).

8.6.2 Gastream for Microalgae Production

During photosynthesis, an enzyme called RuBisCO is responsible for the conversion of CO_2 into organic compounds. However, the CO_2-catalytic center of this enzyme suffers competition with oxygen, which is way more abundant in the air than CO_2. Increasing the content of CO_2 in the cultivation medium would, thus, favor its absorption and the production of organic matter (biomass and bioproducts) (Bhola et al., 2014).

The carbon dioxide can only be assessed by the cells if it is solubilized, which depends greatly on the medium pH, contact surface between the gaseous and liquid phase (bubbles size), the gas-liquid contact time and medium salinity. Alkaline medium is more favorable to CO_2 solubilization because alkalis and carbon dioxide reacts instantaneously producing carbonates and bicarbonates which can be then used by microalgae as carbon source (Sydney et al., 2019). On the other hand, in acid medium the dominant solubilized form is CO_2(aq) which is less stable than its salt form. In eukaryotic microalgae, CO_2 is consumed in the form of bicarbonate (HCO_3^-) and the H^+ left in the medium is transported to neutralize the OH^- produced intracellularly, removing H^+ from the culture medium and increasing the culture pH. Considering this pH increase during microalgae biomass production and that constant addition of CO_2-rich gas results in medium acidification that can cause system collapse, carbon dioxide (or CO_2-rich streams) may be injected during short periods associated with the control of medium's pH (Pourjamshidian et al., 2019).

Because carbon dioxide is the main carbon source for autotrophic production of microalgae, a successful biomass or microalgae-based bioproducts production will be achieved if they access this gas. The extremely low content of CO_2 in the atmosphere of ~410 parts per million (ppm) cannot be considered an optimum source. On the other hand, injecting pure (or high concentrated) CO_2 impacts greatly in process economics. Industrial gas streams are generally rich in carbon (CO_2) (coal-fired power plant: 10–15% vol/vol; natural gas flue gases: 5–6% vol/vol (Vuppaladadiyam et al., 2018); fermentation tanks for bioethanol production: >95%) and can be integrated to a microalgae production facility, decreasing the costs associated with the injection of pure CO_2 and offering environmental advantages and (gas) waste valorization. It has been indicated that the injection of CO_2 in concentrations up to 10% is beneficial to most microalgae, but higher concentrations can significantly reduce the efficiency of the photosystem II (Rinanti, 2016). Many studies have been directed towards the screening of CO_2-tolerant species/strains with high growth rate.

Industrial gas streams also contain SOx and NOx, which can both work as nitrogen and sulphur sources or inhibit the growth of some microalgae strains. NOx, especially the high soluble form NO_2, dissolves forming HNO_3 and HNO_2, which are then accessible by the microalgae. The rate of NOx removal from the gas streams is, thus, directly related to its composition. The use of complexing agent is also described as a strategy to increase NO solubilization. Sulphur is commonly present in SO_2 and SO_3 forms, which form H_2SO_3 an H_2SO_4, respectively, in water solutions. SO_2 can hinder the photosynthetic apparatus of microalgae by competitive inhibition of RuBisCO, especially in acidic environments (Zhang, 2015).

8.6.3 INTEGRATION OF MICROALGAE TECHNOLOGY IN THE SUSTAINABLE AGRI-FOREST SECTOR

The world population growth and increased industrialization results in the continuous pressure on agri-forest sector to produce food and raw materials. At the same time, the sector produces large amounts of organic-matter rich wastes, especially during the processing of the raw materials in the agro-industry, whose environmental impacts should be mitigated to meet the goals for a sustainable development. Ideally, these wastes should not only be treated but reused to produce bioproducts. The waste of food, unoptimized use of the natural resources, waste production during processing, fossil fuel-based transportation, production of animal protein, carbon emissions from cradle-to-gate are and water demand are some of the great challenges in building a more sustainable agri-forest sector. Microalgae technology emerges as a possible strategy to reuse industrial and agriculture liquid (or liquified) and gas wastes to produce bioproducts that could be reintegrated to the production chain (circularity) and replace traditional fossil-based products.

The ability of microalgae to fixate CO_2 from the atmosphere is the result of a different metabolic apparatus than that from heterotrophic bacteria and fungi. Despite microalgae cells being also built of proteins, lipids, carbohydrates and nucleic acids, specific compounds and arrangements of these macromolecules occurs because of specific metabolic pathways. Antimicrobial compounds, growth hormones, polysaccharides, micronutrients, protein hydrolysates and the microalgae biomass itself stimulate both plants and the soil microbiome, improve the soil chemical and bio-logical properties and/or control biological agents and plant diseases, as described in Sections 8.2 and 8.3.

The quality and quantity of bioproducts in microalgae extracts depends greatly on the microalgae species used. However, the large-scale microalgae production is not a widespread technology yet, despite important advances being made in the last few years. Very few species are produced at commercial scale: *Arthrospira* (Spirulina) *platensis* and *A. maxima, Dunaliella salina, Nannochloropsis oculata, C. vulgaris, Haematococcus pluvialis, Isocrysis sp., Tetraselmis sp., Phaeodactylum sp., Pavlova sp.* and *Thalassiosira sp.* (Raja et al., 2018). Among these, *Arthrospira* (Jamal Uddin et al., 2019), *D. salina and Phaeodactylum tricornutum* (Guzmán-Murillo et al., 2013), *N. oculata* (Coppens et al., 2016), *C. vulgaris* (Hajnal-Jafari et al., 2020), *Pavlova lutheri* (Ahmed et al., 2015) have been proved to improve metabolic aspects and plant growth parameters. *Arthrospira* and *Chlorella* are the dominant microalgae produced at commercial scale, are recognized as safe (GRAS) (Niccolai et al., 2019), so their use in integrated systems using agro-forest wastewaters and gas streams will be preferentially discussed.

The scientific literature is rich in studies showing the use of microalgae technology integrated to wastewater treatment. Among the most studied agro-industrial effluents is vinasse, a liquid effluent from the distillation columns of ethanol production. It is a dark acid residue, generally poor in carbohydrates (which were consumed during the anaerobic fermentation) but with high COD that is produced at a rate of 12–15L per liter of sugarcane ethanol (Santana et al., 2017). Vinasse is generally used in fertilization of crop fields, but environmental impacts such as soil desertification and groundwater contamination have been noticed (Carrilho et al., 2016). Moreover, the large volume of production opens opportunities for development of new technologies to give a more rational destination to this effluent. Several microalgae strains have been shown to be capable of growing in vinasse, but limitations regarding its turbidity require its dilution. Sydney et al. (2019) have carried a chemical treatment that allow microalgae to produce microalgae biomass at 100% vinasse medium. Different bioproducts derived from microalgae biomass produced in vinasse have been described, such as carbohydrates, lipids, proteins and pigments. Despite the direct application of microalgae biomass-rich vinasse as fertilizer or biostimulant was not yet evaluated, it seems possible that a potentiated effect can occur.

The presence of toxic compounds in agro-industrial wastewaters are not common and this is an advantage of the use of these wastes for microalgae growth. One exception is the cassava processing wastewater (CPW), generated at 600L/ton of processed root during sedimentation and centrifugation of the cassava juice (Peres et al., 2019). CPW has an average organic load of 20 g/L COD but can reach up to 60 (de Carvalho et al., 2018). It contains cyanogenic glycosides (linamarin and lotaustralin) that limit its direct application as fertilizer (Bolarinwa et al., 2016). However, these compounds can be easily hydrolyzed to cyanide spontaneously or enzymatically. Due to the presence of carbohydrates, microalgae able to develop heterotrophic or mixotrophic metabolism should be preferentially considered to be chosen. *Spirulina* and *Chlorella* have been produced at laboratory scale in untreated CPW reaching biomass concentration comparable to synthetic media, but they can also be produced in treated and biodigested CPW (de Carvalho et al., 2018; Xinyi et al., 2016).

An example of the use of a forest-based wastewater for microalgae growth is described in patent BR 10 2018 074711 8 (Novak et al., 2018). The authors described the use of a black liquor from the Kraft production of paper to produce *Spirulina*. Black liquor is the result of the alkaline cooking of wood, where sodium hydroxide (NaOH) and sodium sulfide (Na_2S) are used for the dissolution of fiber and lignin and is generally burned for electricity production (M'hamdi et al., 2017). Due to the alkaline pH, the *Spirulina* production would be possible without the need of acidification. Concentrations of 2–5% of black liquor were optimum to produce *Spirulina* biomass, but when the cells were immobilized using sodium alginate a black liquor content of up to 35% were possible, with optimized conditions at 20%.

The fermentation of raw materials to produce food, biochemicals, microbial biomass and bioenergy results in the production of high-quality CO_2, with concentration >95%, which is generally released to the atmosphere. Because this is a biogenic CO_2 it does not contribute to increase in its concentration in the atmosphere, but its reuse could contribute for negative carbon footprint. Sugarcane ethanol production is considered a major example of biorefinery because almost all by-products of sugarcane juice extraction and fermentation are somehow reused. However, the recycling of CO_2 is still a challenge. Microalgae technology has long been studied as an alternative, especially because some strains can accumulate intracellular carbohydrates that could be fermented to ethanol (Hirano et al., 1997). However, the current available technologies for the production of biofuel from microalgae still result in very high production costs (Getachew et al., 2020). Another example about CO_2 recycling from agro-industrial activities is the biomethane production. During the purification of biogas (50–75% CH_4) to biomethane (94–99.9% CH_4) (ETIP Bioenergy, 2020), carbon dioxide must be removed. Microalgae technology have been evaluated both as an upgrading strategy (biogas is bubbled in a microalgae tank where carbon dioxide is fixated) (Meier et al., 2019) or in the reuse of the CO_2 removed during physical/chemical separation of CO_2 and CH_4. Microalgae can then be used to produce extracts with biostimulating or biofertilizer properties, or recirculated in the biodigester increasing the biogas production (Bose et al., 2020).

8.7 CONCLUDING REMARKS

Microalgae possess a great potential to transform the agriculture sector, minimizing environmental and harmful health problems in comparison to conventional agrochemicals. Awareness of the use of microalgae bio-based agricultural products is increasing and there is expressive market potential for them. Nonetheless, to meet the challenges necessary for commercialization, one must invest in large-scale field research and in the development of cost-effective production technologies.

In this context, some specific issues must be answered: How does microalgal biomass influence plants compared to the traditional fertilization? Which crops will be most likely to receive its benefits? What will differ (if does) according to the plant species? Will inoculation of microalgae biomass reduce heavy metal uptake? How does microalgal biomass influence soil microbial community arrangement and functions? How does microalgal biomass impact the physical and/or chemical parameters of the recovered soil (pH, C/N ratio, mineralizable C/N, soil organic matter – SOM and water-holding capacity – WHC)? What about the influence on ecosystem in general (greenhouse gas emissions, leaching and runoff)? What are the economic and life cycle implications of microalgal biofertilizers, biostimulants, biopesticides and biocontrol agents compared to the conventional ones? Microalgae could be the future of the environmental development and the sustainable agriculture. As with many of the planet's existential problems, the solutions are out there.

REFERENCES

Abdel-Raouf, N., Al-Homaidan, A. A., & Ibraheem, I. B. M. (2012). Agricultural importance of algae. *African Journal of Biotechnology, 11* (54), 11648–11658. https://doi.org/10.5897/ajb11.3983

Ahmed, F., Zhou, W., & Schenk, P. M. (2015). Pavlova Lutheri is a high-level producer of phytosterols. *Algal Research, 10,* 210–217. https://doi.org/10.1016/j.algal.2015.05.013

Ahmed, M. A., Ahmed, G., Mohamed, H., & Tawfik, M. M. (2011). Integrated effect of organic and biofertilizers on wheat productivity in new reclaimed sandy soil. *Research Journal of Agriculture and Biological Sciences, 7*(1), 105–114.

Alavanja, M. C. R., Hoppin, J. A., & Kamel, F. (2004). Health effects of chronic pesticide exposure: Cancer and neurotoxicity. *Annual Review of Public Health, 25,* 155–197. https://doi.org/10.1146/annurev.publhealth.25.101802.123020

Aldrich, J. R., & Cantelo, W. W. (1999). Suppression of Colorado potato beetle infestation by pheromone-mediated augmentation of the predatory spined soldier bug, Podisus maculiventris (Say) (Heteroptera: Pentatomidae). *Agricultural and Forest Entomology, 1*(3), 209–217. https://doi.org/10.1046/j.1461-9563.1999.00026.x

Arif, M., Khan, M., Akbar, H., & Sajjad, S. A. (2006). Prospects of wheat as a dual response crop and its impact on weeds. *Pakistan Journal of Weed Science Research, 12*(1–2), 13–17.

Arroussi, H. E., Benhima, R., Elbaouchi, A., Sijilmassi, B., El Mernissi, N., Aafsar, A., Meftah-Kadmiri, I., Bendaou, N., & Smouni, A. (2018). *Dunaliella salina* exopolysaccharides: a promising biostimulant for salt stress tolerance in tomato (Solanum lycopersicum). *6TH Congress of the International Society for Applied Phycology.* https://doi.org/10.1007/s10811-017-1382-1

Babu, S., Prasanna, R., Bidyarani, N., & Singh, R. (2014). Analysing the colonisation of inoculated cyanobacteria in wheat plants using biochemical and molecular tools. *Journal of Applied Phycology, 27*(1), 327–338. https://doi.org/10.1007/s10811-014-0322-6

Banu, J. R., Preethi, J., Kavitha, S., Gunasekaran, M., & Kumar, G. (2020). Microalgae based biorefinery promoting circular bioeconomy-techno economic and life-cycle analysis. *Bioresource Technology, 302,* 122822. https://doi.org/10.1016/j.biortech.2020.122822

Bardin, M., Ajouz, S., Comby, M., Lopez-Ferber, M., Graillot, B., Siegwart, M., & Nicot, P. C. (2015). Is the efficacy of biological control against plant diseases likely to be more durable than that of chemical pesticides? *Frontiers in Plant Science, 6*(July), 1–14. https://doi.org/10.3389/fpls.2015.00566

Barone, V., Baglieri, A., Stevanato, P., Broccanello, C., Bertoldo, G., Bertaggia, M., Cagnin, M., Pizzeghello, D., Moliterni, V. M. C., Mandolino, G., Fornasier, F., Squartini, A., Nardi, S., & Concheri, G. (2018). Root morphological and molecular responses induced by microalgae extracts in sugar beet (Beta vulgaris L.). *Journal of Applied Phycology, 30*(2), 1061–1071. https://doi.org/10.1007/s10811-017-1283-3

Barone, V., Puglisi, I., Fragalà, F., Lo Piero, A., Giuffrida, F., & Baglieri, A. (2018). Novel bioprocess for the cultivation of microalgae in hydroponic growing system of tomato plants. *Journal of Applied Phycology, 31*(1), 465–470. https://doi.org/10.1007/s10811-018-1518-y

Barros, M. V., Salvador, R., de Francisco, A. C., & Piekarski, C. M. (2020). Mapping of research lines on circular economy practices in agriculture: From waste to energy. *Renewable and Sustainable Energy Reviews, 131*, 109958. https://doi.org/10.1016/j.rser.2020.109958

Battacharyya, D., Babgohari, M. Z., Rathor, P., & Prithiviraj, B. (2015). Seaweed extracts as biostimulants in horticulture. *Scientia Horticulturae, 196*, 39–48. https://doi.org/10.1016/j.scienta.2015.09.012

Bhola, V., Swalaha, F., Ranjith Kumar, R., Singh, M., & Bux, F. (2014). Overview of the potential of microalgae for CO2 sequestration. *International Journal of Environmental Science and Technology, 11*(7), 2103–2118. https://doi.org/10.1007/s13762-013-0487-6

Biondi, N., Piccardi, R., Margheri, M. C., Rodolfi, L., Smith, G. D., & Tredici, M. R. (2004). Evaluation of *Nostoc* strain ATCC 53789 as a potential source of natural pesticides. *Applied and Environmental Microbiology, 70*(6), 3313–3320. https://doi.org/10.1128/AEM.70.6.3313

Bolarinwa, I. F., Oke, M. O., Olaniyan, S. A., & Ajala, A. S. (2016). A review of cyanogenic glycosides in edible plants. In Soloneski, S., Larramendy, M. L. (Eds.) *Toxicology - New Aspects to This Scientific Conundrum* (pp. 181–191) [Internet]. London: IntechOpen; https://doi.org/10.5772/64886

Bose, A., O'Shea, R., Lin, R., & Murphy, J. D. (2020). A perspective on novel cascading algal biomethane biorefinery systems. *Bioresource Technology, 304*, 123027. https://doi.org/10.1016/j.biortech.2020.123027

Broadley, M., Brown, P., Cakmak, I., Ma, J. F., Rengel, Z., & Zhao, F. (2012). Beneficial elements. In *Marschner's Mineral Nutrition of Higher Plants (3rd ed)* (pp. 249–269). Elsevier, Waltham, MA. http://dx.doi.org/10.1016/b978-0-12-384905-2.00008-x

Bueno, A. F., Sosa-Gómez, D. R., Corrêa-Ferreira, B. S. Moscardi, F., & Bueno, R. C. O. F. (2012). Natural enemies of soybean pests. In C. B. Hoffmann-Campo, B. S. Corrêa-Ferreira, & F. Moscardi (Eds.), *Soy: Integrated Management of Insects and Other Pest Arthropods* (pp. 493–629). Embrapa, Brasília, Brazil.

Cabanelas, I. T. D., Ruiz, J., Arbib, Z., Chinalia, F. A., Garrido-Pérez, C., Rogalla, F., Nascimento, I. A., & Perales, J. A. (2013). Comparing the use of different domestic wastewaters for coupling microalgal production and nutrient removal. *Bioresource Technology, 131*, 429–436. https://doi.org/10.1016/j.biortech.2012.12.152

Carrilho, E. N. V. M., Labuto, G., & Kamogawa, M. Y. (2016). Destination of vinasse, a residue from alcohol industry. In M. N. V. Prasad, & Kaimin Shih (Eds.), *Environmental Materials and Waste* (pp. 21–43). Elsevier, Waltham, MA. https://doi.org/10.1016/B978-0-12-803837-6.00002-0

Carvalho, F. P. (2017). Pesticides, environment, and food safety. *Food and Energy Security, 6*(2), 48–60. https://doi.org/10.1002/fes3.108

Castro, J. de S., Calijuri, M. L., Assemany, P. P., Cecon, P. R., de Assis, I. R., & Ribeiro, V. J. (2017). Microalgae biofilm in soil: Greenhouse gas emissions, ammonia volatilization and plant growth. *Science of the Total Environment, 574*, 1640–1648. https://doi.org/10.1016/j.scitotenv.2016.08.205

Chandel, S. T. (2009). Nematicidal activity of the Cyanobacterium, Aulosira fertilissima on the hatch of Meloidogyne triticoryzae and Meloidogyne incognita. *Archives of Phytopathology and Plant Protection, 42*(1), 32–38. https://doi.org/10.1080/03235400600914363

Chiaiese, P., Corrado, G., Colla, G., Kyriacou, M. C., & Rouphael, Y. (2018). Renewable sources of plant biostimulation: Microalgae as a sustainable means to improve crop performance. *Frontiers in Plant Science, 9.* https://doi.org/10.3389/fpls.2018.01782

Coppens, J., Grunert, O., Van Den Hende, S., Vanhoutte, I., Boon, N., Haesaert, G., & De Gelder, L. (2016). The use of microalgae as a high-value organic slow-release fertilizer results in tomatoes with increased carotenoid and sugar levels. *Journal of Applied Phycology, 28*(4), 2367–2377. https://doi.org/10.1007/s10811-015-0775-2

Crofcheck, C., Monstross, M., Xinyi, E., Shea, A., Crocker, M., & Andrews, R. (2012). Influence of media composition on the growth rate of *Chlorella vulgaris* and Scenedesmus acutus utilized for CO2 mitigation. *American Society of Agricultural and Biological Engineers Annual International Meeting, 1*(12), 532–549. https://doi.org/10.13031/2013.41734

de Carvalho, J. C., Borghetti, I. A., Cartas, L. C., Woiciechowski, A. L., Soccol, V. T., & Soccol, C. R. (2018). Biorefinery integration of microalgae production into cassava processing industry: Potential and perspectives. *Bioresource Technology, 247,* 1165–1172. https://doi.org/10.1016/j.biortech.2017.09.213

Dineshkumar, R., Duraimurugan, M., Sharmiladevi, N., Lakshmi, L. P., Rasheeq, A. A., Arumugam, A., & Sampathkumar, P. (2020). Microalgal liquid biofertilizer and biostimulant effect on green gram (Vigna radiata L) an experimental cultivation. *Biomass Conversion and Biorefinery.* https://doi.org/10.1007/s13399-020-00857-0

Dineshkumar, R., Subramanian, J., Arumugam, A., Rasheeq, A. A., & Sampathkumar, P. (2018). Exploring the microalgae biofertilizer effect on onion cultivation by field experiment. *Waste and Biomass Valorization, 11*(1), 77–87. https://doi.org/10.1007/s12649-018-0466-8

Dineshkumar, R., Subramanian, J., Gopalsamy, J., Jayasingam, P., Arumugam, A., Kannadasan, S., & Sampathkumar, P. (2017). The impact of using microalgae as biofertilizer in maize (Zea mays L.). *Waste and Biomass Valorization, 10*(5), 1101–1110. https://doi.org/10.1007/s12649-017-0123-7

Drever, J. I., & Stillings, L. L. (1997). The role of organic acids in mineral weathering. *Colloids and Surfaces A: Physicochemical and Engineering Aspects, 120,* 167–181.

El-Baky, H. H. A., El-Baz, F. K., & Baroty, G. S. E. (2010). Enhancing antioxidant availability in wheat grains from plants grown under seawater stress in response to microalgae extract treatments. *Journal of the Science of Food and Agriculture, 90*(2), 299–303. https://doi.org/10.1002/jsfa.3815

El-Mougy, N. S., & Abdel-Kader, M. M. (2013). Effect of commercial cyanobacteria products on the growth and antagonistic ability of some bioagents under laboratory conditions. *Journal of Pathogens, 2013,* 1–11. https://doi.org/10.1155/2013/838329

ETIP Bioenergy. (2020). INPUT PAPER Biomethane. Brussels. https://www.etipbioenergy.eu/images/ETIP_B_Factsheet_Biomethane

Faheed, F. A., & Fattah, Z. A. (2008). Effect of *Chlorella vulgaris* as bio-fertilizer on growth parameters and metabolic aspects of lettuce plant. *Journal of Agricultural and Food Chemistry, 4*(1965), 1813–2235. http://www.fspublishers.org

Falaise, C., François, C., Travers, M. A., Morga, B., Haure, J., Tremblay, R., Turcotte, F., Pasetto, P., Gastineau, R., Hardivillier, Y., Leignel, V., & Mouget, J. L. (2016). Antimicrobial compounds from eukaryotic microalgae against human pathogens and diseases in aquaculture. *Marine Drugs, 14*(9), 1–27. https://doi.org/10.3390/md14090159

Garcia-Gonzalez, J., & Sommerfeld, M. (2016). Biofertilizer and biostimulant properties of the microalga *Acutodesmus dimorphus. Journal of Applied Phycology, 28*(2), 1051–1061. https://doi.org/10.1007/s10811-015-0625-2

Gazzoni, D. L. (2012). Pest management perspectives. In C. B. C. Campo, B. S. Corrêa-Ferreira, & F. Moscardi (Eds.), *Soy: Integrated Management of Insects and Other Pest Arthropods* (1st ed., pp. 789–829). Embrapa.

Getachew, D., Mulugeta, K., Gamechu, G., & Murugesan, K. (2020). Values and drawbacks of biofuel production from microalgae. *Journal of Applied Biotechnology Reports*, *7*(1), 1–6. https://doi.org/10.30491/jabr.2020.105917

Glare, T., Caradus, J., Gelernter, W., Jackson, T., Keyhani, N., Köhl, J., Marrone, P., Morin, L., & Stewart, A. (2012). Have biopesticides come of age?. *Trends in Biotechnology*, *30*(5), 250–258. https://doi.org/10.1016/j.tibtech.2012.01.003

Gomiero, T. (2018). Agriculture and degrowth: State of the art and assessment of organic and biotech-based agriculture from a degrowth perspective. *Journal of Cleaner Production*, *197*, 1823–1839. https://doi.org/10.1016/j.jclepro.2017.03.237

Gross, M., Henry, W., Michael, C., & Wen, Z. (2013). Development of a rotating algal biofilm growth system for attached microalgae growth with in situ biomass harvest. *Bioresource Technology*, *150*, 195–201. https://doi.org/10.1016/j.biortech.2013.10.016

Grzesik, M., Romanowska-Duda, Z., & Kalaji, H. M. (2017). Effectiveness of cyanobacteria and green algae in enhancing the photosynthetic performance and growth of willow (Salix viminalis L.) plants under limited synthetic fertilizers application. *Photosynthetica*, *55*(3), 510–521. https://doi.org/10.1007/s11099-017-0716-1

Gupta, K., & Baruah, P. P. (2020). Cypermethrin toxicity to rice field cyanobacterium *Calothrix* sp. *Vegetos*, *33*(3), 401–408. https://doi.org/10.1007/s42535-020-00114-9

Gupta, S., & Dikshit, A. K. (2010). Biopesticides: An ecofriendly approach for pest control. *Journal of Biopesticides*, *3*(1), 186–188.

Guzmán-Murillo, M. A., Ascencio, F., & Larrinaga-Mayoral, J. A. (2013). Germination and ROS detoxification in bell pepper (Capsicum annuum L.) under NaCl stress and treatment with microalgae extracts. *Protoplasma*, *250*(1), 33–42. https://doi.org/10.1007/s00709-011-0369-z

Gyarmati, G. (2017). "The consumption of organic products according to a survey," Proceedings of FIKUSZ 2017, In Monika Fodor (ed.), Proceedings of FIKUSZ '17, pages 125-139, Óbuda University, Keleti Faculty of Business and Management. URL: https://ideas.repec.org/h/pkk/sfyr17/125-139.html

Hajnal-Jafari, T., Seman, V., Stamenov, D., & Đurić, S. (2020). Effect of *Chlorella vulgaris* on growth and photosynthetic pigment content in Swiss Chard (Beta vulgaris L. subsp. cicla). *Polish Journal of Microbiology*, *69*(2), 235–238. https://doi.org/10.33073/pjm-2020-023

Hastings, K. L., Smith, L. E., Lindsey, M. L., Blotsky, L. C., Downing, G. R., Zellars, D. Q., Downing, J. K., & Corena-McLeod, M. (2014). Effect of microalgae application on soil algal species diversity, cation exchange capacity and organic matter after herbicide treatments. *F1000Research*, *3*, 281. https://doi.org/10.12688/f1000research.4016.1

Heimpel, G. E., & Mills, N. J. (2017). *Biological Control: Ecology and Applications*. Cambridge University Press , Cambridge, UK. DOI: 10.1017/9781139029117.

Hellal, F. A., Taalab, A. S., & Safaa, A. M. (2009). Influence of nitrogen and boron nutrition on nutrient balance and Sugar beet yield grown in calcareous soil. *Ozean Journal of Applied Science*, *2*, 1–10.

Hernández-Carlos, B., & Gamboa-Angulo, M. M. (2011). Metabolites from freshwater aquatic microalgae and fungi as potential natural pesticides. *Phytochemistry Reviews*, *10*(2), 261–286. https://doi.org/10.1007/s11101-010-9192-y

Hirano, A., Ueda, R., Hirayama, S., & Ogushi, Y. (1997). CO2 fixation and ethanol production with microalgal photosynthesis and intracellular anaerobic fermentation. *Energy*, *22*(2–3), 137–142. https://doi.org/10.1016/S0360-5442(96)00123-5

Huang, J., Hu, R., Rozelle, S., & Pray, C. (2005). Insect-resistant GM rice in farmers' fields: Assessing productivity and health effects in China. *Science*, *308*(5722), 688–690. https://doi.org/10.1126/science.1108972

Jäger, K., Bartók, T., Ördög, V., & Barnabás, B. (2010). Improvement of maize (Zea mays L.) anther culture responses by algae-derived natural substances. *South African Journal of Botany*, 76(3), 511–516. https://doi.org/10.1016/j.sajb.2010.03.009

Jochum, M., Moncayo, L. P., & Jo, Y. K. (2018). Microalgal cultivation for biofertilization in rice plants using a vertical semi-closed airlift photobioreactor. *PLoS One*, 13(9), e0203456. https://doi.org/10.1371/journal.pone.0203456

Jamal Uddin, A. F. M., Rakibuzzaman, M., Wasin, E. W. N., Husna, M. A., & Mahato, A. K. (2019). Foliar application of *Spirulina* and *Oscillatoria* on growth and yield of okra as bio-fertilizer. *Journal of Bioscience and Agriculture Research*, 22(2), 1840–1844. https://doi.org/10.18801/jbar.220219.227

Kawalekar, J. S. (2013). Role of biofertilizers and biopesticides for sustainable agriculture. *Journal of Bio Innovation*, 3(2277–8330), 73–78.

Khan, S. A., Hussain, M. Z., Prasad, S., & Banerjee, U. C. (2009). Prospects of biodiesel production from microalgae in India. *Renewable and Sustainable Energy Reviews*, 13(9), 2361–2372. https://doi.org/10.1016/j.rser.2009.04.005

Khan, S. A., Sharma, G. K., Malla, F. A., Kumar, A., Rashmi, & Gupta, N. (2019). Microalgae based biofertilizers: A biorefinery approach to phycoremediate wastewater and harvest biodiesel and manure. *Journal of Cleaner Production*, 211, 1412–1419. https://doi.org/10.1016/j.jclepro.2018.11.281

Khan, Z., Park, S. D., & Choi, Y. E. (1999). Three new species of predatory nematodes (Dorylaimida: Actinolaimoidea) from Korea. *Journal of Asia-Pacific Entomology*, 2(1), 7–13. https://doi.org/10.1016/S1226-8615(08)60025-4

Khan, Z., Park, S. D., Shin, S. Y., Bae, S. G., Yeon, I. K., & Seo, Y. J. (2005). Management of Meloidogyne incognita on tomato by root-dip treatment in culture filtrate of the blue-green alga, Microcoleus vaginatus. *Bioresource Technology*, 96(12), 1338–1341. https://doi.org/10.1016/j.biortech.2004.11.012

Kumar, M., Prasanna, R., Bidyarani, N., Babu, S., Mishra, B. K., Kumar, A., Adak, A., Jauhari, S., Yadav, K., Singh, R., & Saxena, A. K. (2013). Evaluating the plant growth promoting ability of thermotolerant bacteria and cyanobacteria and their interactions with seed spice crops. *Scientia Horticulturae*, 164, 94–101. https://doi.org/10.1016/j.scienta.2013.09.014

Kumari, R., Kaur, I., & Bhatnagar, A. K. (2011). Effect of aqueous extract of Sargassum johnstonii Setchell & Gardner on growth, yield and quality of Lycopersicon esculentum Mill. *Journal of Applied Phycology*, 23, 623–633. https://doi.org/10.1007/s10811-011-9651-x

Lachman, J., Orsak, M., Pivec, V., & Kratochvilova, D. (2001). Anthocyanins and carotenoids-major pigments of roses. *Hort Science*, 28, 33–39.

Lorentz, J. F., Calijuri, M. L., Assemany, P. P., Alves, W. S., & Pereira, O. G. (2020). Microalgal biomass as a biofertilizer for pasture cultivation: Plant productivity and chemical composition. *Journal of Cleaner Production*, 276, 124130. https://doi.org/10.1016/j.jclepro.2020.124130

Lv, J., Liu, S., Feng, J., Liu, Q., Guo, J., Wang, L., Jiao, X., & Xie, S. (2020). Effects of microalgal biomass as biofertilizer on the growth of cucumber and microbial communities in the cucumber rhizosphere. *Turkish Journal of Botany*, 44(2), 167–177. https://doi.org/10.3906/bot-1906-1

M'hamdi, A. I., Kandri, N. I., Zeroual, A., Blumberga, D., & Gusca, J. (2017). Life cycle assessment of paper production from treated wood. *Energy Procedia*, 128, 461–468. https://doi.org/10.1016/j.egypro.2017.09.031

Manchanda, T., Tyagi, R., Nalla, V. K., Chahar, S., & Sharma, D. K. (2018). Power generation by algal microbial fuel cell along with simultaneous treatment of sugar industry wastewater. *Journal of Bioprocessing & Biotechniques*, 8(03). https://doi.org/10.4172/2155-9821.1000323

Manjunath, M., Kanchan, A., Ranjan, K., Venkatachalam, S., Prasanna, R., Ramakrishnan, B., Hossain, F., Nain, L., Shivay, Y. S., Rai, A. B., & Singh, B. (2016). Beneficial cyanobacteria and eubacteria synergistically enhance bioavailability of soil nutrients and yield of okra. *Heliyon*, *2*(2). https://doi.org/10.1016/j.heliyon.2016.e00066

Medina-Pastor, P., & Triacchini, G. (2020). The 2018 European Union report on pesticide residues in food. *EFSA Journal*, *18*(4), 1–103. https://doi.org/10.2903/j.efsa.2020.6057

Meier, L., Martínez, C., Vílchez, C., Bernard, O., & Jeison, D. (2019). Evaluation of the feasibility of photosynthetic biogas upgrading: Simulation of a large-scale system. *Energy*, *189*, 116313. https://doi.org/10.1016/j.energy.2019.116313

Melanson, D., & Wells, J. (2016). Algal Research Hitting the Ground in China. University of Kentucky Center for Applied Energy Research (CAER). http://uknow.uky.edu/research/centers-and-institutes/center-applied-energy-research-caer/uk-caer-algal-research-hitting

Misra, S., & Kaushik, D. (1989). Growth promoting substances of Cyanobacteria. I: Vitamins and their influence on rice plant. *Proceedings of the Indian National Science Academy*, *5*, 499–504.

Mohsenpour, S. F., Hennige, S., Willoughby, N., Adeloye, A., & Gutierrez, T. (2021). Integrating micro-algae into wastewater treatment: A review. *Science of The Total Environment*, *752*, 142168. https://doi.org/10.1016/j.scitotenv.2020.142168

Mulbry, W., Kondrad, S., & Pizarro, C. (2007). Biofertilizers from algal treatment of dairy and swine manure effluents: Characterization of algal biomass as a slow release fertilizer. *Journal of Vegetable Science*, *12*(4), 107–125. https://doi.org/10.1300/J484v12n04_08

Mutale-joan, C., Redouane, B., Najib, E., Yassine, K., Lyamlouli, K., Laila, S., Zeroual, Y., & Hicham, E. A. (2020). Screening of microalgae liquid extracts for their bio stimulant properties on plant growth, nutrient uptake and metabolite profile of *Solanum lycopersicum* L. *Scientific Reports*, *10*(1), 1–12. https://doi.org/10.1038/s41598-020-59840-4

Nesamma, A. A., Kashif, M. S., & Jutur, P. P. (2015). Genetic engineering of microalgae for production of value added ingredients. In Se-Known Kim (Ed.), *Handbook of Marine Microalgae: Biotechnology Advances* (1st ed., pp. 405–414). Elsevier Science, Waltham, MA. http://dx.doi.org/10.1016/b978-0-12-800776-1.00026-1

Niccolai, A., Zittelli, G. C., Rodolfi, L., Biondi, N., & Tredici, M. R. (2019). Microalgae of interest as food source: Biochemical composition and digestibility. *Algal Research*, *42*, 101617. https://doi.org/10.1016/j.algal.2019.101617

Novak, A. C., Almeida, A. C. O., Furtado, I. F. S. P. C., & Sydney, E. B. (2018). *Biomass production process from the treatment of black liquor through the cultivation of cyanobacteria and microalgae and use of the biomass obtained* (Patent No. BR 10 2018 074711 8).

Oliveira, J., Mógor, G., & Mógor, A. (2013). Beet productivity due to foliar application of biofertilizers. *VIII Brazilian Congress of Agroecology*, *8*(2), 2–5.

Pacheco, D., Rocha, A. C., Pereira, L., & Verdelhos, T. (2020). Microalgae water bioremediation: Trends and hot topics. *Applied Sciences*, *10*(5), 1886. https://doi.org/10.3390/app10051886

Paudel, Y., & Pradhan, S. (2012). Role of blue green algae in rice productivity. *Agriculture and Biology Journal of North America*, *3*(8), 332–335. https://doi.org/10.5251/abjna.2012.3.8.332.335

Peres, S., Monteiro, M. R., Ferreira, M. L., do Nascimento Junior, A. F., & Palh, M. A. P. F. (2019). Anaerobic digestion process for the production of biogas from cassava and sewage treatment plant sludge in Brazil. *BioEnergy Research*, *12*(1), 150–157. https://doi.org/10.1007/s12155-018-9942-z

Pourjamshidian, R., Abolghasemi, H., Esmaili, M., Amrei, H. D., Parsa, M., & Rezaei, S. (2019). Carbon dioxide biofixation by *Chlorella* sp. in a bubble column reactor at different flow rates and CO2 concentrations. *Brazilian Journal of Chemical Engineering*, *36*(2), 639–645. https://doi.org/10.1590/0104-6632.20190362s20180151

Poyo, T. L. (2013). *Practical aspects of microalgae production: Objectives and needs* [Universidad de Almería]. http://hdl.handle.net/10835/1868

Prasanna, R., Nain, L., Tripathi, R., Gupta, V., Chaudhary, V., Middha, S., Joshi, M., Ancha, R., & Kaushik, B. D. (2008). Evaluation of fungicidal activity of extracellular filtrates of cyanobacteria - Possible role of hydrolytic enzymes. *Journal of Basic Microbiology*, *48*(3), 186–194. https://doi.org/10.1002/jobm.200700199

Priya, H., Prasanna, R., Ramakrishnan, B., Bidyarani, N., Babu, S., Thapa, S., & Renuka, N. (2015). Influence of cyanobacterial inoculation on the culturable microbiome and growth of rice. *Microbiological Research*, *171*, 78–89. https://doi.org/10.1016/j.micres.2014.12.011

Priya, M., Gurung, N., Mukherjee, K., & Bose, S. (2014). Microalgae in removal of heavy metal and organic pollutants from soil. In D. Surajit (Ed.), *Microbial Biodegradation and Bioremediation* (pp. 519–537). Elsevier, Waltham, MA. https://doi.org/10.1016/B978-0-12-800021-2.00023-6

Raja, R., Coelho, A., Hemaiswarya, S., Kumar, P., Carvalho, I. S., & Alagarsamy, A. (2018). Applications of microalgal paste and powder as food and feed: An update using text mining tool. *Beni-Suef University Journal of Basic and Applied Sciences*, *7*(4), 740–747. https://doi.org/10.1016/j.bjbas.2018.10.004

Ranjan, K., Priya, H., Ramakrishnan, B., Prasanna, R., Venkatachalam, S., Thapa, S., Tiwari, R., Nain, L., Singh, R., & Shivay, Y. S. (2016). Cyanobacterial inoculation modifies the rhizosphere microbiome of rice planted to a tropical alluvial soil. *Applied Soil Ecology*, *108*, 195–203. https://doi.org/10.1016/j.apsoil.2016.08.010

Renuka, N., Guldhe, A., Prasanna, R., Singh, P., & Bux, F. (2018). Microalgae as multi-functional options in modern agriculture: Current trends, prospects and challenges. *Biotechnology Advances*, *36*(4), 1255–1273. https://doi.org/10.1016/j.biotechadv.2018.04.004

Rinanti, A. (2016). Biotechnology carbon capture and storage by microalgae to enhance CO_2 removal efficiency in closed-system photobioreactor. In N. Thajuddin (Ed.), *Algae – Organisms for Imminent Biotechnology*. InTechOpen, London. https://doi.org/10.5772/62915

Ronga, D., Biazzi, E., Parati, K., Carminati, D., Carminati, E., & Tava, A. (2019). Microalgal biostimulants and biofertilisers in crop productions. *Agronomy*, *9*(4), 1–22. https://doi.org/10.3390/agronomy9040192

Santana, H., Cereijo, C. R., Teles, V. C., Nascimento, R. C., Fernandes, M. S., Brunale, P., Campanha, R. C., Soares, I. P., Silva, F. C. P., Sabaini, P. S., Siqueira, F. G., & Brasil, B. S. A. F. (2017). Microalgae cultivation in sugarcane vinasse: Selection, growth and biochemical characterization. *Bioresource Technology*, *228*, 133–140. https://doi.org/10.1016/j.biortech.2016.12.075

Santillan-Jimenez, E., Pace, R., Marques, S., Morgan, T., McKelphin, C., Mobley, J., & Crocker, M. (2016). Extraction, characterization, purification and catalytic upgrading of algae lipids to fuel-like hydrocarbons. *Fuel*, *180*, 668–678. https://doi.org/10.1016/j.fuel.2016.04.079

Schreiber, C., Schiedung, H., Harrison, L., Briese, C., Ackermann, B., Kant, J., Schrey, S. D., Hofmann, D., Singh, D., Ebenhöh, O., Amelung, W., Schurr, U., Mettler-Altmann, T., Huber, G., Jablonowski, N. D., & Nedbal, L. (2018). Evaluating potential of green alga *Chlorella vulgaris* to accumulate phosphorus and to fertilize nutrient-poor soil substrates for crop plants. *Journal of Applied Phycology*, *30*(5), 2827–2836. https://doi.org/10.1007/s10811-018-1390-9

Shaaban, M. M. (2001a). Green microalgae water extract as foliar feeding to wheat plants. *Journal of Biological Sciences*, *4*, 628–632.

Shaaban, M. M. (2001b). Nutritional status and growth of maize plants as affected by green microalgae as soil additives. *Journal of Biological Sciences*, *1*(6), 475–479. https://doi.org/10.3923/jbs.2001.475.479

Siddeeg, S. M., Tahoon, M. A., & Ben Rebah, F. (2019). Agro-industrial waste materials and wastewater as growth media for microbial bioflocculants production: A review. *Materials Research Express, 7*(1), 012001. https://doi.org/10.1088/2053-1591/ab5980

Singh, J. S., Kumar, A., & Singh, M. (2019). Cyanobacteria: A sustainable and commercial bio-resource in production of bio-fertilizer and bio-fuel from waste waters. *Environmental and Sustainability Indicators, 3–4*(May), 100008. https://doi.org/10.1016/j.indic.2019.100008

Singh, R., Gautam, N., Mishra, A., & Gupta, R. (2011). Heavy metals and living systems: An overview. *Indian Journal of Pharmacology, 43*(3), 246–253. https://doi.org/10.4103/0253-7613.81505

Singh, V. P., & Trehan, K. (1973). Effect of extracellular products of *Aulosira fertilissima* on the growth of rice seedlings. *Plant Soil, 38*, 457–464.

Supraja, K. V., Behera, B., & Balasubramanian, P. (2020). Efficacy of microalgal extracts as biostimulants through seed treatment and foliar spray for tomato cultivation. *Industrial Crops and Products, 151*(January), 112453. https://doi.org/10.1016/j.indcrop.2020.112453

Sydney, E. B., da Silva, T. E., Tokarski, A., Novak, A. C., de Carvalho, J. C., Woiciecohwski, A. L., Larroche, C., & Soccol, C. R. (2011). Screening of microalgae with potential for biodiesel production and nutrient removal from treated domestic sewage. *Applied Energy, 88*(10). https://doi.org/10.1016/j.apenergy.2010.11.024

Sydney, E. B., Neto, C. J. D., de Carvalho, J. C., Vandenberghe, L. P. S., Sydney, A. C. N., Letti, L. A. J., Karp, S. G., Soccol, V. T., Woiciechowski, A. L., Medeiros, A. B. P., & Soccol, C. R. (2019). Microalgal biorefineries: Integrated use of liquid and gaseous effluents from bioethanol industry for efficient biomass production. *Bioresource Technology, 292*, 121955. https://doi.org/10.1016/j.biortech.2019.121955

Uysal, O., Uysal, F. O., & Ekinci, K. (2015). Evaluation of microalgae as microbial fertilizer. *European Journal of Sustainable Development, 4*(2), 77–82. https://doi.org/10.14207/ejsd.2015.v4n2p77

Valev, D., Santos, H. S., & Tyystjärvi, E. (2020). Stable wastewater treatment with *Neochloris oleoabundans* in a tubular photobioreactor. *Journal of Applied Phycology, 32*(1), 399–410. https://doi.org/10.1007/s10811-019-01890-x

van Lenteren, J. C., Bolckmans, K., Köhl, J., Ravensberg, W. J., & Urbaneja, A. (2018). Biological control using invertebrates and microorganisms: Plenty of new opportunities. *BioControl, 63*(1), 39–59. https://doi.org/10.1007/s10526-017-9801-4

Vasconcelos, Y. (2018). *Pesticides in the balance.* https://revistapesquisa.fapesp.br/en/pesticides-in-the-balance/

Victor, T. J., & Reuben, R. (2000). Effects of organic and inorganic fertilisers on mosquito populations in rice fields of southern India. *Medical and Veterinary Entomology, 14*(4), 361–368. https://doi.org/10.1046/j.1365-2915.2000.00255.x

Vuppaladadiyam, A. K., Yao, J. G., Florin, N., George, A., Wang, X., Labeeuw, L., Jiang, Y., Davis, R. W., Abbas, A., Ralph, P., Fennell, P. S., & Zhao, M. (2018). Impact off flue gas compounds on microalgae and mechanisms for carbon assimilation and utilization. *ChemSusChem, 11*(2), 334–355. https://doi.org/10.1002/cssc.201701611

Wang, J., Zhang, T., Dao, G., Xu, X., Wang, X., & Hu, H. (2017). Microalgae-based advanced municipal wastewater treatment for reuse in water bodies. *Applied Microbiology and Biotechnology, 101*(7), 2659–2675. https://doi.org/10.1007/s00253-017-8184-x

Wilson, M. H., Groppo, J., Placido, A., Graham, S., Morton, S. A., Santillan-Jimenez, E., Shea, A., Crocker, M., Crofcheck, C., & Andrews, R. (2014). CO_2 recycling using microalgae for the production of fuels. *Applied Petrochemical Research, 4*(1), 41–53. https://doi.org/10.1007/s13203-014-0052-3

Wuang, S. C., Khin, M. C., Chua, P. Q. D., & Luo, Y. D. (2016). Use of Spirulina biomass produced from treatment of aquaculture wastewater as agricultural fertilizers. *Algal Research, 15*, 59–64. https://doi.org/10.1016/j.algal.2016.02.009

Xiao, R., & Zheng, Y. (2016). Overview of microalgal extracellular polymeric substances (EPS) and their applications. *Biotechnology Advances*, *34*(7), 1225–1244. https://doi.org/10.1016/j.biotechadv.2016.08.004

Xinyi, E., Crofcheck, C., & Crocker, M. (2016). Application of recycled media and algae-based anaerobic digestate in *Scenedesmus* cultivation. *Journal of Renewable and Sustainable Energy*, *8*(1). https://doi.org/10.1063/1.4942782

Yilmaz, E., & Sönmez, M. (2017). The role of organic/bio–fertilizer amendment on aggregate stability and organic carbon content in different aggregate scales. *Soil and Tillage Research*, 168, 118–124. https://doi.org/10.1016/j.still.2017.01.003

Zhang, J., Wang, X., & Zhou, Q. (2017). Co-cultivation of *Chlorella* spp and tomato in a hydroponic system. *Biomass and Bioenergy*, *97*, 132–138. https://doi.org/10.1016/j.biombioe.2016.12.024

Zhang, X. (2015). *Microalgae Removal of CO2 from Flue Gas*. IEA Clean Coal Centre, London, UK.

Zhou, W., Li, Y., Gao, Y., & Zhao, H. (2017). Nutrients removal and recovery from saline wastewater by *Spirulina platensis*. *Bioresource Technology*, *245*(June), 10–17. https://doi.org/10.1016/j.biortech.2017.08.160

Zulpa, G., Zaccaro, M. C., Boccazzi, F., Parada, J. L., & Storni, M. (2003). Bioactivity of intra and extracellular substances from cyanobacteria and lactic acid bacteria on "wood blue stain" fungi. *Biological Control*, *27*(3), 345–348. https://doi.org/10.1016/S1049-9644(03)00015-X

9 Heavy Metal Remediation
The Microbial Approach

Shilpi Srivastava[1] and Atul Bhargava[2]
[1]Amity University Uttar Pradesh, Lucknow
Campus, Lucknow, India
[2]Mahatma Gandhi Central University, Bihar, India

CONTENTS

9.1 INTRODUCTION

Environmental pollution is defined as the disequilibrium condition in an environment due to accumulation of toxic substances in the surroundings that reduce the ability of the contaminated sites to support life. Human zeal for rapid industrial development has exerted a substantial strain on the environment and in many cases has been detrimental to it (Cramer and Cole 2017). Although environmental pollution is an age-old phenomenon and has occurred throughout the human history, the process has been accelerated since the advent of industrialisation (Freitas et al. 2004; Shigaki 2020). Increased human population coupled with a concomitant interference in the environment through activities like industrialisation, burning of fossil fuels, urbanisation, technological revolution, deforestation, modern agricultural practices, mining and exploration has resulted in the deterioration in the quality of the natural environment in many parts of the world and has emerged as one of the most serious global challenges (Bhargava et al. 2012a; Ukaogo et al. 2020). The situation is aggravated by the misuse of environmental resources and improper disposal of wastes (Pushpanathan et al. 2014). Environmental pollution has become a global problem that is increasingly affecting ecosystems, growth and productivity of different life

DOI: 10.1201/9781003110477-9

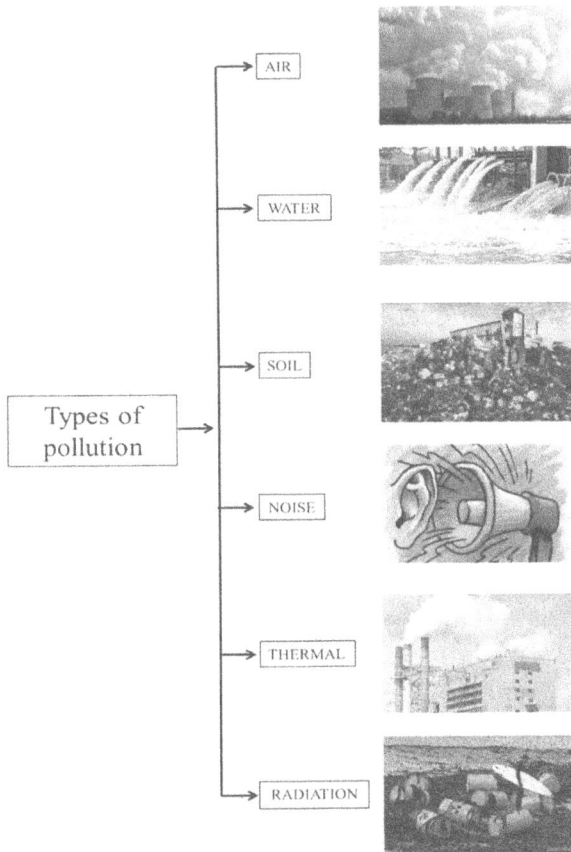

FIGURE 9.1   Different types of pollution. (Koul and Taak 2018.)

forms along with a decline in environmental quality (Rai 2016). Pollution also negatively affects biodiversity and is a major cause for its depletion (McNeely 1992;
Backhaus et al. 2012; Sordello et al. 2019). A pollutant can be defined as any chemical or physical substance released by human beings into the natural environment
beyond the permissible limit which can have disastrous consequences. Different
types of pollution are known namely air pollution, water pollution, soil/land pollu-
tion, thermal pollution, radioactive pollution and noise pollution (Figure 9.1) (Koul
and Taak 2018). The problem of pollution is much serious in developing countries
due to illiteracy, poverty and poor legislation.

9.2   HEAVY METALS

Heavy metals constitute a group of naturally occurring ill-defined subset of metals
and metalloids that have relatively higher density as compared to water and are toxic
even at parts per billion (ppb) levels. The term 'heavy metal' is referred to metals
having high atomic weights ranging from 63.5 to 200.6 g/mol and densities greater

than 5 g/cm^3 (Srivastava and Majumder 2008). These comprise transition metals, some metalloids, lanthanides and actinides (Babula et al. 2008). According to the density criteria mentioned above, 53 elements are usually considered as heavy metals. Of these some are crucial for survival of life forms while others are fatal. Some heavy metals like cadmium (Cd), lead (Pb), uranium (U), thallium (Tl), chromium (Cr), silver (Ag), and mercury (Hg) are toxic, and directly or indirectly affect different living organisms including plants (Srivastava and Bhargava 2015). Depending on their requirement and utility for living forms, heavy metals are categorised into essential and non-essential metals. In plants, they play an important role in the physiological and biochemical processes like mitochondria and chloroplast functioning, acting as cofactors for numerous enzymes, DNA synthesis, lignin formation, protein modifications and trafficking, sugar metabolism, oxidative stress responses, cell wall metabolism, hormone signalling and nodulation and nitrogen fixation (Bhargava et al. 2012a; Morkunas et al. 2018; Pietrini et al. 2019; Alejandro et al. 2020). Table 9.1 provides information on the various roles these elements play in the plant's well-being. Heavy metals are present in the soil naturally in low amounts, but their concentration greatly increases due to numerous geological and anthropogenic activities

TABLE 9.1
The Importance of Heavy Metals for Plants (Srivastava and Bhargava 2015)

Metal	Beneficial Effects of Heavy Metals	Reference
Cu	Important role in CO_2 assimilation and ATP synthesis.	Thomas et al. (1998)
	Component of plastocyanin and cytochrome oxidase	Demirevska-kepova et al. (2004)
Fe	Synthesis of chlorophyll	Miller et al. (1995); Spiller et al. (1982)
	Component of cytochromes	Soetan et al. (2010)
Zn	Synthesis of cytochrome	Tisdale et al. (1984)
	Synthesis of tryptophan and auxin	Alloway (2004); Brennan (2005)
	Reduce the adverse effects of short periods of heat and salt stress	Disante et al. (2010); Tavallali et al. (2010)
Co	Inhibition of ethylene production	Lau and Yang (1976)
	Role in salt tolerance	Ibrahim et al. (1989)
Mn	Activation of enzymes like decarboxylating malate dehydrogenase, isocitrate dehydrogenase, and nitrate reductase	Mukhopadhay and Sharma (1991)
Mo	Regulatory component in the maintenance of nitrogen fixation in legumes	Kaiser et al. (2005); Soetan et al. (2010)
	Integral part of molybdenum co-factor (Moco) which binds to molybdenum-requiring enzymes.	Bittner et al. (2001); Mendel and Haensch (2002)
Ni	Cofactor of enzymes involved in DNA biosynthesis and amino acid metabolism	Arinola et al. (2008)
	Component of the enzyme urease	Aydinalp and Marinova (2009)
Hg	No beneficial effect reported	

like industrialisation, mining, foundries and smelters, indiscriminate use of agro-chemicals, inefficient waste disposal and waste incineration (Chibuike and Obiora 2014; Mishra et al. 2017). Heavy metals are also transported over long distances in gaseous and particulate form leading to their accumulation in different living systems (Adriano et al. 2005). Pollution by heavy metals in environments like soil, water and runoff water is increasing day by day and has become a serious threat to different living forms in the ecosystems since it affects crop yields, soil biomass and fertility, leading to bioaccumulation of metals in the food chain (Rajkumar et al. 2009; Bhargava et al. 2012b; Siddiquee et al. 2015; Mishra 2017). Heavy metal toxic-ity is increasing at an alarming pace since heavy metals are nondegradable and have a tendency of bioaccumulation along the food chain (Gautam et al. 2014).

Plants have a tendency to absorb and accumulate heavy metals in both aboveg-round and belowground parts when grown in soils having high concentrations of these elements. Some heavy metals are beneficial in minor amounts, and in fact improve plant growth and development. However, higher uptake directly or indi-rectly affects plant health causing various morphological, physiological and meta-bolic abnormalities (Amari et al. 2017; Hassan et al. 2017). Table 9.2 provides an exhaustive account of the different deleterious effects of heavy metals on plant health. These toxic effects range from reduction in morphological and yield param-eters to destruction of photosynthetic pigments, inhibition of physiologically active enzymes, disturbance in mineral metabolism, genomic instability and damage to cell structures as a fallout of oxidative stress (Jalmi et al. 2018; Ghosh and Roy 2019; Sinegovskaya et al. 2020). Heavy metals also induce detrimental effects in microbial populations that include hinderance in cellular processes, reduced oxygen uptake, elongated lag phase, growth retardation, reduced biomass, denaturation of proteins, enzyme deactivation, destruction of nucleic acids, alteration in transcription process, hindering cell division and even mortality of microbial cells (Ding et al. 2017; Igiri et al. 2018). Humans get exposed to heavy metals through skin contact, inhalation and contaminated foodstuffs. Once inside in appreciable amounts, the heavy metal lowers the energy levels and induces immense damage to vital organs like the brain, lungs, kidney, and liver, besides altering the blood composition (Figure 9.2). Long-term exposure to heavy metals of their salts leads to progressive physical, muscular, and neurological degenerative processes that imitate disorders like Parkinson's dis-ease (Anyanwu et al. 2018; Bjorklund et al. 2019), Alzheimer's disease (Lee et al. 2018), epigenetic alterations (Caffo et al. 2014), cardiovascular disorders (Alissa and Ferns 2011), muscular dystrophy (Chin-Chan et al. 2015), macular degenera-tion (Pamphlett et al. 2020), fatty liver disease (Lin et al. 2017) and cancer of lungs, liver, prostate, skin and bladder (Kim et al. 2015; Vella et al. 2017). Thus, we see that heavy metals released from various sources are toxic not only to plant systems but also constitute a significant risk to public health.

9.3 BIOREMEDIATION

Removal of heavy metals from contaminated sites is urgently required for the pro-tection and conservation of the environment (Glick 2010). Several physical, chemical and biological techniques have been used for this purpose. The physical methods

TABLE 9.2
Adverse Effects of Some Heavy Metals on Plant Growth and Development

Metal	Toxic Effect	Reference
Zn	Decrease in chlorophyll and carotenoids	Sagardoy et al. (2009)
	Lower activity of RuBisCO	Mukhopadhyay et al. (2013)
	Growth retardation in roots and shoots	Jain et al. (2010)
Hg	Seed injuries and reduced seed viability	Patra et al. (2004)
	Reduction in leghemoglobin, root nodule size and number	Mondal et al. (2015)
Cu	Reduction in plant height and fresh weight	Chiou and Hsu (2019)
	Disruption of the root cuticle, severe deformation of root structure	Sheldon and Menzies (2005)
	Reduction in chlorophyll	Singh et al. (2007); Karimi et al. (2012)
Co	Adverse effect on shoot growth and biomass	Li et al. (2009)
	Chlorosis and necrosis	Gopal et al. (2003)
Mn	Chlorosis, puckering and crinkling of leaves, necrosis with purple-colored veins.	Fernando and Lynch (2015); Santos et al. (2017)
	Cytoplasmic injures; plasma membrane rupturing	Santandrea et al. (1997, 1998)
Pb	Reduced seed germination and plant growth	Zulfiqar et al. (2019)
	Inhibition of carotenoid and plastoquinone production	Mitra et al. (2020)
	Inhibition of enzyme activities and ATP production	Pourrut et al. (2011)
	Inhibition of root growth and root dry mass	Fahr et al. (2013)
Cr	Delayed and reduced seed germination	Datta et al. (2011); Stambulska et al. (2018)
	Chlorosis and reduced photosynthesis	
	Reduction in uptake of essential minerals	Kumar et al. (2019a)
		Sharma et al. (2020)
Ni	Induction of senescence, leaf and meristem chlorosis	Ahmad and Ashraf (2011)
	depletion of low molecular weight proteins	Kukkola et al. (2000)
	Reduces the translocation of N from root to shoot	Ameen et al. (2019)
Cd	Inhibition of respiration	Llamas et al. (2000)
	Inhibition of photosynthesis	Dias et al. (2013)
	Reduction in carotenoid content	Castagna et al. (2013)
	Inhibition of DNA replication, gene expression, and cell division.	Ghosh and Roy (2019)
Mo	Weakened nitrogen fixation capacity	Yang et al. (2020)
	Reduction In shoot and root yield	Kevresan et al. (2001)
Se	Damage to chloroplast structure	Guo et al. (2016)
	Increased oxidative stress	Gupta and Gupta (2017)
	Decrease in germination and seedling growth	Lapaz et al. (2019)

include adsorption, electroremediation, nano-photocatalysis, excavation and replacement of soil, soil washing with strong acids, vitrification, heat treatment, and use of membrane adsorption techniques (Khulbe and Matsuura 2018; Sharma et al. 2018; Tahir et al. 2019; Akhtar et al. 2020), while the chemical methods comprise of precipitation, leaching, chemical extraction and oxidation, floatation, ion exchange, coagulation, flocculation and use of soil amendments (Wuana and Okieimen 2011;

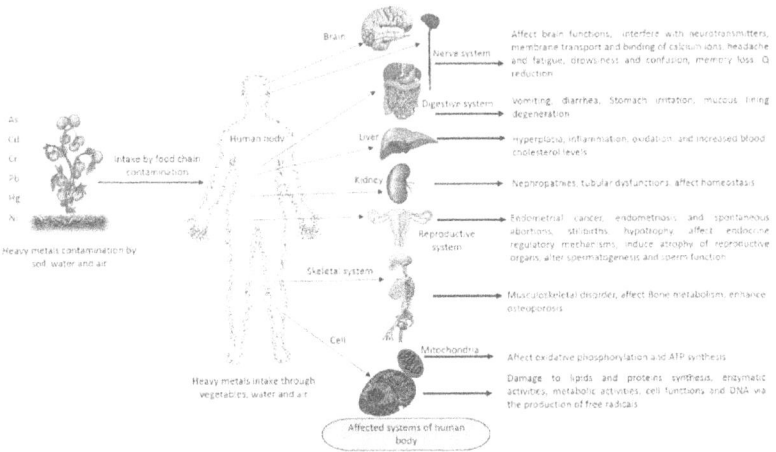

FIGURE 9.2   Effect of heavy metals on human health. (Kumar et al. 2019b.)

Sharma et al. 2018; Akhtar et al. 2020). However, most of the above-mentioned remediation technologies are complex, costly, labour intensive and cause environmental disturbances, and have thus found limited acceptability among the communities (Bhargava et al. 2012b). Bioremediation is a highly promising, cost-effective and efficient emerging technology that utilises the bioaccumulation and biodegradative potential of organisms for efficient management of environmental pollutants. This noninvasive and cost-effective approach involves the use of living organisms for enhancing the efficacy of natural degradation process by conversion of pollutants into less toxic or nontoxic form (Chandra et al. 2013). Common bioremediation techniques include landfarming, bioaugmentation, biosparging, bioventing, biostimulaton, bioslurping, biopiling, composting and the use of biofilters and bioreactors (Azubuike et al. 2016).

9.4   MICROBES IN HEAVY METAL REMEDIATION

Microorganisms are ubiquitously present in the biosphere and play a major role in the regulation of primary elements (Tapia-Torres et al. 2016; Jacoby et al. 2017), recycling of wastes and detoxification (Wei and Zimmermann 2017; Ru et al. 2020) and suppression of plant diseases (Pineda et al. 2017). The degradation potential of microorganisms ranges from synthetic substances such as plastics, radionuclides, heavy metals, pharmaceutical products and agrichemicals to polyaromatic hydrocarbons (Ojuederie and Babalola 2017; Malla et al. 2018) (Table 9.3). The mechanisms involved in bioremediation mediated by microbes involve bioaccumulation, bioleaching, biosorption, biotransformation, bioaccumulation, biomineralisation and metal-microbe interactions (Verma and Kuila 2019). Microbes like bacteria, fungi, algae and yeast also play a major role in the bioremediation of heavy metals since they have the capability to endure metal toxicity utilizing different metabolic processes. Cleaning heavy metal-contaminated sites using microorganisms, a process successfully used for decades, utilises the metabolic activity of microorganisms for reduction,

TABLE 9.3
Catabolic Genes Reported in Bacteria Responsible for Bioremediation (Das and Dash 2014)

Potential Bacteria	Target Substance	Catabolic Genes	References
Pseudomonas aeruginosa	Organic and Inorganic mercury	*merA, merB*	De et al. (2008); Dash and Das (2012)
Cycloclasticus sp.	PAH	*phnA1, phnA2, phnA3, and phnA4*	Kasai et al. (2002)
Pseudomonas sp.	Phenol	*dmpN*	Selvaratnam et al. (1997)
Staphylococcus aureus	Chromate	*chrB*	Aguilar-Barajas et al. (2008)
Bacillus subtilis, Bacillus cereus	Cobalt-Zinc-Cadmium	*czcD*	Abdelatey et al. (2011)
Pseudomonas sp., Bordetella sp.	Nickel-Cobalt-Cadmium	*nccA*	Abou-Shanab et al. (2003)

removal, transformation and degradation of heavy metals. This process facilitates heavy metal decomposition or immobilisation by utilisation of the metabolic potential of microorganisms with novel catabolic functions resulting from selection or by introduction of genes encoding such functions. With respect to heavy metals, it has been observed that the microorganisms do not convert toxic heavy metals into harmless simpler forms. Considering their unique specificity towards contaminants, the microbes are selected primarily on the basis of their metabolic activity. The vast diversity of microbes enables them in selective utilisation of heavy metals for their growth and development. Those microbes that are capable of surviving in metal rich environments have a diverse set of resistance mechanisms either acquired during the evolutionary history or obtained by successful gene transfer from plasmids. These resistance mechanisms must provide the ability to microbes for tolerating the toxicity of heavy metals at high concentrations and at the same allow the entrance of essential metals required for maintenance of homeostasis (Castro et al. 2019). Among the microorganisms microalgae, bacteria and fungi have been the favourite choice for bioremediation of heavy metals.

9.4.1 ALGAE

The use of microalgae or 'wonder organism' in bioremediation of heavy metals has recently gained attention (Zeraatkar et al. 2016). These microscopic entities are capable of effectively carrying out bioremediation by two mechanisms, viz., bioassimilation and biosorption (Sreekumar et al. 2020). Microalgae can grow in contaminated water as 'algal blooms' and absorb pollutants in the biomass which can be harvested and processed for use as an efficient biosorbent. A number of metal-binding groups such as carboxylate, amine, imidazole, phosphate, sulfhydryl, sulphate and hydroxyl present in the cell wall of algae aid in the binding of metals to the algal wall. Blue-green algae or Cyanobacteria, a diverse group of nitrogen fixing,

TABLE 9.4
Algae Used in Bioremediation of Heavy Metals

Algal Class	Algae	Heavy Metal	References
Blue green algae/	*Anabaena subcylindrica*	Lead, Manganese	El-Sheekh et al. (2005)
Cyanobacteria	*Gloeothece* sp.	Copper	Micheletti et al. (2008b)
	Nostoc commune	Cadmium	Hwang et al. (2018)
	Nostoc muscorum	Lead, Manganese,	El-Sheekh et al. (2005);
	Oscillatoria angustissima	Chromium	Gupta and Rastogi
	Spirulina platensis	Cobalt, Copper, Zinc	(2008a)
		Chromium	Mohapatra and Gupta
			(2005)
			Finocchio et al. (2010)
Green algae	*Chlamydomonas angulosa*	Cadmium	Hwang et al. (2018)
	Chlamydomonas reinhardtii	Chromium	Arıca et al. (2005)
	Chlorella miniata	Chromium	Han et al. (2007)
	Chlorella sorokiniana	Nickel	Akhtar et al. (2004)
		Chromium	Akhtar et al. (2008)
	Chlorella vulgaris	Cadmium	Aksu (2001)
		Chromium	Fraile et al. (2005)
		Copper	Mehta and Gaur (2001)
		Lead	El-Naas et al. (2007)
		Nickel	Aksu (2002)
	Scenedesmus obliquus	Cadmium	Chen et al. (2012)
	Spirogyra sp.	Chromium	Bishnoi et al. (2007)
		Copper	Gupta et al. (2006)
		Lead	Gupta and Rastogi (2008b)

oxygenic photosynthetic prokaryotes need special mention here. This group is abundantly found in natural environments and confer numerous advantages over other microbes in bioremediation of heavy metals due to unique cell wall composition, high mucilage volume with high-binding affinity, large surface area, simple nutrient requirement, ease in culturing and low cost (Micheletti et al. 2008a, 2008b). Several cyanobacterial species have been efficiently utilised in remediation of heavy metal-polluted sites (Table 9.4). The metal sorption capability of certain cyanobacterial species has been found to be better than bacteria and fungi in treatment of contaminated sites.

    i. Bioaccumulation by viable algae, and
   ii. Biosorption by non-living, non-growing biomass or biomass products.

Of these two, biosorption by non-viable cells is more advantageous since it affords greater capacity to treat large volumes, greater selectivity and specificity for heavy metals, removal of many metals simultaneously, no requirement of growth media, nutrients and expensive reagents, good results over a wide range of physic-chemical conditions, low capital investment and operational costs, no limitation of metal toxicity and significant recovery of the contaminant (heavy metal) (Suresh Kumar et al. 2015). Table 9.4 reveals the bioremediation of heavy metals by microalgae belonging primarily to green algae. Microalgae afford the following advantages with respect to phytoremediation of heavy metals (Suresh Kumar et al. 2015):

i. Performs efficiently at low level of contaminants
ii. Rarely generate toxic sludge
iii. Affords easy culturing
iv. Good binding affinity
v. Ideal at both small scale and large scale
vi. Efficacy and low cost

Intensive research needs to be carried out in microalgae with reference to growth parameters, cellular structure, pretreatment and recovery of heavy metals from algal biomass which can go a long way in cutting down costs and pave the way for their increased utilisation in bioremediation programmes.

9.4.2 FUNGI

Mycoremediation refers to the use of fungus (live or dead) for the removal of contaminants from a polluted site. Among different microorganism, fungi are considered as most efficient in removal of heavy metals due to their relatively higher tolerance to heavy metals, large surface to volume ratio and robust nature in comparison to bacteria and algae (Akhtar and Mannan 2020). Fungi use mechanisms such as bioaccumulation, biovolatilisation, biosorption, biomineralisation, bioreduction, biooxidation, extracellular precipitation and intracellular precipitation in the remediation of heavy metals. However, this capability varies on the fungal species, dose of inoculum, time, pH, temperature, nature and concentration of heavy metals and the shaking rate (Kumar and Dwivedi 2021a). A large number of fungi play a crucial role in the precipitation of metals like insoluble oxalates, oxides, carbonates, and phosphates (Liang and Gadd 2017; Suyamud et al. 2020) (Table 9.5). Of the diverse fungi that immobilise heavy metals, the urease-positive fungi need special mention since the mechanism of immobilisation is linked with urea degradation. These group of fungi when grown in urea-rich medium hydrolyse urea-producing ammonia and free carbonate resulting in the precipitation of metals as carbonates such as barium carbonate ($BaCO_3$), cadmium carbonate ($CdCO_3$), cobalt(II) carbonate ($CoCO_3$), basic copper carbonate ($Cu_2(OH)_2CO_3$), lanthanum carbonate ($La_2(CO_3)_3$), and nickel carbonate ($NiCO_3$) (Li and Gadd 2017; Liang and Gadd 2017; Liu et al. 2019; Li et al. 2019, 2020). Extracellular proteins, amino acids, and polysaccharides also play an important role in toxic metal immobilisation.

9.4.3 YEAST

Yeasts have also been used as biosorbents for recovery of a range of heavy metals like silver, gold, cadmium, cobalt, chromium, copper, nickel, lead, uranium and

TABLE 9.5
Fungi and Yeast Used in Bioremediation of Heavy Metals

Fungi	Heavy Metal	References
Arthrinium malaysianum	Chromium	Majumder et al. (2017)
Aspergillus flavus	Cadmium, Lead, Mercury	Mahmoud et al. (2017)
Aspergillus fumigatus	Lead	Kumar Ramasamy et al. (2011)
Aspergillus versicolor	Chromium, Copper, Nickel	Taştan et al. (2010)
Beauveria bassiana	Lead	Gola et al. (2018)
Botrytis cinerea	Zinc	Tunali and Akar (2006)
Fusarium solani	Silver	El Sayed and El-Sayed (2020a)
	Zinc	El Sayed and El-Sayed (2020b)
Gloeophyllum sepiarium	Chromium	Achal et al. (2011)
Hypocrea lixii	Copper	Salvadori et al. (2013)
Mucor rouxii	Cadmium, Nickel, Zinc	Yan and Viraraghavan (2003)
Paneibacillus sp.	Copper, lead, Zinc	Govarthanan et al. (2016)
Penicillium polonicum	Lead	Xu et al. (2020)
Pleurotus ostreatus	Cadmium, Nickel	Anacletus et al. (2017)
Pythium sp.	Chromium	Kavita et al. (2011)
Rhizopus arrhizus	Zinc	Fourest et al. (1994)
Rhizopus oryzae	Chromium	Sukumar (2010)
Trichoderma brevicompactum	Copper, Lead	Zhang et al. (2020)
Trichoderma koningiopsis	Copper	Salvadori et al. (2014)
Trichoderma lixii	Copper	Kumar and Dwivedi (2021b)
Yeast		
Candida tropicalis	Cadmium	Rehman and Anjum (2011)
Cryptococcus laurentii	Chromium, Gallium, Iron	Rusinova-Videva et al. (2014)
Kodamaea transpacifica	Chromium	Campaña-Pérez et al. (2019)
Saccharomyces cerevisiae	Cadmium	Talos et al. (2009)
	Chromium	De Rossi et al. (2018)
	Lead	Lívia de et al. (2015)
Saccharomyces pastorianus	Cadmium, Copper, Lead	Kordialik-Bogacka and Diowksz (2014)

zirconium (Podgorskiĭ et al. 2004; Rehman and Anjum 2011) (Table 9.5). Baker's yeast (*Saccharomyces cerevisiae*) has been the yeast of choice. In a recent study, S. *cerevisiae* cells exhibited high adsorption capacities for several heavy metals along with a number of rare earth elements (Ojima et al. 2019).

9.4.4 BACTERIA

Bacteria are considered as important biosorbents due to their ubiquitous nature, miniature size, ability to flourish in varied environments and resilience to stressful conditions (Srivastava et al. 2015; Tarekegn et al. 2020). The great biosorption ability reported by certain bacteria is due to their high surface-to-volume ratios and potential active chemisorption sites on the cell wall (Mosa et al. 2016). It has been proved that bacteria are quite stable and exhibit good survival capacity especially in mixed cultures (Sannasi et al. 2006). Therefore, consortia of bacterial cultures are far more efficient in the biosorption of heavy metals as compared to single species (Kader et al. 2007; Tarekegn et al. 2020). Among bacterial biosorbents, Gram-positive bacteria in general bind larger quantities of metals as compared to Gram-negative bacteria (Prabhakaran et al. 2016). Table 9.6 shows the different bacterial species used in the bioremediation of heavy metals.

TABLE 9.6
Bacteria Used in Bioremediation of Heavy Metals

Bacteria	Heavy Metal	Reference
Aeromonas hydrophila	Chromium	Ranjan et al. (2009)
Arthrobactor sp.	Lead	Wang et al. (2018)
Arthrobacter viscosus	Zinc	Malkoc et al. (2016)
Azotobacter chroococcum	Cadmium, Chromium, Copper	Rizvi et al. (2020)
Bacillus cereus	Cadmium	Huang et al. (2013)
	Chromium	Dong et al. (2013)
Bacillus gibsonii	Lead	Zhang et al. (2013)
Bacillus salmalaya	Chromium	Dadrasnia et al. (2015)
Bacillus subtilis	Chromium	Vullo et al. (2008)
	Iron	Krishna Kanamarlapudi and
	Lead	Muddada (2019)
		Rizvi et al. (2020)
Bacillus thuringiensis	Manganese	Huang et al. (2020)
Cupriavidus gilardii	Copper	Yang et al. (2017)
Enterobacter cloacae	Chromium	Rahman et al. (2015)
Klebsiella sp.	Lead	Wei et al. (2016)
Methylococcus capsulatus	Chromium	Hasin et al. (2010)
Pseudomonas	Chromium	Chang et al. (2019)
	Uranium	Sar and D'Souza (2001)
Pseudomonas aeruginosa	Cadmium, Chromium, Nickel	Rizvi et al. (2020)
Pseudomonas alcaliphila	Chromium	El-Naggar et al. (2020)
Pseudomonas koreensis	Cadmium, Chromium, Lead	Ayangbenro et al. (2019)
Pseudomonas putida	Chromium	Vullo et al. (2008)
Pseudomonas veronii	Cadmium, copper, zinc	Vullo et al. (2008)
Ralstonia pickettii	Manganese	Huang et al. (2018)
Rhodococcus sp.	Lead	Hu et al. (2020)
Serratia sp.	Lead	Chen et al. (2019)
Sporosarcina ginsengisoli	Arsenic	Achal et al. (2012)
Streptomyces ciscaucasicus	Zinc	Li et al. (2010)
Variovorax paradoxus	Zinc	Malkoc et al. (2016)

9.5 FUTURE PERSPECTIVES

Biological remediation using microorganisms is far superior to physical and chemical methods of contaminant removal. However, bioremediation is effective only when the environmental conditions permit microbial growth and activity. Also, a lot needs to be done with respect to deciphering the mechanism involved in various processes involved in the remediation of heavy metals using microbes. Apart from conventional bioremediation techniques, recombinant DNA technology has opened new avenues to engineer genetically modified microorganisms (GEMs) that show greater resilience, adaptability and remediation capacity for heavy metals. Another upcoming strategy is the utilisation of nanotechnology in remediation of contaminated sites, a process termed as nanobioremediation. Nanoscale materials (1–100 nm) like zeolites, metal oxides, dendrimers, carbon nanotubes, chitosan, and graphene nanosheets can be synthesized using microorganisms to reduce metal ions where fungi offer many advantages as compared to bacteria.

REFERENCES

Abdelatey, L. M., W. K. B. Khalil, T. H. Ali and K. F. Mahrous. 2011. Heavy metal resistance and gene expression analysis of metal resistance genesin Gram-positive and Gram-negative bacteria present in Egyptian soils. Journal of Applied Sciences in Environmental Sanitation. 6: 201–11.

Abou-Shanab, R. I., T. A. Delorme, J. S. Angle, R. L. Chaney, K. Ghanem, H. Moawad and H. A. Ghozlan. 2003. Phenotypic characterization of microbes in the rhizosphere of *Alyssum murale*. International Journal of Phytoremediation. 5: 367–80.

Achal, V., D. Kumari and X. Pan. 2011. Bioremediation of chromium contaminated soil by a brown-rot fungus, *Gloeophyllum sepiarium*. Research Journal of Microbiology. 6: 166.

Achal, V., X. Pan, Q. Fu and D. Zhang. 2012. Biomineralization based remediation of As (III) contaminated soil by *Sporosarcina ginsengisoli*. Journal of Hazardous Materials. 201: 178–84.

Adriano, D. C., N. S. Bolan, J. Vangronsveld and W. W. Wenzel. 2005. Heavy metals. In: Hillel D. (Ed.), Encyclopedia of Soils in the Environment. Elsevier, Amsterdam. 175–82.

Aguilar-Barajas, E., E. Paluscio, C. Cervantes and C. Rensing. 2008. Expression of chromate resistance genes from *Shewanella* sp. strain ANA-3 in *Escherichia coli*. FEMS Microbiology Letters. 285: 97–100.

Ahmad, M. S. and M. Ashraf. 2011. Essential roles and hazardous effects of nickel in plants. Reviews in Environmental Contamination and Toxicology. 214: 125–67.

Akhtar, F. Z., K. M. Archana, V. G. Krishnaswamy and R. Rajagopal. 2020. Remediation of heavy metals (Cr, Zn) using physical, chemical and biological methods: a novel approach. SN Applied Sciences. 2: 267.

Akhtar, N., J. Iqbal and M. Iqbal. 2004. Removal and recovery of nickel(II) from aqueous solution by loofa sponge-immobilized biomass of *Chlorella sorokiniana*: characterization studies. Journal of Hazardous Materials. 108: 85–94.

Akhtar, N., M. Iqbal, S. I. Zafar and J. Iqbal. 2008. Biosorption characteristics of unicellular green alga *Chlorella sorokiniana* immobilized in loofa sponge for removal of Cr(III). Journal of Environmental Sciences (China). 20: 231–39.

Akhtar, N. and M. A. Mannan. 2020. Mycoremediation: expunging environmental pollutants. Biotechnology Reports. 26: e00452.

Aksu, Z. 2001. Equilibrium and kinetic modelling of cadmium(II) biosorption by *C. vulgaris* in a batch system: effect of temperature. Separation and Purification Technology. 21: 285–94.

Aksu, Z. 2002. Determination of the equilibrium, kinetic and thermodynamic parameters of the batch biosorption of nickel(II) ions onto *Chlorella vulgaris*. Process Biochemistry. 38: 89–99.

Alejandro, S, S. Höller, B. Meier and E. Peiter. 2020. Manganese in plants: from acquisition to subcellular allocation. Frontiers in Plant Science. 11: 300.

Alissa, E. M. and G. A. Ferns. 2011. Heavy metal poisoning and cardiovascular disease. Journal of Toxicology. 2011: 870125.

Alloway, B. J. 2004. Zinc in soil and crop nutrition. International Zinc Association. Belgium, Brussels.

Amari, T., T. Ghnaya and C. Abdelly. 2017. Nickel, cadmium and lead phytotoxicity and potential of halophytic plants in heavy metal extraction. South African Journal of Botany. 111: 99–110.

Ameen, N., M. Amjad, B. Murtaza, G. Abbas, M. Shahid, M. Imran, M. A. Asif, and N. K. Niazi. 2019. Biogeochemical behavior of nickel under different abiotic stresses: toxicity and detoxification mechanisms in plants. Environmental Science and Pollution Research. 26: 10496–514.

Anacletus, F., K. Nwauche and C. Ighorodje-Monago. 2017. Mineral and heavy metal composition of crude oil polluted soil amended with non-ionic surfactant (Triton X-100) and white rot fungus (*Pleurotus ostratus*). Journal of Environmental and Analytical Toxicology. 7: 449–51.

Anyanwu, B. O., A. N. Ezejiofor, Z. N. Igweze and O. E. Orisakwe. 2018. Heavy metal mixture exposure and effects in developing nations: an update. Toxics. 6: 65.

Arıca, M. Y., İ. Tüzün, E. Yalçna, Ö. İnce and G. Bayramoğlu. 2005. Utilisation of native, heat and acid-treated microalgae *Chlamydomonas reinhardtii* preparations for biosorption of Cr(VI) ions. Process Biochemistry. 40: 2351–58.

Arinola, O. G., S. O. Nwozo J. A. Ajiboye and A. H. Oniye. 2008. Evaluation of trace elements and total antioxidant status in Nigerian cassava processors. Pakistan Journal of Nutrition. 7: 770–72.

Ayangbenro, A. S., O. O. Babalola, O. S. Aremu. 2019. Bioflocculant production and heavy metal sorption by metal resistant bacterial isolates from gold mining soil. Chemosphere. 231: 113–20.

Aydinalp, C. and S. Marinova. 2009. The effect of heavy metals on seed germination and plant growth on alfalfa plant (*Medicago sativa*). Bulgarian Journal of Agricultural Science. 15: 347–50.

Azubuike, C. C., C. B. Chikere and G. C. Okpokwasili. 2016. Bioremediation techniques–classification based on site of application: principles, advantages, limitations and prospects. World Journal of Microbiology and Biotechnology. 32: 180.

Babula, P., V. Adam, R. Opatrilova, J. Zehnalek, L. Havel and R. Kizek. 2008. Uncommon heavy metals, metalloids and their plant toxicity: a review. Environmental Chemistry Letters. 6: 189–213.

Backhaus, T., J. Snape and J. Lazorchak. 2012. The impact of chemical pollution on biodiversity and ecosystem services: The need for an improved understanding. Integrated Environmental Assessment and Management. 8: 575–76.

Bhargava, A., V. K. Gupta, A. K. Singh and R. Gaur. 2012a. Microbes for heavy metal remediation. In: Gaur, R., S. Mehrotra and R. R. Pandey. (Eds.), Microbial Applications. IK International Publishing, New Delhi. 167–77.

Bhargava, A., F. F. Carmona, M. Bhargava and S. Srivastava. 2012b. Approaches for enhanced phytoextraction of heavy metals. Journal of Environmental Management. 105: 103–20.

Bishnoi, N. R., R. Kumar, S. Kumar and S. Rani. 2007. Biosorption of Cr(III) from aqueous solution using algal biomass *Spirogyra* spp. Journal of Hazardous Materials. 145: 142–47.

Bittner, F., M. Oreb and R. R. Mendel. 2001. ABA3 is a molybdenum cofactor sulfurase required for activation of aldehyde oxidase and xanthine dehydrogenase in *Arabidopsis thaliana*. Journal of Biological Chemistry. 276: 40381–384.

Bjorklund, G., T. Hofer, V. M. Nurchi and J. Aaseth. 2019. Iron and other metals in the pathogenesis of Parkinson's disease: toxic effects and possible detoxification. Journal of Inorganic Biochemistry. 199: 110717.

Brennan, R. F. 2005. Zinc application and its availability to plants. PhD dissertation, School of Environmental Science, Division of Science and Engineering, Murdoch University.

Caffo, M., G. Caruso, G. L. Fata, V. Barresi, M. Visalli, M. Venza and I. Venza. 2014. Heavy metals and epigenetic alterations in brain tumors. Current Genomics. 15: 457–63.

Campaña-Pérez, J. F., P. Portero Barahona, P. Martín-Ramos and E. J. Carvajal Barriga. 2019. Ecuadorian yeast species as microbial particles for Cr(VI) biosorption. Environmental Science and Pollution Research. 26: 28162–172.

Castagna, A., D. Di Baccio, R. Tognetti, A. Ranieri and L. Sebastiani. 2013. Differential ozone sensitivity interferes with cadmium stress in poplar clones. Biologia Plantarum. 57: 313–24.

Castro, C., M. S. Urbieta, J. Plaza Cazón and E. R. Donati. 2019. Metal biorecovery and bioremediation: whether or not thermophilic are better than mesophilic microorganisms. Bioresource Technology. 279: 317–26.

Chandra S., R. Sharma, K. Singh and A. Sharma. 2013. Application of bioremediation technology in the environment contaminated with petroleum hydrocarbon. Annals of Microbiology. 63: 417–31.

Chang, J., S. Deng, Y. Liang and J. Chen. 2019. Cr(VI) removal performance from aqueous solution by *Pseudomonas* sp. strain DC-B3 isolated from mine soil: characterization of both Cr(VI) bioreduction and total Cr biosorption processes. Environmental Science and Pollution Research. 26: 28135–145.

Chen, H., J. Xu, W. Tan and L. Fang. 2019. Lead binding to wild metal-resistant bacteria analyzed by ITC and XAFS spectroscopy. Environmental Pollution. 250: 118–26.

Chen, C. Y., H. W. Chang, P. C. Kao, J. L. Pan and J. S. Chang. 2012. Biosorption of cadmium by CO_2-fixing microalga *Scenedesmus obliquus* CNW-N. Bioresource Technology. 105: 74–80.

Chibuike, G. U. and S. C. Obiora. 2014. Heavy metal polluted soils: effect on plants and bioremediation methods. Applied and Environmental Soil Science. 2014: 752708.

Chin-Chan, M., J. Navarro-Yepes, B. Quintanilla-Vega, V. Campos-Peña and R. Weissert. 2015. Environmental pollutants as risk factors for neurodegenerative disorders: Alzheimer and Parkinson diseases environmental factors in neurodegeneration. Frontiers in Cellular Neuroscience. 9: 124.

Chiou, W. Y. and F. C. Hsu. 2019. Copper toxicity and prediction models of copper content in leafy vegetables. Sustainability. 11: 22.

Cramer, A. J. and J. M. Cole. 2017. Removal or storage of environmental pollutants and alternative fuel sources with inorganic adsorbents *via* host-guest encapsulation. Journal of Material Chemistry A. 5: 10746–771.

Dadrasnia, A., K. S. Chuan Wei, N. Shahsavari, M. S. Azirun and S. Ismail. 2015. Biosorption potential of *Bacillus salmalaya* strain 139SI for removal of Cr (VI) from aqueous solution. International Journal of Environmental Research and Public Health. 12: 15321–338.

Das, H. and H. R. Dash. 2014. Microbial bioremediation: a potential tool for restoration of contaminated areas. In: Das S. (Ed.), Microbial Degradation and Bioremediation. Elsevier, Amsterdam. 1–21.

Dash, H. R. and S. Das. 2012. Bioremediation of mercury and importance of bacterial mer genes. International Biodeterioration and Biodegradation. 75: 207–13.

Datta, J. K., A. Bandhyopadhyay, A. Banerjee and N. K. Mondal. 2011. Phytotoxic effect of chromium on the germination, seedling growth of some wheat (*Triticum aestivum* L.) cultivars under laboratory condition. Journal of Agricultural Technology. 7: 395–402.

De Rossi, A., M. R. Rigon, M. Zaparoli, R. D. Braido, L. M. Colla, G. L. Dotto and J. S. Piccin. 2018. Chromium (VI) biosorption by *Saccharomyces cerevisiae* subjected to chemical and thermal treatments. Environmental Science and Pollution Research. 25: 19179–186.

De, J., N. Ramaiah and L. Vardanyan. 2008. Detoxification of toxic heavy metals by marine bacteria highly resistant to mercury. Marine Biotechnology. 10: 471–77.

Demirevska-kepova, K., L. Simova-Stoilova, Z. Stoyanova, R. Holzer and U. Feller. 2004. Biochemical changes in barely plants after excessive supply of copper and manganese. Environmental and Experimental Botany. 52: 253–66.

Dias, M. C., C. Monteiro, J. Moutinho-Pereira, C. Correia, B. Goncalves and C. Santos. 2013. Cadmium toxicity affects photosynthesis and plant growth at different levels. Acta Physiologia Plantarum. 35: 1281–89.

Ding, Z., J. Wu, A. You, B. Huang and C. Cao. 2017. Effects of heavy metals on soil microbial community structure and diversity in the rice (*Oryza sativa* L. subsp. *Japonica*, food crops Institute of Jiangsu Academy of Agricultural Sciences) rhizosphere. Soil Science and Plant Nutrition. 63: 75–83.

Disante, K. B., D. Fuentes and J. Cortina. 2010. Response to drought of Zn-stressed *Quercus suber* L. seedlings. Environmental and Experimental Botany. 70: 96–103.

Dong, G., Y. Wang, L. Gong, M. Wang, H. Wang, N. He, Y. Zheng and Q. Li. 2013. Formation of soluble Cr (III) end-products and nanoparticles during Cr (VI) reduction by *Bacillus cereus* strain XMCr-6. Biochemical Engineering Journal. 70: 166–72.

El-Naas, M. H., F. A. Al-Rub, I. Ashour and M. Al Marzouqi. 2007. Effect of competitive interference on the biosorption of lead (II) by Chlorella vulgaris. Chemical Engineering and Processing: Process Intensification. 46: 1391–99.

El-Naggar, N. E., A. Y. El-Khateeb, A. A. Ghoniem, M. S. El-Hersh and W. I. A. Saber. 2020. Innovative low-cost biosorption process of Cr^{6+} by *Pseudomonas alcaliphila* NEWG-2. Scientific Reports. 10: 14043.

El Sayed, M. T. and A. S. A. El-Sayed. 2020a. Tolerance and mycoremediation of silver ions by *Fusarium solani*. Heliyon. 6: e03866.

El Sayed, M. T. and A. S. A. El-Sayed. 2020b. Bioremediation and tolerance of zinc ions using *Fusarium solani*. Heliyon. 6: e05048.

El-Sheekh, M. M., W. A. El-Shouny, M. E. H. Osman and E. W. E. El-Gammal. 2005. Growth and heavy metals removal efficiency of *Nostoc muscorum* and *Anabaena subcylindrica* in sewage and industrial wastewater effluents. Environmental Toxicology and Pharmacology. 19: 357–65.

Fahr, M., L. Laplaze, N. Bendaou, V. Hocher, M. El Mzibri, D. Bogusz and A. Smouni. 2013. Effect of lead on root growth. Frontiers in Plant Science. 4: 175.

Fernando, D. R. and J. P. Lynch. 2015. Manganese phytotoxicity: new light on an old problem. Annals of Botany. 116: 313–19.

Finocchio, E., A. Lodi, C. Solisio and A. Converti. 2010. Chromium (VI) removal by methylated biomass of *Spirulina platensis*: the effect of methylation process. Chemical Engineering Journal. 156: 264–69.

Fourest, E., C. Canal and J. C. Roux. 1994. Improvement of heavy metal biosorption by mycelial dead biomasses (*Rhizopus arrhizus, Mucor miehei* and *Penicillium chrysogenum*): pH control and cationic activation. FEMS Microbiology Reviews. 14: 325–32.

Fraile, A., S. Penche, F. Gonzalez, M. L. Blazquez, J. A. Munoz and A. Ballester. 2005. Biosorption of copper, zinc, cadmium and nickel by *Chlorella vulgaris*. Chemical Ecology. 21: 61–75.

Freitas, H., M. N. V. Prasad and J. Pratas. 2004. Plant community tolerant to trace elements growing on the degraded soils of Sao Domingos mine in the south east of Portugal: environmental implications. Environment International. 30: 65–72.

Gautam, R. K., S. Soni and M. C. Chattopadhyaya. 2014. Functionalized magnetic nanoparticles for environmental remediation. Handbook of Research on Diverse Applications of Nanotechnology in Biomedicine, Chemistry, and Engineering, IGI Global, PA, USA. 518–51.

Ghosh, R. and S. Roy. 2019. Cadmium toxicity in plants: unveiling the physicochemical and molecular aspects. In: Hasanuzzaman, M., M. N. Vara Prasad and K. Nahar. (Eds.), Cadmium Tolerance in Plants. Academic Press, Amsterdam. 223–46.

Glick, B. R. 2010. Using soil bacteria to facilitate phytoremediation. Biotechnology Advances. 28: 367–74.

Gola, D., A. Malik, M. Namburath and S. Z. Ahammad. 2018. Removal of industrial dyes and heavy metals by *Beauveria bassiana*: FTIR, SEM, TEM and AFM investigations with Pb(II). Environmental Science and Pollution Research. 25: 20486–496.

Gopal, R., B. K. Dube, P. Sinha and C. Chatterjee. 2003. Cobalt toxicity effects on growth and metabolism of tomato. Communications in Soil Science and Plant Analysis. 34: 619–28.

Govarthanan, M., R. Mythili, T. Selvankumar, S. Kamala-Kannan, A. Rajasekar and Y. C. Chang. 2016. Bioremediation of heavy metals using an endophytic bacterium *Paenibacillus* sp. RM isolated from the roots of *Tridax procumbens*. 3 Biotech. 6: 242.

Guo, H., C. Hong, X. Chen, Y. Xu, Y. Liu, D. Jiang and B. Zheng. 2016. Different growth and physiological responses to cadmium of the three *Miscanthus* species. PLoS ONE. 11: e0153475.

Gupta, M. and S. Gupta. 2017. An overview of selenium uptake, metabolism, and toxicity in plants. Frontiers in Plant Sciences. 7: 2074.

Gupta, V. K., A. Rastogi, V. K. Saini and N. Jain. 2006. Biosorption of copper(II) from aqueous solutions by *Spirogyra* species. Journal of Colloid and Interface Science. 296: 59–63.

Gupta, V. K. and A. Rastogi. 2008a. Sorption and desorption studies of chromium(VI) from nonviable cyanobacterium *Nostoc muscorum* biomass. Journal of Hazardous Materials. 154: 347–54.

Gupta, V. K. and A. Rastogi. 2008b. Biosorption of lead from aqueous solutions by green algae *Spirogyra* species: kinetics and equilibrium studies Journal of Hazardous Materials. 152: 407–14.

Han, X., Y. S. Wong, M. H. Wong and N. F. Tam. 2007. Biosorption and bioreduction of Cr(VI) by a microalgal isolate, *Chlorella miniata*. Journal of Hazardous Materials. 146: 65–72.

Hasin, A. A., S. J. Gurman, L. M. Murphy, A. Perry, T. J. Smith, and P. E. Gardiner. 2010. Remediation of chromium (VI) by a methane-oxidizing bacterium. Environmental Science and Technology. 44: 400–05.

Hassan, T. U., A. Bano and I. Naz. 2017. Alleviation of heavy metals toxicity by the application of plant growth promoting rhizobacteria and effects on wheat grown in saline sodic field. International Journal of Phytoremediation. 19: 522–29.

Hu, X., J. Cao, H. Yang, D. Li, Y. Qiao, J. Zhao, Z. Zhang and L. Huang. 2020. Pb^{2+} biosorption from aqueous solutions by live and dead biosorbents of the hydrocarbon-degrading strain *Rhodococcus* sp. HX-2. PLoS One. 15: e0226557

Huang, F., Z. Dang, C. L. Guo, G. N. Lu, R. R. Gu, H. J. Liu and H. Zhang. 2013. Biosorption of Cd(II) by live and dead cells of *Bacillus cereus* RC-1 isolated from cadmium-contaminated soil. Colloids and Surfaces B: Biointerfaces. 107: 11–18.

Huang, H., Y. Zhao, Z. Xu, Y. Ding, W. Zhang and L. Wu. 2018. Biosorption characteristics of a highly Mn(II)-resistant *Ralstonia pickettii* strain isolated from Mn ore. PLoS One. 13(8): e0203285.

Huang, H., Y. Zhao, Z. Xu, Y. Ding, X. Zhou and M. Dong. 2020. A high Mn(II)-tolerance strain, *Bacillus thuringiensis* HM7, isolated from manganese ore and its biosorption characteristics. Peer J. 8: e8589.

Hwang, K., G. J. Kwon, J. Yang, M. Kim, W. J. Hwang, W. Youe and D. Y. Kim. 2018. *Chlamydomonas angulosa* (Green Alga) and *Nostoc commune* (Blue-Green Alga) microalgae-cellulose composite aerogel beads: manufacture, physicochemical characterization, and Cd (II) adsorption. Materials (Basel). 11: 562.

Ibrahim, A., S. El-Abd and A. S. El-Beltagy. 1989. A possible role of cobalt in salt tolerance of plant. Egyptian Journal of Soil Science. Special Issue: 359–70.

Igiri, B. E., S. I. R. Okoduwa, G. O. Idoko, E. P. Akabuogu, A. O. Adeyi and I. K. Ejiogu. 2018. Toxicity and bioremediation of heavy metals contaminated ecosystem from tannery wastewater: a review. Journal of Toxicology. 2018: 1–16

Jacoby, R., M. Peukert, A. Succurro, A. Koprivova and S. Kopriva. 2017. The role of soil microorganisms in plant mineral nutrition-current knowledge and future directions. Frontiers in Plant Science. 8: 1617.

Jain, R., S. Srivastava, S. Solomon, A. Shrivastava and A. Chandra. 2010. Impact of excess zinc on growth parameters, cell division, nutrient accumulation, photosynthetic pigments and oxidative stress of sugarcane (*Saccharum* spp.). Acta Physiologia Plantarum. 32: 979–986.

Jalmi, S. K., P. K. Bhagat, D. Verma, S. Noryang, S. Tayyeba, K. Singh, D. Sharma and A. K. Sinha. 2018. Traversing the links between heavy metal stress and plant signaling. Frontiers in Plant Science. 9: 12.

Kader, J., P. Sannasi, O. Othman, B. S. Ismail and S. Salmijah. 2007. Removal of Cr (VI) from aqueous solutions by growing and non-growing populations of environmental bacterial consortia. Global Journal of Environmental Research. 1: 12–17.

Kaiser, B. N., K. L. Gridley, J. N. Brady, T. Phillips and S. D. Tyerman. 2005. The role of molybdenum in agricultural plant production. Annals of Botany. 96: 745–754.

Karimi, P., R. A. Khavari-Nejad, V. Niknam, F. Ghahremaninejad and F. Najafi. 2012. The effects of excess copper on antioxidative enzymes, lipid peroxidation, proline, chlorophyll, and concentration of Mn, Fe, and Cu in *Astragalus neo-mobayenii*. The Scientific World Journal. 2012: 615670.

Kasai, Y., H. Kishira and S. Harayama. 2002. Bacteria belonging to the genus *Cycloclasticus* play a primary role in the degradation of aromatic hydrocarbons released in a marine environment. Applied Environmental Microbiology. 68: 5625–33.

Kavita, B., J. Limbachia and H. Keharia. 2011. Hexavalent chromium sorption by biomass of chromium tolerant *Pythium* sp. Journal of Basic Microbiology. 51: 173–82.

Kevresan, S., N. Petrovic, M. Popovic and J. Kandrac. 2001. Nitrogen and protein metabolism in young pea plants as affected by different concentrations of nickel, cadmium, lead, and molybdenum. Journal of Plant Nutrition. 24: 1633–44.

Khulbe, K. C. and T. Matsuura. 2018. Removal of heavy metals and pollutants by membrane adsorption techniques. Applied Water Science. 8: 19–49.

Kim, H. S., Y. J. Kim and Y. R. Seo. 2015. An overview of carcinogenic heavy metal: molecular toxicity mechanism and prevention. Journal of Cancer Prevention. 20: 232–40.

Kordialik-Bogacka, E. and A. Diowksz. 2014. Metal uptake capacity of modified *Saccharomyces pastorianus* biomass from different types of solution. Environmental Science and Pollution Research. 21: 2223–29.

Koul, B. and P. Taak. 2018. Soil pollution: causes and consequences. In: Koul, B. and P. Taak (Eds.), Biotechnological Strategies for Effective Remediation of Polluted Soils. Springer, Singapore. 1–37.

Krishna Kanamarlapudi, S. L. R. and S. Muddada. 2019. Structural changes of *Bacillus subtilis* biomass on biosorption of iron (ii) from aqueous solutions: isotherm and kinetic studies. Polish Journal of Microbiology. 68: 549–58.

Kukkola, E., P. Rautio and S. Huttunen. 2000. Stress indications in copper and nickel-exposed scots pine seedlings. Environmental and Experimental Botany. 43: 197–210.

Kumar, P., J. Tokas and H. R. Singal. 2019a. Amelioration of chromium VI toxicity in Sorghum (*Sorghum bicolor* L.) using glycine betaine. Scientific Reports. 9: 16020.

Kumar, S., S. Prasad, K. K. Yadav, M. Shrivastava, N. Gupta, S. Nagar, Q. V. Bach, H. Kamyab, S. A. Khan, S. Yadav and L. C. Malav. 2019b. Hazardous heavy metals contamination of vegetables and food chain: role of sustainable remediation approaches – a review. Environmental Research. 179: 108792.

Kumar, V. and S. K. Dwivedi. 2021a. Mycoremediation of heavy metals: processes, mechanisms, and affecting factors. Environmental Science and Pollution Research. 28: 10375–412.

Kumar, V. and S. K. Dwivedi. 2021b. Bioremediation mechanism and potential of copper by actively growing fungus *Trichoderma lixii* CR700 isolated from electroplating wastewater. Journal of Environmental Management. 277: 111370.

Kumar Ramasamy, R., S. Congeevaram and K. Thamaraiselvi. 2011. Evaluation of isolated fungal strain from e-waste recycling facility for effective sorption of toxic heavy metal Pb (II) ions and fungal protein molecular characterization – a mycoremediation approach. Asian Journal of Experimental Biological Science. 2: 342–47.

Lapaz, A.M., L. F. M. Santos, C. H. P. Yoshida, R. Heinrichs, M. Campos and A. R. Reis 2019. Physiological and toxic effects of selenium on seed germination of cowpea seedlings. Bragantia. 78: 498–508.

Lau, O. and S. F. Yang. 1976. Inhibition of ethylene production by cobaltous ion. Plant Physiology. 58: 114–117.

Lee, H. J., M. K. Park and Y. R. Seo. 2018. Pathogenic mechanisms of heavy metal induced-Alzheimer's disease. Toxicology and Environmental Health Science. 10: 1–10.

Li, H. F., C. Gray, C. Mico, F. J. Zhao and S. P. McGrath. 2009. Phytotoxicity and bioavailability of cobalt to plants in a range of soils. Chemosphere. 75: 979–86.

Li, H., Y. Lin, W. Guan, J. Chang, L. Xu, J. Guo and G. Wei. 2010. Biosorption of Zn(II) by live and dead cells of *Streptomyces ciscaucasicus* strain CCNWHX 72-14. Journal of Hazardous Materials. 179: 151–59.

Li, Q. and G. M. Gadd. 2017. Fungal nanoscale metal carbonates and production of electrochemical materials. Microbiology and Biotechnology. 10: 1131–36.

Li, Q., D. Liu, C. Chen, Z. Shao, H. Wang, J. Liu, Q. Zhang and G. M. Gadd. 2019. Experimental and geochemical simulation of nickel carbonate mineral precipitation by carbonate-laden ureolytic fungal culture supernatants. Environmental Science: Nano. 6: 1866–75.

Li, Q., J. Liu and G. M. Gadd. 2020. Fungal bioremediation of soil co-contaminated with petroleum hydrocarbons and toxic metals. Applied Microbiology and Biotechnology. 104: 8999–9008.

Liang, X. and G. M. Gadd. 2017. Metal and metalloid biorecovery using fungi. Microbiology and Biotechnology. 10: 1199–205.

Lin, Y., I. Lian, C. Kor, C. C. Chang, P. Y. Su, W. T. Chang, Y. F. Liang, W. W. Su and M. S. Soon. 2017. Association between soil heavy metals and fatty liver disease in men in Taiwan: A cross sectional study. BMJ Open. 7: e014215.

Liu, F., L. Csetenyi and G. M. Gadd. 2019. Amino acid secretion influences the size and composition of copper carbonate nanoparticles synthesized by ureolytic fungi. Applied Microbiology and Biotechnology. 103: 7217–30.

Lívia de C. F., H. B. Mario and C. Benedito. 2015. Potential application of modified *Saccharomyces cerevisiae* for removing lead and cadmium. Journal of Bioremediation and Biodegradation. 6: 2.

Llamas, A., C. I. Ullrich and A. Sanz. 2000. Cd2+ effects on transmembrane electrical potential, respiration and membrane permeability of rice (Oryza sativa) roots. Plant and Soil. 219: 21–8.

Mahmoud, M. E., G. M. El Zokm, A. E. M. Farag and M. S. Abdelwahab. 2017. Assessment of heat-inactivated marine *Aspergillus flavus* as a novel biosorbent for removal of Cd(II), Hg(II), and Pb(II) from water. Environment Science and Pollution Research. 24: 18218–28.

Majumder, R., L. Sheikh, A. Naskar, Vineeta, M. Mukherjee and S. Tripathy. 2017. Depletion of Cr(VI) from aqueous solution by heat dried biomass of a newly isolated fungus *Arthrinium malaysianum*: a mechanistic approach. Scientific Reports. 7: 11254.

Malkoc, S., E. Kaynak and K. Guven. 2016. Biosorption of zinc(II) on dead and living biomass of *Variovorax paradoxus* and *Arthrobacter viscosus*. Desalination and Water Treatment. 57: 15445–54.

Malla, M. A., A. Dubey, S. Yadav, A. Kumar, A. Hashem and E. F. Abd Allah. 2018. Understanding and designing the strategies for the microbe-mediated remediation of environmental contaminants using omics approaches. Frontiers in Microbiology. 9: 1132.

McNeely, J. A. 1992. The sinking ark: pollution and the worldwide loss of biodiversity. Biodiversity Conservation. 1: 2–18.

Mehta, S. K. and J. P. Gaur. 2001. Removal of Ni and Cu from single and binary metal solutions by free and immobilized *Chlorella vulgaris*. European Journal of Parasitology. 37: 261–71.

Mendel, R. R. and R. Haensch. 2002. Molybdoenzymes and molybdenum cofactor in plants. Journal of Experimental Botany. 53: 1689–98.

Micheletti, E., G. Colica, C. Viti, P. Tamagnini, and R. De Philippis. 2008a. Selectivity in the heavy metal removal by exopolysaccharide-producing cyanobacteria. Journal of Applied Microbiology. 105: 88–94.

Micheletti, E., S. Pereira, F. Mannelli, P. Moradas-Ferreira, P. Tamagnini and R. De Philippis. 2008b. Sheathless mutant of cyanobacterium *Gloeothece* sp. strain PCC 6909 with increased capacity to remove copper ions from aqueous solutions. Applied Environmental Microbiology. 74: 2797–804.

Miller, W. L., D. W. King, J. Lin and D. R. Kester. 1995. Photo-chemical redox cycling of iron in coastal seawater. Marine Chemistry. 50: 63–77.

Mishra, G. K. 2017. Microbes in heavy metal remediation: a review on current trends and patents. Recent Patents in Biotechnology. 11: 188–96.

Mishra, J., R. Singh and N. K. Arora. 2017. Alleviation of heavy metal stress in plants and remediation of soil by rhizosphere microorganisms. Frontiers in Microbiology. 8: 1706.

Mitra, A., S. Chatterjee, A. V. Voronina, C. Walther and D. K. Gupta. 2020. Lead toxicity in plants: a review. In: Gupta D., S. Chatterjee and C. Walther. (Eds.), Lead in Plants and the Environment. Radionuclides and Heavy Metals in the Environment. Springer, Cham. 99–116.

Mohapatra, H. and R. Gupta. 2005. Concurrent sorption of Zn(II), Cu(II) and Co(II) by *Oscillatoria angustissima* as a function of pH in binary and ternary metal solutions. Bioresource Technology. 96: 1387–98.

Mondal, N. K., C. Das and J. K. Datta. 2015. Effect of mercury on seedling growth, nodulation and ultrastructural deformation of *Vigna radiata* (L) Wilczek. Environmental Monitoring and Assessment. 187: 241.

Morkunas, I., A. Woźniak, V. C. Mai, R. Rucińska-Sobkowiak and P. Jeandet. 2018. The role of heavy metals in plant response to biotic stress. Molecules. 23: 2320.

Mosa, K. A., I. Saadoun, K. Kumar, M. Helmy and O. P. Dhankher. 2016. Potential biotechnological strategies for the cleanup of heavy metals and metalloids. Frontiers in Plant Science. 7: 303.

Mukhopadhay, M. J. and A. Sharma. 1991. Manganese in cell metabolism of higher plants. Botanical Reviews. 57: 117–49.

Mukhopadhyay, M., A. Das, P. Subba, P. Bantawa, B. Sarkar, P. D. Ghosh and T. K. Mondal. 2013. Structural, physiological and biochemical profiling of tea plants (*Camellia sinensis* (L.) O. Kuntze) under zinc stress. Biologia Plantarum. 57: 474–80.

Ojima, Y., S. Kosako, M. Kihara, N. Miyoshi, K. Igarashi and M. Azuma. 2019. Recovering metals from aqueous solutions by biosorption onto phosphorylated dry baker's yeast. Scientific Reports. 9: 225.

Ojuederie, O. B. and O. O. Babalola. 2017. Microbial and plant-assisted bioremediation of heavy metal polluted environments: A review. International Journal of Environmental Research and Public Health. 14: 1504.

Pamphlett, R., S. Cherepanoff, L. K. Too, S. Kum Jew, P. A. Doble and D. P. Bishop. 2020. The distribution of toxic metals in the human retina and optic nerve head: implications for age-related macular degeneration. PLoS One. 15: e0241054.

Patra, M., N. Bhowmik, B. Bandopadhyay and A. Sharma. 2004. Comparison of mercury, lead and arsenic with respect to genotoxic effects on plant systems and the development of genetic tolerance. Environmental and Experimental Botany. 52: 199–223.

Pietrini, F., M. Carnevale, C. Beni, M. Zacchini, F. Gallucci and E. Santangelo. 2019. Effect of different copper levels on growth and morpho-physiological parameters in giant reed (*Arundo donax* L.) in semi-hydroponic mesocosm experiment. Water. 11: 1837.

Pineda, A., I. Kaplan and T. M. Bezemer. 2017. Steering soil microbiomes to suppress aboveground insect pests. Trends in Plant Science. 22: 770–78.

Podgorskiĭ, V. S., T. P. Kasatkina and O. G. Lozovaia. 2004. Drozhzhi–biosorbenty tiazhelykh metallov [Yeasts–biosorbents of heavy metals]. Mikrobiol Z. 66: 91–103.

Pourrut, B., M. Shahid, C. Dumat, P. Winterton and E. Pinelli. 2011. Lead uptake, toxicity, and detoxification in plants. Reviews in Environmental and Contamination Toxicology. 213: 113–36.

Prabhakaran, P., M. A. Ashraf and W. S. Aqma. 2016. Microbial stress response to heavy metals in the environment. RSC Advances. 6: 109862–77.

Pushpanathan, M., S. Jayashree, P. Gunasekaran and J. Rajendhran. 2014. Microbial bioremediation – a metagenomic approach. In: Das S. (Ed.), Microbial Biodegradation and Bioremediation. Elsevier, Amsterdam. 407–19.

Rahman, A., N. Nahar, N. N. Nawani, J. Jass, K. Hossain, Z. A. Saud, A. K. Saha, S. Ghosh, B. Olsson and A. Mandal. 2015. Bioremediation of hexavalent chromium (VI) by a soil-borne bacterium, *Enterobacter cloacae* b2-dha. Journal of Environmental Science and Health Part A. 50: 1136–47.

Rai, P. K. 2016. Biomagnetic Monitoring of Particulate Matter. Elsevier, Amsterdam.

Rajkumar, M., M. N. V. Prasad, H. Freitas and N. Ae 2009. Biotechnological applications of serpentine bacteria for phytoremediation of heavy metals. Critical Reviews in Biotechnology. 29: 120–30.

Ranjan, D., P. Srivastava, M. Talat and S. H. Hasan. 2009. Biosorption of Cr(VI) from water using biomass of *Aeromonas hydrophila*: Central composite design for optimization of process variables. Applied Biochemistry and Biotechnology. 2009. 158: 524–39.

Rehman, A. and M. S. Anjum. 2011. Multiple metal tolerance and biosorption of cadmium by *Candida tropicalis* isolated from industrial effluents: glutathione as detoxifying agent. Environmental Monitoring and Assessment. 174: 585–95.

Rizvi, A., B. Ahmed and A. Zaidi. 2020. Biosorption of heavy metals by dry biomass of metal tolerant bacterial biosorbents: an efficient metal clean-up strategy. Environment Monitoring and Assessment. 192: 801.

Ru J., Y. Huo and Y. Yang. 2020. Microbial degradation and valorization of plastic wastes. Frontiers in Microbiology. 11: 442.

Rusinova-Videva, S., K. Pavlova and K. Georgieva. 2014. Effect of different carbon sources on biosynthesis of exopolysaccharide from Antarctic strain *Cryptococcus Laurentii* AL62. Biotechnology and Biotechnological Equipment. 25: 80–84.

Sagardoy, R., F. Morales, A. F. López-Millán, A. Abadía and J. Abadía. 2009. Effects of zinc toxicity on sugar beet (*Beta vulgaris* L.) plants grown in hydroponics. Plant Biology. 11: 339–50.

Salvadori, M. R., L. F. Lepre, R. A. Ando, C. A. Oller do Nascimento and B. Corrêa. 2013. Biosynthesis and uptake of copper nanoparticles by dead biomass of *Hypocrea lixii* isolated from the metal mine in the Brazilian Amazon Region. PLoS One. 8: e80519.

Salvadori, M. R., R. A. Ando, C. A. Oller do Nascimento and B. Corrêa. 2014. Intracellular biosynthesis and removal of copper nanoparticles by dead biomass of yeast isolated from the wastewater of a mine in the Brazilian Amazonia. PLoS ONE 9: e87968.

Sannasi, P., J. Kader, O. Othman and S. Salmijah. 2006. Single and multi-metal removal by an environmental mixed bacterial isolate. In: Mendez-Vilas, A. (Ed.), Modern Multidisciplinary Applied Microbiology: Exploiting Microbes and Their Interactions. Wiley-VCH Verlag GmbH & Co. KGaA, Germany. 136–41.

Santandrea, G., S. Schiff and A. Bennici. 1997. Manganese toxicity to different growth processes in vitro in *Nicotiana*. Plant Cell Tissue and Organ Culture. 50: 125–29.

Santandrea, G., S. Schiff and A. Bennici. 1998. Effects of manganese on *Nicotiana* species cultivated in vitro and characterization of regenerated Mn-tolerant tobacco plants. Plant Science. 132: 71–78.

Santos, E. F., J. M. Kondo Santini, A. P. Paixão, E. F. Júnior, J. Lavres, M. Campos and A. R. dos Reis. 2017. Physiological highlights of manganese toxicity symptoms in soybean plants: Mn toxicity responses. Plant Physiology and Biochemistry. 113: 6–19.

Sar, P. and S. F. D'Souza. 2001. Biosorptive uranium uptake by *Pseudomonas* strain: Characterization and equilibrium studies. Journal of Chemical Technology and Biotechnology. 76: 1286–94.

Selvaratnam, S., B. A. Schoedel, B. L. McFarland and C. F. Kulpa. 1997. Application of the polymerase chain reaction (PCR) and reverse transcriptase/PCR for determining the fate of phenol degrading *Pseudomonas putida* ATCC 11172 in a bioaugmented sequencing batch reactor. Applied Microbiology and Biotechnology. 47: 236–40.

Sharma, S., S. Tiwari, A. Hasan, V. Saxena and L. M. Pandey. 2018. Recent advances in conventional and contemporary methods for remediation of heavy metal-contaminated soils. 3 Biotech. 8: 216.

Sharma, A., D. Kapoor, J. Wang, B. Shahzad, V. Kumar, A. S. Bali, S. Jasrotia, B. Zheng, H. Yuan and D. Yan. 2020. Chromium bioaccumulation and its impacts on plants: an overview. Plants. 9. 100.

Sheldon, A. R. and N. W. Menzies. 2005. The effect of copper toxicity on the growth and root morphology of rhodes grass (*Chloris gayana* Knuth.) in resin buffered solution culture. Plant and Soil. 278: 341–49.

Shigaki, T. 2020. Health effects of environmental pollutants. In: Li, X. and P. Liu. (Eds.), Gut Remediation of Environmental Pollutants. Springer, Singapore. 1–29.

Siddiquee, S., K. Rovina and S. A. Azad. 2015. Heavy metal contaminants removal from wastewater using the potential filamentous fungi biomass: a review. Journal of Microbial and Biochemical Technology. 7: 384–93.

Sinegovskaya, V. T., O. A. Terekhova, S. I. Lavrentyeva, L. E. Ivachenko and K. S. Golokhvast. 2020. Effect of heavy metals on oxidative processes in soybean seedlings. Russian Agricultural Science. 46: 28–32.

Singh, D., K. Nath and Y. K. Sharma. 2007. Response of wheat seed germination and seedling growth under copper stress. Journal of Environmental Biology. 28: 409–14.

Soetan, K. O., C. O. Olaiya and O. E. Oyewole. 2010. The importance of mineral elements for humans, domestic animals and plants: a review. African Journal of Food Science. 4: 200–22.

Sordello, R., F. Flamerie De Lachapelle, B. Livoreil and S. Vanpeene. 2019. Evidence of the environmental impact of noise pollution on biodiversity: a systematic map protocol. Environmental Evidence. 8: 8.

Spiller, S. C., A. M. Castelfranco and P. A. Castelfranco. 1982. Effects of iron and oxygen on chlorophyll biosynthesis: I. In vivo observations on iron and oxygen deficient plants. Plant Physiology. 69: 107–11.

Sreekumar, N., A. Udayan and S. Srinivasan. 2020. Algal bioremediation of heavy metals. In: Shah, M. P. (Ed.), Removal of Toxic Pollutants Through Microbiological and Tertiary Treatment. Elsevier, Amsterdam. 279–307.

Srivastava, N. K. and C. B. Majumder. 2008. Novel biofiltration methods for the treatment of heavy metals from industrial wastewater. Journal of Hazardous Materials. 151: 1–8.

Srivastava, S., S. B. Agrawal and M. K. Mondal. 2015. A review on progress of heavy metal removal using adsorbents of microbial and plant origin. Environmental Science and Pollution Research. 22: 15386–415.

Srivastava, S. and A. Bhargava. 2015. Genetic diversity and heavy metal stress in plants. In: Ahuja, M. R. and S. M. Jain. (Eds.), Sustainable Development and Biodiversity. Springer, Switzerland. 241–70.

Stambulska, U. Y., M. M. Bayliak and V. I. Lushchak. 2018. Chromium(VI) toxicity in legume plants: modulation effects of rhizobial symbiosis. BioMed Research International. 2018: 8031213.

Sukumar, M. 2010. Reduction of hexavalent chromium by *Rhizopus oryzae*. African Journal of Environmental Science and Technology. 4: 412–18.

Suresh Kumar, K., H.-U. Dahms, E.-J. Won, J.-S. Lee and K.-H. Shin. 2015. Microalgae – A promising tool for heavy metal remediation. Ecotoxicology and Environmental Safety. 113: 329–52.

Suyamud, B., J. Ferrier, L. Csetenyi, D. Inthorn and G. M. Gadd. 2020. Biotransformation of struvite by *Aspergillus niger*: phosphate release and magnesium biomineralization as glushinskite. Environmental Microbiology. 22: 1588–602.

Tahir, M. B., H. Kiran and T. Iqbal. 2019. The detoxification of heavy metals from aqueous environment using nano-photocatalysis approach: a review. Environmental Science and Pollution Research. 26: 10515–28.

Talos, K., C. Pager, S. Tonk, C. Majdik, B. Kocsis, F. Kilar and T. Pernyeszi. 2009. Cadmium biosorption on native *Saccharomyces cerevisiae* cells in aqueous suspension. Acta Universitatis Sapientiae Agriculture and Environment. 1: 20–30.

Tapia-Torres, Y., M. D. Rodríguez-Torres, J. J. Elser, A. Islas, V. Souza and F. García-Oliva. 2016. How to live with phosphorus scarcity in soil and sediment: lessons from bacteria. Applied Environmental Microbiology. 82: 4652–62.

Tarekegn, M. M., F. Z. Salilih, A. I. Ishetu and F. Yildiz. 2020. Microbes used as a tool for bioremediation of heavy metal from the environment. Cogent Food and Agriculture. 6: 1.

Taştan, B. E., S. Ertuğrul and G. Dönmez. 2010. Effective bioremoval of reactive dye and heavy metals by *Aspergillus versicolor*. Bioresource Technology. 101: 870–76.

Tavallali, V., M. Rahemi, S. Eshghi, B. Kholdebarin and A. Ramezanian. 2010. Zinc alleviates salt stress and increases antioxidant enzyme activity in the leaves of pistachio (*Pistacia vera* L. 'Badami') seedlings. Turkish Journal of Agriculture and Forestry. 34: 349–59.

Thomas, F., C. Malick, E. C. Endreszl and K. S. Davies. 1998. Distinct responses to copper stress in the halophyte, *Mesembryanthemum crystallium*. Physiologia Plantarum. 102: 360–68.

Tisdale, S. L., W. L. Nelson and J. D. Beaten. 1984. Zinc in Soil Fertility and Fertilizers. Macmillan Publishing Company, New York.

Tunali, S. and T. Akar. 2006. Zn(II) biosorption properties of *Botrytis cinerea* biomass. Journal of Hazardous Materials. 131: 137–45.

Ukaogo, P. O., U. Ewuzie and C. V. Onwuka. 2020. Environmental pollution: causes, effects, and the remedies. In: Chowdhary, P., A. Raj, D. Verma and Y. Akhter. (Eds.), Microorganisms for Sustainable Environment and Health. Elsevier, Amsterdam. 419–29.

Vella, V., R. Malaguarnera, R. Lappano, M. Maggiolini and A. Belfiore. 2017. Recent views of heavy metals as possible risk factors and potential preventive and therapeutic agents in prostate cancer. Molecular Cell Endocrinology. 457: 57–72.

Verma, S. and A. Kuila. 2019. Bioremediation of heavy metals by microbial process. Environmental Technology and Innovation. 14: 100369.

Vullo, D. L., H. M. Ceretti, M. A. Daniel, S. A. Ramírez and A. Zalts. 2008. Cadmium, zinc and copper biosorption mediated by *Pseudomonas veronii* 2e. Bioresource Technology. 99: 5574–81.

Wang, T., J. Yao, Z. Yuan, Y. Zhao, F. Wang and H. Chen. 2018. Isolation of lead-resistant *Arthrobacter* strain GQ-9 and its biosorption mechanism. Environmental Science and Pollution Research. 25: 3527–38.

Wei, R. and W. Zimmermann. 2017. Microbial enzymes for the recycling of recalcitrant petroleum-based plastics: how far are we? Microbial Biotechnology. 10: 1308–22.

Wei, W., Q. Wang, A. Li, J. Yang, F. Ma, S. Pi and D. Wu. 2016. Biosorption of Pb (II) from aqueous solution by extracellular polymeric substances extracted from *Klebsiella* sp. J1: Adsorption behaviour and mechanism assessment. Scientific Reports. 6: 31575.

Wuana, R. A. and F. E. Okieimen. 2011. Heavy metals in contaminated soils: A review of sources, chemistry, risks and best available strategies for remediation. ISRN Ecology. 2011: 402647.

Xu, X., R. Hao, H. Xu and A. Lu. 2020. Removal mechanism of Pb(II) by *Penicillium polonicum*: Immobilization, adsorption, and bioaccumulation. Scientific Reports. 10: 9079.

Yan, G. and T. Viraraghavan. 2003. Heavy-metal removal from aqueous solution by fungus *Mucor rouxii*. Water Research. 37: 4486–96.

Yang, J., Z. Song, J. Ma and H. Han. 2020. Toxicity of molybdenum-based nanomaterials on the soybean–rhizobia symbiotic system: implications for nutrition. ACS Applied NanoMaterials. 3: 5773–82.

Yang, Y., M. Hu, D. Zhou, W. Fan, X. Wang and M. Huo. 2017. Bioremoval of Cu^{2+} from CMP wastewater by a novel copper-resistant bacterium *Cupriavidus gilardii* CR3: characteristics and mechanisms. RSC Advances. 7: 18793–802.

Zeraatkar, A. K., H. Ahmadzadeh, A. F. Talebi, N. R. Moheimani and M. P. McHenry. 2016. Potential use of algae for heavy metal bioremediation, a critical review. Journal of Environmental Management. 181: 817–31.

Zhang, B., R. Fan, Z. Bai, S. Wang, L. Wang and J. Shi. 2013. Biosorption characteristics of *Bacillus gibsonii* S-2 waste biomass for removal of lead (II) from aqueous solution. Environmental Science and Pollution Research. 20: 1367–73.

Zhang, D., C. Yin, N. Abbas, Z. Mao and Y. Zhang. 2020. Multiple heavy metal tolerance and removal by an earthworm gut fungus *Trichoderma brevicompactum* QYCD-6. Scientific Reports. 10: 6940.

Zulfiqar, U., M. Farooq, S. Hussain, M. Maqsood, M. Hussain, M. Ishfaq, M. Ahmad and M. Z. Anjum. 2019. Lead toxicity in plants: impacts and remediation. Journal of Environmental Management. 250: 109557.

10 Microbial Bioremediation
A Sustainable Approach for Restoration of Contaminated Sites

Anamika Harshvardhan[1] and Purabi Saikia[2]
[1]Institute of Wood Science and Technology, Karnataka, India
[2]Central University of Jharkhand, Jharkhand, India

CONTENTS

10.1 INTRODUCTION

Incessant increase of contaminants due to various human-induced activities over the last few decades have resulted in deterioration of the ecosystem that harms human beings, livestock, crops, indigenous organisms, biodiversity, hence disrupting sustainable development (Arora, 2018). Xenobiotics have also increased tremendously due to the expansion of industries and agricultural intensification (Das and Adholeya, 2012). The hazardous materials adversely affect both biotic and abiotic components of the ecosystem. Microbial bioremediation of contaminated sites relies on breaking down of target pollutants by microbial strains and their enzymes. It is an innovative method that utilizes naturally occurring microorganisms and/or their derivatives to clean up and degrade various types of contaminants generated due to various anthropogenic

DOI: 10.1201/9781003110477-10

activities such as intensive agriculture, industrialization, and urbanization (Kumar et al., 2018). It is involved in degrading, removing, transforming, immobilizing, or detoxifying various physico-chemical and hazardous wastes from the soil, air, and water through the enzymatic pathways of microorganisms (Tekere, 2019).

The application of bioremediation as a biotechnological tool involves microorganisms for eliminating the harmful toxic contaminants from a polluted environment that operates as a substitute and a very applicable strategy (Muthuirulan et al., 2014). Microorganisms due to their impressive metabolic ability, their global distribution, the capability of multiplying at any extreme temperature, and environmental conditions are being used for bioremediation of contaminated sites (Tang et al., 2007). It involves the process of recycling waste into a usable form and decontaminates contaminated environments by exploiting metabolic microorganisms which further depends upon the ability of a particular microorganism to convert it into a less toxic form (Strong and Burgess, 2008). Microorganisms produce enzymes in appropriate environmental conditions that spell the contaminants and convert them into harmless and less/non-toxic products (Kumar et al., 2017). Microbial bioremediation is advantageous over traditional methods by considering pollutant degradation capacity; it has public acceptance, economic feasibility, very few or no by-products, reusability, and eco-friendly process (Chikere et al., 2012; Dell Anno et al., 2012; Tangahu et al., 2011). Microorganisms are highly diverse with metabolic versatility and can use many harmful components as their nutrient source (Das and Dash, 2014). Microbial bioremediation is a natural process in which complete destruction of target organic pollutants is possible and degrades it into carbon dioxide and water and converts hazardous inorganic pollutants into less toxic forms (Jain and Bajpai, 2012; Tekere, 2019). Biofuel can also be produced by coupling microbial bioremediation with biofuel production (Thomas et al., 2016; Waghmare et al., 2014).

There are certain disadvantages associated with microbial bioremediation as it is a highly specific process that requires appropriate environmental conditions and proper levels of nutrients for the bioremediation of harmful, poisonous, and lethal wastes. Besides, it is a time-intensive process, requires skilled manpower, and limited to only biodegradable compounds. Sometimes, products of bioremediation may be more noxious than the original material (Abatenh et al., 2017).

Thus, this chapter emphasized on common methods and factors affecting the microbial bioremediation, as well as microbes available for bioremediation of contaminated sites. Novel microbial species with great remediation potential must be explored in the future to expedite the process of microbial bioremediation.

10.2   BIOREMEDIATION STRATEGIES

Microbial bioremediation is a natural process in which microorganisms or their enzymes degrade, detoxify, immobilize, break down or alter toxic pollutants into CO_2, H_2O, microbial biomass, mineral salts, and other less toxic metabolites (Chakraborty et al., 2012). It can be carried out both onsite (*in-situ* bioremediation) without causing any damage and alteration to the natural environment and offsite (*ex-situ* bioremediation) based on their ability to transport and remove pollutants (Figure 10.1) (Kumar et al., 2011).

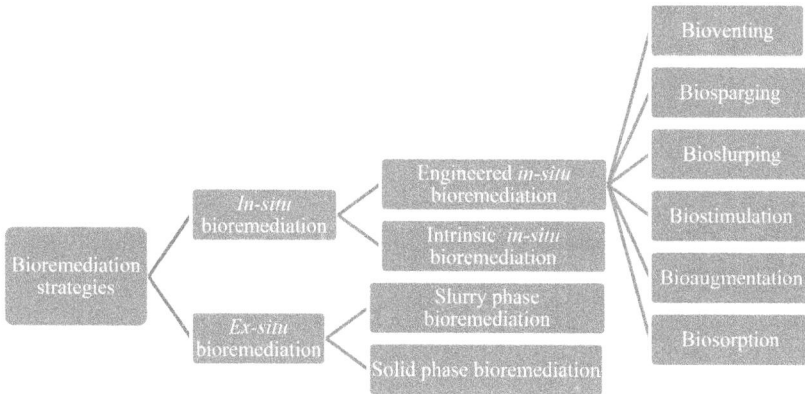

FIGURE 10.1 Types of bioremediation strategies.

10.2.1 *IN-SITU* BIOREMEDIATION

In-situ bioremediation is a procedure that is carried out at its site of contamination (USEPA, 2006; 2012), in which chemotaxis plays a fundamental role, as microbes have chemotactic ability to move towards contaminated areas. In case of the *in-situ* bioremediation process, oxygen and nutrients are added to the contaminated sites to hasten the microbial multiplication and accelerate the process of bioremediation (Hazen, 2010). It is considered to provide greater cost-benefit than *ex-situ* bioremediation because the removal process is considered to be more useful than the immobilization of contaminants. It also reduces the contact of workers with the contaminated medium (Thomé et al., 2018). *In-situ* bioremediation is further bifurcated into intrinsic *in-situ* bioremediation and engineered *in-situ* bioremediation (Hazen, 2010).

10.2.1.1 Intrinsic *in-situ* Bioremediation

Intrinsic *in-situ* bioremediation is also known as passive bioremediation or natural attenuation that degrades organic compounds by indigenous microorganisms present in the contaminated site (USEPA, 2000a; 2000b; 2006). It is a cost-effective strategy that exterminates pollutants and creates fewer disturbances to the nearby areas of contaminated sites. It basically relies upon indigenous microorganisms to degrade pollutants without any artificial augmentation (Kumar et al., 2018) and is also limited to the depth at which microorganisms can degrade contaminants (Banerjee et al., 2016). It is good for improving soil quality, texture, structure, and nitrogen fixation that helps in enhancing plant growth. The major prerequisites for successful intrinsic *in-situ* bioremediation includes adequate population of microorganisms, availability of nutrients and ideal conditions for the growth of microbes, and adequate time to naturally deplete the pollutants at the contaminated site (Kumar et al., 2018).

10.2.1.2 Engineered *in-situ* Bioremediation

Engineered *in-situ* bioremediation is also known as accelerated or attenuated *in-situ* bioremediation, which accelerates the biodegradation process by augmenting microbes to contaminated sites and improving its physico-chemical characteristics (Hazen, 2010). It is deduced that genetically altered free-living bacteria might have fewer chances of survival in the stressed environmental conditions (Kumar et al., 2018). Hence, the selection of genetically modified bacteria which are fast growing, having high metabolic flexibility, high bioremediation prospective without any environmental risk will help in attaining a secure and sustainable environment in a very short span of time (Kumar et al., 2018). Engineered *in-situ* bioremediation process is of various types including bioventing, biosparging, bio-slurping, bio-stimulation, and bioaugmentation.

10.2.1.2.1 Bioventing

In bioventing, oxygen flow and nutrients are provided with the help of wells to the anaerobic contaminated sites that acts as an electron acceptor for power generation to revive the growth of microorganisms (Abatenh et al., 2017; Das and Adholeya, 2012; Lim et al., 2016) and also to initiate biodegradation process. It is best applicable in sites of high temperature and where the water level is generally low. It degrades hydrocarbons and volatile compounds present beneath the surface of the Earth generally in the vadose zone or unsaturated zone (Lee et al., 2006). Bioventing is one of the dominantly used *in-situ* bioremediation procedures for treating volatile pollutants with the addition of oxygen to contaminated sites. Bioventing can be carried out at the site of contamination by the addition of oxygen directly in the unsaturated zone for the removal of semi-volatile and non-volatile contaminants (Maier, 2000).

10.2.1.2.2 Biosparging

It is an *in-situ* bioremediation process, in which nutrients and air are injected into a saturated zone to enhance the remediation process and biological activities of indigenous organisms (Kumar et al., 2018). It is commonly implemented in areas of contamination that are readily volatile such as gasoline, petroleum products and need to be eliminated immediately by biosparging. Soil porosity plays a key role in the usefulness of this process (Mohapatra, 2008; Vidali, 2001).

10.2.1.2.3 Bio-Slurping

Bio-slurping is a technique to remediate hydrocarbon-contaminated sites through the application of vacuum-enhanced dewatering technologies (Parker, 1996) that help in recovering free product and remediate the vadose zone by combining vacuum-assisted free-product recovery with bioventing and soil vapor-extraction (Anonymous, 1998).

10.2.1.2.4 Bio-Stimulation

In the process of bio-stimulation, the contaminated site is supplemented with nutrients, electron donors, or an electron acceptor (Abdulsalam et al., 2011; Tyagi et al., 2011). It involves the addition of nitrogen and phosphate-rich nutrients to stimulate the growth of microorganisms and accelerate the process of degradation (Boopathy, 2000). It is often employed under ideal conditions of pH, temperature, water content, and nutrients for remediating wide varieties of xenobiotics (Kumar et al., 2011). Microorganisms used to assimilate the hydrocarbons that were present in the con-taminated soil in a large quantity that entered from the leakage of underground stor-age tanks, ponds, and landfills (Eweis et al., 1998). Bio-stimulation is also known as 'enhanced bioremediation' as it is often accompanied by bioaugmentation.

10.2.1.2.5 Bioaugmentation

Bioaugmentation requires the addition of microbial strains to the contaminated sites, usually in combination with bio-stimulation (Boopathy, 2000) that is involved in the improvement of the performance of microorganisms by adding genetically modified organisms (Abdulsalam et al., 2011). This technique primarily focuses to improve the insufficient biodegradative activities of indigenous microbes of the polluted sites by adding genetically altered microorganisms having preferred catabolic abilities to enhance the bioremediation potential of polluted sites and degenerate unmanageable compounds (El Fantroussi and Agathos, 2005).

10.2.1.2.6 Biosorption

Biosorption is a microbial bioremediation process that employed microorganisms for the removal of various positively charged heavy metals (HMs) ions, such as Pb, Cr, As, and Cd, from aqueous solutions as a result of their attraction to negatively charged microbial cell membranes (Sari and Tuzen, 2009).

10.2.2 Ex-situ BIOREMEDIATION

Ex-situ bioremediation is a process in which contaminated soil or water is trans-ferred from the site of its origin to elsewhere for its quality improvements (Song et al., 2017). It is basically of two types, *viz.*, slurry phase bioremediation and solid phase bioremediation.

10.2.2.1 Slurry Phase Bioremediation

It is one of the biological processes within a bioreactor, contaminated sludge is mixed with water to mechanically breakdown the contaminants in soil by the process of abrasion (Kumar et al., 2018). It contains a bioreactor supplemented with oxygen and nutrients to establish ideal conditions for microbial growth and to degrade the specific pollutants. In the bioreactor, the contaminated materials are mixed with water and chemicals to create three phases, that is, solid, liquid, and gas. It is a con-tinuous process so that the microbes can be kept in close association with hazardous contaminants (Kumar et al., 2018). After the completion of this process, the water is extracted from the soil and the soil is reloaded in the environment for reuse after quality checking and improvement (USEPA, 2006).

10.2.2.2 Solid Phase Bioremediation

It is an *in-situ* approach that treat the contaminated materials above ground. The major drawback associated with solid phase bioremediation is that it requires more space and is a comparatively time-intensive process (Kumar et al., 2018).

10.3 PHYSICO-CHEMICAL FACTORS AFFECTING MICROBIAL BIOREMEDIATION

Environmental circumstances such as temperature, pH, salinity, availability of oxygen, and nutrients, affect the growth, metabolic activities, and biodegradation potential of microorganisms and to some extent the solubility and volatility of the pollutants and mobility of metal ions in soil (Dermont et al., 2008; Naik and Duraphe, 2012). Light availability has a positive impact on degradation and detoxification of pollutants by photosynthetic microorganisms like algae (Muñoz et al., 2003) as it helps in degrading petroleum products and hydrocarbons through their photochemical actions. Bacteria can maintain the metabolic activity in temperature fluctuations; however, seasonal temperature fluctuations used to affect the pollutant degradation rate (Palmisano et al., 1991) as it influences the detoxification and degradation of contaminants by changing the physico-chemical properties of oils, biochemical properties of microbes, and the solubility of hydrocarbons (Tabatabaee et al., 2005). The rate of HMs absorption affects the growth and development of microorganisms at ambient temperature (Fang et al., 2011) and the most suitable temperature range is 25–35°C for adsorption of HMs by microorganisms (Gan et al., 2012; Goyal et al., 2003; Hu et al., 2010). On the other hand, hydrocarbon degradation used to occur over a range of temperatures from close to 0°C up to more than 30°C (Kumar et al., 2018). Highest rates of biodegradation used to occur at neutral pH or slightly alkaline (Ibrahim, 2016). The optimum pH of aerobic microorganisms differs from anaerobic microorganisms (Fang et al., 2011). An unsuitable pH may cause an adverse impact on microbial growth (Mohsenzadeh et al., 2010) by affecting the enzyme activities in microorganisms and also by adsorption of HMs (Morto-Bermea et al., 2002).

Although, the degradation of hydrocarbons takes place both in aerobic and anaerobic conditions, aerobic conditions are considered more efficient as oxygenases, the primary enzymes needed for biodegradation of hydrocarbons used to be more effective in the presence of oxygen (Kumar et al., 2018). The optimum nutrient levels enhance the microbial bioremediation process and it can be boosted by supplementing the essential nutrients in the contaminated sites. In case of nitrogen acute petroleum oil spills, C, N, and P are added in the ratio of ~100:10:1 (Speight and El-Gendy, 2018), while the optimum C:N and C:P ratios for enhanced bioremediation are 10:1 and 30:1, respectively (Azad et al., 2014). Besides, temperature, soil texture, moisture content, nutrient status, and soil organic matter content also affect the microbial bioremediation of contaminated sites (Azad et al., 2014). The concentration of HMs in the contaminated sites also affects the microbial adsorption rate (Brunetti et al., 2012; Ehrlich, 1997; Tyagi et al., 2014). The degradation of pollutants is often achieved through complex microbial population interactions where the efficiency also depends on the concentration and physico-chemical characteristics of pollutants and the polluted sites, and their accessibility to microorganisms (El Fantroussi and Agathos, 2005). The chemical nature of the contaminants, its physical state (solid, liquid, gases), concentration, and chemical bond type are also influencing bioremediation (Azad et al., 2014).

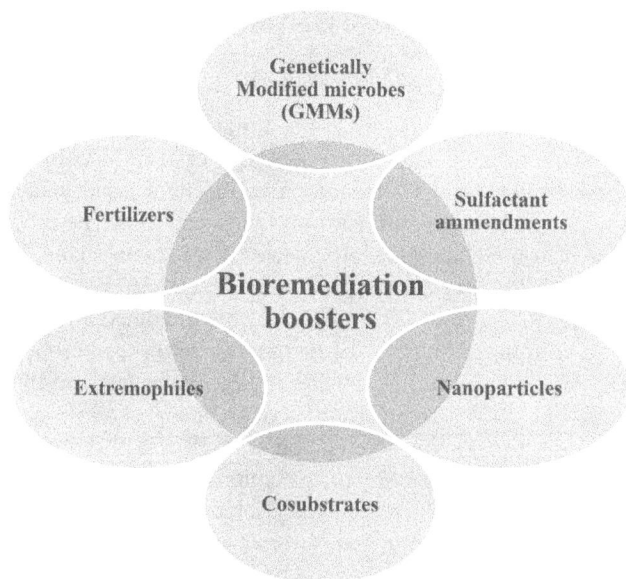

FIGURE 10.2 Bioremediation boosters help to enhance the degradation of various organic and inorganic pollutants.

There are several bioremediation boosters including the GMMs, fertilizers, nanoparticles, extremophiles, surfactant amendments, and co-substrate that enhance the bioremediation potential of microorganisms (Figure 10.2) (Kumari and Singh, 2016). GMMs are proficient in destroying definite pollutants (Wasilkowski et al., 2012) by combining genes artificially that do not occur together naturally. Similarly, C:N ratio significantly affect the microbial biodegradation process and addition of nutrients is an important aspect to achieve the balance of C:N ratio for efficacious biodegradation of organic and biodegradable contaminants (Jin and Fallgren, 2007). On the other hand, biodegradation potential can be improved through the addition of surfactants that help in improved bioavailability of pollutants (Chang et al., 2008) through enhancing the solubility of hydrophobic hydrocarbons. Besides, supplement of metabolites like salicylate and phenanthrene as inducer of co-metabolic degradation help in boosting the biodegradation of polyaromatic hydrocarbons (PAHs) with high molecular weight (Somtrakoon et al., 2008). Metallic nanoparticles, oxide nanoparticles, and magnetic nanoparticles are another group of bioremediation boosters that can be bio-synthesized by microorganisms (Li et al., 2011). It is being used to degrade the petroleum hydrocarbon from air, water, or soil at a faster rate through trapping and reducing the metal ions on the surface of microbial cells or by transporting it into the microbial cell to produce nanoparticles in the presence of various microbial enzymes (Zhang et al., 2011). Microbes may degrade the petroleum hydrocarbons in extreme environmental conditions including low temperature (Wang et al., 2011) through nitrate-reducing activity, and also in high salinity (Fathepure, 2014).

10.4 MICROBIAL AGENTS RESPONSIBLE FOR BIOREMEDIATION

10.4.1 NATURALLY OCCURRING MICROBIAL AGENTS

Different species of bacteria, cyanobacteria, fungi, and algae have been reported for the bioremediation of contaminated sites. Bacteria are usually considered as the best option for microbial bioremediation due to their rapid metabolic rates, a range of degradation and metabolic pathways, and can easily be modified genetically to increase their bioremediation efficiencies (Prakash and Irfan, 2011). Species of *Pseudomonas* and *Bacillus* are the most widely used bacteria for bioremediation of toxic compounds (Chen et al., 2015). At least 60% bioremediation used to be performed by using bacteria or fungi as the main tool and these microbes can be used alone or in association with other organisms, mainly plants (e.g., mycorrhizal fungi) (Vieira and Stefenon, 2017). The involvement of green algae, brown algae, and blue-green algae (cyanobacteria) in the degradation of PAHs was also reported (Cerniglia and Gibson, 1977). Various microorganisms commonly involved in bioremediation are *Bacillus* (Wierzba, 2015), *Pseudomonas* (Bojórquez and Voltolina, 2016; Vullo et al., 2008), *Aspergillus* (Kapoor and Viraraghavan, 1997; Tastan et al., 2010), *Streptomyces* (Ehrlich, 1997), *Rhizopus* (Abd-Alla et al., 2012), and *Penicillium* (Sarret et al., 1998). However, the use of consortium rather than monoculture has higher biodegradation efficiency (Kadali et al., 2012). Microbial consortium of five different fungi (*Phanerochaete chrysosporium*, *Cunninghamella* sp., *Alternaria alternate* [Fr.] Keissler, *Penicillium chrysogenum*, and *Aspergillus niger*) and three bacteria (*Bacillus* sp., *Zoogloea* sp., and *Flavobacterium* sp.) improved the rate of biodegradation of contaminants by 41.3% (Li et al., 2009).

10.4.2 GENETICALLY MODIFIED (GM) MICROBIAL AGENTS

In certain oil-polluted sites, indigenous microorganisms are unable to degrade the pollutants thus genetically modified microorganisms (GMMs) are incorporated to degrade pollutants. GMMs have been relatively easy to construct through the reshuffling of genes that enhance the performance of *in-situ* bioremediation (Coelho et al., 2015). In this technique, desired genes for the production of a protein are inserted into the genome of the microorganism, where it degrades the contaminants by producing suitable protein (Jain et al, 2010). The University of Tennessee in collaboration with Oak Ridge National Laboratory released the first GEM for field trial (Sayler et al., 1999). Use of GM bacteria represent an efficient technology for the deletion and detoxification of HMs and recalcitrant compounds from the polluted sites (Muhammad et al., 2008) that have been used for the elimination of HMs such as Cd, Hg, Ni, Cu, As, and Fe (Verma and Singh, 2005). GM *Escherichia coli* strain JM109 has the aptitude to eradicate Hg from metal-contaminated water, soil, and sediment (Chen and Wilson, 1997). Besides, transgenic bacteria expressing metallothioneins and polyphosphate kinase containing *merA* genes have been used for the bioremediation of Hg (De et al., 2006; Ruiz et al., 2011). On the other hand, *E. coli* containing the ArsR gene can stimulate the bioaccumulation of As (Kostal et al., 2004). Microorganisms are unable to destroy inorganic metals but they can change the oxidation states of inorganic metals which are less toxic through their microbial reduction system. The metal regulatory gene of bacteria are able to transform toxic HMs to a less toxic forms (Hasin et al., 2010). GM bacteria expressing metallothioneins (MT) can accelerate the accumulation of HMs (Pazirandeh et al., 1995). There are certain difficulties associated with the deployment of GMM-assisted bioremediation because of the bureaucratic barriers and the poor survival of GMMs in the metal-contaminated sites (Abhilash et al., 2009). A list of microorganisms used for bioremediation of a range of pollutants are given in Table 10.1.

TABLE 10.1

Indigenous, Extremophiles, and Genetically Modified Microorganisms (GMMs) Used for Remediation of Various Pollutants

Pollutants	Microbial Group	Microorganism(s)	References
Indigenous Microorganisms			
Heavy metals (Pb, Hg, Ni, Cd, Zn, U)		*Saccharomyces cerevisiae*	
PAHs	Bacteria	*Pseudomonas, Sphingomonas, Cycloclasticus, Burkholderia, Polaromonas, Neptunomonas, Janibacter, Rhodococcus rhodochrous, Enterobacter cloaceae*	Iwabuchi et al., 2002; Iyer et al., 2006; Kumar et al., 2018
Dibenzothiophene, fluoranthene, pyrene, and chrysene	Bacteria	*Bjerkandera adusta*	Valentýn et al., 2007
Phenanthrene, anthracene, fluoranthene, pyrene, chrysene, benzo (β) fluoranthene, benzo (k) fluoranthene, benzo (α) pyrene, dibenzo (a, h) anthracene, benzo (g, h, i) perylene, and indeno (1, 2, 3 - c, d) pyrene	Bacteria	*Irpex lacteus*	Leonardi et al., 2007
Naphthalene	Fungi	*Cunninghamella elegans*	Cerniglia and Gibson, 1977
Petroleum hydrocarbons	Fungi	*Amorphoteca, Neosartorya, Talaromyces, Candida, Yarrowia, Pichia*	Chaillan et al., 2004
Crude oil hydrocarbons	Fungi	*Aspergillus, Cephalosporium, Pencillium*	Singh, 2006

(Continued)

TABLE 10.1 *(Continued)*
Indigenous, Extremophiles, and Genetically Modified Microorganisms (GMMs) Used for Remediation of Various Pollutants

Pollutants	Microbial Group	Microorganism(s)	References
Indigenous Microorganisms			
Petroleum compounds	Fungi	*Candida lipolytica, Rhodotorula mucilaginosa, Geotrichum* sp., *Trichosporon mucoides*	Boguslawska-Was and Dabrowski, 2001
Crude oil and mixed hydrocarbon substrates	Achlorophyllous Alga	*Prototheca zopfii*	Walker et al., 1975
Fluoranthene, pyrene, and a mixture of fluoranthene and pyrene	Green Algae	*Chlorella vulgaris, Scenedesmus platydiscus, Scenedesmus quadricauda, Selenastrum capricornutum*	Lei et al., 2007
Naphthalene and phenanthrene	Cyanobacteria	*Oscillatoria* spp., *Agmenellum quadruplicatum* strainPR-6.	Cerniglia et al., 1980; Narro et al., 1992
Extremophiles			
PAHs	Thermophilic bacteria	*Nocardia, Bacillus, Mycobacterium, Stenotrophomonas, Pasteurell*	Kumar et al., 2018
Petroleum hydrocarbon	Cold climatic bacteria	*Pseudomonas, Thiobacillus, Geobacter*	Yeung et al., 2013
Crude oil	Halophilic bacteria	*Haloferax* sp., *Halobacterium* sp.	Al-Mailem et al., 2010
Phenol and benzoate	Halophilic bacteria	*Halomonas organivorans*	Moreno et al., 2011
Genetically Modified Microorganisms (GMMs)			
Oil	Genetically modified bacteria	*Pseudomonas putida* PaW85	Jussila et al., 2007
Naphthalene	Genetically modified bacteria	*Pseudomonas fluorescens* HK44	Ripp et al., 2000; Sayler and Ripp, 2000
Hg	Genetically modified bacteria	*Escherichia coli* JM109	Chen and Wilson, 1997
Ni	Genetically modified bacteria	*E. coli* SE5000	Fulkerson et al., 1998
Cd	Genetically modified bacteria	*Caulobacter* spp. JS4022/ p723-6H	Patel et al., 2010
Naphthalene, toluene and biphenyl	Genetically modified bacteria	*P. putida* S12	Marconi et al., 1997

10.5 MICROBIAL BIOREMEDIATIONS OF ORGANIC AND INORGANIC POLLUTANTS

Fossil fuels are at the forefront in promoting major positive changes in global economic growth and are also responsible for major environmental concerns and challenges. The persistent environmental concern caused by the fossil fuels in areas of oil exploitation and majority of environmental accidents occurred during its transportation and uses in industries and automobiles (Maier et al., 2000). Besides, the substantial use of the synthetic compounds to fulfill the human needs in the past few decades has resulted in long-term environmental problems. Specifically, with extensive use of inorganic fertilizers, pesticides, weedicides in agriculture and synthetic drugs and medicines in public health sectors throughout the world have resulted in frequent contamination of agricultural soils, surface and groundwater resources, and the overall environment. Dominant organic contaminants include xenobiotics, volatile organic, phenolic, and halogenated hydrocarbon and inorganic contaminants include nitrate, phosphate, salt along with HMs such as As, Cu, Zn, Hg, and so on. PAHs are a dominant group of hydrophobic organic pollutants that used to be present in the environment for a longer period and are toxic, carcinogenic, and hence considered to be hazardous to human beings and the environment (Kuiper et al., 2004). PAHs are widely found in the environment being produced by both natural and anthropogenic sources and the natural sources include the products produced by plants and termites (Azuma et al., 1996; Krauss et al., 2005; Wilcke et al., 2000). A variety of bacterial strains, for example, *Pseudomonas aeruginosa*, *Mycobacterium* sp., and *Haemophilus* sp. have been identified to play a role in PAHs degradation or its transformation (Simarro et al., 2013; Seo et al., 2009).

Varieties of toxic organic compounds such as pesticides, synthetic dyes, polychlorinated biphenyls (PCBs), and PAHs have been reported to degrade by several bacterial strain (Kafilzadeh et al., 2011) belonging to different genera such as *Bacillus*, *Corynebacterium*, *Staphylococcus*, *Pseudomonas*, and *Sphingomonas* under laboratory conditions (Chaudhry et al., 2005; Glick, 2010; Hussain et al., 2013; Hussain et al., 2011; Imran et al., 2015). Several bacterial strains have been identified for their potential to degrade a range of organic compounds under laboratory conditions, but such transformations to be achieved in the natural environment is quite difficult (Glick, 2010). Desorption of PAHs from soil has been found to be enhanced by microorganisms through the production of surfactant (Makkar and Rockne, 2003). PCBs also known as 'congeners' are a group of 209-member compounds having a range of applications from an insulator in transformer production to an extender in insecticide (Ang et al., 2005). Various types of microorganisms are needed to bioremediate numerous types of organic compounds present at contaminated sites. *Pseudomonas putida* is the first patent for microbial bioremediation (Prescott et al., 2002). Accidental spills during transportation, production, refining, lead to seepage of an estimated amount of 600.000 MT Yr^{-1} (Kvenvolden and Cooper, 2003). Leakage of oil into water bodies not only causes water pollution but also damage to our biodiversity by causing the death of marine plants and animals (Kothari et al., 2013). Accidental leakage has indirectly hampered various industries such as fishing, tourism, and also to oil industries (Kothari et al., 2013). Leakage in underground storage tanks may lead to contamination to the land ecosystem by entering into the soil (Eweis et al., 1998). Some examples of disastrous oil spills are the *Torrey Canyon* of southern England with the spill of ~117,000 mg occurred in the year 1967, the *Arrow* incident of Nova Scotia with the spill of ~11,000 mg occurred in the year 1970, the *Metula* in Estrecho de Magallanes of Strait of Magellan with the spill of ~53,000 mg occurred in the year 1973 (Freedman, 1995). Phenol is the major pollutant that is released into the environment from oil industries as a principal constituent of wastewater. The temperature around 10–25°C influences the bioremediation of phenol in which phenol containing wastewater is treated by psychrotrophic *Pseudomonas putida* (Margesin and Schinner, 2001; Onysko et al., 2000).

A huge amount of non-biodegradable HMs is added in nature due to rapid population growth, associated development, globalization, enhanced industrialization, and anthropogenic activities that affect human health (Jarup, 2003) as well as plant and animal metabolism once they reach beyond the natural admissible limit. Various industries including mining, surface finishing, fuel-energy production, iron and steel, fertilizer and pesticide, electric appliance manufacturing industries, metallurgy, electroplating, leather manufacturing, photography, aerospace, atomic energy installation (Mahmood et al., 2012), and other compulsory industrial/ household products have directly or indirectly brought a significant change in the presence and amount of HMs concentration in the environment, leading to severe environmental pollution and biodiversity loss (Anyanwu et al., 2018; Volesky, 1990). Most of the HMs even at very low concentrations are toxic, carcinogenic, mutagenic, and teratogenic in nature that induces oxidative stress because of the generation of reactive oxygen species (ROS) (Ali et al., 2019). The pollution caused by toxic and hazardous HMs including Cr(VI), Pb(II), Cd(II), Al(III), Ni(II), Co(II), U(VI) (McIntyre, 2003) is a disastrous environmental problem because of their antagonistic effects on soil, air, water, and human health. Humans are exposed to a range of environmental pollutant through the inhalation of suspended particulate matters and biomagnification of contaminants through the food chain (Dadar et al., 2016). Therefore, the reclamation of HMs polluted sites with the help of microorganisms is of absolute necessity for the safety of the human and the ecosystem health.

Microbial bioremediation is an economically feasible procedure showing successful results in removing HMs and organic compounds such as petroleum hydrocarbons (Das and Adholeya, 2012) from contaminated sites. Nonbiodegradable inorganic compounds especially HMs cannot be destroyed biologically, but only bioaccumulate or transform from one oxidation state or organic complex to another and one toxic form to a less toxic form (Abatenh et al., 2017; Tekere, 2019). Microorganisms have established the proficiencies to protect themselves from HMs toxicity actively through uptake (bioaccumulation), and/or passively through adsorption, methylation of frequently volatile methylated compounds, oxidation, and reduction (Abatenh et al., 2017). Microorganisms have evolved mechanisms such as sequestration of proteins through which they detoxify or resist these HMs to survive metal stressed conditions (Rehan and Alsohim, 2019). The interface between the microorganisms with the toxic and hazardous contaminants in microbial bioremediation leads to immobilization and assortment rather than their destruction and eradication from the environment (Muthuirulan et al., 2014). Soil amendments can escalate the removal of HMs by microorganisms and the concentration of amendments have fluctuating effects on the leaching rate of HMs (Tyagi et al., 2014). Removal rates are higher in addition to additives such as $FeSO_4.7H_2O$ than when these additives are added individually (Race, 2017).

10.6 CONCLUSIONS AND FUTURE RESEARCH PROSPECTS

Microbial bioremediation involves both onsite and offsite cleaning of contaminants present in soil, groundwater, and marine environment with the assistance of naturally occurring and GMMs including bacteria, cyanobacteria, fungi, and algae by absorbing toxic compounds or by metabolizing it. It uses microorganisms to detoxify and transform environmental contaminants into comparatively less toxic forms with less or no by-products and is thus considered as an environment-friendly, cost effective, and less labor-intensive approach. Therefore, it is acknowledged as an attractive option and the best substitute to traditional physico-chemical methods to remediating, cleaning, managing, and recovering polluted environments through inherent microbial metabolic activity. Novel microbial species with great remediation potential must be explored in the future to expedite the process of microbial bioremediation. In addition, the use of microbial genetic transformation is a quite powerful tool, which should be further exploited to design hyper-resistant microorganisms. Molecular studies can also help in developing specific probes to identify and monitor specific degradative microorganisms in the environment. Further, the biodegradation potential can significantly be boosted up using various biodegradation boosters including GMMs, fertilizers, surfactant amendments, nanoparticles, and so on in an integrated manner.

REFERENCES

Abatenh, E., Gizaw, B., Tsegaye, Z. (2017). Application of microorganisms in bioremediation-review. *J. Environ. Microbiol.*, 1(1), 02–09.

Abd-Alla, M.H., Morsy, F.M., El-Enany, A.W.E., Ohyama, T. (2012). Isolation and characterization of a heavy-metal-resistant isolate of Rhizobium leguminosarum bv. viciae potentially applicable for biosorption of Cd^{2+} and Co^{2+}. *Int. Biodeter. Biodegr.*, 67, 48–55.

Abdulsalam, S., Bugaje, I., Adefila, S., Ibrahim, S. (2011). Comparison of biostimulation and bioaugmentation for remediation of soil contaminated with spent motor oil. *Int. J. Environ. Sci. Technol.*, 8, 187–194.

Abhilash, P.C., Jamil, S., Singh, N. (2009). Transgenic plants for enhanced biodegradation and phytoremediation of organic xenobiotics. *Biotechnol. Adv.*, 27, 478–488.

Ali, H., Khan, E., Ilahi, I. (2019). Environmental chemistry and ecotoxicology of hazardous heavy metals: Environmental persistence, toxicity, and bioaccumulation. *J. Chem.*, 2019, 1–14. https://doi.org/10.1155/2019/6730305

Al-Mailem, D.M., Sorkhoh, N.A., Al-Awadhi, H., Eliyas, M., Radwan, S.S. (2010). Biodegradation of crude oil and pure hydrocarbons by extreme halophilic archaea from hypersaline coasts of the Arabian Gulf. *Extremophiles*, 14, 321–328.

Ang, E.L., Zhao, H., Obbard, J.P. (2005). Recent advances in the bioremediation of persistent organic pollutants via biomolecular engineering. *Enzyme Microb. Technol.*, 37, 487–496.

Anonymous. (1998). Application Guide for Bioslurping, Vol. I, Summary of the Principles and Practices of Bioslurping, Technical Memorandum, TM-2300-ENV, Naval Facilities Engineering Service Center, Port Hueneme, California, pp. 1–16. https://clu-in.org/download/techfocus/mpe/Bioslurp-app-guide-V1.pdf

Anyanwu, B.O., Ezejiofor, A.N., Igweze, Z.N., Orisakwe, O.E. (2018). Heavy metal mixture exposure and effects in developing nations: An update. *Toxics*, 6, 65.

Arora, N.K. (2018). Bioremediation: A green approach for restoration of polluted ecosystems. *Environ. Sustainability*, 1, 305–307.

Azad, M.A.K., Amin, L., Sidik, N.M. (2014). Genetically engineered organisms for bioreme-diation of pollutants in contaminated sites. *Chin. Sci. Bull.*, 59(8), 703–714.

Azuma, H., Toyota, M., Asakawa, Y., Kawano, S. (1996). Naphthalene-a constituent of *Magnolia* flowers. *Phytochemistry*, 42(4), 999–1004.

Banerjee, A., Roy, A., Dutta, S., Mondal, S. (2016). Bioremediation of hydrocarbon a review. *Int. J. Adv. Res.*, 4(6), 1303–1313.

Boguslawska-Was, E., Dabrowski, W. (2001). The seasonal variability of yeasts and yeast-like organisms in water and bottom sediment of the Szczecin Lagoon. *Int. J. Hyg. Environ. Health*, 203, 451–458.

Bojórquez, C., Voltolina, D. (2016). Removal of cadmium and lead by adapted strains of Pseudomonas aeruginosa and Enterobacter cloacae. *Rev. Int. Contam. Ambie.*, 32, 407–412.

Boopathy, R. (2000). Factors limiting bioremediation technologies. *Bioresour. Technol.*, 74(1), 63–67.

Brunetti, G., Farrag, K., Soler-Rovira, P., Ferrara, M., Nigro, F., Senesi, N. (2012). The effect of compost and Bacillus licheniformis on the phytoextraction of Cr, Cu, Pb and Zn by three Brassicaceae species from contaminated soils in the Apulia region, Southern Italy. *Geoderma*, 170, 322–330.

Cerniglia, C.E., Gibson, D.T. (1977). Metabolism of naphthalene by *Cunninghamella elegans*. *Appl. Environ. Microbiol.*, 34, 366–370.

Cerniglia, C.E., Gibson, D.T., Baalen, C.V. (1980). Oxidation of naphthalene by cyanobacteria and microalgae. *J. Gen. Microbiol.*, 116, 495–500.

Chaillan, F., Le Fleche, A., Bury, E., Phantavong, Y., Grimont, P., Saliot, A., Oudot, J. (2004). Identification and biodegradation potential of tropical aerobic hydrocarbon degrading microorganisms. *Res. Microbiol.*, 155(7), 587–595.

Chakraborty, R., Wu, C.H., Hazen, T.C. (2012). Systems biology approach to bioremediation. *Curr. Opin. Biotechnol.*, 23, 1–8.

Chang, M.W., Holoman, T.P., Yi, H. (2008). Molecular characterization of surfactantdriven microbial community changes in anaerobic phenantherene-degrading cultures under methanogenic conditions. *Biotechnol. Lett.*, 30, 1595–1601.

Chaudhry, Q., Blom-Zandstra, M., Gupta, S., Joner, E.J. (2005). Utilizing the synergy between plants and rhizosphere microorganisms to enhance breakdown of organic pollutants in the environment. *Environ. Sci. Pollut. Res.*, 12, 34–48.

Chen, M., Xu, P., Zeng, G., Yang, C., Huang, D., Zhang, J. (2015). Bioremediation of soils con-taminated with polycyclic aromatic hydrocarbons, petroleum, pesticides, chlorophenols and heavy metals by composting: Applications, microbes and future research needs. *Biotechnol. Adv.*, 33, 745–755.

Chen, S.L., Wilson, D.B. (1997). Genetic engineering of bacteria and their potential for Hg2? bioremediation. *Biodegradation*, 8, 97–103.

Chikere, C.B., Chikere, B.O., Okpokwasili, G.C. (2012). Bioreactor-based bioremediation of hydrocarbon polluted Niger Delta marine sediment, *Nigeria Biotech.*, 2(1), 53–66.

Coelho, L.M., Rezende, H.C., Coelho, L.M., de Sousa, P.A., Melo, D.F., Coelho, N.M. (2015). Bioremediation of polluted waters using microorganisms. In: *Advances in Bioremediation of Wastewater and Polluted Soil*, Shiomi, N. (ed.). IntechOpen. https://doi.org/10.5772/60770

Dadar, M., Adel, M., Saravi, H.N., Dadar, M. (2016). A comparative study of trace metals in male and female Caspian kutum (Rutilus frisii kutum) from the southern basin of Caspian Sea. *Environ. Sci. Pollut. Res.*, 23, 24540–24546.

Das, M., Adholeya, A. (2012). Role of microorganisms in remediation of contaminated soil. In: *Microorganisms in Environmental Management*, Satyanarayana, T., Johri, B.N. (eds.), pp. 81–111. Dordrecht: Springer, Netherlands.

Das, S., Dash, H.R. (2014). Microbial bioremediation: A potential tool for restoration of contaminated areas. In: *Microbial Biodegradation and Bioremediation*, Das, S. (ed.), pp. 1–21. Elsevier B.V.: Elsevier Inc. https://www.elsevier.com/books/microbial-biodegradation-and-bioremediation/das/978-0-12-800021-2

De, J., Sarker, A., Rahman, N.S. (2006). Bioremediation of toxic substances by mercury resistant marine bacteria. *Ecotoxicology*, 15, 385–389.

Dell Anno, A., Beolchini, F., Rocchetti, L., Luna, G.M., Danovaro, R. (2012). High bacterial biodiversity increases degradation performance of hydrocarbons during bioremediation of contaminated harbor marine sediments. *Environ Pollut.*, 167, 85–92.

Dermont, G., Bergeron, M., Mercier, G., Richerlaflèche, M. (2008). Soil washing for metal removal: A review of physical/chemical technologies and field applications. *J. Hazard. Mater.*, 152, 1–31.

Ehrlich, H.L. (1997). Microbes and metals. *Appl. Microbiol. Biotechnol.*, 48, 687–692.

El Fantroussi, S., Agathos, S.N. (2005). Is bioaugmentation a feasible strategy for pollutant removal and site remediation? *Curr. Opin. Microbiol.*, 8, 268–275.

Eweis, J.B., Ergas, S.J., Chang, D.P.Y., Schroeder, E.D. (1998). *Bioremediation principles*, pp. 296, Singapore: McGraw-Hill Companies, Inc.

Fang, L., Zhou, C., Cai, P., Chen, W., Rong, X., Dai, K., Liang, W., Gu, J., Huang, Q. (2011). Binding characteristics of copper and cadmium by cyanobacterium Spirulina platensis. *J. Hazard. Mater.*, 190, 810–815.

Fathepure, B.Z. (2014). Recent studies in microbial degradation of petroleum hydrocarbons in hypersaline environment: A review article. *Front. Microbiol.*, 5, 1–16.

Freedman, B. (1995). *Environmental Ecology: The Ecological Effects of Pollution, Disturbance, and other Stresses.* Elsevier B.V.: Elsevier Inc. https://www.elsevier.com/books/environmental-ecology/freedman/978-0-08-050577-0

Fulkerson, J.F., Garner, R.M., Mobley, H.L.T. (1998). Conserved residues and motifs in the NixA protein of Helicobacter pylori are critical for the high affinity transport of nickel ions. *J. Biol. Chem.*, 273, 235–241.

Gan, W.J., Yue, H.E., Zhang, X.F., Shan, Y.H., Zheng, L.P., Lin, Y.S. (2012). Speciation analysis of heavy metals in soils polluted by electroplating and effect of washing to the removal of the pollutants. *J. Ecol. Rural Environ.*, 28, 82–87.

Glick, B.R. (2010). Using soil bacteria to facilitate phytoremediation. *Biotechnol. Adv.*, 28, 367–374.

Goyal, N., Jain, S.C., Banerjee, U.C. (2003). Comparative studies on the microbial adsorption of heavy metals. *Adv. Environ. Res.*, 7, 311–319.

Hasin, A.A., Gurman, S.J., Murphy, L.M., et al. (2010). Remediation of chromium (VI) by a methane-oxidizing bacterium. *Environ Sci Technol.*, 44, 400–405.

Hazen, T.C. (2010). In situ: Groundwater bioremediation. In: *Handbook of Hydrocarbon and Lipid Microbiology*, Timmis, K.N. (ed.), pp. 2583–2594. Berlin: Springer.

Hu, N., Luo, Y., Song, J. (2010). Influence of soil organic matter, pH and temperature on adsorption by four soils from Yangtze river delta. *Acta Pedol. Sin.*, 44, 437–443.

Hussain, S., Devers-Lamrani, M., El-Azahari., N, Martin-Laurent, F. (2011). Isolation and characterization of an isoproturon mineralizing *Sphingomonas* sp. strain SH from a French agricultural soil. *Biodegradation*, 22, 637–650.

Hussain, S., Maqbool, Z., Ali, S., Yasmin, T., Imran, M., Mehmood, F., Abbas, F. (2013). Biodecolorization of Reactive Black-5 by a metal and salt tolerant bacterial strain *Pseudomonas* sp. RA20 isolated from Paharang drain effluents in Pakistan. *Ecotoxicol. Environ. Saf.*, 98, 331–338.

Ibrahim, H.M.M. (2016). Biodegradation of used engine oil by novel strains of Ochrobactrum anthropi HM-1 and Citrobacter freundii HM-2 isolated from oil-contaminated soil. *3 Biotech.*, 6, 226.

Imran, M., Arshad, M., Khalid, A., Hussain, S., Mumtaz, M.W., Crowley, D.E. (2015). Decolorization of reactive black-5 by Shewanella sp. in the presence of metal ions and salts. *Water Environ. Res.*, 87, 579–586.

Iwabuchi, N., Sunairi, M., Urai, M., Itoh, C., Anzai, H., Nakajima, M., Haryama, S. (2002). Extracellular polysaccharides of *Rhodococcus rhodochrous* S-2 stimulate the degradation of aromatic components in crude oil by indigenous marine bacteria. *Appl. Environ. Microbiol.*, 68, 2337–2343.

Iyer, A., Mody, K., Jha, B. (2006). Emulsifying properties of a marine bacterial exopolysaccharide. *Enzyme Microb. Technol.*, 38, 220–222.

Jain, P.K., Bajpai, V. (2012). Biotechnology of bioremediation – A review. *Int. J. Environ. Sci.*, 3(1), 535–549.

Jain, P.K., Gupta, V.K., Gaur, R.K, Bajpai, V., Gautam, N., Modi, D.R. (2010). Fungal enzymes: Potential tools of environmental processes. In: *Fungal Biochemistry and Biotechnology*, Gupta, V.K., Tuohy, M., Gaur, R.K. (eds.), pp. 44–56. Germany: LAP Lambert Academic Publishing.

Jarup, L. (2003). Hazards of heavy metal contamination. *Br. Med. Bull.*, 68, 167–182.

Jin, S., Fallgren, P.H. (2007). Site-specific limitations of using urea as nitrogen source in biodegradation of petroleum wastes. *Soil Sediment Cont.*, 16(5), 497–505.

Jussila, M.M., Zhao, J., Suominen, L., Lindström, K. (2007). TOL plasmid transfer during bacterial conjugation in vitro and rhizoremediation of oil compounds in vivo. *Environ. Pollut.*, 146(2), 510–524.

Kadali, K.K., Simons, K.L., Sheppard, P.J., Ball, A.S. (2012). Mineralisation of weathered crude oil by a hydrocarbonoclastic consortia in marine mesocosms. *Water Air Soil Poll.*, 223, 4283–4295.

Kafilzadeh, F., Sahragard, P., Jamali, H., Tahery, Y. (2011). Isolation and identification of hydrocarbons degrading bacteria in soil around Shiraz Refinery. *Afr. J. Microbiol. Res.*, 4(19), 3084–3089.

Kapoor, A., Viraraghavan, T. (1997). Heavy metal biosorption sites in Aspergillus niger. *Bioresour. Technol.*, 61, 221–227.

Kostal, J.R.Y., Wu, C.H., Mulchandani, A. et al. (2004). Enhanced arsenic accumulation in engineered bacterial cells expressing ArsR. *Appl. Environ. Microbiol.*, 70, 4582–4587.

Kothari, V., Panchal, M., Srivastava, N. (2013). Microbial degradation of hydrocarbons. https://www.researchgate.net/profile/Vijay-Kothari-2/publication/261613425_Microbial_Degradation_of_Hydrocarbons/links/00b7d534d0ae4005fd000000/Microbial-Degradation-of-Hydrocarbons.pdf

Krauss, M., Wilcke, W., Martius, C., Bandeira, A.G., Garcia, M.V., Amelung, W. (2005). Atmospheric versus biological sources of polycyclic aromatic hydrocarbons (PAHs) in a tropical rainforest environment. *Environ. Pollut.*, 135(1), 143–154.

Kuiper, Lagendijk, E.L., Bloemberg, G.V., Lugtenberg, B.J.J. (2004). Rhizoremediation: A beneficial plant-microbe interaction. *Mol. Plant-Microb. Interact.*, 17, 6–15.

Kumar, A., Bisht, B.S., Joshi, V.D., Dhewa, T. (2011). Review on bioremediation of polluted environment: A management tool. *Int. J. Environ. Sci.*, 1(6), 1079–1093.

Kumar, M., Prasad, R., Goyal, P., Teotia, P., Tuteja, N., Varma, A., Kumar, V. (2017). Environmental biodegradation of xenobiotics: role of potential microflora. In: *Xenobiotics in the Soil Environment, Soil Biology*, Hashmi, M., Kumar, V., Varma, A. (eds.), 319–334. Cham: Springer.

Kumar, V., Shahi, S.K., Singh, S. (2018). Bioremediation: An eco-sustainable approach for restoration of contaminated sites. In: *Microbial Bioprospecting for Sustainable Development*, Singh, J., Sharma, D., Kumar, G., Sharma, N. (eds.), pp. 115–136. Singapore: Springer.

Kumari, B., Singh, D.P. (2016). A review on multifaceted application of nanoparticles in the field of bioremediation of petroleum hydrocarbons. *Ecol. Eng.*, 97, 98–105.

Kvenvolden, K.A., Cooper, C.K. (2003). Natural seepage of crude oil into the marine environment. *Geo-Mar Lett.*, 23, 140–146.

Lee, T.H., Byun, I.G., Kim, Y.O., Hwang, I.S., Park, T.J. (2006). Monitoring biodegradation of diesel fuel in bioventing processes using in situ respiration rate. *Water Sci. Technol.*, 53, 263.

Lei, A.P., Hu, Z.L., Wong, Y.S., Tam, N.F.Y. (2007). Removal of fluoranthene and pyrene by different microalgal species. *Bioresour. Technol.*, 98(2), 273–280.

Leonardi, V., Sasek, V., Petruccioli, M., Annibale, A.D., Erbanova, P., Cajthaml, T. (2007). Bioavailability modification and fungal biodegradation of PAHs in aged industrial soils. *Int. Biodeter. Biodeg.*, 60, 165–170.

Li, X., Lin, X., Li, P., Liu, W., Wang, L., Ma, F., Chukwuka, K.S. (2009). Biodegradation of the low concentration of polycyclic aromatic hydrocarbons in soil by microbial consortium during incubation. *J. Hazard Mater.*, 172(2–3), 601–605.

Li, X., Xu, H., Chen, Z.S., Chen, G. (2011). Biosynthesis of nanoparticles by microorganisms and their applications. *J. Nanomat.*, 2011, 1–16. https://doi.org/10.1155/2011/270974

Lim, M.W., Lau, E.V., Poh, P.E. (2016). A comprehensive guide of remediation technologies for oil contaminated soil - present works and future directions. *Mar. Pollut. Bull.*, 109, 14–45.

Mahmood, Q., Rashid, A., Ahmad, S.S., Azim, M.R., Bilal, M. (2012). Current status of toxic metals addition to environment and its consequences. In: *The Plant Family Brassicaceae*, Anjum, N.A., Iqbal Ahmad, I., Pereira, M. E., Duarte, A. C., Umar, S., Khan, N.A. (eds.), pp. 35–69. Dordrecht, Netherland: Springer.

Maier, R.M. (2000). Microorganisms and organic pollutants. In: *Environmental Microbiology*, Maier, R.M., Pepper, L.I., Gerba, P.C. (eds.), chapter 16, pp. 363–400. CA: Elsevi.

Maier, R.M., Pepper, I.L., Gerba, C.P. (2000). *A Textbook of Environmental Microbiology*. San Diego, CA: Academic.

Makkar, R.S., Rockne, K.J. (2003). Comparison of synthetic surfactants and biosurfactants in enhancing biodegradation of polycyclic aromatic hydrocarbons. *Environ. Toxicol. Chem.*, 22, 2280–2292.

Marconi, A.M., Kieboom, J., deBont, J.A.M. (1997). Improving the catabolic functions in the toluene-resistant strain *Pseudomonas putida* S12. *Biotechnol Lett.*, 19, 603–606.

Margesin, R., Schinner, F. (2001). Biodegradation and bioremediation of hydrocarbons in extreme environments. *Appl. Microbiol. Biotechnol.*, 56, 650–663.

McIntyre, T. (2003). Phytoremediation of heavy metals from soils. *Adv. Biochem. Eng. Biotechnol.*, 78, 97–123.

Mohapatra, P.K. (2008). *Textbook of Environmental Microbiology.* New Delhi: I.K. International Publishing House.

Mohsenzadeh, F., Nasseri, S., Mesdaghinia, A., Nabizadeh, R., Zafari, D., Khodakaramian, G., Chehregani, A. (2010). Phytoremediation of petroleum-polluted soils: Application of Polygonum aviculare and its root-associated (penetrated) fungal strains for bioremediation of petroleum-polluted soils. *Ecotoxicol. Environ. Saf.*, 73, 613–619.

Moreno, M.D.L., Sanchez-Porro, C., Piubeli, F., Frias, L., Garcia, M.T., Mellado, E. (2011). Cloning, characterization and analysis of cat and ben genes from the phenol degrading halophilic bacterium *Halomonas organivorans*. *PLoS ONE.*, 6, e21049.

Morto-Bermea, O., Hernández, A.E., Gaso, I., Segovia, N. (2002). Heavy metal concentrations in surface soils from Mexico City. *Bull. Environ. Contam. Toxicol.*, 68, 383–388.

Muhammad, S., Muhammad, S., Sarfraz, H. (2008). Perspectives of bacterial ACC deaminase in phytoremediation. *Trends Biotechnol.*, 25, 356–362.

Muñoz, R., Guieysse, B., Mattiasson, B. (2003). Phenanthrene biodegradation by an algal-bacterial consortium in two-phase partitioning bioreactors. *Appl. Microbiol. Biotechnol.*, 61(3), 261–267.

Muthuirulan, P., Sathyanarayanan, J., Paramasamy, G., Jeyaprakash, R. (2014). Microbial bioremediation: A metagenomic approach. In: *Microbial Biodegradation and Bioremediation*, Das, S. (ed.), pp. 407–419. Elsevier Inc.

Naik, M.G., Duraphe, M.D. (2012). Review paper on-parameters affecting bioremediation. *Int. J. Life Sci. Pharma Res.*, 2(3), L77–L80.

Narro, M.L., Cerniglia, C.E., Baalen, C.V., Gibson, D.T. (1992). Metabolism of phenanthrene by the marine cyanobacterium *Agmenellum quadruplicatum* strainPR-6. *Appl. Environ. Microbiol*, 58, 1351–1359.

Onysko, K.A., Budman, H.M., Robinson, C.W. (2000). Effect of temperature on the inhibition kinetics of phenol biodegradation by Pseudomonas putida Q5. *Biotech. Bio.*, 70(3), 291–299.

Palmisano, A.C., Schwab, B.S., Maruscik, D.A., Ventullo, R.M. (1991). Seasonal changes in mineralization of xenobiotics by stream microbial communities. *Can. J. Microbiol.*, 37(12), 939–948.

Parker, J.C. (1996). Evaluating the effectiveness of product recovery, bioventing, and bioslurping systems. *Environ. Sys. Technol.*, 4 (Winter 95), 4–6.

Patel, J., Zhang, Q., Michael, R., et al. (2010). Genetic engineering of *Caulobacter crescentus* for removal of cadmium from water. *Appl. Biochem. Biotechnol.*, 160, 232–243.

Pazirandeh, M., Chrisey, L.A., Mauro, J.M. et al. (1995). Expression of the Neurospora crassa metallothionein gene in Escherichia coli and its effect on heavy-metal uptake. *Appl. Microbiol. Biotechnol.*, 43, 1112–1117.

Prakash, B., Irfan, M. (2011). Pseudomonas aeruginosa is present in crude oil contaminated sites of Barmer region (India). *J. Bioremediat. Biodeg*, 2, 129.

Prescott, L.M., Harley, J.P., Klein, D.A. (2002). *Microbiology*, 5th edn. New York: McGraw-Hill.

Race, M. (2017). Applicability of alkaline precipitation for the recovery of EDDS spent solution. *J. Environ. Manag.*, 203, 358–363.

Rehan, M., Alsohim, A.S. (2019). Bioremediation of heavy metals. In: *Environmental Chemistry and Recent Pollution Control Approaches*, Saldarriaga-Noreña, H., Murillo-Tovar, M.A., Farooq, R., Dongre, R., Riaz, S. (eds.). IntechOpen. https://doi.org/10.5772/intechopen.88339

Ripp, S., Nivens, D.E., Ahn, Y., Werner, C., Jarrel, J., Easter, J.P., Cox, C.D., Burlage, R.S., Sayler, G.S. (2000). Controlled field release of a bioluminescent genetically engineered microorganism for bioremediation process monitoring and control. *Environ. Sci. Technol.*, 34, 846–853.

Ruiz, O.N., Alvarez, D., Gongalez-Ruiz, G. et al (2011). Characterization of mercury bioremediation by transgenic bacteria expressing metallothionein and polyphosphate kinase. *BMC Biotechnol.*, 11, 1–8.

Sari, A., Tuzen, M. (2009). Kinetic and equilibrium studies of biosorption of Pb(II) and Cd(II) from aqueous solution by macrofungus (*Amanita rubescens*) biomass. *J. Hazard Mater.*, 164, 1004–1111.

Sarret, G., Manceau, A., Spadini, L., Roux, J.C., Hazemann, J.L., Soldo, Y., Eybert-Bérard, L., Menthonnex, J. (1998). Structural determination of Zn and Pb binding sites in penicillium chrysogenum cell walls by EXAFS spectroscopy. *Environ. Sci. Technol.*, 32, 1648–1655.

Sayler, G.S., Cox, C.D., Burlage, R., Ripp, S., Nivens, D.E. (1999). Field application of a genetically engineered microorganism for polycyclic aromatic hydrocarbon bioremediation process monitoring and control. In: *Novel Approaches for Bioremediation of Organic Pollution*, Fass, R., Flashner, Y., Reuveny, S. (eds.), pp. 241–254. New York: Kluwer Academic/Plenum Publishers.

Sayler, G.S., Ripp, S. (2000). Field applications of genetically engineered microorganisms for bioremediation processes. *Curr. Opin. Biotechnol.*, 11, 286–289.

Seo, J.S., Keum, Y.S., Li, Q.X. (2009). Bacterial degradation of aromatic compounds. *Int. J. Environ. Res. Public Health*, 6, 278–309.

Simarro, R., González, N., Bautista, L.F., Molina, M.C. (2013). Biodegradation of high-molecular-weight polycyclic aromatic hydrocarbons by a wood-degrading consortium at low temperatures. *FEMS Microbiol. Ecol.*, 83, 438–449.

Singh, H. (2006). *Mycoremediation: Fungal Bioremediation*. New York: Wiley Interscience.

Somtrakoon, K., Suanjit, S., Pokethitiyook, P., Kruatrachue, M., Lee, H., Upatham, S. (2008). Phenanthrene stimulates the degradation of pyrene and fluoranthene by *Burkholderia* sp. VUN10013. *World J. Microbiol. Biotechnol.*, 24, 523–531.

Song, B., Zeng, G., Gong, J., Liang, J., Xu, P., Liu, Z., Zhang, Y., Zhang, C., Cheng, M., Liu, Y., Ye, S.,Yi, H., Ren, X. (2017). Evaluation methods for assessing effectiveness of in situ remediation of soil and sediment contaminated with organic pollutants and heavy metals. *Environ. Int.*, 105, 43–55.

Speight, J.G., El-Gendy, N.S. (2018). Bioremediation of contaminated soil. In: *Introduction to Petroleum Biotechnology*, chapter 10, pp. 361–417. https://doi.org/10.1016/B978-0-12-805151-1.00010-2.

Strong, P.J., Burgess, J.E. (2008). Treatment methods for wine-related ad distillery wastewaters: A review. *Biorem. Journal*, 12, 70–87.

Tabatabaee, A., Assadi, M.M., Noohi, A.A., Sajadian, V.A. (2005). Isolation of biosurfactant producing bacteria from oil reservoirs. *Iran J. Environ. Health Sci. Eng.*, 2(1), 6–12.

Tang, C.Y., Criddle, C.S., Leckie, J.O. (2007). Effect of flux (transmembrane pressure) and membrane properties on fouling and rejection of reverse osmosis and nanofiltration membranes treating perfluorooctane sulfonate containing waste water. *Environ. Sci. Technol.*, 41, 2008–2014.

Tangahu, B.V., Abdullah, A.R.S., Basri, H., Idris, M., Anuar, N., Mukhlisin, M. (2011). A review on heavy metals (As, Pb, and Hg) uptake by plants through phytoremediation. *Int. J. Chem. Eng.*, 2011, 31.

Tastan, B.E., Ertugrul, S., Dönmez, G. (2010). Effective bioremoval of reactive dye and heavy metals by *Aspergillus versicolor*. *Bioresour. Technol.*, 101, 870–876.

Tekere, M. (2019). Microbial bioremediation and different bioreactors designs applied. In: *Biotechnology and Bioengineering*, E. Jacob-Lopes and L. Queiroz Zepka (eds.), IntechOpen. DOI: 10.5772/intechopen.83661.

Thomas, D.G., Minj, N., Mohan, N., Rao, P.H. (2016). Cultivation of microalgae in domestic wastewater for biofuel applications–An upstream approach. *J. Algal Biomass Utln.*, 7(1), 62–70.

Thomé, A., Reginatto, C., Vanzetto, G., Braun, A.B. (2018). Remediation technologies applied in polluted soils: New perspectives in this field. In: *The International Congress on Environmental Geotechnics*, pp. 186–203. Singapore: Springer.

Tyagi, M., da Fonseca, M.M.R., de Carvalho, C.C.C.R. (2011). Bioaugmentation and bio-stimulation strategies to improve the effectiveness of bioremediation processes. *Biodegradation*, 22, 231–241.

Tyagi, S., Kumar, V., Singh, J., Teotia, P., Bisht, S., Sharma, S. (2014). Bioremediation of pulp and paper mill effluent by dominant aboriginal microbes and their consortium. *Int. J. Environ. Res.*, 8, 561–568.

United States Environmental Protection Agency (USEPA). (2000a). Engineered approaches to in situ bioremediation of chlorinated solvents: fundamentals and field applications, EPA-542-R-00-008.

United States Environmental Protection Agency (USEPA). (2000b). Ground water issue. EPA 540/S-01/500.

United States Environmental Protection Agency (USEPA). (2006). In situ and ex situ biodegradation technologies for remediation of contaminated sites. EPA/625/R-06/015.

United States Environmental Protection Agency (USEPA) (2012). A citizen guide to bioremediation. EPA 542-F-12-003.

Valentýn, L., Lu-Chau, T.A., Lopez, C., Feijoo, G., Moreira, M.T., Lema, J.M. (2007). Biodegradation of dibenzothiophene, fluoranthene, pyrene and chrysene in a soil slurry reactor by the white-rot fungus *Bjerkandera* sp. BOS55. *Process Biochem.*, 42, 641–648.

Verma, N., Singh, M. (2005). Biosensors for heavy metals. *J. Biomet*, 18, 121–129.

Vidali, M. (2001). Bioremediation: An overview. *Pure Appl. Chem.*, 73(7), 1163–1172.

Vieira, J.D., Stefenon, V.M. (2017). Soil bioremediation in heavy metal contaminated mining areas: A microbiological/biotechnological point of view. *J. Adv. Microbiol.*, 4(1), 1–10.

Volesky, B. (1990). Biosorption and biosorbents. In: *Biosorption of Heavy Metals*, Volesky, B. (ed.), pp. 3–5. FL: CRC Press.

Vullo, D., Ceretti, H., Alejandra, D.M., Ramírez, S., Zalts, A. (2008). Cadmium, zinc and copper biosorption mediated by pseudomonas veronii 2E. *Bioresour. Technol.*, 99, 5574–5581.

Waghmare, P.R., Kadam, A.A., Saratale, G.D., Govindwar, S.P. (2014). Enzymatic hydrolysis and characterization of waste lignocellulosic biomass produced after dye bio-remediation under solid state fermentation. *Bioresour. Technol.*, 168, 136–141.

Walker, J.D., Colwell, R.R., Vaituzis, Z., Meyer, S.A. (1975). Petroleum-degrading achlorophyllous alga *Prototheca zopfii*. *Nature*, 254, 423–424.

Wang, S.J., Wang, X., Lu, G.L., Wang, Q.H., Li, F.S., Guo, G.L. (2011). Bioremediation of petroleum hydrocarbon-contaminated soils by cold-adapted microorganisms: Research advance. *J. Appl. Ecol.*, 22(4), 1082–1088.

Wasilkowski, D., Swedziol, Z., Mrozik, A. (2012). The applicability of genetically modified microorganism in bioremediation of contaminated environments. *Chemik*, 66(8), 817–826.

Wierzba, S. (2015). Biosorption of lead(II), zinc(II) and nickel(II) from industrial wastewater by Stenotrophomonas maltophilia and Bacillus subtilis. *Pol. J. Chem. Technol.*, 17, 79–87.

Wilcke, W., Amelung, W., Martius, C., Garcia, M.V., Zech, W. (2000). Biological sources of polycyclic aromatic hydrocarbons (PAHs) in the Amazonian rainforest. *J. Plant Nutr. Soil Sci.*, 163(1), 27–30.

Yeung, C.W., Van Stempvoort, D.R., Spoelstra, J., Bickerton, G., Voralek, J., Greer, C.W. (2013). Bacterial community evidence for anaerobic degradation of petroleum hydrocarbons in cold climate groundwater. *Cold Reg. Sci. Technol.*, 86, 55–68.

Zhang, X., Yan, S., Tyagi, R.D., Surampalli, R.Y. (2011). Synthesis of nanoparticles by microorganisms and their application in enhancing microbiological reaction rates. *Chemosphere*, 82(4), 489–494.

11 Endophytes as Potential Plant Growth Promoters in Forestry
Recent Advances and Perspectives

*Vinay Kumar¹, Lata Jain¹ Sorabh Chaudhary²,
and Ravindra Soni³*
¹National Institute of Biotic Stress Management,
Raipur, Chhattisgarh, India
²Central Potato Research Institute (Regional Station),
Meerut, Uttar Pradesh, India
³Indira Gandhi Krishi Vishwa Vidyalaya,
Raipur, Chhattisgarh, India

CONTENTS

11.1   INTRODUCTION

Plants are sessile in nature and need to cope with adverse environmental and edaphic conditions with their intrinsic biological mechanisms. Microorganisms are one of most natural inhabitants of diverse environments having enormous metabolic potential to mitigate biotic and abiotic stresses. Bacterial and fungal endophytes ubiquitously live inside the plant tissues without causing any negative impact and their presence is often beneficial for the host, as they enhance the growth, improve tolerance to various environmental stresses, and can and suppress pathogen colonization and modulate immune response in plants (Kumar et al. 2019; Dini-Andreote 2020; Kumar et al. 2020c). A tropical forest is considered one of the most diverse terrestrial ecosystems, having the biggest number of microbial endophytes, and known for storage of many molecules with diverse biological activity (Roy and Banerjee 2018). Like all other plants, trees are vulnerable to attack by many pests and pathogens (Rabiey et al. 2019). The tree activates its physiological defence pathways through producing chemical messengers (volatile and non-volatile) to counteract the negative effect of resin production during wounding (Khan et al. 2016). Trees endophytes consists of ascomycetes and some basidiomycetes belong to the group of endophytes (Petrini 1986; Rodriguez et al. 2009).The host-endophytes interaction may play an important role in plants' growth and development and make them resistant to pest and pathogens (Santoyo et al. 2016; Goel et al. 2018b; Kumar et al. 2020b). Previous reports suggested that the fungal endophytes associated with woody perennial plants can be complex and labile both in ecological/environmental and evolutionary time (Saikkonen 2007).

Carbon sequestration is one of the most major ecosystem services provided by forest trees (Quine et al. 2011). Trees have long lives, endophytes play a vital role in preparing their hosts to counter extreme environmental conditions and pest-pathogen attacks. These endophytic microbes interact with their plant host and elicit positive responses to herbivore, plant pathogens, environmental stresses and synthesize bioactive metabolites/compounds (Omomowo and Babalola 2019; Kumar et al. 2020c). Recent studies have shown the presence of endophytes in plants in forest ecosystem and also shown the significance of endophytes in shaping the plant diversity in the forest. These microbes have potential applications in biocontrol of forest diseases, synthesis of secondary metabolites, enzymes for various industrial and biotechnological applications. Among the endophytic microbes, fungal endophytes are dominant and untapped, in the habitat adaptation of plants. Further studies in fungal diversity of plant and plant tissues using a multi-omics approach may be helpful to enhance the activity of drug, pharmaceutical and clinical applications (Mane and Vedamurthy 2019). Endophytes facilitate the fermentation procedure used as bio-fertilizers to increase the soil fertility and crop yield, and produce economically important plants and plant products (Chutulo and Chalannavar 2018). Endophytic microbes (bacterial and fungi) play a significant role in agriculture, food industry, pharmaceuticals, natural cycling, as biofertilizers and many other ways in the daily life of human beings. Agricultural microbial biotechnology through the integration of beneficial plant microbes may offer a promising strategy to improve plant health and agricultural productivity in a sustainable manner (Timmusk et al. 2017).

11.2 ISOLATION OF ENDOPHYTES

11.2.1 FUNGAL ENDOPHYTES

Several studies have focused on studying the diversity of fungal endophytes and its importance to the host and environment (Kumar et al. 2019). Fungal diversity in tree forests is highly affected due to samples, sampling strategy, isolation methods, temperature, seasonal changes and climatic conditions (Suryanarayanan et al. 2011; Singh et al. 2017). Studies were conducted to observe the diversity and biological activities of endophytic fungi from *Terminalia spp.* that showed the variation between one species to another, tissue and organ level (Toghueo and Boyom 2019). A survey was conducted to study the association of endophytic fungi in the mixed forest system containing *Betula platyphylla*, *Ulmus macrocarpa* and *Quercus liaotungensis* which showed all three tree species had similar overall endophyte colonization rates but infection rates was significantly higher in twigs as compared to leaf tissues (Sun et al. 2012). Many fungal endophytes have been reported to found in more than one host with different percentage of occurrence among the hosts in the same habitat indicated the host preference of microbes (Zhou and Hyde 2001). Among the fungal endophytes isolated from the *Terminalia* sp. includes *Aspergillus flavus*, *Alternaria sp.*, *Diaporthe arengae* and *Lasiodiplodia theobromae* were most predominantly present in the bark and leaves tissues (Patil et al. 2014). Fungal endophytes were isolated from the various plant tissues of *Eucalyptus globulus* (Lupo et al. 2001), roots and leaves tissues of *Lepanthes* and healthy sugar beet (Bayman et al. 1997; Shi et al. 2016). Andhale et al. (2019) isolated fungal endophytes, *Alternaria sp.* from the *Plumbago zeylanica* L an important medicinal plant with anticancer properties and showed the association of *Alternaria sp.* led to enhanced plumbagin content. Plumbagin is having anticancer, antifungal and antibacterial activities (Aziz et al. 2008). A list of endophytic microbes (fungi and Bacteria) isolated from different plants/trees and their functional attributes is mentioned in Table 11.1. Turbat et al. (2020) isolated endophytic fungi from different parts of *Sophora flavescen*, a medicinal plant represented by *Alternaria*, *Didymella*, *Fusarium* and *Xylogone* major genera's. Endophytes recovered from different tissues of Neem, *Azadirachta indica*, from different locations generated huge diversity of *Periconia*, *Drechslera* and *Stenella* endophytes (Verma et al. 2007).

11.2.2 BACTERIAL ENDOPHYTES

Bacterial endophytes colonize and live inside the plants different plant tissues and have been found to play an essential role in the functioning of the host plants (Kumar et al. 2020c). They are involved in imparting tolerance to various biotic and abiotic stresses to the host by producing anti-microbial metabolites, growth regulators/phyto-hormones and competition with host for nutrient and space (Rosenblueth and Martínez-Romero 2006; Mercado-Blanco and Lugtenberg 2014). The composition of microbial endophytes varied among the genotypes, tissues, organs of the plants, soil and geographical locations (Hardoim et al. 2015). Munir et al. (2020) isolated native bacterial endophytes healthy, symptomatic and asymptomatic citrus

TABLE 11.1

List of Endophytes Isolated from Forestry/Agroforestry Plant with Their Probable Functional Attributes

S. No.	Name of the Endophytic Microbes	Bacterial/ Fungal	Host Plants	Plant Tissues	Location /Country	PGP Activities / Antimicrobial	Reference
1.	*Bipolaris tetramera*,*Petriella* sp. *Penicillium*, *Trichophaea* and *Ulocladium* sp.	Fungal	*Pinus roxburgii*	Twigs or Stems	Jammu & Kashmir, India	Antimycotic activity	Qadri et al. 2013
2.	*Serendipita indica*	Fungal	*Ocimum basilicum*	Root tissues	Erfurt, Germany	Growth attributes and Heavy Metal remediation	Sabra et al. 2018
3.	*Alternaria alternate*, *Geotrichum albida*, *Thielaviopsis basicola* and *Penicillium frequentans*	Fungal	*Pinus roxburghii*	Spike	Uttrakhand, India	antimicrobial agents, pharmaceutical compounds	Bhardwaj et al. 2015
4.	*Rhizoctonia* sp.	Fungal	*Cynodon dactylon*	Leaf tissues	Jiangsu province china	Antibacterial activities	Ma et al. 2004
5.	*Colletotrichum gloeosporioides*	Fungal	*Vitex negundo*	Leaf tissues	Tamilnadu, India	Antibacterial activities	Arivudainambi et al. 2011
6.	Six numbers of fungal endophytes	Fungal	Jamun, *Syzygium aromaticum*	Stem Leaf, and internal stem tissues	Uttrakhand, India	Production of Bioactive compounds, enzymes	Kapoor et al. 2018
7.	*Alternaria alternate* and *Aspergillus niger* were more dominated and *Aspergillus fumigatus A. japonicus*, *A. niger* and *Chaetomium globosum*,	Fungal	*Eugenia jambolana* (*Sczhigium cumini*)	Stem, Leaf and Petiole	Rohtak, India	Antibacterial activity against MDR strains	Yadav et al. 2016

(Continued)

TABLE 11.1 (Continued)
List of Endophytes Isolated from Forestry/Agroforestry Plant with Their Probable Functional Attributes

S. No.	Name of the Endophytic Microbes	Bacterial/ Fungal	Host Plants	Plant Tissues	Location /Country	PGP Activities / Antimicrobial	Reference
8.	*Aspergillus fumigatus, Penicillium oxalicum, Preussia sp., Peyronellaea eucalyptica, P. sancta and Alternaria tenuissima.*	Fungal	*Juniperus procera*	Twigs or Stems	Saudi Arabia	Antimicrobial agents/ Antibacterial activities	Gherbawy and Elhariry 2014
9.	*Apiosordaria otanii, Penicillium oxalicum, Daldinia fissa and Polyporus arcularius*	Fungal	*Abies pindrow*	Twigs or Stems	Jammu & Kashmir, India	Antimicrobial activities	Qadri et al. 2013
10.	*Beauveria bassiana*	Fungal	*Grapevine, Vitis vinifera,*	Hardwood cuttings	Germany	entomopathogenic against Grape leafhopper	Rondot and Reineke 2016
11.	*Alternaria alternata, A. brassicae, Cladosporium cladospor ioides, Diaporthe helianthi, Fusarium proliferatum, Lasiodiplodia theobromae and Glomerella acutata*	Fungal	*Rauwolfia serpentine,* Sarpgandha	Stems or twigs	Jammu & Kashmir, India	Antimicrobial activities	Qadri et al. 2013
12.	*Chaetomium globosum, Diaporthe phaseolorum, Thielavia subthermophila and Valsa sordid*	Fungal	*Picrorhiza kurroa*	Stems or twigs	Jammu & Kashmir, India	Antimicrobial activities	Qadri et al. 2013

(Continued)

TABLE 11.1 (Continued)
List of Endophytes Isolated from Forestry/Agroforestry Plant with Their Probable Functional Attributes

S. No.	Name of the Endophytic Microbes	Bacterial/ Fungal	Host Plants	Plant Tissues	Location /Country	PGP Activities / Antimicrobial	Reference
13.	Penicillium chrysogenum, Alternaria alternata	Fungal	Asclepias sinaica	Leaf tissues	Germany	Plant growth promoting traits	Foudaet al. 2015
14.	Colletotrichum sps., Chaetomium coarctatum, Meyerozyma guilliermondii Fusarium proliferatum, and Penicillium crustosum	Fungal	Indian Medicinal plant, Ocimum sanctum	leaf and stem tissues	India	Anti-phytopathogenic activity against	Chowdhary and Kaushik 2015
15.	Aspergillus, Alternaria,Chaetomium globosum,Fusarium sp.	Fungal	Aegle marmelos	Bark, Leaf and root samples	Varanasi, India	Tissue specific diversity of endophytic microbes	Gond et al. 2007
16.	Gliocladium catenulatum	Fungal	Theobroma cacao	Branches of cacao	Bahia State, Brazil	Antibiosis and mycoparasitism	Rubini et al. 2005
17.	Twenty endophytic fungal isolates, Colletotrichum, Alternaria and Chaetomium	Fungal	Catharanthus roseus	Bark, leaf and stem	Coastal region of India	cytotoxic potential against the HeLa and MCF-7 cells,	Dhayanithy et al. 2019
18.	Fusarium proliferatum and Aspergillus fumigates	Fungal	Oxalis corniculata	Root tissues	Swat, Pakistan	Plant-growth-promoting	Bilal et al. 2018
19.	Aspergillus sp.	Fungal	Cynodon dactylon	Leaf tissues	Sheyang port on Yellow sea	Antibacterial activities	Li et al. 2005
20.	Chloridium sps.	Fungal	Azadirachata indica	Root tissues	Varanasi, India	Antibacterial activities	Kharwar et al. 2009

(Continued)

TABLE 11.1 (Continued)
List of Endophytes Isolated from Forestry/Agroforestry Plant with Their Probable Functional Attributes

S. No.	Name of the Endophytic Microbes	Bacterial/ Fungal	Host Plants	Plant Tissues	Location /Country	PGP Activities / Antimicrobial	Reference
21.	Bacillus, Paenibacillius, Pseudomonas, Rhizobium, Rummeliibacillus and Methylobacterium	Bacteria	Handroanthus impetiginosus 'pink lapacho	Leaf and roots tissues	South and Mesoamerica.	PGP traits	Yarte et al. 2022
22.	Bacillus amyloliquefaciens	Bacteria	Citrus spp.	Leaf and fruit tissue	Thailand	Biocontrol of citrus canker in lime.	Daungfu et al. 2019
23.	Bacillus subtilis Bacillus pumilus Bacillus amyloliquefaciens	Bacteria	Pennicetum glaucum	Root, Leaf, and Stem tissues	Uttar Pradesh, India	PGP and abiotic stress tolerance	Kushwaha et al. 2019
24.	Pantoea, Pseudomonas, Enterobacter Bacillus subtilis	Bacteria	Eleusine coracana			Plant-growth-promoting	Misganaw et al. 2019
25.	Pseudomonas, Chryseobacterium sp. and Serratia sp	Bacteria	Persea americana Avocado trees	Root samples	Michoacán, Mexico.	Biocontrol of avocado thrips	Tzec-Interián et al. 2020
26.	Pseudomonas sp. and Bacillus sp.	Bacteria	Malus domestica	Buds	Lithuania, Europe	Shoot growth and response to pathogens	Tamošiūnė et al. 2018
27.	Enterobacter sp.	Bacteria	Populus deltoidesxnigra	Cuttings	United States	Endophyte-assisted phytoremediation	Doty et al. 2017
28.	Conexibacter, Gemmatimonas, Holophaga, Luteolibacter, Methylophilus	Bacteria	Panax notoginseng	Flower, Leaf, stem, root and fiber tissues	China	Important metabolites production	Dong et al. 2018
29.	Acinetobacter sp., Klebsiella sp., Pseudomonas sp.	Bacteria	Mussaenda roxburghii	Stem, Leaf and Root tissues	Eastern Himalayan province, India	PGP and antimicrobial activities against fungal phytopathogens	Pandey et al. 2015

leaves from different citrus-growing areas showed the presence of dominant genera's namely *Bacillus subtilis, Bacillus sp., B. velezensis, B. methylotrophicus, B. pumilus, B. vallismortis, Pantoea eucrina, Proteus mirabilis* and *Sphingobium*. Endophytic bacterial diversity varied across the plant tissues, genotypes, healthy versus stressed tissues and tissue-specific endophytes have been tested for antimicrobial activities and plant growth-promoting traits were summarized (Soni et al. 2017; Chutulo and Chalannavar 2018; Kumar et al. 2018; Kumar et al. 2020c). In comparison between endophytes isolated from healthy and disease plant tissues of citrus, it was found that the *B. subtilis* was dominated species in the healthy plants and showed significant differences among the healthy and diseased with and without symptomatic trees (Munir et al. 2020).

11.3 SIGNIFICANT ROLE/BENEFITS OF MICROBIAL ENDOPHYTES IN THE HOST PLANTS

Plants have evolved to form complex, beneficial relationships with the microorganisms in their surroundings and associated with plant microbiome. Among them, endophytes are the microbes that live inside the plant tissue, causing any negative symptoms or disease in the host. These microbes are known to involve various plant developmental processes starting from the germination of seeds and development seedling, adaptation of plant under the environmental conditions, growth and development of plant. In addition to these, microbes can transfer vertically from seed to the progeny, as similar to the transfer of hereditary traits and protecting the host from the adverse effect of biotic and abiotic stresses (Goel et al. 2018a; Lata et al. 2018; Kumar et al. 2019; Kumar et al. 2020c; Dubey et al. 2020; Kumar et al. 2020a). PGP bacteria infects roots confers nutrient acquisition to the host plants (Backer et al. 2018). Production of phytohormone which are essential for plant growth and solubilization of minerals antimicrobial activities against pest and pathogens (Berg 2009; Goel et al. 2017; Turbat et al. 2020). Higher plants have been reported to the host bacterial and fungal endophytes in a symbiotic plant-microbe interaction where associated microbes provide assured benefits to their hosts in exchanging the nutrients derived from the host. Such interactions are called as symbiotic interactions/relations (Lugtenberg et al. 2002; Helpern et al. 2015; Bamisile et al. 2018; Khare et al. 2018; Goel et al. 2018b).

The bacterial endophytes transmission may occur from parent plant to seed and then to the seedlings and this process is called vertical transmission (Robinson et al. 2016). Identifying such trait-specific microbes can directly be inoculated to provide desired traits like resistance by bio-priming of seeds with desired endophytic microbes. Fungal endophytes namely *Penicillium chrysogenum, Phoma sp.,* and *Trichoderma koningii* isolated from *Opuntia spp.* are involved in seed dormancy breaking and inducing seed germination (Delgado-Sánchez et al. 2013).

Recent advances in plant microbiome research led to holobiont theory which stated that the appearance of phenotype or traits in plant is due to the genes present in the plant genome and also due to the presence of associated endophytic microbes second genome in the host which plays a significant role in the plant growth and development. A holobiont refers to an assemblage of the plant and its symbionts living and functioning as a unit of biological organization, have ability to replicate and pass on its genetic composition regarded as a unit of selection (van Opstal and Bordenstein 2015; Theis et al. 2016). Due to vertical transmission nature of plant-associated endophytes, these may be utilized in microbial-assisted breeding (Gopal and Gupta 2016).

Endophytic microbes influence plant growth and development either directly or indirectly (i) directly PGP endophytic microbes help in the acquisition of nutrients such as nitrogen, phosphorous, iron and synthesis of phytohormones. While (ii) indirectly through production of siderophore, metabolites, HCN and antibiotics against pathogenic microbes and solubilization of inorganic phosphate (Tzec-Interian et al. 2020). The role of endophytic microbes in the host plants is depicted in Figure 11.1.

Alternatively, endophytic microbes could be explored as they colonize and reside within the interior tissues of plants that are stable and protected. Their interaction with a plant can grow into a longer relationship and contribute to the networks of role in the plant growth development and health.

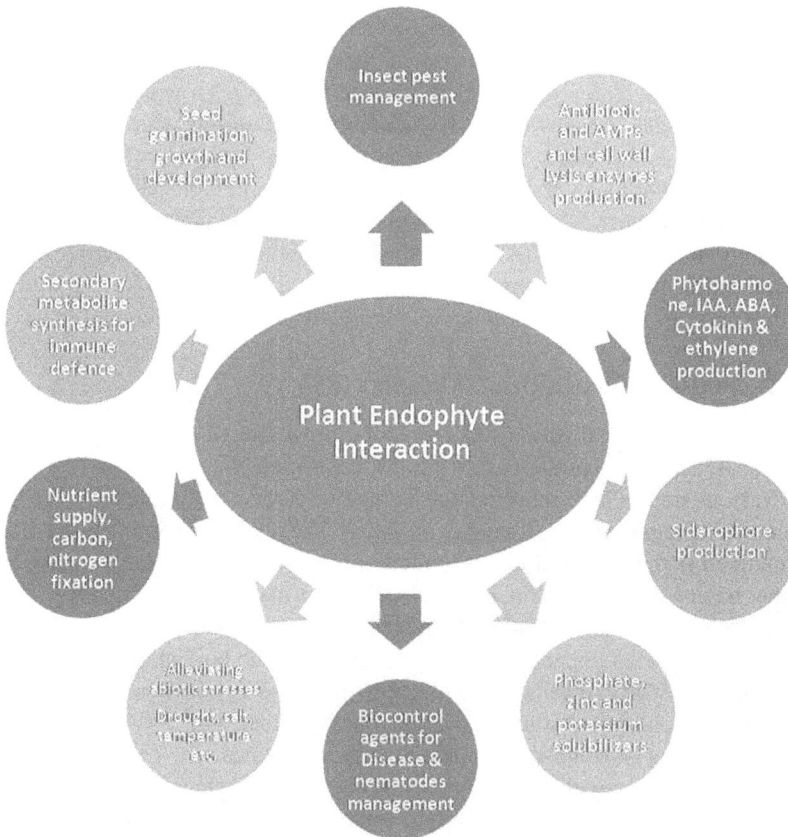

FIGURE 11.1 Schematic representation of role of endophytes and plant endophyte interaction.

11.4 PLANT GROWTH-PROMOTING ACTIVITIES

11.4.1 Nutrient Acquisition

Soil fertility or soil health is significantly influenced by soil microorganisms (Gianinazzi and Schuepp 1994). Plants serve as the primary and major source of carbon for soil microbes, through carbon-based root exudates formed during the photosynthesis or through plant residue. This process involves the transfer of nutrient transfer between symbiont and the host. Endophytic microbes include bacteria, fungi, actinorhizal and rhizobial symbioses that bring nutrients to host plants like fixing atmospheric nitrogen in plant tissues, phosphates solubilization (White et al. 2019). Nutrient transfer in plant-fungal symbiosis has been well described (Behie and Bidochka 2014). The ability of fungal endophytes to transfer nutrients/minerals to host is quite a recent discovery that needs more focused attention. In recent studies, bacterial endophytes namely *Bacillus, Paenibacillus, Rhizobium, Pseudomonas, Rummeliibacillus* and *Methylobacterium* genera have been isolated from *Handroanthus impetiginosus* a medicinal, ornamental and forestal tree native of the Americas, distributed throughout forests showed production of IAA and indole-like compounds (Yarte et al. 2022). Similarly, supply of minerals is also involved in plant growth promotion including the synthesis of uptake system of siderophores (Katiyar and Goel 2004).

Plants have the ability to absorb two inorganic forms nitrogen viz., nitrate and ammonium directly from soils and are essential for crop growth and yield (Hao et al. 2020). Fungal endophytes have a symbiotic relationship with the host plants and ammonium is the major form of nitrogen released by symbiotic fungi (Behie and Bidochka 2014; Chen et al. 2018). Recent advances in molecular microbiology showed that plants grown in flooded or acidic soils have a major presence of ammonium; ammonium transporter (AMT)-mediated acquisition of ammonium from soils is crucial for the nitrogen requirement of the plants.

11.4.2 Phosphate-Solubilizing Microbes (PSM)

Phosphorus plays a vital role in the metabolic processes such as macromolecular biosynthesis, cellular respiration, signal transduction and photosynthesis (Yarte et al. 2022) Various studies have reported that about 49.3 percent of cultivated land is deficient in available phosphate (Zhang et al. 2018). The applied phosphorus fertilizers become fixed in the soil resulting in limited or non-accessible to plants. Endophytic microbes having phosphate-solubilizing activities normally enhance the accessibility of phosphorus (P) released from soil to the plants particularly in phosphorus-limiting soils. These microbes enhances phosphate uptake by the host plant and providing protection against microbial plant pathogens, inducing systemic resistance in the host plant. The phosphate-solubilizing microbes enhance soil fertility due to their potential to convert insoluble phosphate to soluble phosphate by producing organic acids, ion exchange and chelation (Nahas 1996). The endophytic microbes isolated from tree species have shown potential to solubilize available phosphate to the forestry plants. It was noticed that the bacterial endophytes isolated from *Handroanthus impetiginosus*, a medicinal forest tree, have been found to solubilize the phosphate and zinc (Yarte et al. 2022). Other studies have reported the importance of bacterial endophytes in solubilizing phosphate, and endophytes *Bacillus* and *Enterobacter* showed a potential to solubilize phosphate in rice (Kumar et al. 2019; Kumar et al. 2020c).

11.4.3 PHYTOHORMONE PRODUCTION/SYNTHESIS

There are five major classes of phytohormones, *viz.*, auxins, cytokinins, gibberellins, ethylene and abscisic acid. Auxins and cytokinins play a crucial role in initiation and development of root architecture, lateral roots, flower morphogenesis, apical dominance and regulation of vascular differentiation (Aloni et al. 2006). Auxin also plays a crucial role in the establishment and maintenance of beneficial plant-endophytic microbe interaction (Yarte et al. 2022). While ethylene and abscisic acid are considered as stress hormones required for plant to mitigate environmental stresses. These hormones are normally synthesized inside the plant tissues during the different plant developmental stages. In addition, endophytic microbes which have the ability to produce phytohormones inside the host plants in turn help in the growth and development of the host plant. These microbes have been identified to produce different phytohormone such as auxin, cytokinins, gibberellins and other defence-related hormones and bioactive compounds produced in the plants. Several microbes synthesize ethylene, a potent regulator of plant growth and development, and they promote plant growth by keeping the level of ethylene in plant low, thus preventing harmful effects of high ethylene concentrations such as inhibition of root elongation in legume crops (Glick 1995; Gamalero and Glick 2015). Recent studies have further proved the role of endophytic microbe in synthesis of phytohormones. Endophytes Fungi *Fusarium*, *Didymella and Alternaria* isolated from *Sophora flavescen* showed significantly produced indole acetic acid (IAA) either *via* the Tryptophan-dependent or Tryptophan-independent pathways (Turbat et al. 2020). Endophytic fungus *Preussia* sp. isolated from *Boswellia sacra*, an economically important frankincense-producing tree that showed enzymes and IAA production, helped in the promotion of plant growth through increasing length of shoots, internodes and leaves (Khan et al. 2016).

11.4.4 SIDEROPHORE PRODUCTION

Siderophore production is a main attribute of PGP endophytes and enables plant growth under iron-limiting conditions through iron sequestration. An extracellular siderophores may be produced by plant-associated microbes in both mutualistic and antagonistic manner. Bacterial endophytes known to secrete siderophores in soil while interacting with host plants, where siderophores chelate iron from the environment for use by plant and microbial cells (Hardoim et al. 2015). Siderophore production contributes to an uptake of enhanced nutrient by the host plants. Siderophores also help in controlling the growth of pathogenic microbes by depriving them iron (Saha et al. 2016).

11.4.5 PRODUCTION OF BIOACTIVE COMPOUNDS

Microorganisms are well known for the key source of biologically active compounds/ small molecules and able to produce >20,000 compounds/molecules, which influences the fitness, performance and survival of other organisms (Demain and Sanchez 2009). Secondary metabolites are very useful and potential source of immunosuppressive, antidiabetic, antioxidant, antifungal, antiviral agents and anticancer drugs (Gunatilaka 2006). Several studies have shown that endophytic microbes living inside the plants can produce novel biologically active compounds and hence have a high potential for agricultural, pharmaceutical, and other industrial applications (Köberl et al. 2013). Majority of the bioactive compounds produced by endophytic microbes are essential for enhancing the adaptability of both endophytic partners (bacteria and fungi) and their host plants, tolerance to environmental stresses, defence signalling and establishment of symbiosis and their genetic regulation (Sun et al. 2014; Hardoim et al. 2015; Kumar et al. 2020b; Kumar et al. 2020a). The plant-associated microbes emits volatile compounds further restricting the growth and proliferation of fungal pathogens or phytopathogens (Bailly and Weisskopf 2017; Garbeva and Weisskopf 2020). Recently, it was found that *Bacillus cereus* and *B. subtilis* isolated from grapevine produced volatile compounds which inhibited the growth of fungal pathogen, *Phytophthora infestans* (Bruisson et al. 2019). In addition to the direct effect of bioactive compounds, they have been found to induce or provoke synthesis of novel bioactive compounds/metabolites or enhance the synthesis of secondary metabolites which may be exploited as important medicinal resources for pharmaceutical purposes (Zhang et al. 2006; Firáková et al. 2007; Rodriguez et al. 2009). The study by Teimoori-Boghsani et al. (2020) showed the endophytes associated with *Salvia abrotanoides* found to induce synthesis of cryptotanshinone by the plant and independent of the host. Pandey et al. (2016) showed that the bacterial endophytes of *Papaver somniferum* namely *Marmoricola sp.* and *Acinetobacter sp.* upregulate the expression of genes involved in the alkaloid benzylisoquinoline biosynthesis. Host plants can be induced to produce desired metabolites of agricultural or pharmaceutical importance like drugs for treating cancer (Khare et al. 2018). It is evident from several studies that the endophytes have the ability to hijack the host's metabolic setups for the production of bioactive compounds in higher quantities and providing a basis for production of secondary metabolites of agricultural and pharmaceutical importance (Morelli et al. 2020). Endophytic fungi belonging to the genera, namely *Alternaria, Botryodiplodia, Botrytis, Aspergillus, Cladosporium* isolated from *Taxus baccata*, have taxol-producing ability being used for production of taxol, an anticancer drug (Heinig and Jennewein 2009).

11.5 RESISTANCE/TOLERANCE MECHANISM IN PLANTS

11.5.1 BIOTIC STRESS TOLERANCE

A very complex network governs stress and developmental responses in the plant system. Biotic stress includes all the living organisms including bacteria, fungi, virus, and nematodes, which are causing disease in crop plants. Plant usually adopt defence mechanism that includes Systemic Acquired Resistance (SAR) and Induced

Systemic Resistance (ISR) to combat the adverse effect of pathogens. SAR is induced by the plant pathogenic microbes whereas the ISR is mediated by the beneficial rhizospheric microbiota like *Pseudomonas, Paenibacillius* and *Trichoderma spp.* Endophytes play a crucial role in mitigating biotic stresses by inducing ISR against pathogen and/or suppressing the growth of phytopathogens, improving agricultural productivity (Gunatilaka 2006; Kloepper and Ryu 2006). Plant hormones namely Jasmonic acid and Salicylic acid play crucial roles during plant stress response (Naseem et al. 2015; Khare et al. 2016). Recent studies have shown the endophyte-mediated resistance (EMR) which seems to be different from ISR and SAR, as salicylic acid, jasmonate and ethylene are not involved (Pieterse et al. 2014; Constantin et al. 2019).

PGPR, endophytic microbes and mycorrhizaa is widely used to control or manage the insect pest and diseases. Biocontrol is a process in which one or more beneficial microorganisms in the rhizosphere unfavourably affect the survival or activity of a plant pathogen. Endophytic microbes typically cover the same ecological niches as occupied by plant phytopathogens and proposed as biocontrol agents that could be applied as an alternative to chemical pesticides (Goel at al. 2018b; Dash et al. 2019). Biocontrol of pathogens and insect pests may be controlled through competition for nutrients, the production of antimicrobial metabolites and activation of immune responses in plant (Islam et al. 2015; Desgarennes et al. 2020). Previous studies have reported the role of tree endophytes for potential as biological control agents (Arnold et al. 2003; Clay 2004). Endophytic microbes mainly adopted different strategies to combat against pathogenic microbes which includes (i) competition; (ii) production of cell wall lysis enzyme; (iii) production of antibiotics and (iv) production of lipopeptides or antimicrobial peptides (AMPs). Antimicrobial compounds play an important role in the suppression of phytopathogens by antagonistic microorganisms (Raaijmakers et al. 2002).The endophytic bacterial are known to produce lipopeptides as an antimicrobial substance. Kumar et al. (2020c) identified the presence of surfactin and iturin in the rice bacterial endophytes having antimicrobial activities. Microbial endophytes are considered as potential strategies for sustainable agriculture.

11.5.2 Abiotic Stress Tolerance

Abiotic stresses or environmental stresses include drought, heat, salinity, heavy metal, and oxidative stress, which are considered as severe threats to agro-ecosystems (Khare and Arora 2015; Goel et al. 2018a; Goel et al. 2018b). Endophytic microbes' association with plants offers numerous benefits ranging from seed germination, promotion of plant growth to development and tolerance of biotic and abiotic stresses. These stresses cause several morpho-biochemicals, physiological and molecular alterations which ultimately lead to loss of productivity and degradation of ecosystems (Wang et al. 2003). The plant-beneficial microbe interaction has mutual effect on both the partners (Phillips et al. 2004). Studies have reported the role of endophytes in stress tolerance. Yin et al. (2014) showed the endophytes *Epichloe spp.* to play an important role in increasing host plants' resistance to abiotic stresses. Several studies have also reported that plants under salt in the presence of endophytes showed an enhanced antioxidant enzyme activity which facilitates plants to tolerate salt stress (Hashem et al. 2014; Ahmad et al. 2015; Yasmeen and Siddiqui 2017). Several studies have confirmed the involvement of bacterial and fungal endophytes in mitigating environmental stresses by producing certain anti-oxidants, metabolites, enzymes, synthesis of stress hormones like ethylene, ABA and compatible solutes in the host plants. They are also found to induce or enhance the expression of genes related to stress-mitigating pathways.

11.6 CONCLUSION

Due to ever-increasing global warming and excessive use of chemical and synthetic pesticides led to environmental pollution. Microbial-based bio-fertilizers and bio-pesticide and their formulations seem to be the key components to improve crop productivity in sustainable eco-friendly agro-ecosystems. The usage of microbial endophytes has been proposed as a promising biotechnological tool for improvement of quality and yields of agricultural and forest crops under biotic and abiotic stress. These is an urgent need to explore endophytes from extreme, diverse habitats, high altitude and forest regions and grasses to identify their potential in producing phytohormone, producing natural products like antibiotics, lipopeptides and other bioactive compounds which can be effective against multi-drug-resistant bacteria.

REFERENCES

Aloni, R., E. Aloni, M. Langhans, and C. I. Ullrich. 2006. Role of auxin in regulating Arabidopsis flower development. *Planta*. 223(2): 315–328.
Andhale, N. B., M. Shahnawaz, and A. B. Ade. 2019. Fungal endophytes of *Plumbago zeylanica L.* enhances plumbagin content. *Botan. Stud.* 60(1): 21. https://doi.org/10.1186/s40529-019-0270-1.
Arivudainambi, U. S., T. D. Anand, V. Shanmugaiah, C. Karunakaran, and A. Rajendran 2011. Novel bioactive metabolites producing endophytic fungus *Colletotrichum gloeosporioides* against multidrug-resistant *Staphylococcus aureus*. *FEMS Immunol. Med. Microbiol.* 61: 340–345.
Arnold, A. E., L. C. Mejia, D. Kyllo, E. I. Rojas, Z. Maynard, N. Robbins, and E. A. Herre. 2003. Fungal endophytes limit pathogen damage in a tropical tree. *Proc. Natl. Acad. Sci. USA* 100: 15649–15654. doi: 10.1073/pnas.2533483100.
Aziz, M. H., N. E. Dreckschmidt, and A. K. Verma. 2008. Plumbagin, a medicinal plant derived naphthoquinone, is a novel inhibitor of the growth and invasion of hormone-refractory prostate cancer. *Cancer Res.* 68: 9024–9032.
Backer, R., J. S. Rokem, G. Ilangumaran, J. Lamont, D. Praslickova, E. Ricci, S. Subramanian, and D. L. Smith. 2018. Plant growth-promoting rhizobacteria: Context, mechanisms of action, and roadmap to commercialization of biostimulants for sustainable agriculture. *Front. Plant Sci.* 9: 1473. doi: 10.3389/fpls.2018.01473.
Bailly, A., and L. Weisskopf. 2017. Mining the volatilomes of plant-associated microbiota for new biocontrol solutions. *Front. Microbiol.* 8: 1638. doi: 10.3389/fmicb.2017.01638.
Bamisile, B. S., C. K. Dash, K. S. Akutse, R. Keppanan, and L. Wang. 2018. Fungal endophytes: Beyond herbivore management. *Front. Microbiol.* 9: 1–11.
Bayman, P., L. L. Lebron, and R. L. Tremblay. 1997. Variation in endophytic fungi from roots and leaves of *Lepanthes* (*Orchidaceae*). *New Phytol.* 135: 143e–149.
Behie, S. W., and M. J. Bidochka. 2014. Nutrient transfer in plant–fungal symbioses. *Trends Plant Sci.* 19: 734–740.

Berg, G. 2009. Plant-microbe interactions promoting plant growth and health: perspectives for controlled use of microorganisms in agriculture. *Appl. Microbiol. Biotechnol.* 84: 11–18.

Bhardwaj, A., D. Sharma, N. Jadon, and P. K. Agrawal. 2015. Antimicrobial and phytochemical screening of endophytic fungi isolated from spikes of *Pinus roxburghii. Arch. Clin. Microbiol.* 6(3): 1–9.

Bilal, L., S. Asaf, M. Hamayun, H. Gul, A. Iqbal, I. Ullah, I. J. Lee, and A. Hussain. 2018. Plant growth promoting endophytic fungi *Aspergillus fumigatus* TS1 and *Fusarium proliferatum* BRL1 produce gibberellins and regulates plant endogenous hormones. *Symbiosis.* 76: 117–127.

Bruisson, S., M. Zufferey, F. L'Haridon, E. Trutmann, A. Anand, A. Dutartre, M. De-Vrieze, and L. Weisskopf. 2019. Endophytes and epiphytes from the Grapevine leaf microbiome as potential biocontrol agents against Phytopathogens. *Front. Microbiol.* 10: 2726. doi: 10.3389/fmicb.2019.02726.

Chen, A., M. Gu, S. Wang, J. Chen, and G. Xu. 2018. Transport properties and regulatory roles of nitrogen in arbuscular mycorrhizal symbiosis. *Semin. Cell Dev. Biol.* 74: 80–88.

Chowdhary, K., and N. Kaushik. 2015. Fungal endophyte diversity and bioactivity in the Indian medicinal plant *Ocimum sanctum* linn. *PLoS ONE.* 10(11): e0141444. doi: 10.1371/journal.pone.0141444.

Chutulo, E. C., and R. K. Chalannavar. 2018. Endophytic Mycoflora and their bioactive compounds from *Azadirachta Indica*: A comprehensive review. *J Fungi (Basel).* 4(2): 42.

Clay, K. 2004. Fungi and the food of the gods. *Nature.* 427: 401–402.

Constantin, M.E., F.J. de Lamo, B.V. Vlieger, M. Rep, and Takken, F.L.W. 2019. Endophyte-mediated resistance in Tomato to *Fusarium oxysporum* is independent of ET, JA, and SA. *Front. Plant Sci.* 10:979. doi: 10.3389/fpls.2019.00979.

Dash, B., R. Soni, V. Kumar, D. C. Suyal, D. Dash, and R. Goel. 2019. Mycorrhizosphere: Microbial interactions for sustainable agricultural production. In: *Mycorrhizosphere and Pedogenesis.* A. Varma, and D. Choudhary (eds.), pp. 321–338. Singapore: Springer. doi: https://doi.org/10.1007/978-981-13-6480-8_18.

Daungfu, O., S. Youpensuk, and S. Lumyong. 2019. Endophytic bacteria isolated from citrus plants for biological control of citrus canker in lime plants. *Trop. Life Sci. Res.* 30: 73.

Delgado-Sánchez, P., J. F. Jiménez-Bremont, M. L. Guerrero-González, and J. Flores. 2013. Effect of fungi and light on seed germination of three Opuntia species from semiarid lands of central Mexico. *J. Plant Res.* 126: 643–649. doi: 10.1007/s10265-013-0558-2.

Demain, A. L., and S. Sanchez. 2009. Microbial drug discovery: 80 years of progress. *J. Antibiot.* 62: 5–16.

Desgarennes, D., G. Carrio, J. L. Monribot, J. A. Tzec-interia, O. Ferrera-rodri, and D. L. Santos. 2020. Characterization of plant growth-promoting bacteria associated with avocado trees (Persea americana Miller) and their potential use in the biocontrol of Scirtothrips perseae (avocado thrips). *PLoS ONE.* 15(4): e0231215. https://doi.org/10.1371/journal.pone.0231215.

Dhayanithy, G., K. Subban, and J. Chelliah. 2019. Diversity and biological activities of endophytic fungi associated with *Catharanthus roseus. BMC Microbiol.* 19: 22.

Dini-Andreote, F. 2020. Endophytes: The second layer of plant defense. *Trends Plant Sci.* 25(4): 319–322. doi: 10.1016/j.tplants.2020.01.007.

Dong, L., R. Cheng, L. Xiao, F. Wei, G. Wei, J. Xu, Y. Wang, X. Guo, Z. Chen, and S. Chen. 2018. Diversity and composition of bacterial endophytes among plant parts of *Panax notoginseng. Chin Med.* 13: 41.

Doty, S. L., J. L. Freeman, C. M. Cohu, J. G. Burken, A. Firrincieli, A., Simon et al. (2017). Enhanced degradation of TCE on a superfund site using endophyte assisted poplar tree phytoremediation. *Environ. Sci. Technol.* 51: 10050–10058. doi: 10.1021/acs.est.7b01504.

Dubey, P., V. Kumar, K. Ponnusamy, R. Sonwani, A. K. Singh, D. C. Suyal, and R. Soni. 2020. Microbe assisted plant stress management. In: *The Recent Advancements in Microbial Diversity*, 1st edition. S. De Mandal and P. Bhatt (eds.), pp. 351–378. Cambridge, MA: Academic Press.

Firáková, S., M. Šturdíková, and M. Múčková. 2007. Bioactive secondary metabolites produced by microorganisms associated with plants. *Biologia*. 62: 251–257.

Fouda, A. H., S. E. D. Hassan, A. Mohamed Eid, E. El-Din Ewais. 2015. Biotechnological applications of fungal endophytes associated with medicinal plant *Asclepias sinaica* (Bioss.), *Ann. Agric. Sci.* 60(1): 95–104. https://doi.org/10.1016/j.aoas.2015.04.001.

Hao, D. L., J. Y. Zhou, S. Y. Yang, W. Qi, K. J. Yang, and Y. H. Su. 2020. Function and regulation of ammonium transporters in plants. *Int. J. Mol. Sci.* 21(10): 3557. https://doi.org/10.3390/ijms21103557.

Hashem, A., A. A. Alqarawi, R. Radhakrishnan, A. F. Al-Arjani, H. A. Aldehaish, D. Egamberdieva, and E. F. Abd Allah. 2018. Arbuscular mycorrhizal fungi regulate the oxidative system, hormones and ionic equilibrium to trigger salt stress tolerance in *Cucumis sativus L. Saudi J Biol. Sci.* 25(6): 1102–1114. https://doi.org/10.1016/j.sjbs.2018.03.009.

Gamalero, E., and B. R. Glick. 2015. Bacterial modulation of plant ethylene levels. *Plant Physiol.* 169: 13–22.

Garbeva, P., and L. Weisskopf. 2020. Airborne medicine: bacterial volatiles and their influence on plant health. *New Phytol.* 226(1): 32–43. doi: 10.1111/nph.16282.

Gherbawy, Y. A., and H. M. Elhariry. 2014. Endophytic fungi associated with high-altitude *Juniperus* trees and their antimicrobial activities. *Plant Biosystems.* 131–140. https://doi.org/10.1080/11263504.2014.984011.

Gianinazzi, S., and H. Schuepp. 1994. *Impact of Arbuscular Mycorrhizas on Sustainable Agriculture and Natural Ecosystems*, p. 226. Basel: Birkhäuser Verlagp.

Glick, B. R. 1995. The enhancement of plant growth by free-living bacteria. *Can. J. Microbiol.* 41: 109–117. doi: 10.1139/m95-015.

Goel, R., D. C. Suyal, V. Kumar, L. Jain, and R. Soni. 2018a. Stress-tolerant beneficial microbes for sustainable agricultural production. In: *Microorganisms for Green Revolution. Microorganisms for Sustainability.* D. Panpatte, Y. Jhala, H. Shelat, and R. Vyas (eds.), vol. 7, pp. 141–159. Singapore: Springer. doi: https://doi.org/10.1007/978-981-10-7146-1_8.

Goel, R., V. Kumar, D. C. Suyal, Narayan, and R. Soni. 2018b. Toward the unculturable microbes for sustainable agricultural production. In: *Role of Rhizospheric Microbes in Soil.* V. Meena (ed.). Singapore: Springer. doi: https://doi.org/10.1007/978-981-10-8402-7_4.

Goel, R., V. Kumar, D. K. Suyal, B. Dash, P. Kumar, and R. Soni. 2017. Root-associated bacteria: Rhizoplane and endosphere. In: *Plant-Microbe Interactions in Agro-Ecological Perspectives.* D. Singh, H. Singh, and R. Prabha (eds.). Singapore: Springer. doi: https://doi.org/10.1007/978-981-10-5813-4_9.

Gond, S. K., V. C. Verma, A. Kumar, V. Kumar, and R. N. Kharwar. 2007. Study of endophytic fungal community from different parts of *Aegle marmelos* Correae (*Rutaceae*) from Varanasi (India). *World J. Microbiol. Biotechnol.* 23(10): 1371–1375.

Gopal, M., and A. Gupta. 2016. Microbiome selection could spur next-generation plant breeding strategies. *Front. Microbiol.* 7: 1971.

Gunatilaka, A. A. L. 2006. Natural products from plant-associated micro-organisms: distribution, structural diversity, bioactivity, and implications of their occurrence. *J. Nat. Prod.* 69: 509–526.

Hardoim, P. R., L. S. vanOverbeek, G. Berg, A. M. Pirttila, S. Compant, A. Campisano, M. Doring, and A. Sessitsch. 2015. The hidden world within plants: Ecological and evolutionary considerations for defining functioning of microbial endophytes. *Microbiol. Mol. Biol. Rev.* 79: 293–320.

Heinig, U., and S. Jennewein. 2009. Taxol: A complex diterpenoid natural product with an evolutionarily obscure origin. *Afr. J. Biotechnol.* 8: 1370–1385.

Islam, S., A. M. Akanda, A. Prova, M. T. Islam, and M. M. Hossain. 2015. Isolation and identification of plant growth promoting rhizobacteria from cucumber rhizosphere and their effect on plant growth promotion and dis- ease suppression. *Front. Microbiol.* 6: 1360. https://doi.org/10.3389/fmicb.2015.01360.

Kapoor, N., P. Rajput, M. A. Mushtaque, and L. Gambhir. 2018. Bio-prospecting fungal endophytes of high altitude medicinal plants for commercially imperative enzymes. *Biosci. Biotech. Res. Comm.* 1(3): 370–375.

Katiyar, V., and R. Goel. 2004. Siderophore mediated plant growth promotion at low temperature by mutant of fluorescent pseudomonad. *Plant Growth Regul.* 42: 239–244.

Khan, A. L., A., Al-Harrasi, A. Al-Rawahi, Z. Al-Farsi, A. Al-Mamari, M. Waqas et al. 2016. Endophytic Fungi from frankincense tree improves host growth and produces extracellular enzymes and indole acetic acid. *PLoS ONE.* 11(6): e0158207.

Khare, E. K., and Arora, N. K. 2015. Effects of soil environment on field efficacy of microbial inoculants. In: *Plant Microbes Symbiosis: Applied Facets.* N. K. Arora (ed.), pp. 353–381. Netherland: Springer.

Khare, E., J. Mishra, and N. K. Arora. 2018. Multifaceted interactions between endophytes and plant: developments and prospects *Front. Microbial.* 9: 2732. https://doi.org/10.3389/fmicb.2018.02732.

Khare, E., K. M. Kimand, and K. J. Lee. 2016. Rice OsPBL1 (*Oryza sativa* Arabidopsis PBS1-LIKE 1) enhanced defense of Arabidopsis against *Pseudomonas syringae* DC3000. *Eur. J. Plant Pathol.* 146: 901–910.

Kharwar, R. N., V. C. Verma, A. Kumar, S. K. Gond, J. K. Harper, W. M. Hess, E. Lobkovosky, C, Ma, Y. H. Ren, and G. A. Strobel. 2009. Javanicin, an antibacterial naphthaquinone from an endophytic fungus of Neem, *Chloridium sp. Curr. Microbiol.* 58: 233–238.

Kloepper, J. W., and C. M. Ryu. 2006. Bacterial endophytes as elicitors of induced systemic resistance. In: *Microbial Root Endophytes.* B. J. E. Schulz, C. J. C. Boyle, and T. N. Sieber (eds.), pp. 33–52. Berlin, Germany, Springer-Verlag. doi: 10.1007/3-540-33526-9_3.

Köberl, M., R. Schmidt, E. M. Ramadan, R. Bauer, and G. Berg. 2013. The microbiome of medicinal plants: Diversity and importance for plant growth, quality and health. *Front. Microbiol.* 4: 400. doi: 10.3389/fmicb.2013.00400.

Kumar, V., L. Jain, S. K. Jain, and P. Kaushal. 2018. Metagenomic analysis of endophytic and rhizospheric bacterial diversity of Lathyrus sativus using illumina based sequencing. In: Proceeding of International Conference on Microbiome Research, Pune, with Abstract ID 2018EVM0152.

Kumar, V., L. Jain, P. Kaushal, and R. Soni. (2020a). Fungal endophytes and their applications as growth promoters and biological control agents. In: *Fungi Bio-Prospects in Sustainable Agriculture, Environment and Nano-Technology.* V. K. Sharma, M. P. Shah, S. Parmar, and A. Kumar (eds.), pp. 315–337. Cambridge, MA: Academic Press (Elsevier). doi: https://doi.org/10.1016/B978-0-12-821394-0.00012-3.

Kumar, V., L. Jain, R. Soni, P. Kaushal, and R. Goel. 2020b. Bio-prospecting of endophytic microbes from higher altitude plants: Recent advances and their biotechnological applications. In: *Microbiological Advancements for Higher Altitude Agro-Ecosystems & Sustainability. Rhizosphere Biology.* R. Goel, R. Soni, and D. Suyal (eds.), pp. 375–392. Singapore: Springer. https://doi.org/10.1007/978-981-15-1902-4_18.

Kumar, V, L. Jain, S. K. Jain, S. Chaturvedi, and P. Kaushal. 2020c. Bacterial endophytes of rice (*Oryza sativa* L.) and their potential for plant growth promotion and antagonistic activities. *S. Afr. J. Bot.* 34: 50–63.

Kumar, V., R. Soni, L. Jain, B. Dash, and R. Goel. 2019. Endophytic fungi: Recent advances in identification and explorations. In: *Advances in Endophytic Fungal Research. Fungal Biology*. B. Singh (ed.), pp. 267–281. Cham: Springer. doi: https://doi.org/10.1007/978-3-030-03589-1_13.

Kushwaha, P., P. L. Kashyap, P. Kuppusamy, A. K. Srivastava, and R. K. Tiwari. 2019. Functional characterization of endophytic bacilli from pearl millet (*Pennisetum glaucum*) and their possible role in multiple stress tolerance. *Plant Biosys*. 154(4): 503–514.

Lata, R., S. Chowdhury, S. K. Gond, and J. F. Jr. White. 2018. Induction of abiotic stress tolerance in plants by endophytic microbes. *Lett. Appl. Microbiol*. 66: 268–276.

Li, Y., Y. C. Song, J. Y. Liu, Y. M. Ma, and R. X. Tan. 2005. Anti-helicobacter pylori substances from endophytic fungal cultures. *World J. Microbiol. Biotechnol*. 21: 553–558.

Lugtenberg, B. J. J., T. F. C. Chin-A-Woeng, and G. V. Bloemberg. 2002. Microbe–plant interactions: Principles and mechanisms. *Antonie Van Leeuwenhoek*. 81: 373–383.

Lupo, S., S. Tiscornia, and L. Bettucci. 2001. Endophytic fungi from flowers, capsules and seeds of *Eucalyptus globules*. *Rev. Iberoam. Microl*. 18: 38e41.

Ma, Y. M., Y. Li, J. Y. Liu, Y. C. Song, and R. X. Tan. 2004. Anti-*Helicobacter pylori* metabo- lites from *Rhizoctonia sp*. Cy064, an endophytic fungus in *Cynodon dactylon*. *Fitoterapia*. 75: 451–456.

Mane, R. S. and A. B. Vedamurthy. 2019. The fungal endophytes: Sources and future prospects. *J. Med. Plants Stud*. 6(2): 121–126.

Mercado-Blanco, J., and B. J. J. Lugtenberg. 2014. Biotechnological applications of bacterial endophytes. *Curr. Biotechnol*. 3: 60–75.

Misganaw, G., A. Simachew, and A. Gessesse. 2019. Endophytes of finger millet (*Eleusine coracana*) seeds. *Symbiosis*. 78: 203–213.

Morelli, M., O. Bahar, K. K. Papadopoulou, and D. L. Hopkins. 2020. Role of endophytes in plant health and defense against pathogens. *Front. Plant Sci*. 11: 1–5. https://doi.org/10.3389/fpls.2020.01312.

Munir, S., Y. Li, P. He, M. Huang, P. He, P. He et al. 2020. Core endophyte communities of different citrus varieties from citrus growing regions in China. *Sci. Rep*. 1–12. https://doi.org/10.1038/s41598-020-60350-6.

Nahas, E. 1996. Factors determining rock phosphate solubilization by microorganisms isolated from soil. *World J. Microbiol. Biotech*. 12: 567–572.

Naseem, M., M. Kaltdorf, and T. Dandekar. 2015. The nexus between growth and defence signalling: Auxin and cytokinin modulate plant immune response pathways. *J. Exp. Bot*. 66: 4885–4896. https://doi.org/10. 1093/jxb/erv297.

Omomowo, O. I., and O. O. Babalola. 2019. Bacterial and fungal endophytes: Tiny giants with immense beneficial potential for plant growth and sustainable agricultural productivity. *Microorganisms*. 7(11): 481.

Pandey, P. K., R. Samanta, and R. N. S. Yadav. 2015. Plant beneficial endophytic bacteria from the ethnomedicinal *Mussaenda roxburghii* (Akshap) of Eastern Himalayan Province, India. *Adv. Biol.*,Article ID 580510, 8. Doi:10.1155/2015/580510.

Pandey, S. S., S. Singh, C. S. Babu, K., Shanker, N. K. Srivastava, and A. Kalra. 2016. Endophytes of opium poppy differentially modulate host plant productivity and genes for the biosynthetic pathway of benzylisoquinoline alkaloids. *Planta*. 43: 1097–1114.

Patil, M. P., R. H. Patil, S. G., Patil, and V. L. Maheshwari. 2014. Endophytic mycoflora of Indian medicinal plant, *Terminalia arjuna* and their biological activities. *Int. J. Biotechnol. Wellness Ind*. 3: 53.

Petrini, O. 1986. Taxonomy of endophytic fungi of aerial plant tissues. In: Fokkema NJ, van den Huevel J (eds) Microbiology of the phyllosphere. Cambridge University Press, Cambridge, pp. 175–187

Phillips, D. A., T. C. Fox, M. D. King, T. V. Bhuvaneswari, and L. R Teubner. 2004. Microbial products trigger amino acid exudation from plant roots. *Plant Physiol*. 136: 2887–2894.

Pieterse, C. M. J., C. Zamioudis, R. L. Berendsen, D. M. Weller, S. C. M. Van Wees, and P. A. H. M. Bakker. 2014. Induced systemic resistance by beneficial microbes. *Annu. Rev. Phytopathol.* 52: 347–375.

Qadri, M., S. Johri, and B. A. Shah. et al. 2013. Identification and bioactive potential of endophytic fungi isolated from selected plants of the Western Himalayas. *SpringerPlus.* 2: 8.

Quine, C. P., C. Cahalan, A. Hester, J. Humphrey, K. Kirby, A. Moffat, and G. Valatin. 2011. Chapter 8 Woodlands in UK National Ecosystem Assessment Technical Report. Cambridge: UNEP-WCMC.

Rabiey, M., L. E. Hailey, S. R. Roy, K. Grenz, M. A. S. Al-Zadjali, G. A. Barrett, and R. W. Jackson. 2019. Endophytes vs tree pathogens and pests: Can they be used as biological control agents to improve tree health? *Eur. J. Plant Pathol.* 155: 711–729.

Raaijmakers, J. M., M. Vlami, and J. T. de Souza. 2002. Antibiotic production by bacterial biocontrol agent. *Anton van Leeuwenhoek.* 81: 537–547.

Robinson, R. J., B. A. Fraaije, I. M. Clark, R. W. Jackson, P. R. Hirsch, and T. H. Mauchline. 2016. Wheat seed embryo excision enables the creation of axenic seedlings and Koch's postulates testing of putative bacterial endophytes. *Sci Rep.* 6: 25581. doi: 10.1038/srep25581.

Rondot, Y., and A. Reineke. 2018. Endophytic Beauveria bassiana in grapevine Vitis vinifera (L.) reduces infestation with piercing-sucking insects. *Biol. Control.* 116: 82–89.

Rodriguez, R. J., J. F. Jr. White, A. E. Arnold, and R. S. Redman. 2009. Fungal endophytes: Diversity and functional roles. *New Phytol.* 182: 314–330.

Rosenblueth, M. and E. Martínez-Romero. 2006. Bacterial endophytes and their interactions with hosts. *Mol. Plant-Microbe Interact.* 19: 827–837. doi: 10.1094/MPMI-19-0827.

Roy, S., and D. Banerjee. 2018. Diversity of Endophytes in Tropical Forests. In: *Endophytes of Forest Trees. Forestry Sciences.* A. Pirttilä, and A. Frank (eds.), vol 86, pp. 43–62. Cham: Springer. https://doi.org/10.1007/978-3-319-89833-9_3.

Rubini, M. R., R. T. Silva-Ribeiro, A. W. V. Pomella, C. S. Maki, W. L. Araújo, D. R. Dos Santos et al. 2005. Diversity of endophytic fungal community of cacao (Theobroma cacao L.) and biological control of *Crinipellis perniciosa*, causal agent of Witches' Broom Disease. *Int. J. Biol. Sci.* 1: 2433.

Sabra, M., A. Aboulnasr, P. Franken, E. Perreca, L. P. Wright, and I. Camehl. 2018. Beneficial root endophytic fungi increase growth and quality parameters of sweet basil in heavy metal contaminated soil. *Front Plant Sci.* 9: 1726. doi: 10.3389/fpls.2018.01726.

Saha, M., S. Sarkar, B. Sarkar, B. K., Sharma, S. Bhattacharjee, and P. Tribedi. 2016. Microbial siderophores and their potential applications: A review. *Environ. Sci. Pollut. Res.* 23: 3984–3999.

Saikkonen, K. 2007. Forest structure and fungal endophytes. *Fungal Biol. Rev.* 21: 67–74. https://doi.org/10.1016/j.fbr.2007.05.001.

Santoyo, G., G. Moreno-Hagelsieb, M. del Carmen Orozco-Mosqueda, and B. R. Glick. 2016. Plant growth-promoting bacterial endophytes. *Microbiol. Res.* 183: 92–99.

Shi, Y. W., C. Li, H. M. Yang, T. Zhang, Y. Gao, J. Zeng, Q. Lin, O. Mahemuti, Y. G. Li, X. Huo, and K. Lou. 2016. Endophytic fungal diversity and space-time dynamics in sugar beet. *Eur. J. Soil Biol.* 77: 77–85.

Singh, D. K., V. K. Sharma, J. Kumar, A. Mishra, S. K. Verma, T. N. Sieber et al. 2017. Diversity of endophytic mycobiota of tropical tree *Tectona grandis* Linn.f.: Spatiotemporal and tissue type effects. *Sci. Rep.* 7(1): 3745.

Soni, R., V. Kumar, D. C. Suyal, L. Jain, R. Goel. 2017. Metagenomics of Plant Rhizosphere Microbiome. In: *Understanding Host-Microbiome Interactions—An Omics Approach.* R. Singh, R. Kothari, P. Koringa, and S. Singh (eds.), pp. 193–205. Singapore: Springer. doi: https://doi.org/10.1007/978-981-10-5050-3_12.

Sun, J., Q. Zhang, J. Zhou, and Q. Wei. 2014. Illumina amplicon sequencing of 16S rRNA tag reveals bacterial community development in the rhizosphere of apple nurseries at a replant disease site and a new planting site. *PLoS One*. 9(10): e111744. https://doi.org/10.1371/journal.pone.0111744.

Sun, X., Q. Ding, K. D. Hyde, L. D. Guo, and J. T. White. 2012. Community structure and preference of endophytic fungi of three woody plants in a mixed forest. *Fungal Ecol*. 5: 624–632. https://doi.org/10.1016/j.funeco.2012.04.001.

Suryanarayanan, T. S., M. B. Govinda Rajulu, E. Thirumalai, M. S. Reddy, and N. P. Money. 2011. Agni's fungi: Heat-resistant spores from the Western Ghats, southern India. *Fungal Biol*. 115(9): 833–838.

Tamošiûnë, I., G. Stanienë, P. Haimi, V. Stanys, R. Rugienius, and D. Baniulis. 2018. Endophytic *Bacillus* and *Pseudomonas spp*. modulate apple shoot growth, cellular redox balance, and protein expression under in Vitro conditions. *Front. Plant Sci*. 9: 889. doi: 10.3389/fpls.2018.00889.

Teimoori-Boghsani, Y., A. Ganjeali, T. Cernava, H. Müller, J. Asili, and G. Berg. 2020. Endophytic fungi of native *Salvia abrotanoides* plants reveal high taxonomic diversity and unique profiles of secondary metabolites. *Front. Microbiol*. 10: 3013.doi: 10.3389/fmicb.2019.03013.

Theis, K. R., N. M. Dhelly, J. L. Klassen, R. M. Brucker, J. F. Baines, T. C. G. Bosch, J. F. Cryan, S. F. Gilbert, C. J. Goodnight, E. A. J, Lloyd et al. 2016. Getting the hologenome concept right: An eco-evolutionary framework for hosts and their microbiomes. *mSystems*. 1(2): e00028 -16.doi: 10.1128/mSystems.00028-16.

Timmusk, S., L. Behers, J. Muthoni, A. Muraya, and A. C. Aronsson. 2017. Perspectives and challenges of microbial application for crop improvement. *Front. Plant Sci*. 8: 49. doi: 10.3389/fpls.2017.00049.

Toghueo, R. M. K., and F. F. Boyom. 2019. Endophytic fungi from *Terminalia Species*: A comprehensive review. *J. Fungi*. 5: 43. doi:10.3390/jof5020043.

Turbat, A., D. Rakk, A. Vigneshwari, S. Kocsubé, H. Thu, Á. Szepesi, L. D. Bakacsy, B. Škrbić, E. A. Jigjiddorj, C. Vágvölgyi, and A. Szekeres. 2020. Characterization of the plant growth-promoting activities of endophytic fungi isolated from *Sophora flavescens*. *Microorganisms*. 8(5): 683.

Tzec-Interián, J. A., D. Desgarennes, G. Carrión, J. L. Monribot-Villanueva, J. A. Guerrero-Analco, O. Ferrera-Rodríguez, D. L. Santos-Rodríguez, N. Liahut-Guin, G. E. Caballero-Reyes, and R. Ortiz-Castro. 2020. Characterization of plant growth-promoting bacteria associated with avocado trees (*Persea americana* Miller) and their potential use in the biocontrol of *Scirtothrips perseae* (avocado thrips). *PLoS One*. 15(4): e0231215. doi: 10.1371/journal.pone.0231215.

van Opstal, E. J., and S. R. Bordenstein. 2015. Rethinking heritability of the microbiome. *Science*. 349: 1172–1173.

Verma, V. C., S. K. Gond, A. Kumar, R. N. Kharwar, and G. Strobel. 2007. The endophytic mycoflora of bark, leaf, and stem tissues of *Azadirachta indica* A. Juss (Neem) from Varanasi (India). *Microb. Ecol*. 54: 119–125.

Wang, W. X., Vinocur, B., and A. Altman. 2003. Plant responses to drought, salinity and extreme temperatures: towards genetic engineering for stress tolerance. *Planta*. 218: 1–14.

White, J. F., K. L. Kingsley, Q. Zhang, R. Verma, N. Obi, S. Dvinskikh, M. T. Elmore, S. K. Verma, S. K. Gond, and K. P. Kowalski. 2019. Review: Endophytic microbes and their potential applications in crop management. *Pest. Manag. Sci*. 10: 2558–2565. https://doi.org/10.1002/ps.5527.

Yadav, M., A. Yadav, S. Kumar, and J. P. Yadav. 2016. Spatial and seasonal influences on culturable endophytic mycobiota associated with different tissues of *Eugenia jambolana* Lam. and their antibacterial activity against MDR strains. *BMC Microbiol*. 16: 44.

Yarte, M. E., M. I. Gismondi, B. E. Llorente, and E. E. Larraburu. 2022. Isolation of endo-phytic bacteria from the medicinal, forestal and ornamental tree *Handroanthus impe-tiginosus. Environ. Technol.* 43(8): 1129–1139.

Yasmeen, R., and Z. S. Siddiqui. 2017. Physiological responses of crop plants against Trichoderma harzianum in saline environment. *Acta. Bot. Croat.* doi: 10.1515/botcro-2016-0054.

Yin, L., A. Ren, M. Wei, L. Wu, Y. Zhou, X. Li et al. 2014. Neotyphodium coenophialum-infected tall fescue and its potential application in the phytoremediation of saline soils. *Int. J. Phytoremediat.* 16: 235–246. https://doi.org/10.1080/15226514.2013.773275.

Zhang, H. W., Y. C. Song, and R. X. Tan. 2006. Biology and chemistry of endophytes. *Nat. Prod. Rep.* 23: 753–771.

Zhang, Q, G. Xiong, J. Li, Z. Lu, Y. Li, W. Xu, Y. Wang, C. Zhao, and Z. Tang Xie. 2018. Nitrogen and phosphorus concentrations and allocation strategies among shrub organs: The effects of plant growth forms and nitrogen-fixation types. *Plant Soil.* 427(1–2): 305–319. doi: 10.1007/s11104-018-3655-0.

Zhou, D. Q., and K. D. Hyde. 2001. Host-specificity, host-exclusivity, and host-recurrence in saprobic fungi. *Mycol. Res.* 105: 1449e–1457.

12 Harnessing Beneficial Root Microbiome Alleviates Abiotic Stress Tolerance in Crops

Mehtab Muhammad Aslam[1,2,3],
Muhammad Waseem[4,5]*, Abah Felix*[3]*,*
Eyalira J. Okal[3]*, Witness J. Nyimbo*[3]*,*
and Muhammad Tahir ul Qamar[6]
[1]The Chinese University of Hong Kong,
Hong Kong, China
[2]Yangzhou University, Jiangsu, China
[3]Fujian Agriculture and Forestry University,
Fujian, China
[4]University of Narowal, Narowal, Pakistan
[5]South China Agricultural University,
Guangzhou, China
[6]Guangxi University, Nanning, China

CONTENTS

DOI: 10.1201/9781003110477-12

12.1 INTRODUCTION

Food shortage is becoming a serious concern due to the progressively growing world population and complications arising from the changing climate. Most often, crops are faced with adverse environmental conditions, for example, drought, extreme temperatures, soil salinization, and heavy metals contaminations resulted in reduced crop production (Gong *et al.* 2020; Teixeira *et al.* 2013). Drought is considered to be one of the most substantial abiotic stress reported to affect up to 45% of the farmland worldwide (Wilhite 2016). Abiotic stresses deteriorate plants hence affecting their growth and final productivity. Plants react to stresses by inducing several physiological and molecular variations that include stomatal closure, osmotic adjustment, phytohormone production, for example, abscisic acid (ABA), reprogramming gene expression, and other adaptive growth mechanisms (Peck and Mittler 2020). Developing stress-tolerant plant cultivars is the primary approach employed by scientists to attenuate the negative impacts of abiotic stress on agricultural crop production. The conventional breeding methods are highly appreciated due to their importance in creating stress-tolerant varieties with improved crop yields, although these methods are time-consuming (Blum 1988; Borlaug 1983). The exploitation of genetically modified crops has been another important approach used to mitigate adverse effects of abiotic stress on crop production; however, it comes with its challenges that include policy and regulation issues (Haque *et al.* 2018; Hartung and Schiemann 2014). Notably, both conventional and genetic engineering methods do not appreciate the soil ecosystems' ecological significance and the environment inhabited by the crop.

Apart from plant breeding techniques, different research studies have appreciated the root-associated microbial contribution to improving crop abiotic stress tolerance. Recent studies seek to understand molecular mechanisms that trigger plant–microbiome interactions, the advantageous effects of microbial interactions, and the means through which beneficial functions of these microbes can be harnessed (Meena *et al.* 2017). Plant–microbiome interactions have been shown to induce complex chemical processes in plant cellular systems that may offer protection against abiotic stress. In essence, microbes that inhabit areas of prolong abiotic stresses are assumed to have adaptive tolerant traits, making them potential candidates to be utilized as plant growth promoters in a stress environment. The aptitude of the plant microbiome, especially rhizospheric microorganisms, to tolerate harsh conditions and confer beneficial traits to their host plants has been ascribed due to the presence of various genes, specific hormones, and proteins including Indole-3-acetic acid (3-IAA), nifH, *1-Aminocyclopropane-1-Carboxylate* (ACC) deaminase, and siderophore (Huang *et al.* 2014; Scharf *et al.* 2016). Leveau et al. highlighted varieties of useful reporter genes such as *xylE*, *lacZ*, and *gusA* that modulate the synthesis of transcriptional products in plant-associated microbes as they respond to environmental stresses in their habitat (Leveau *et al.* 2007). Furthermore, studies that employed the genomic and proteomic techniques to track metabolic and signal interactions of rhizospheric microorganisms in *Zea mays*, *Arabidopsis thaliana*, and *Triticum aestivum* outline phytohormone modulation, quorum sensing, close signal and metabolic interactions with host plants (Li *et al.* 2019b).

To understand the advantageous roles of plant root microbes, there is a need to explore specific individual microbes, mechanistic mode of action, and associated host plant. The interaction with other root microbes and their practicability viability in-field experimentation to improve crop production. Various reports have recommended gene engineering of the plant microbes. Attempts to modify *Escherichia coli*, *Pseudomonas putida*, and *Staphylococci* genome to improve heavy metal biosorption capacity (Almaguer-Cantú *et al.* 2011; Wallenstein 2017). Mosa et al. point out the possibility of rhizoremediation exploitation using siderophores-producing bacteria in the soil such as *Pseudomonas sp.*, *Enterobacter*, *Rhizobacteria*, *Azospirillum*, *Bacillus*, and *Agrobacterium*. The phytoextraction ability of these microbes can be significantly improved through modern genetic engineering techniques that target key genes in the plant-associated microorganisms (Mosa *et al.* 2016; Xie *et al.* 2019).

There is close association of plants and related root microbiota in their evolutionary relationship over millions of years. The plant genome was mentioned as a key determinant in the selection of mutual microbes. Schlaeppi *et al.* (2014) reviewed the nature of the soil-micro habitat to impose a higher selective pressure on plant-associated microbiota than the host-plant phylogeny. Therefore, it is important to understand the evolutionary route of plant–microbe interactions and utilizing acquired information to promote plants' adaptability (Escudero-Martinez and Bulgarelli 2019). Besides providing detailed insights on severe environmental stresses that influence plant traits, this chapter also outlines the role of root-associated beneficial microbes to improve the efficiency of host–microbe interactions and the tolerance against abiotic stress.

12.2   PLANT TRAITS AFFECTED BY ABIOTIC STRESSES

Abiotic stress is one of the major environmental stresses that interrupts global plant growth and agricultural productivity. Abiotic factors including water, soil, and temperature challenged plants which disturbs cellular and plant developmental processes (Figure 12.1). Stresses like salinity and drought were demonstrated to induce deterious effects on developmental and physiological processes particularly by disturbing

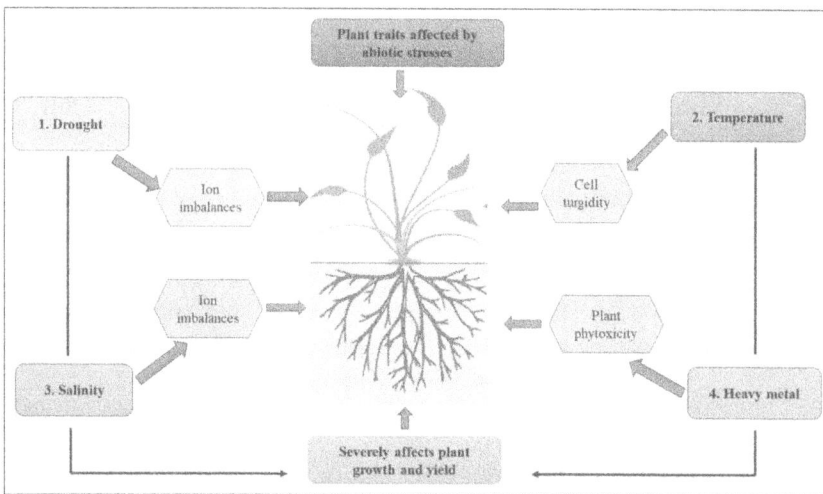

FIGURE 12.1   Plant growth-related traits affected by abiotic stress that leads to severe yield losses.

the ion homeostasis of plant cells (Khan *et al.* 2016). Abiotic stress directly affects vegetative growth before the reproductive stage (Table 12.1), especially in crop species (Uddin *et al.* 2011). Plants growing in the ecosystem are determined to a great extent by abiotic factors (Aslam *et al.* 2021). For example, for the plant to grow in the soil, macro-nutrients (phosphorus, P; potassium, K; and nitrogen, N) and micronutrients (sulfur, S; calcium, Ca; and magnesium, Mg) are required. Climate change and environmental destruction have made abiotic stresses more intense and frequent. Some of the stresses directly affect the plant's above-ground parts, whereas

TABLE 12.1
Major Plant Traits Affected by Abiotic Stresses among Several Crops

Plant	Stress Type	Plant Trait Influenced	Reference
Rice	Drought	Detrimental effect on blooming stage, filling stage, and maturity	Zhang *et al.* 2018
Maize	Drought	Increased reduction in growth and yield	Anjum *et al.* 2016
Soybean	Drought	Reduction in shoot and root biomass	Du *et al.* 2020
Wheat	Drought	Disturb wheat grain filling	Liu *et al.* 2016
Vegetables crops	Drought	Reduced water and nutrient uptake efficiency	Rouphael *et al.* 2012
Rice	Temperature	Delays rice germination, and redue growth yield	Hussain *et al.* 2019
Maize	Temperature	Effect Rubisco and photosynthetic activity	Crafts-Brandner and Salvucci 2002
Wheat	Temperature	Reduced yield and growth	Begum and Nessa 2014; Rosenzweig *et al.* 1996
Maize	Temperature	Reduce plant growth and grain size	Harrison *et al.* 2011
Common bean	Temperature	Delays seed germination and reduce plant growth	Dubal *et al.* 2016
Rice	Salinity	Decrease plant dry matter, germination rate, and sterility	Korres *et al.* 2019
Wheat	Salinity	Reduction in grain size, protein, fiber, and fat content	Abbas *et al.* 2013
Maize	Salinity	Reduce photosynthesis, nutrient and water uptake, and reduces plant yield	Iqbal *et al.* 2020
Barely	Salinity	Decreased seed germination rate	Kanbar 2014
Arabidopsis	Salinity	Delays seed germination and disturb electrolyte imbalance	Demirbas *et al.* 2013
Rice	Heavy metal	Strongly influenced plant growth and yield	Zhou *et al.* 2003
Wheat	Heavy metal	Imposed toxic effect on plant growth and grain size	Athar and Ahmad 2002
Maize	Heavy metal	Reduced plant height and slow down growth	Armienta *et al.* 2019
Barely	Heavy metal	Inhibit shoot growth	Juknys *et al.* 2009
Arabidopsis	Heavy metal	Decreased leaf area, primary root length, disturb auxin accumulation	Wang *et al.* 2015

other stresses negatively affect root structure (Aslam *et al.* 2020; Duque *et al.* 2013). This negative alteration at molecular levels causes damaging effects and abnormal changes in metabolism, which leads to poor growth, development, yield, and sometimes outright death of plants.

12.2.1 DROUGHT

Drought stress effects plant growth and survival, resulting in reduced productivity (Martignago *et al.* 2019). It is proven that the leading cause of drought is the concentration of atmospheric CO_2, leading to a decrease in water availability in the soil with a subsequent effect on carbon absorption and plant growth (Aksu and Altay 2020). Plants experience drought when the water supply to the roots becomes a challenge or evaporation rate from plant leaves becomes very high (Anjum *et al.* 2011). Drought stress generates several physiological and biochemical responses in plants (Galeano *et al.* 2019). Plant responses include stomatal closure, cell development suppression, and photosynthesis. Plant growth is altered with physiological traits of individual fluctuations, including low height, reduced leaf size, fewer leaves, and alteration of the reproductive stages (Adiloğlu 2012). It is revealed that severe drought stress interrupts the mesophyll cell structure, causing chloroplast to break, swollen grana, and abjection of starch granules (Zhang *et al.* 2019).

Plants respond to adjust to water shortage at cellular levels by accusing osmolytes and proteins important in drought tolerance (Shinozaki and Yamaguchi-Shinozaki 2007). Drought stress causes plants to alter water use to uphold its cellular functions and osmotic alteration by the formation and accretion of compatible solutes like free amino acids, sugars, and proline (Abid *et al.* 2018). Meanwhile, their association with beneficial microorganisms improves plant ability to tolerate drought (Zhang *et al.* 2019). Microbes can alter the plant physiological response to drought promoting plant endurance under transpiration by induction of plant growth hormone and increased antioxidase activity together with the photosynthetic rate (Zhang *et al.* 2019).

12.2.2 TEMPERATURE

Heat stress is a rise in soil and air temperatures beyond a certain threshold for a minimum amount of time, resulting in permanent wilting or harm to plant performance (Lamaoui *et al.* 2018). Rising temperatures could have altered geographical distribution and agricultural seasons including earlier crops maturation (Ali *et al.* 2011), impaired fertilization and pollen abortion, premature ripening of fruits and vegetables, leaf senescence and abscission, growth retardation, fruit discoloration and damage (Kaushal *et al.* 2016). During seed development, heat stress affect all its biochemical events reduceing seed size, quantity, quality, and viability. Extreme temperatures significantly affect photosynthetic activities, respiration, water equilibrium, and leaves membrane stability (Kohila and Gomathi 2018). The reproductive stage is also highly sensitive to heat stress (Kaushal *et al.* 2016).

Heat has direct influences on plant growth and yields through growth inhibition, starvation, low ion flux, and reactive oxygen species (ROS) such as single oxygen ($^1O_2$), superoxide radical (O_2), hydrogen peroxide (H_2O_2) and hydroxyl radical (OH^-).

Although breeding programs improved crop management and planting time. Thermotolerant defense mechanisms of plants like phytohormone-regulated processes, secretion of osmolytes, and upregulation of enzymatic or/and non-enzymatic antioxidant for ROS detoxification (Khan *et al.* 2013) has helped cope with temperature stress in plants, evolving low- cost techniques that smallholder crop farmers can quickly adopt is a significant challenge.

Heat stress adversely affects many cereal crops, especially during anthesis and grain-filling stages (Vardharajula *et al.* 2011). It is demonstrated that an increase in temperature causes irreversible damages such as protein denaturation, aggregation, degradation, and membrane lipids fluidity, deactivation of chloroplastic and mitochondrial enzymes, protein biosynthesis (Gao and Lan 2016), and severe cellular injury. The rates of biochemical and enzymatic reactions doubles for every 1°C increase up to 20–30°C. Temperatures extremes reduce reaction rates because protein's tertiary/quaternary structures are disrupted and enzymes become either inactivated, degraded, or denatured. Temperature extremes are among major environmental factors negatively affecting plant growth and development, altered plant morphology and biochemistry (Waraich *et al.*, 2012).

Temperature influences most of the plant development processes (Nievola *et al.* 2017), and sensitive to physiological and biochemical damages at extreme temperature that is more than optimum temperature (Masouleh and Sassine 2020). Generally, elevated greenhouse gasses including CO_2 in the atmosphere increase mean annual temperatures (Wang *et al.* 2016) that significantly decrease seed germination and growth, plant cell turgidity, and water stress (Akter and Islam 2017). High temperature affects the plants internal anatomy at tissue at the cellular and subcellular levels (Naz *et al.* 2018).

The processes used by plants to respond to temperature extremes have proven to be similar, however, the intracellular signalling and physical responses differ (Rodríguez Graña *et al.* 2015). For example, plants established the complex mechanisms to endure the low-temperature stresses by accretion of proteins or/and carbohydrates, resulting in a massive accumulation of amino acids, soluble sugars, and cold-induced proteins (Li *et al.* 2018). However, plants depend on beneficial microbes to ameliorate the effects of temperature stress by producing plant hormones that play roles in increasing tolerance against temperature stress, intensifying biofilm formation, and reducing ABA production (Khan *et al.* 2020).

12.2.3 SALINITY

Salinity is defined as the soil with altered electrical conductivity (EC, approximately 40 mM NaCl) at 25°C with 15% exchangeable sodium (Na^+) (Shrivastava and Kumar 2015). Soil salinity include the accumulation of water-soluble salts in the top soil profile to a level (Rengasamy 2006) that limits plant growth, development, and finally lower yield (Machado and Serralheiro 2017). NaCl is one of the most predominant salt contaminant in the soil, which, when ionized by water, releases Na^+ and chloride (Cl^+) ions (Uddin *et al.* 2009) challenging ionic and osmotic stress at the cellular level (Cha-um and Kirdmanee 2011). Salinity is negatively affecting photosynthesis, plant respiration, and protein synthesis in plant cells (Wang *et al.* 2020).

Plants develop several mechanisms to control osmotic stress; one sub-mechanism is osmoprotectants accumulation inclduing glycine betaine, mannitol, and raffinose (Taji *et al.* 2004). Salt tolerance in plants can be enhanced through the salt-tolerant beneficial microbes such as plant growth-promoting rhizobacteria (PGPR) (Etesami and Beattie 2018; Hahm *et al.* 2017). The beneficial microbes activate plant antioxidant defense machinery by up-regulation the activities of key enzymes, including superoxide dismutase (SOD), catalase (CAT), and peroxidase (POD) that collect the overproduced ROS and preventing salt toxicity in plant (Sharma *et al.* 2016).

12.2.4 HEAVY METALS

Heavy metals are group of toxic metals that are biologically and industrially essential (Khalid *et al.* 2017), and their contamination in soils is now a global apprehension due to potential hazardous impacts on soil, agricultural productivity, and food scurity (Ding *et al.* 2020; Malar *et al.* 2015). Among these metals is one category known as vital micronutrients required for plant growth inclduing Cu, Fe, Mn, Mg, Mo, Ni, and Zn, and the other is non-vital elements such as Ag, As, Co, Cd, Cr, Hg, Pb, Se, and Sb (Ali and Baek 2020). The former is beneficial for human health. Plant growth and development require various metals; however, these heavy metals' extreme quantities can be lethal to plant life (Syed *et al.* 2018). Metals such as As, Au, Ag, Cr, Cd, Hg, Ni, Se, Ur, and Zn are highly harmful heavy metals that pollute the atmosphere, water, and soil with adverse impacts on the soil quality and crop production (Ojuederie *et al.* 2019). It was found that when plants are exposed to these heavy metals, it leads to cellular plant damage (Yadav 2010). The study demonstrates that massive metal releases like Pb affects various plants biological activities like transpiration, root elongation, chlorophyll biosynthesis, seed germination and development, cell division, and alteres cell membrane permeability, hence causing phytotoxicity (Tiwari and Lata 2018). Cadmium metal severely causes a change in various enzyme activities, including the ones taking part in the Malvin Calvin cycle, carbohydrate and phosphorus metabolism, and CO_2 fixation leading to inhibit plant growth, leaf epinastic, chlorosis, alterations in chloroplast ultrastructure (Schützendübel *et al.* 2002). Plants develop various detoxifying mechanisms, mainly based on cellular compartmentalization and chelation to minimize heavy metals toxicity in tissues. However, metal immobilizing and metal-resistant plant microbes have lately been exploited to improve the plant growth and discourage the heavy metals accumulation in the plant tissues (Khanna *et al.* 2019).

12.3 PLANT RESPONSES TO ABIOTIC STRESSES

Although plants sense, respond, and tolerate to various abiotic stresses in a complex and dynamic ways (Khan *et al.* 2014; Marschner and Rengel 2012), beneficial microorganisms have shown effectiveness in supporting plant growth by mitigating against abiotic stresses (Figure 12.2). They achieve this through their enormous metabolic capabilities (Meena *et al.* 2017) to secrete hormones, mobilize nutrients (Chakraborty *et al.* 2015), and produce organic compounds such as 3-IAA, and ACC deaminases solubilizing phosphate (PHO_3^{-2}), and hormone secretion. These compounds deactivate

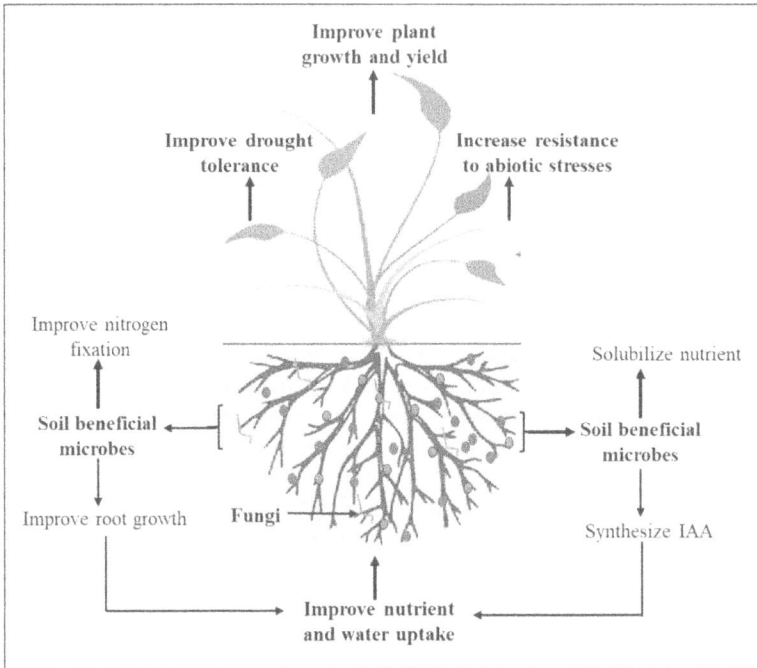

FIGURE 12.2 Contribution of beneficial soil microbiota to improve plant resistance to several abiotic stresses by facilitating nutrient and water uptake.

the synthesis of ROS and ameliorate plants nutrient uptake, thereby stimulating the plant growth and development. Through these, they help and influence drought resistance, extreme temperatures, heavy metals, salinity, and nutrient deficiency hence increasing agricultural productivity.

12.3.1 Drought Tolerance

Drought has reduced global agricultural production, especially in the developing world imposing food security at global and regional levels (Kogan 2019b). Drought is the most damaging environmental challenge with about US$2.0 billion losses of production in 2016 alone (Kogan 2019a), thereby creating food insecurity, hunger, ill-health, and in extreme cases death. Drought conditions cause low productivity and outright mortality of plants by causing salt accumulation and nutrient immobilization in soils, making it unproductive due to dryness and saline (Arora 2019). Recently, only China has lost about US$13.7 billion to drought, while that of Australia was estimated at US$6.0 billion (Kogan 2019a). Since 2001, about $932 billion has been lost by 42 top-ranking agrarian countries to drought (Kogan *et al.* 2019). This indicates that agricultural productivity is lowered to water scarcity. Several mitigation strategies are adapted and employed to reduce the consequence of drought stress. One such strategy is the plant rhizosphere/endo-rhizosphere colonization by beneficial PGPR. These beneficial microbes reduce drought stress by secreting extracellular polysaccharides (EPS), phytohormones, ACC, and volatile compounds. They also induce the accumulation of antioxidants; osmolytes, differential expression of genes and proteins that respond to stress and alteration in root architechture system. Plants facing drought stress can change the behavior of rhizospheric microbial community by changing the cocktail of carbohydrates that secrete from their roots (Gargallo-Garriga *et al.* 2018). This promotes the activity of microorganisms, with the effect of releasing more nutrients that remarkably promote plant growth.

Danielle *et al.* (2018) demonstrated that *Bouteloua gracilis* inoculated with microbial communities shows positive performance under moderate drought. Soil microbes produce enzymes that may stimulate stomatal closure during drought to reduce water loss and hydraulic failure. Extracellular polysaccharides, glycoconjugates, and proteins secreted by microbe and released from roots in exudates influence soil water holding capacity at the microzone, soil profile, soil pH, and promote soil aggregation. They influence the rhizospheric microzone and microbial activities during both well-irrigated and moderate drought conditions, which affect plant physiological responses to subsequent drought in extreme cases (Sanna *et al.* 2019). Mayak *et al.* (2004) reported that *Achromobacter piechaudii* ARV8 could reduce ethylene production in tomato seedlings. Tomato and pepper seedlings exposed to transient water stress showed fresh and dry weights increase when inoculated with the bacterium (Mayak *et al.* 2004). *Pseudomonas fluorescens* Pf1 has also demonstrated effectiveness in enhancing water stress resistance in *Vigna radiata* by secreting CAT that detoxifies compounds accumulated in the leguminous crop during adverse conditions. Generally, plant–microbe interactions improve plant performance amidst drought, especially under moderate drought conditions.

12.3.2 Temperature Tolerance

Climate change and environmental degradation occur due to anthropogenic activities in the globe, the mean temperature of earth has risen by 0.9°C for the past 100 years. This is due to the excess greenhouse gasses (carbon dioxide, CO_2; methane, CH_4;, chlorofluorocarbons, CFCs; and nitrous oxide, N_2O) being released in the atmosphere. It is predicted to increase further to about 1.5°C or more by 2050 (Verma *et al.* 2019) and between 1.5°C and 11°C by 2100 (Vardharajula *et al.* 2011). The unprecedented temperature rise resulted in the increased events of climatic factors: droughts, ice melting at the polar region, floods, water deficit in some areas, infrequent and erratic precipitation, shrinking of lakes (e.g., lake chad), heat waves, and wind storm. This has negatively affected plant growth and agricultural productivity across the world, amidst food demand by the increasing world population expected to be 9.7 billion by 2050 (Arora 2019; Dillard 2019).

PGPR and fungi found in rhizospheric soil play a major role in militating against temperature extremes like in other abiotic stress conditions in the ecological system. They improve plant growth/yield during heat stress by the secretion of hormones (such as auxin, cytokinin, gibberellic acid, brassinosteroids, ethylene, ABA, salicylic acid, and jasmonic acid); involved in biological nitrogen fixation; solubilization and mineralization of inorganic iron and phosphorus; secretion of siderophores ('iron carrier'), ACC, and exopolysaccharide, ammonia, antibiotics, volatile metabolites,

IAA, and antioxidants (Gargallo-Garriga *et al.* 2018). For instance, Rhizobium strains enhance nodulation of N_2 fixation under unfavorable conditions of extremely high or low temperatures or under acidic conditions.

Rhizobacteria and Arbuscular mycorrhiza have shown effectiveness against heat and other abiotic stress factors. If properly harnessed, especially in the developing economies, can improve agricultural productivity among smallholder farmers in the face of climate change and environmental degradation. Some different reports and demonstrations showed the practical effects of PGPR and fungi against temperature stress. It is demonstrated that inoculation of wheat with *P. putia* (strain AKMP7) could increase significantly shoot/root length and dry biomass by 52.93/27.7 cm and 16.8 g/plant, respectively (Grover *et al.* 2011). Its reported growth and biochemical parameters: total sugar contents, proline, carbohydrates, protein contents and chlorophyll, chlorophyll significantly enhanced under heat stress. The rhizospheric microorganisms play a key role in promoting plant growth during abiotic stress than uninoculated plants. The enhancement of biochemical parameters promotes plant growth/yield by increasing photosynthetic activities and suppressing ROS generation on leaves. Thermotolerant PGPR, *Bacillus cereus* proved effective in stimulating growth in tomatoes under heat-stressed conditions. Tomato augmented with ACC producing *B. cereus* enhanced the plant biochemical and physiological parameters (Mukhtar *et al.* 2020). Other agronomic treats such as shoot-to-root length, total leaf surface area, plant fresh to dry weights were significantly increased.

Cold weather vulnerable maize plants due to oxidative damage have also shown promising results when augmented with PGPR in cold stress conditions. It is reported that the treatment of maize plants suffering cold stress with *B. simplex* strain R41 *B. amyloliquefaciens* subsp. *plantarum, Pseudomonas* sp. DSMZ 13134 with micronutrients (Zn/Mn) or seaweed extracts showed improved growth and yield. (Ho *et al.* 2017). A similar result was also observed when tomato (*Solanum lycopersicum*) seeds were subjected to the same treatment using bacteria from the genera *Arthrobacter, Flavimonas, Flavobacterium, Pseudomonas,* and *Pedobacter* with significant increment in the height of the plant and root length (Ho *et al.* 2017). Another study showed a cold-tolerant bacterium, *Methylobacterium phyllosphaerae* (IARI-HHS2-67), isolated from the phyllosphere of *Triticum aestivum*, exhibited improved growth, efficient nutrient uptake, and survival compared to a non-inoculated under cold weather conditions. The chilling resistance of the grapevine plantlets showed enhanced chilling resistance when inoculated with rhizobacteria, *Burkholderia phytofirmans* (PsJN strain) (Barka *et al.* 2006). An increase in CO_2 fixation and oxygen evolution was observed. The root length and plantlet biomass also increased. Despite the hike in temperature due to climate change menace, agricultural production can be improved upon if the opportunity presented by thermotolerant PGPR and Arbuscular mycorrhiza are properly utilized.

12.3.3 SALINITY RESISTANCE

Salinity alone deteriorates 80 million hectares of arable lands and is among the harshest environmental stress that hampers crop production. The production yields are lost (~20–50%) due to drought and salinity alone. Munns and Tester reported that world's irrigable lands (~50%) are contaminated with salts (Munns and Tester 2008). Plants are sensitive to salinity especially at the early stage of their life span. A high salt concentration challenge water availability in soil for plant causing altered osmotic pressure and ionic stress/toxicity leading to physiological drought. Cations (Na^+, Ca^{2+}, and Mg^{2+}) and anions (Cl^{2-}, SO_4^{2-}, HCO_3^{2-}, CO_3^{2-}, and NO_3^{2-}) are the prevalent ions in saline soils. Other constituents include B, SrO^{2+}, Mo, Ba^{2+}, and Al^{3+}. Therefore, saline soils contain dissolved $CaSO_4$, KCl, Na_2SO_4, NaCl, Na_2CO_3, $MgSO_4$, and $MgCl_2$ (Tavakkoli *et al.* 2010) in high concentration. Ca^{2+}, K^+, Mg^{2+}, and NH^{4+} are essential to plant growth while AL^{3+}, H^+, and Na^+ alter soil pH. The negatively charged soil colloids are attracted to and react with these cations. Water-soluble salts on the topsoil and subsoil contribute to the salinization of soils, negatively influencing agricultural productivity, environmentally healthy, and economic growth (Sha Valli Khan *et al.* 2014).

Osmotic stress brings about physiological drought or cellular dehydration as plants are constrained from taking up adequate water. The plants under these conditions suffer the secondary effect of salinization, including mechanical and oxidative as a result of nutritional imbalances (Khan *et al.* 2014) effects arise due to salinity include stunted growth and development, poor yield, ecological imbalance, soil erosion, residual effects of elements (e.g., B, F, and Se) on human health in the food chain (Hu and Schmidhalter 2005; Parida and Das 2005) and, consequently, food insecurity and low income for farmers.

Some bacteria and fungi have been demonstrated to negate the effects of soil salinization. The symbiotic association between plants and microorganisms is greatly beneficial against abiotic salt stress in *Zea mays* upon *Rhizobium* inoculation. *Pseudomonas* is helpful in decreasing electrolyte leakage triggered by salinity maintaining leaf water content. *P. stutzeri P. aeruginosa*, and *P. fluorescens* were reported to ameliorate sodium chloride stress in tomato plants, which result in an increase in root and shoot length (Ho *et al.* 2017). It has been demonstrated that *Oryza sativa* inoculated with a mixture of *P. pseudoalcaligenes* and *B. pumilus* resulted in higher glycine concentrations and betaine-like quaternary compounds, assisting in salinity tolerance of the plant. The beneficial rhizobacteria bacteria could serve as an instrument against salinity in rice plants (Jha *et al.* 2011). A proteomic investigation has shown that halotolerant PGPR *Dietzia natronolimnaea* STR1 relief wheat plants against salinity by modulating a stress-responsive gene (Qurashi and Sabri 2011). *Oceanobacillus profundus* (Pmt2) and *Staphylococcus saprophyticus* (ST1) bacteria were able to form a biofilm, produce EPS and accumulate endogenous osmolytes (proline and glycine betaine) that enhanced the development of lentil (*Lens culinaris* Var. Masoor-93) under salt stress.

12.3.4 HEAVY METAL SEQUESTRATION

Heavy metals are the major contributor in environmental pollutants degrading agroecosystem, water, and air for years due to urbanization and industrialization. These metals (and metalloids): Cr, Cu, Hg, Cd, Zn, Mn, and Ni pollute the environment and are considered as a significant source of poisoning (Briffa *et al.* 2020). In their high concentration, heavy metals adversely limit seed germination, nutrient mobilization, and plant growth (Sethy and Ghosh 2013). They can be found in various forms in the soil, natural water, and air. Pesticides, paints, and fertilizers contain some of these heavy metals (Moghal *et al.* 2020). Anthropogenic activities degrading agricultural lands and ecological systems are increasing that cause environmental and metal pollution in soils. Such human activities include mining and smelting, indiscriminate and unwholesome horticultural and agricultural practices, sewage sludge, fossil fuel combustion, and oil spillage (Sha Valli Khan *et al.* 2014). They are highly persistent in the environment and are majorly released into the environment through anthropogenic sources (Ojedokun and Bello 2016). Toxic heavy metals accumulate in water bodies and aquatic creatures therein. Apart from their hazardous effect on human health (Chisti *et al.* 2018), their high concentrations cause toxicity in crop plants and adversely influence plant growth and protein content (Ho *et al.* 2017). The elevated levels of toxic metal ions in the environment give rise to many deleterious effects on living organisms. Although micronutrients are essential for metabolism, enzymatic activities and sometimes structural functions, their excess can be problematic. Heavy toxic metals (Hg and Cd) harmful to plants even at the lowest concentrations and cause the production of ROS that blocks functional groups of biomolecules resulting in oxidative injury in plants. Metal-contaminated soils reduce not only crop yields but the consumption of foods grown on such soils also pose health hazards to the populace because of their residual effect.

It has been demonstrated that 5–20 mg/kg soil mercury in *Lycopersicum esculentum* plant reduces the rate of survival, plant height, flowering, seed viability, and pollen viability (Shekar *et al.* 2011). Apart from different defensive mechanism in which plants respond to heavy metals and metalloid toxicity (Ahsan and Fukuhara 2010), microorganisms play a vital role in minimizing the harmful effects of toxic heavy metals and metalloids on plants. For instance, *P. putida* allows for healthy growth and has been reported to increase shoot-to-root length, plant biomass, and chlorophyll content of *Eruca sativa* (Ho *et al.* 2017) under Cd stress. It had shown that *Brassica napus* seedlings augmented with *Kluyvera ascorbata* SUD165 could be protected against nickel toxicity. *In planta* studies, Ho et al. demonstrated that with either phenol or catechol, inoculation of *A. thaliana* with *Achromobacter xylosoxidans* strain F3B showed positive growth result as they aid rate of pollutant removal (Ho *et al.* 2012). This endophytic strain helps plants to tolerate the aromatic compound stress and improve phytoremediation. A mercury-reducing bacterium strain MELD1, *Photobacterium halotolerans*, significantly increased the root length, yield, biomass, and restricted mercury uptake by *A. thaliana* (Mathew *et al.* 2015a). This bacterium has also been proved effective against Hg (about 25 ppm) in *Vigna Unguiculata Sesquipedalis*. This improves yield and reduces the translocation of Hg to the bean pods. Therefore, the association between plant and their associated beneficial microbes can enhance the photoprotection, phytostabilization, and phytoremediation of mercury (Mathew *et al.* 2015b).

12.3.5 Nutrient Deficiency

Plants need essential nutrient elements in order to complete their life cycle. These elements are categorized into macroelements (N, P, and K), microelements (Mg, Ca, and S), and trace elements (Cl, Br, Fe, Mn, Cu, Zn, Ni, and Mb). Macronutrients are required for structural entities. In contrast, the trace elements in addition to have structural role in plants are also required in relatively small quantities as regulator and catalytic activators of enzymes (George *et al.* 1995). Deficiency or toxicity in soil of any of these mineral elements limit plant growth and final yield.

Microorganisms in a symbiotic association that modulate local and systemic mechanisms offering defense against adverse environmental conditions including nutrient deficiency (Ahemad and Kibret 2014). For instance, rhizobacteria and fungi facilitate plant growth by assisting nutrient acquisition (Perner *et al.* 2008) and modulate plant hormone secretion (Table 12.2). Physiological and molecular research has shown that AM fungi enhance plant P uptake through hyphae network in the surrounding soil to absorb nutrients transferred to and taken up by plant roots (Marschner *et al.* 2011a). This increases the area of microzone from where nutrients are taken from. Phosphorus absorbed by the hyphae and transported to host plant roots via fungus-plant interface at root epidermis and/or root hairs, in turn the fungi get their organic carbon supply from the host plant (Smith and Smith 2011). Plants exudates and microbial mucilage secretion in the soil can solubilize insoluble inorganic P by releasing OH, protons, or CO_2, and organic acid like oxalate, citrate, and malate (Marschner *et al.* 2011a). Meanwhile, the root morphology, soil structure, P supply, fungal species, and light intensity are some of the factors that determine the level of benefits that plants may derive from the fungi. Some soil organisms promote plant growth through growth hormones or other mechanisms (Smith and Smith 2011;

TABLE 12.2
Contribution of Soil Beneficial Microbes to Alleviates Several Abiotic Stresses

Plant	Microorganism	Effect on Plant	References
Rice	*Enterobacteriaceae*	Improve drought resistance with increased rhizosheath	Zhang *et al.* 2020
Wheat	*Arthrobacter, Devosia,* and *Bacillus*	Improve plant growth and N_2 availability	Chen *et al.* 2019
Maize	*Bacillus subtilis*	Reduce heavy metal uptake by plant	Li *et al.* 2019a
Barely and several other crops	*Piriformospora indica*	Improves P uptake, plant growth, and stress resistance to biotic and abiotic stresses	Achatz *et al.* 2010; Aslam *et al.* 2019
Peanut	*Arthrocnemum indicum*	Improves plant growth and salinity resistance	Sharma *et al.* 2016
Arabidopsis	*Actinobacteria*	Improves plant root and microbial interaction to facilitate plant growth	Lundberg *et al.* 2012
Soybean	*Sinorhizobia,* and *bradyrhizobia*	Improve saline and alkali resistance	Han *et al.* 2020
Several crops	*Pseudomonas aeruginosa*	Mobilize P, N_2, and K	Rashid *et al.* 2016
Pea	*P. syringae*	Improve plant tolerance to abiotic stresses in field	Martín-Sanz *et al.* 2012

Sylvia *et al.* 2005). The microbial biomass release nutrients in the root zone, from where plant with a high nutrient uptake capacity. Root growth may act as a mechanism by which plants absorb nutrients, mobilized by rhizospheric microorganisms of root (Marschner *et al.* 2011b).

Endomycorrhiza also plays a key role in the nutrition of plants, when soils suffering nutrients starvation. For instance, endomycorrhizal fungus, Endosome improve soil structure by stabilizing soil aggregates, serving as bioremediation agents of contaminated soils (Kennedy 2005), and dune stabilization (sand aggregation), thereby improving plant nutrition. The plant's ability to withstand stress conditions is mostly dependent on the efficiency of its association with root microorganisms. For example, (Dommergues 1978) *Alnus tenuifolia* invaded nitrogen-deficient soil of glacier bay, Alaska; *Myrica asplenifolia* colonized areas have been a severe disturbance resulting from road construction; *Podocarpus lawrencei* invades exposed rocky subalpine and alpine sites in New South Wales. Plant roots were associated with N2-fixing microorganisms in all three examples: Actinomycetes-like organisms in *Alnus tenuifolia* and *Myrica asplenifolia* system; bacteria associated with mycorrhizae in the podocarps system (Dommergues 1978). The ability to fix N_2 by these organisms enables the hosts in an unfavorable environment colonization process. On black wastes from anthracite mining in Pennsylvania, the only stressful original colonists of nitrogen-deficient wastes were either N_2-fixing plants or certain ectotrophic mycorrhizal species (Dommergues 1978). It was only ectomycorrhizal fungi that were able to withstand such adverse conditions in such wastelands with high acidity, high temperature, and very low available nitrogen content.

Under Fe-deficiency stress, microorganisms secrete siderophores that chelate Fe^3 in iron-deficient soils (Jin *et al.* 2014). Microbes produce different siderophores such as fungi synthesize ferrochrome and enterobactin, ferrioxamines and pyoverdine by bacteria (Hider 1984). In iron deficient soil, roots are stimulated to produce more exudates, leading to an increase in microbial community and activities around the root zone. This alteration benefits plant iron acquisition via secretion and accumulation of siderophores and protons around the root zone, improving iron bioavailability in the microzone (Dommergues 1978; Jin *et al.* 2014). Rhizobium nodulation is another complex process that involves rhizobium and host plant. The interaction between rhizobia and leguminous plant in flavonoid presence leads to the formation of nodules that improve plant Fe and nitrogen uptake capacity. AM fungal association enhances root length and the nutrient acquisition area of the root system and increases the production of Fe^{3+} chelators and protons (Diem and Dommergues 1980; Jin *et al.* 2014). Photosynthetic organisms and plants in particular particularly require iron for cellular and metabolic processes, such as oxygen transport, photosynthetic electron transport, cellular respiration, chlorophyll biosynthesis, thylakoid biogenesis, and chloroplast development, reduction of ROS induced damage, sulfur and nitrogen metabolism, and cofactor assembly (Kroh and Pilon 2020). In general, symbionts in the rhizosphere and photosphere help the efficacy of nutrient uptake and allow host plants to persist in a low nutrient environment. Naturally, different plant species share the broadly specific mycorrhizal fungi that modulate nutrients and carbon cycles (provide up to 80% of nitrogen and phosphorus for plants), and functions of ecosystem (van der Heijden *et al.* 2015).

12.5   CONCLUSION AND FUTURE PROSPECTIVE

In summary, the plant microbiome exhibits a high potential of being harnessed to help crops thrive against abiotic stress and enhance production. Microbe such as *Enterobacter sp.*, *Pseudomonas sp.*, *Azospirillum*, *Rhizobacteria*, *Bacillus*, and *Agrobacterium* produce hormones and siderophore, which enable plants to survive in water-deficient conditions and heavy metal-contaminated soil, respectively. Furthermore, rhizobium bacteria promote root nodulation of legumes, while mycorrhizal fungi enhance the root's length and the nutrient attainment area of the root system in host plants. Diverse microbes that inhabit plant roots secrete a range of secretions like siderophore and phytohormones, respectively, enhancing microbial community diversity in the roots and controlling plant growth. PGPBs can, therefore, be harnessed and utilized to enhance crop development and production in a stressed abiotic environment. The PGPB can be incorporated with manure or developed as biofertilizers known for their less toxic effects on the soil and ecosystem.

REFERENCES

Abbas G, Saqib M, Rafique Q, Rahman AU, Akhtar J, Haq MAU, Nasim M. 2013. Effect of salinity on grain yield and grain quality of wheat (Triticum aestivum L.). *Pakistan Journal of Botany*, **50**, 185–189.

Abid M, Ali S, Qi LK, Zahoor R, Tian Z, Jiang D, Snider JL, Dai T. 2018. Physiological and biochemical changes during drought and recovery periods at tillering and jointing stages in wheat (Triticum aestivum L.). *Scientific Reports*, **8**, 1–15.

Achatz B, von Rüden S, Andrade D, Neumann E, Pons-Kühnemann J, Kogel K-H, Franken P, Waller F. 2010. Root colonization by *Piriformospora indica* enhances grain yield in barley under diverse nutrient regimes by accelerating plant development. *Plant and Soil*, **333**, 59–70.

Adiloğlu S. 2012. Determination of some trace element nutritional status of Cherry laurel (Prunus laurocerasus L.) with leaf analysis which grown natural conditions in Eastern Black Sea region of Turkey. *Scientific Research and Essays*, **7**, 1237–1243.

Ahemad M, Kibret M. 2014. Mechanisms and applications of plant growth promoting rhizobacteria: current perspective. *Journal of King Saudi University of Science*, **26**, 1–20.

Ahsan A, Fukuhara T. 2010. Mass and heat transfer model of tubular solar still. *Solar Energy*, **84**, 1147–1156.

Aksu G, Altay H. 2020. The effects of potassium applications on drought stress in sugar beet. *Sugar Tech*, **22**, 1–11.

Akter N, Islam MR. 2017. Heat stress effects and management in wheat. A review. *Agronomy for Sustainable Development*, **37**, 37.

Ali M, Baek K-H. 2020. Jasmonic acid signaling pathway in response to abiotic stresses in plants. *International Journal of Molecular Sciences*, **21**, 621.

Ali S Z, Sandhya V, Grover M, Linga V R, Bandi V. 2011. Effect of inoculation with a thermotolerant plant growth promoting Pseudomonas putida strain AKMP7 on growth of wheat (Triticum spp.) under heat stress. *Journal of Plant Interaction*, **6**, 239–246.

Almaguer-Cantú V, Morales-Ramos LH, Balderas-Rentería I. 2011. Biosorption of lead (II) and cadmium (II) using Escherichia coli genetically engineered with mice metallothionein I. *Water Science and Technology*, **63**, 1607–1613.

Anjum SA, Tanveer M, Ashraf U, Hussain S, Shahzad B, Khan I, Wang L. 2016. Effect of progressive drought stress on growth, leaf gas exchange, and antioxidant production in two maize cultivars. *Environmental Science and Pollution Research*, **23**, 17132–17141.

Anjum SA, Xie X-y, Wang L-c, Saleem MF, Man C, Lei W. 2011. Morphological, physiological and biochemical responses of plants to drought stress. *African Journal of Agricultural Research*, **6**, 2026–2032.

Armienta MA, Beltrán M, Martínez S, Labastida I. 2019. Heavy metal assimilation in maize (Zea mays L.) plants growing near mine tailings. *Environmental Geochemistry and Health*, **42**, 1–15.

Arora NK. 2019. Impact of climate change on agriculture production and its sustainable solutions. *Environmental Sustainability*, **2**, 95–96.

Aslam MM, Akhtar K, Karanja JK, Haider FU. 2020. Understanding the adaptive mechanisms of plant in low phosphorous soil. In: *Plant Stress Physiology*. London, UK: IntechOpen. https://doi.org/10.5772/intechopen.91873.

Aslam MM, Karanja J, Bello SK. 2019. Piriformospora indica colonization reprograms plants to improved P-uptake, enhanced crop performance, and biotic/abiotic stress tolerance. *Physiological and Molecular Plant Pathology*, **106**, 232–237.

Aslam MM, Karanja JK, Yuan W, Zhang Q, Zhang J, Xu W. 2021. Phosphorus uptake is associated with the rhizosheath formation of mature cluster roots in white lupin under soil drying and phosphorus deficiency. *Plant Physiology and Biochemistry*, **166**, 531–539.

Athar R, Ahmad M. 2002. Heavy metal toxicity: effect on plant growth and metal uptake by wheat, and on free living Azotobacter. *Water Air and Soil Pollution*, **138**, 165–180.

Barka EA, Nowak J, Clément C. 2006. Enhancement of chilling resistance of inoculated grapevine plantlets with a plant growth-promoting rhizobacterium, Burkholderia phytofirmans strain PsJN. *Applied and Environmental Microbiology*, **72**, 7246–7252.

Begum F, Nessa A. 2014. Effects of temperature on some physiological traits of wheat. *Journal of Bangladesh Academy of Sciences*, **38**, 103–110.

Blum A. 1988. *Plant Breeding for Stress Environments*. Boca Raton: CRC Press. https://doi.org/10.1201/9781351075718.

Borlaug NE. 1983. Contributions of conventional plant breeding to food production. *Science*, **219**, 689–693.

Briffa J, Sinagra E, Blundell R. 2020. Heavy metal pollution in the environment and their toxicological effects on humans. *Heliyon*, **6**, e04691.

Cha-um S, Kirdmanee C. 2011. Remediation of salt-affected soil by the addition of organic matter: an investigation into improving glutinous rice productivity. *Scientia Agricola*, **68**, 406–410.

Chakraborty U, Chakraborty B, Dey P, Chakraborty AP. 2015. Role of microorganisms in alleviation of abiotic stresses for sustainable agriculture. *Abiotic Stresses in Crop Plants*, 232–253.

Chen S, Waghmode TR, Sun R, Kuramae EE, Hu C, Liu B. 2019. Root-associated microbiomes of wheat under the combined effect of plant development and nitrogen fertilization. *Microbiome*, **7**, 136.

Chisti HTN, Rangreez TA, Bashir R, Mobin R, Najar AA. 2018. Synthesis and characterization of graphene Th (IV) phosphate composite cation exchanger: analytical application as lead ion-selective membrane electrode. *Desalination and Water Treatment*, **57**, 23893–23902.

Crafts-Brandner SJ, Salvucci ME. 2002. Sensitivity of photosynthesis in a C4 plant, maize, to heat stress. *Plant Physiology*, **129**, 1773–1780.

Demirbas S, Vlachonasios K, Acar O, Kaldis A. 2013. The effect of salt stress on Arabidopsis thaliana and Phelipanche ramosa interaction. *Weed Research*, **53**, 452–460.

Diem H, Dommergues Y. 1980. Significance and improvement of rhizospheric N2 fixation. *Recent Advances in Biological Nitrogen Fixation*, 190–226.

Dillard HR. 2019. Global food and nutrition security: from challenges to solutions. *Food Security*, **11**, 249–252.

Ding Z, Kheir AM, Ali MG, Ali OA, Abdelaal AI, Zhou Z, Wang B, Liu B, He Z. 2020. The integrated effect of salinity, organic amendments, phosphorus fertilizers, and deficit irrigation on soil properties, phosphorus fractionation and wheat productivity. *Scientific Reports*, **10**, 1–13.

Dommergues YR. 1978. The plant–microorganism system. In: *Developments in Agricultural and Managed Forest Ecology*, Edited by YR Dommergues and SV Krupa, vol **4**, pp. 1–475. Netherlands: Elsevier.

Du Y, Zhao Q, Chen L, Yao X, Zhang W, Zhang B, Xie F. 2020. Effect of drought stress on sugar metabolism in leaves and roots of soybean seedlings. *Plant Physiology and Biochemistry*, **146**, 1–12.

Dubal I, Troyjack C, Koch F, Aisenberg G, Szareski VJ, Pimentel JR, Nardino M, Carvalho I, Olivoto T, Souza V, Villela F, Aumonde T. 2016. Effect of temperature on bean seed germination: vigor and isozyme expression. *American Journal of Agricultural Research*, **1**, 1–9.

Duque AS, de Almeida AM, da Silva AB, da Silva JM, Farinha AP, Santos D, Fevereiro P, de Sousa Araújo S. 2013. Abiotic stress responses in plants: unraveling the complexity of genes and networks to survive. In: *Abiotic stress-plant responses and applications in agriculture*, Edited by Kourosh Vahdati and Charles Leslie, 49–101. IntechOpen. https://doi.org/10.5772/45842.

Escudero-Martinez C, Bulgarelli D. 2019. Tracing the evolutionary routes of plant–microbiota interactions. *Current Opinion in Microbiology*, **49**, 34–40.

Etesami H, Beattie GA. 2018. Mining halophytes for plant growth-promoting halotolerant bacteria to enhance the salinity tolerance of non-halophytic crops. *Frontiers in Microbiology*, **9**, 148.

Galeano E, Vasconcelos TS, Novais de Oliveira P, Carrer H. 2019. Physiological and molecular responses to drought stress in teak (Tectona grandis Lf). *PloS One*, **14**, e0221571.

Gao J, Lan T. 2016. Functional characterization of the late embryogenesis abundant (LEA) protein gene family from Pinus tabuliformis (Pinaceae) in Escherichia coli. *Scientific Reports*, **6**, 19467.

Gargallo-Garriga A, Preece C, Sardans J, Oravec M, Urban O, Peñuelas J. 2018. Root exudate metabolomes change under drought and show limited capacity for recovery. *Scientific Reports*, **8**, 12696.

George E, Marschner H, Jakobsen I. 1995. Role of arbuscular mycorrhizal fungi in uptake of phosphorus and nitrogen from soil. *Critical Reviews in Biotechnology*, **15**, 257–270.

Gong Z, Xiong L, Shi H, Yang S, Herrera-Estrella LR, Xu G, Chao D-Y, Li J, Wang P-Y, Qin F. 2020. Plant abiotic stress response and nutrient use efficiency. *Science China Life Sciences*, **63**, 635–674.

Grover M, Ali SZ, Sandhya V, Rasul A, Venkateswarlu B. 2011. Role of microorganisms in adaptation of agriculture crops to abiotic stresses. *Journal of Microbiology and Biotechnology*, **27**, 1231–1240.

Hahm M-S, Son J-S, Hwang Y-J, Kwon D-K, Ghim S-Y. 2017. Alleviation of salt stress in pepper (Capsicum annum L.) plants by plant growth-promoting rhizobacteria. *Journal of Microbiology and Biotechnology*, **27**, 1790–1797.

Han Q, Ma Q, Chen Y, Tian B, Xu L, Bai Y, Chen W, Li X. 2020. Variation in rhizosphere microbial communities and its association with the symbiotic efficiency of rhizobia in soybean. *ISME Journal*, **14**, 1915–1928.

Haque E, Taniguchi H, Hassan M, Bhowmik P, Karim MR, Śmiech M, Zhao K, Rahman M, Islam T. 2018. Application of CRISPR/Cas9 genome editing technology for the improvement of crops cultivated in tropical climates: recent progress, prospects, and challenges. *Frontiers in Plant Science*, **9**, 617.

Harrison L, Michaelsen J, Funk C, Husak G. 2011. Effects of temperature changes on maize production in Mozambique. *Climate Research*, **46**, 211–222.

Hartung F, Schiemann J. 2014. Precise plant breeding using new genome editing techniques: opportunities, safety and regulation in the EU. *The Plant Journal*, **78**, 742–752.

Hider RC. 1984. Siderophore mediated absorption of iron. In: *Siderophores from Microorganisms and Plants*, vol **58**, pp. 25–87. Heidelberg, Berlin: Springer. https://doi.org/10.1007/BFb0111308.

Ho Y-N, Mathew DC, Hsiao S-C, Shih C-H, Chien M-F, Chiang H-M, Huang C-C. 2012. Selection and application of endophytic bacterium Achromobacter xylosoxidans strain F3B for improving phytoremediation of phenolic pollutants. *Journal of Hazardous Materials*, **219**, 43–49.

Ho Y-N, Mathew DC, Huang C-C. 2017. Plant-microbe ecology: interactions of plants and symbiotic microbial communities. In: Plant Ecology, Edited by Zubaida Yousaf, vol 93, p. 119. London, UK: IntechOpen.

Hu Y, Schmidhalter U. 2005. Drought and salinity: a comparison of their effects on mineral nutrition of plants. *Journal of Plant Nutrition and Soil Science*, **168**, 541–549.

Huang X-F, Chaparro JM, Reardon KF, Zhang R, Shen Q, Vivanco JM. 2014. Rhizosphere interactions: root exudates, microbes, and microbial communities. *Botany*, **92**, 267–275.

Hussain S, Khaliq A, Ali B, Hussain HA, Qadir T, Hussain S. 2019. Temperature extremes: impact on rice growth and development. In: *Plant Abiotic Stress Tolerance*, Edited by M Hasanuzzaman, K Hakeem, K Nahar and H Alharby, pp. 153–171. Cham: Springer. https://doi.org/10.1007/978-3-030-06118-0_6.

Iqbal S, Hussain S, Qayyaum MA, Ashraf M. 2020. The response of maize physiology under salinity stress and its coping strategies. In: *Plant Stress Physiology*, Edited by Akbar Hossain. London: IntechOpen. https://doi.org/10.5772/intechopen.92213.

Jha Y, Subramanian R, Patel S. 2011. Combination of endophytic and rhizospheric plant growth promoting rhizobacteria in *Oryza sativa* shows higher accumulation of osmo-protectant against saline stress. *Acta Physiologiae Plantarum*, **33**, 797–802.

Jin CW, Ye YQ, Zheng SJ. 2014. An underground tale: contribution of microbial activity to plant iron acquisition via ecological processes. *Annals of Botany*, **113**, 7–18.

Juknys R, Račaitė M, Vitkauskaitė G, Venclovienė J. 2009. The effect of heavy metals on spring barley (Hordeum vulgare L.). *Agriculture*, **96**, 111–124.

Kanbar A. 2014. Effect of salinity stress on germination and seedling growth of barley (Hordeum vulgare L.) varieties. *Advances in Environmental Biology*, **8**, 244–248.

Kaushal N, Bhandari K, Siddique KH, Nayyar H. 2016. Food crops face rising temperatures: an overview of responses, adaptive mechanisms, and approaches to improve heat tolerance. *Cogent Food & Agriculture*, **2**, 1134380.

Kennedy, A. 2005. Soil biota in the rhizosphere. In: Sylvia, D.M., Fuuhrman, J.J., Hartel P.G., Zuberer, D.A., (eds.). *Principles and Applications of Soil Microbiology*. Upper Sadle River, NJ: Prentice Hall. p. 242–262.

Khalid S, Shahid M, Niazi NK, Murtaza B, Bibi I, Dumat C. 2017. A comparison of technologies for remediation of heavy metal contaminated soils. *Journal of Geochemical Exploration*, **182**, 247–268.

Khan MA, Asaf S, Khan AL, Jan R, Kang S-M, Kim K-M, Lee I-J. 2020. Thermotolerance effect of plant growth-promoting Bacillus cereus SA1 on soybean during heat stress. *BMC Microbiology*, **20**, 1–14.

Khan MIR, Iqbal N, Masood A, Per TS, Khan NA. 2013. Salicylic acid alleviates adverse effects of heat stress on photosynthesis through changes in proline production and ethylene formation. *Plant Signaling & Behavior*, **8**, e26374.

Khan MS, Khan MA, Ahmad D. 2016. Assessing utilization and environmental risks of important genes in plant abiotic stress tolerance. *Frontiers in Plant Science*, **7**, 792.

Khan PSV, Nagamallaiah G, Rao MD, Sergeant K, Hausman J. 2014. Abiotic stress tolerance in plants: Insights from proteomics. In: *Emerging Technologies and Management of Crop Stress Tolerance*, Edited by Parvaiz Ahmad and Saiema Rasool, pp. 23–68. Elsevier Academic Press. https://doi.org/10.1016/B978-0-12-800875-1.00002-8.

Khanna K, Jamwal VL, Gandhi SG, Ohri P, Bhardwaj R. 2019. Metal resistant PGPR lowered Cd uptake and expression of metal transporter genes with improved growth and photosynthetic pigments in Lycopersicon esculentum under metal toxicity. *Scientific Reports*, **9**, 1–14.

Kogan F. 2019a. Climate change and food security current and future. In: *Remote Sensing for Food Security*, pp. 191–224. Cham: Springer. https://doi.org/10.1007/978-3-319-96256-6.

Kogan F. 2019b. Monitoring drought from space and food security. In: *Remote Sensing for Food Security*, pp. 75–113. Cham: Springer. https://doi.org/10.1007/978-3-319-96256-6.

Kogan F, Guo W, Yang W. 2019. Drought and food security prediction from NOAA new generation of operational satellites. *Geomatics Natural Hazards and Risk*, **10**, 651–666.

Kohila S, Gomathi R. 2018. Adaptive physiological and biochemical response of sugarcane genotypes to high-temperature stress. *Indian Journal of Plant Physiology*, **23**, 245–260.

Korres NE, Varanasi VK, Slaton NA, Price AJ, Bararpour T. 2019. Effects of salinity on rice and rice weeds: Short-and long-term adaptation strategies and weed management. In: *Advances in Rice Research for Abiotic Stress Tolerance*, Edited by Mirza Hasanuzzaman, Masayuku Fujita, Kamrun Nahar and Jiban Biswas, pp. 159–176. Woodhead Publishing.

Kroh GE, Pilon M. 2020. Regulation of iron homeostasis and use in chloroplasts. *International Journal of Molecular Science*, **21**, 3395.

Lamaoui M, Jemo M, Datla R, Bekkaoui F. 2018. Heat and drought stresses in crops and approaches for their mitigation. *Frontiers in Chemistry*, **6**, 26.

Leveau JH, Loper JE, Lindow SE. 2007. Reporter gene systems useful in evaluating in situ gene expression by soil-and plant-associated bacteria. In: *Manual of Environmental Microbiology*, Edited by Christon J. Hurst, Ronald L. Crawford, Jay L. Garland, David A. Lipson, Aaron L. Mills and Linda D. Stetzenbach, 3rd edition, pp. 734–747. American Society of Microbiology (ASM) Press.

Li S, Yang Y, Zhang Q, Liu N, Xu Q, Hu L. 2018. Differential physiological and metabolic response to low temperature in two zoysiagrass genotypes native to high and low latitude. *Plos One*, **13**, e0198885.

Li X, Cai Y, Liu D, Ai Y, Zhang M, Gao Y, Zhang Y. 2019a. Occurrence, fate, and transport of potentially toxic metals (PTMs) in an alkaline rhizosphere soil-plant (Maize, Zea mays L.) system: the role of Bacillus subtilis. *Environmental Science and Pollution Research*, **26**, 5564–5576.

Li Z, Yao Q, Guo X, Crits-Christoph A, Mayes MA, Lebeis SL, Banfield JF, Hurst GB, Hettich RL, Pan C. 2019b. Genome-resolved proteomic stable isotope probing of soil microbial communities using 13CO2 and 13C-methanol. *Frontiers in Microbiology*, **10**, 2706.

Liu Y, Liang H, Lv X, Liu D, Wen X, Liao Y. 2016. Effect of polyamines on the grain filling of wheat under drought stress. *Plant Physiology and Biochemistry*, **100**, 113–129.

Lundberg DS, Lebeis SL, Paredes SH, Yourstone S, Gehring J, Malfatti S, Tremblay J, Engelbrektson A, Kunin V, Del Rio TG. 2012. Defining the core Arabidopsis thaliana root microbiome. *Nature*, **488**, 86–90.

Machado RMA, Serralheiro RP. 2017. Soil salinity: effect on vegetable crop growth. Management practices to prevent and mitigate soil salinization. *Journal of Horticulture*, **3**, 30.

Malar S, Sahi S, Favas P. 2015. Assessment of mercury heavy metal toxicity-induced physiochemical and molecular changes in Sesbania grandiflora L. *International Journal of Environment Science*, **12**, 3273–3282.

Marschner P, Crowley D, Rengel Z. 2011a. Rhizosphere interactions between microorganisms and plants govern iron and phosphorus acquisition along the root axis–model and research methods. *Soil Biology and Biochemistry*, **43**, 883–894.

Marschner P, Rengel Z. 2012. Nutrient availability in soils. In: *Marschner's Mineral Nutrition of Higher Plants*, Edited by Horst Marschner, 3rd edition, pp. 315–330. Academic Press.

Marschner P, Umar S, Baumann K. 2011b. The microbial community composition changes rapidly in the early stages of decomposition of wheat residue. *Soil Biology and Biochemistry*, **43**, 445–451.

Martignago D, Rico-Medina A, Blasco-Escaméz D, Fontanet-Manzaneque JB, Caño-Delgado AI. 2019. Drought resistance by engineering plant tissue-specific responses. *Frontiers in Plant Science*, **10**, 1676.

Martín-Sanz A, Pérez de la Vega M, Caminero C. 2012. Resistance to Pseudomonas syringae in a collection of pea germplasm under field and controlled conditions. *Plant Pathology*, **61**, 375–387.

Masouleh SSS, Sassine Y N. 2020. Molecular and biochemical responses of horticultural plants and crops to heat stress. *Ornamental Horticulture*, **26**, 148–158.

Mathew DC, Ho Y-N, Gicana RG, Mathew GM, Chien M-C, Huang C-C. 2015a. A rhizosphere-associated symbiont, Photobacterium spp. strain MELD1, and its targeted synergistic activity for phytoprotection against mercury. *PloS One*, **10**, e0121178.

Mathew DC, Mathew GM, Gicana RG, Huang C-C. 2015b. Genome Sequence of Photobacterium halotolerans MELD1, with mercury reductase (merA), isolated from Phragmites australis. *Genome Announcements*, **3**, e00530–15.

Mayak S, Tirosh T, Glick BR. 2004. Plant growth-promoting bacteria that confer resistance to water stress in tomatoes and peppers. *Plant Science*, **166**, 525–530.

Meena KK, Sorty AM, Bitla UM, Choudhary K, Gupta P, Pareek A, Singh DP, Prabha R, Sahu PK, Gupta VK. 2017. Abiotic stress responses and microbe-mediated mitigation in plants: the omics strategies. *Frontiers in Plant Science*, **8**, 172.

Moghal AAB, Ashfaq M, Al-Shamrani MA, Al-Mahbashi A. 2020. Effect of heavy metal contamination on the compressibility and strength characteristics of chemically modified semiarid soils. *Journal of Hazard Toxic Radioactive Waste*, **24**, 04020029.

Mosa KA, Saadoun I, Kumar K, Helmy M, Dhankher OP. 2016. Potential biotechnological strategies for the cleanup of heavy metals and metalloids. *Frontiers in Plant Science*, **7**, 303.

Mukhtar T, Smith D, Sultan T, Seleiman MF, Alsadon AA, Ali S, Chaudhary HJ, Solieman TH, Ibrahim AA, Saad MA. 2020. Mitigation of heat stress in solanum lycopersicum L. by ACC-deaminase and exopolysaccharide producing bacillus cereus: effects on biochemical profiling. *Sustainibility*, **12**, 2159.

Munns R, Tester M. 2008. Mechanisms of salinity tolerance. *Annual Review of Plant Biology*, **59**, 651–681.

Naz N, Durrani F, Shah Z, Khan N, Ullah I. 2018. Influence of heat stress on growth and physiological activities of potato (Solanum tuberosum L.). *Phyton-International Journal of Experimental Botany*, **87**, 225.

Nievola CC, Carvalho CP, Carvalho V, Rodrigues E. 2017. Rapid responses of plants to temperature changes. *Temperature (Austin)*, **4**, 371–405.

Ojedokun AT, Bello OS. 2016. Sequestering heavy metals from wastewater using cow dung. *Water Resources and Industry*, **13**, 7–13.

Ojuederie OB, Olanrewaju OS, Babalola OO. 2019. Plant growth promoting rhizobacterial mitigation of drought stress in crop plants: implications for sustainable agriculture. *Agronomy*, **9**, 712.

Parida AK, Das AB. 2005. Salt tolerance and salinity effects on plants: a review. *Ecotoxicology and Environmental Safety*, **60**, 324–349.

Peck S, Mittler R. 2020. Plant signaling in biotic and abiotic stress. *Journal of Experimental Botany*, **71**, 1649–1651.

Perner H, Rohn S, Driemel G, Batt N, Schwarz D, Kroh LW, George E. 2008. Effect of nitrogen species supply and mycorrhizal colonization on organosulfur and phenolic compounds in onions. *Journal of Agricultural and Food Chemistry*, **56**, 3538–3545.

Qurashi AW, Sabri AN. 2011. Osmoadaptation and plant growth promotion by salt tolerant bacteria under salt stress. *African Journal of Microbiology Research*, **5**, 3546–3554.

Rashid MI, Mujawar LH, Shahzad T, Almeelbi T, Ismail IM, Mohammad O. 2016. Bacteria and fungi can contribute to nutrients bioavailability and aggregate formation in degraded soils. *Microbiological Research*, **183**, 26–41.

Rengasamy P. 2006. World salinization with emphasis on Australia. *Journal of Experimental Botany*, **57**, 1017–1023.

Rodríguez Graña VM, Soengas Fernández MDP, Alonso-Villaverde Iglesias V, Sotelo Pérez T, Cartea González ME, Velasco Pazos P. 2015. Effect of temperature stress on the early vegetative development of Brassica oleracea L. *BMC Plant Biology*, **15**, 145.

Rosenzweig C, Tubiello FN. 1996. Effects of changes in minimum and maximum temperature on wheat yields in the central US: a simulation study. *Agricultural and Forest Meteorology*, **80**, 215–230.

Rouphael Y, Cardarelli M, Schwarz D, Franken P, Colla G. 2012. Effects of drought on nutrient uptake and assimilation in vegetable crops. In: *Plant Responses to Drought Stress*, pp. 171–195. Heidelberg, Berlin: Springer.

Sanna S, Ryan M, Johansen RB, Dunbar JM. 2019. Plant-microbe interactions before drought influence plant physiological responses to subsequent severe drought. *Scientific Reports*, **9**, 1–10.

Scharf BE, Hynes MF, Alexandre GM. 2016. Chemotaxis signaling systems in model beneficial plant–bacteria associations. *Plant Molecular Biology*, **90**, 549–559.

Schlaeppi K, Dombrowski N, Oter RG, van Themaat EVL, Schulze-Lefert P. 2014. Quantitative divergence of the bacterial root microbiota in Arabidopsis thaliana relatives. *Proceedings of the National Academy of Sciences*, **111**, 585–592.

Schützendübel A, Nikolova P, Rudolf C, Polle A. 2002. Cadmium and H2O2-induced oxidative stress in Populus × canescens roots. *Plant Physiology and Biochemistry*, **40**, 577–584.

Sethy SK, Ghosh S. 2013. Effect of heavy metals on germination of seeds. *Journal of Natural Science, Biology, and Medicine*, **4**, 272.

Sha Valli KP, Nagamallaiah GV, Dhanunjay Rao M, Sergeant K, Hausman JF. 2014. Abiotic stress tolerance in plants: insights from proteomics. In: *Emerging Technologies and Management of Crop Stress Tolerance*, pp. 23–68. Cambridge: Academic Press.

Sharma S, Kulkarni J, Jha B. 2016. Halotolerant rhizobacteria promote growth and enhance salinity tolerance in peanut. *Frontiers in Microbiology*, **7**, 1600.

Shekar CC, Sammaiah D, Shasthree T, Reddy KJ. 2011. Effect of mercury on tomato growth and yield attributes. *International Journal of Pharma and Bio Science*, **2**, B358–B364.

Shinozaki K, Yamaguchi-Shinozaki K. 2007. Gene networks involved in drought stress response and tolerance. *Journal of Experimental Botany*, **58**, 221–227.

Shrivastava P, Kumar R. 2015. Soil salinity: a serious environmental issue and plant growth promoting bacteria as one of the tools for its alleviation. *Saudi Journal of Biological Sciences*, **22**, 123–131.

Smith SE, Smith FA. 2011. Roles of arbuscular mycorrhizas in plant nutrition and growth: new paradigms from cellular to ecosystem scales. *Annual Review of Plant Biology*, **62**, 227–250.

Syed R, Kapoor D, Bhat AA. 2018. Heavy metal toxicity in plants: a review. *Plant Archives*, **18**, 1229–1238.

Sylvia DM, Fuhrmann JJ, Hartel PG, Zuberer DA. 2005. *Principles and Applications of Soil Microbiology*. New Jersey: Pearson.

Taji T, Seki M, Satou M, Sakurai T, Kobayashi M, Ishiyama K, Narusaka Y, Narusaka M, Zhu J-K, Shinozaki K. 2004. Comparative genomics in salt tolerance between Arabidopsis and Arabidopsis-related halophyte salt cress using Arabidopsis microarray. *Journal of Plant Physiology*, **135**, 1697–1709.

Tavakkoli E, Rengasamy P, McDonald GK. 2010. High concentrations of Na+ and Cl–ions in soil solution have simultaneous detrimental effects on growth of faba bean under salinity stress. *Journal of Experimental Botany*, **61**, 4449–4459.

Teixeira EI, Fischer G, Van Velthuizen H, Walter C, Ewert F. 2013. Global hot-spots of heat stress on agricultural crops due to climate change. *Agricultural and Forest Meteorology*, **170**, 206–215.

Tiwari S, Lata C. 2018. Heavy metal stress, signaling, and tolerance due to plant-associated microbes: an overview. *Frontiers in Plant Science*, **9**, 452.

Uddin K, Juraimi AS, Ismail MR, Othman R, Rahim AA. 2009. Growth response of eight tropical turfgrass species to salinity. *African Journal of Biotechnology*, **8**, 5799–5806.

Uddin MK, Juraimi AS, Ismail MR, Hossain MA, Othman R, Rahim AA. 2011. Effect of salinity stress on nutrient uptake and chlorophyll content of tropical turfgrass species. *Australian Journal of Crop Science*, **5**, 620.

van der Heijden MG, Martin FM, Selosse MA, Sanders IR. 2015. Mycorrhizal ecology and evolution: the past, the present, and the future. *New Phytologist*, **205**, 1406–1423.

Vardharajula S, Zulfikar Ali S, Grover M, Reddy G, Bandi V. 2011. Drought-tolerant plant growth promoting Bacillus spp.: effect on growth, osmolytes, and antioxidant status of maize under drought stress. *Journal of Plant Interaction*, **6**, 1–14.

Verma M, Mishra J, Arora NK. 2019. Plant growth-promoting rhizobacteria: diversity and applications. In: *Environmental Biotechnology: For Sustainable Future*, Edited by CS Ranbir, KA Naveen and K Richa, pp 129–173. Singapore: Springer. https://doi.org/10.1007/978-981-10-7284-0.

Wallenstein MD. 2017. Managing and manipulating the rhizosphere microbiome for plant health: a systems approach. *Rhizosphere*, **3**, 230–232.

Wang D, Heckathorn SA, Mainali K, Tripathee R. 2016. Timing effects of heat-stress on plant ecophysiological characteristics and growth. *Frontiers in Plant Science*, **7**, 1629.

Wang R, Wang J, Zhao L, Yang S, Song Y. 2015. Impact of heavy metal stresses on the growth and auxin homeostasis of Arabidopsis seedlings. *Biometals*, **28**, 123–132.

Wang X, Sun R, Tian Y, Guo K, Sun H, Liu X, Chu H, Liu B. 2020. Long-term phytoremediation of coastal saline soil reveals plant species-specific patterns of microbial community recruitment. *Msystems*, **5**, e00741–19.

Wilhite DA. 2016. Global drought detection and impact assessment from space. In: *Droughts*, pp 226–239. Routledge.

Xie K, Guo L, Bai Y, Liu W, Yan J, Bucher M. 2019. Microbiomics and plant health: an interdisciplinary and international workshop on the plant microbiome. *Molecular Plant*, **12**, 1–3.

Yadav S. 2010. Heavy metals toxicity in plants: an overview on the role of glutathione and phytochelatins in heavy metal stress tolerance of plants. *South African Journal of Botany*, **76**, 167–179.

Zhang J, Zhang S, Cheng M, Jiang H, Zhang X, Peng C, Lu X, Zhang M, Jin J. 2018. Effect of drought on agronomic traits of rice and wheat: a meta-analysis. *International Journal of Environmental Research and Public Health*, **15**, 839.

Zhang W, Xie Z, Zhang X, Lang D, Zhang X. 2019. Growth-promoting bacteria alleviates drought stress of G. uralensis through improving photosynthesis characteristics and water status. *Journal of Plant Interaction*, **14**, 580–589.

Zhang Y, Du H, Xu F, Ding Y, Gui Y, Zhang J, Xu W. 2020. Root-bacteria associations boost rhizosheath formation in moderately dry soil through ethylene responses. *Plant Physiology*, **183**, 780–792.

Zhou Q, Wang X, Liang R, Wu Y. 2003. Effects of cadmium and mixed heavy metals on rice growth in Liaoning, China. *Soil and Sediment Contamination*, **12**, 851–864.

13 Plant–Microbe Interactions Significance in Sustainable Agriculture

Gowardhan Kumar Chouhan, Saurabh Singh,
Arpan Mukherjee, Anand Kumar Gaurav,
and Jay Prakash Verma
Banaras Hindu University, Uttar Pradesh, India

CONTENTS

13.1 INTRODUCTION

In nature, plants and animals always interact with various microbes during their life cycle. From an early time of evolution, humans are spending their lives associated with diverse microbes that help them adapt to a normal healthy life (Kirjavainen *et al.*, 2019). Similarly, plants also cohabit with various microbes such as bacteria, fungi, archaea, and protists together called microbiota (Bulgarelli *et al.*, 2013; Das *et al.*, 2021). The interaction of microbes with their host plants may either be positive (interaction of microbes with their host plant conferring stress tolerance – biotic and abiotic; promote plant growth and development) or negative (an interaction between microbes and their host plant leading to disease development) may play a significant role in sustainable agriculture (Newton *et al.*, 2010; Chouhan *et al.*, 2021a). The plant provides various flavonoids to microbes as a signal that make

DOI: 10.1201/9781003110477-13

effective communication between plant and microbes for their growth and development (Hu¨ckelhoven, 2007). The establishment of effective communication between plants and microbes may affect the plant both positively and negatively. In positive relationships, they show interactions such as symbiosis, mutualism, and commensalism to enhance plant immunity (Badri *et al.*, 2009). In the same way, in negative relationships, microbes release various chemicals that can alter the composition of root exudates that can negatively change the physiology of plants (Pritchard and Birch, 2011). Plants release a large number of root exudates and deposits (accounting up to 40% of plant photosynthesized carbons) such as sugar and amino acids, organic acids, and polymerized sugars to the soil to tackle pathogenic microbes and promote the growth of beneficial microbes (Badri *et al.*, 2009; Glick, 2014; Chouhan *et al.*, 2021b; Liu *et al.*, 2021). In general, root exudates contain various primary and secondary metabolites, which facilitate communication between plants and microbes and play a significant role for the better growth and development of plants (Kumar *et al.*, 2014; Kumar and Verma, 2019). Shoot also serve as a deck for the interaction of these microbes by providing various secondary metabolites. Microbes associated with shoot can also synthesize secondary metabolites. In addition, microbes associated with the plant have the potential for bioremediation. Soil pollutants such as heavy metals, polyaromatic hydrocarbons, surfactants, emulsifiers, herbicides/insecticides, and other complex organic and inorganic contaminants are major factors that cause serious environmental hazards such as decreased soil fertility causing plant tissue damages (Santillo and Johnston, 2003). Phytosiderophore produced by the plant under Fe- or Zn-deficient conditions chelate with Fe or Zn and help in sequestration of these metals' ions and soil bioremediation. Currently, many studies have been done on plant growth-promoting microbes, which can be used as consortia to mitigate the problems of soil pollution. Moreover, a number of studies have been done in the field of molecular biology such as genomic, metagenomic, proteomic, metabolomic and transcriptomic that can reveal the process of plant–pathogen interaction (Kaul *et al.*, 2016; Verma *et al.*, 2017). In addition, studies on how the plant defence system reacts when the pathogens attack and how pathogens overcome the plant defence system leading to the emergence of the disease have also been done. The overall mechanism of pathogen attack and the plant defence system against diverse pathogenic microbial species is a challenging task for humankind (Li *et al.*, 2013). Environmental factors such as light, temperature, and carbon flux are also important factors that can significantly affect the plant–microbe interaction (Hua, 2013). Physical, chemical, and biological factors also regulate the plant–microbe relation governing a complex biological activity such as biofertilization, bioremediation, and biocontrol activity (Singh *et al.*, 2017). Thus, an understanding of the mechanism between microbes and their host plant with the changing environmental conditions may give the overall idea about the interrelationship, either beneficial or harmful and have a significant impact on sustainable agriculture. In this chapter, we discuss an overview of the significance of plant–microbe interaction and its mechanism. In addition, we discuss the use of potential microbes, which can be used in the form of consortia as biofertilizers, in rhizoremediation, and as biocontrol agents that can enhance agriculture production under sustainable agriculture, including the future research directions.

13.2 PLANT–MICROBE INTERACTION FOR INCREASED CROP YIELD UNDER INCREASING DEMAND FOR FOOD

In nature, every organism for its growth and development has to manage association with neighbours. From the very beginning, coevolution between hosts and their microbial associations plays a critical role in host fitness and health that provide a dynamic effect on agricultural systems (Chisholm et al., 2006; Jones and Dangl, 2006). Plant for their better growth and survival make an association with other plants and microbes for nutrient acquisition, assimilation, and development of the immune system against various biotic and abiotic factors (Mukherjee et al., 2019; 2021). In general, microbes may have a different range of associations with their host plants such as mutualism, interspecific cooperation, commensal and disease-causing, and parasitism (Hu¨ckelhoven, 2007; Links et al., 2014,). These establishments form an array of interactions in which both partners (plant and their associated microbes) may affect either negative or positive. Although the percentage of pathogenic microbe in the associated microbial community is low, they significantly influence plant health and development, leading to low agriculture production. The first level of the innate immune response system of plants responds initially to microbe-associated molecular patterns which may follow the same pattern for pathogenic and pathogenic microbes (Dodds and Rathjen, 2010). This interactive structure of the plant immune response system and microbe-associated molecular patterns provide unique molecular insight for cell recognition, cell biology, and evolutionary change across the biological kingdoms. This mechanism gives us an idea to understand plant immune system functions and how we deal with them by using the novel biotechnological approach to crop improvement, protection, and their production under sustainable agriculture. An agriculture point of view, microbial behaviour, including mutualistic or pathogenic describes important attributes between plant and microbe relationship. Understanding the specific interaction mechanism between them can significantly enhance crop production under sustainable agriculture (Li et al., 2013; Chouhan et al., 2021c). A number of studies show that plant growth-promoting rhizobacteria (PGPR) have the potential that can be used as biofertilizer, rhizoremediation, and biocontrol agents (Krishna et al., 2021). Manipulation of these microbes has a significant potential to enhance agriculture production and reduce plant diseases, chemical fertilizers, and pesticides.

13.2.1 PLANT–MICROBE INTERACTIONS

13.2.1.1 Beneficial versus Pathogenic Microbes

A diverse number of microbes form microbial communities associated with the host plant, showing either positive or negative effects on a host plant. Plant perceives both types of microbes including beneficial and pathogenic as intruders and thus mount defence reactions against them. But the successful interaction of microbes with their host depends on strategy prepared by microbes so that can successfully infect the plant tissues and surpass the plant defence system (Soto et al., 2006). Plant–microbe interaction mainly depends upon the ability of both candidates to reconcile their respective preference of a mode of action that may be pathogenic or mutualistic depending upon chemical signalling between them (Soto *et al.*, 2009). Harmful microbes firstly colonize in apoplast or xylem of plant and release some toxins and hydrolytic enzymes lead to plant disease. In contrast, beneficial microbes provide beneficial interaction based on a nutrient exchange (De La Fuente and Burdman, 2011, Kafle *et al.*, 2018). In general, a plant recognized microbes by two different processes such as plant transmembrane pattern recognition receptors (PRRs) recognize pathogen-associated molecular patterns (PAMPs) and resistance-genes (R-Genes) products such as polymorphic nucleotide-binding proteins and leucine-rich repeat proteins (NB-lRR). The primary encounter with pathogen effectors by R-Genes products activates the active defence responses of the host plant (Dodds and Rathjen, 2010). The "zigzag" model is one of the most accepted models of pathogen–plant interactions that categorize four different phase interactions between pathogen and their host plant (Jones and Dangl 2006). In the first phase, transmembrane PRRs recognize PAMPs and initiate PAMP-triggered immunity (PTI) to suppress the pathogenic microbes and develop resistance against future pathogenic invaders. In the second phase, if pathogenic microbes enter in cells of plant tissue then they release effector substances that can interfere with PTI and lead to nonfunctioning of its activity resulting in effector-triggered susceptibility (ETS). In the third phase, in some cases, these effector substances of pathogenic microbes recognized by R-gene products resulting in effector-triggered immunity (ETI) that triggers hypersensitive response leads to cell death thereby no further damage from pathogenic microbial attack. Finally, in the fourth phase, to reduce the effect of ETI of host plant against pathogens, either some potential pathogenic microbes modify previously recognized effector molecules or produce evolved effector molecules. Recent molecular studies reveal that symbiotic microbes act as "intelligent pathogens." Global transcription analysis of leguminous plants such as *Lotus japonicus* (Deguchi *et al.*, 2007) and *Medicago truncatula* (Lohar *et al.*, 2006) show the upregulation of defence-related genes when plant seed is treated with beneficial microbes. These beneficial microbes decline in number with nodule development to accord with the notion that a very fine line separates pathogenic microbes from beneficial symbiotic microbes (Djordjevic *et al.*, 1987). There are mainly two types of secretion systems including type III or type IV secretion systems (T3SSs or T4SSs) that decide the movement of effector molecules from the cytoplasm of microbes to plant cytoplasm that results in mutualism or pathogenesis. A number of studies were done on various leguminous crops such as *Rhizobium* sp. NGR234 (Zehner *et al.*, 2008), *Mesorhizobium loti* R7A (Hubber *et al.*, 2004), and *Mesorhizobium loti* MAFF303099 (Hubber *et al.*, 2004) show that the cluster of genes follows type III secretion systems (T3SSs) (Pueppke and Broughton, 1999) that decide the phenomena of mutualistic behaviour of microbes with their host plant. The pathogenic microbes suppress the plant immunity by releasing effector molecules that trigger the PAMPs altering the conformational and biochemical changes in host protein, RNA metabolism, and activity of kinase enzyme involved in plant protection against plant pathogens (Block *et al.*, 2008). Rhizobial microbes secrete effector molecules that follow the same kind of function as proteins secreted by pathogens to establish a respective interaction with plants (Dai *et al.*, 2008; Kambara *et al.*, 2009). Till date, these interactions are multifarious and our current understanding on interactions of symbiotic or pathogenic

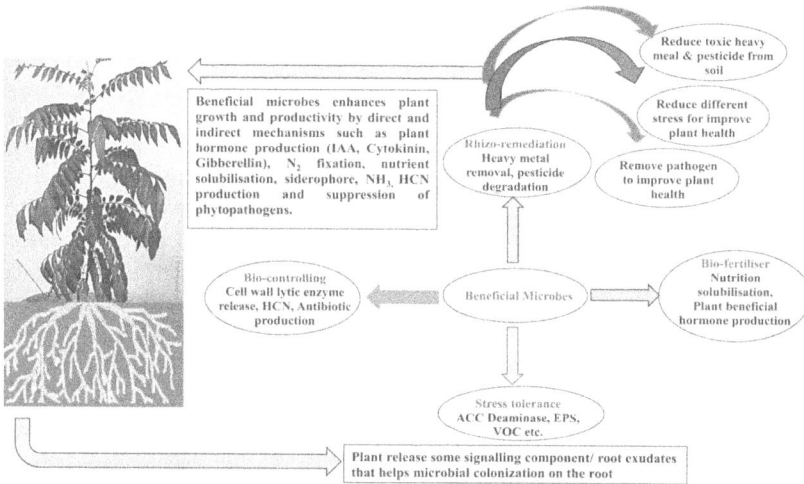

FIGURE 13.1 Plant growth-promoting activities of beneficial microbes as Biofertilizers, biocontrolling agents, stress tolerance and rhizoremediation) for direct and indirect promoting plant growth and development.

microbes in respect to various environmental factors such as biotic and abiotic factors remain poorly known (Deakin and Broughton, 2009). Microbes associated with their host plant show complex interactions with them, mutualistic and pathogenic association. Microbes with beneficial interaction with the plant are described in Figure 13.1.

13.2.1.2 Biofertilizer Application: A Sustainable and Eco-friendly Agriculture Practice

Indiscriminate use of chemical fertilizers and pesticides for agriculture production can have negative health effects on human health and other biotic components of ecosystems, which is a great concern in the context of sustainable agriculture. Apart from being costly, these synthetic chemicals contaminate the environment, cause soil salinity, decreasing the nutritional quality of food produce and eutrophication of surface water (Youssef and Eissa 2014). These agrochemicals definitely increase crop production to some extent but in long-term use, they show negative effects on human health (Mukherjee et al., 2021). Indiscriminate use of these chemicals in agroecosystem leads to decreased soil fertility and increased salt content in the agriculture field (Aggani 2013) that may increase greenhouse gas emission as well as acidification of soil and water (Kumar *et al.* 2018; Singh et al., 2020). They make it more vulnerable to plant disease by weakening its root system. The symbiotic aspect between microbes and their host plant can play a remarkable role as a biofertilizer that has the potential to boost agriculture production without harming the environmental system. In symbiosis, the microbes utilize simple nitrogen from the atmosphere and convert them into a simple form such as ammonium and nitrate that can easily be available to the plant, in turn, they obtain carbon sources as food material from their host plant. Symbiotic microbes such as *Rhizobium*, *Azorhizobium*, and *Sinorhizobium* genera act as biofertilizers, estimated at about 65% of total nitrogen supply to agri-crops world (Prasad *et al.*, 2014). Most potential symbiotic PGPR microbes such as *Rhizobium*, *Mesorhizobium*, *Sinorhizobium*, *Azorhizobium*, *Bradyrhizobium*, and *Allorhizobium* are frequently used worldwide for enhancing agriculture production. Currently, a huge number of biofertilizers have been used based on a symbiotic rela-tionship between microbes and their host plant that can be a good alternative to chemical fertilizer and boost crop production under sustainable agriculture (Kumar *et al.*, 2009). However, further studies are needed at different adverse biotic and abiotic conditions such as drought, salinity, high soil temperatures, adverse soil pH, presence of organic acids to improve the fertility of agriculture soil and overall effi-ciency of the biofertilization process. An elaborate description of the interactions and their role is given in Table 13.1.

13.2.1.3   Rhizoremediation: Microbes-Assisted Phytoremediation

Rhizoremediation is a subprocess of phytoremediation where plants and their asso-ciated microbes work together to extract toxic metals from soil (Jing *et al.* 2007; Hao *et al.*, 2014). This is a beneficial approach between plants and microbes where microbes make metals bioavailable to plant; in turn, plants help in the extraction of such toxic waste from contaminated sites (Chaudhry *et al.* 2005). Over the last century, an unprecedented increase of industrialization led to imposing the burden of chemicals, metals, pesticides, and salt in the agroecosystem (Huettermann *et al.*, 2009). These anthropogenic chemicals show highly negative effects on agriculture soil, crop productivity, quality of food material, and human health. Based on recent studies, plant–microbe interaction has been exploited to sort out these problems, which are not only cheaper but also eco-friendly. Citric acid and Oxalic acid from a wild grass *Echinochloa crus-galli* significantly increase both translocation and bioaccumulation of metals such as Cd, Cu, and Pb, which indicate that organic acids have the potential to chelate such toxic metals (Kim *et al.* 2010). Genoproteomic machinery of both partners (plant and microbes) helps them in their survival and growth in changing environmental conditions. The microbial population around the rhizospheric region is directly proportional to root exudates produced by plant root, leading to the dynamic interaction between them (Kumar et al., 2021). Root exudates released by a plant in the rhizospheric region help them survive in distressed condi-tions either by affecting other microbes or plants or by extracting toxic waste from contaminated sites (Luo *et al.*, 2012). Since pollutants present in the region directly or indirectly affect the plant growth, its removal by means of natural alternative could be a good alternative to enhance crop production under sustainable conditions.

13.2.1.4   Biocontrol Activity

Worldwide, the primary cause of loss of agriculture production is plant disease. According to FAO, 25% of loss of crop production is mainly due to plant disease and pests. Therefore, an alternative eco-friendly natural system needs to control plant disease and pests (Savary *et al.* 2012). The use of biocontrol agents to prevent plant disease and pests is one of the most prominent techniques, which could be an alter-native to synthetic chemicals. In the past few decades, this phenomenon has been widely used all over the world to control plant disease under the umbrella of the

TABLE 13.1

Plant–Microbe Interaction and Factors Regulated in Different Studies

Plant Studied	Microbes Used	Factors Regulated	Isolation Source	References
M. micrantha	*Mikania micrantha*	Potassium solubilization, Increase bioavailability of potassium to plants, 2-fold increase in K contents in plants	Rhizospheric soil	Sun *et al.*, 2020
Avicennia marina	*Bacillus altitudinis, B. anthracis and B. marisflavi, Penicillium citrinum, Aspergillus quadrilineatus and Gibberella intermedia*	Heavy metal accumulation and Zn solubilization	Wastewater treated mangrove soil	Kayalvizhi and Kathiresan, 2019
Malus domestica L. Borkh	*Paenibacillus mucilaginosus*	Potassium solubilizing, Increased phytohormones and organic acids production.	NA	Chen *et al.*, 2020
Temperate forest	*Pseudomonas*	Increase in Si, P solubilization followed by decrease in pH which consequently related to secretion of organic acids.	NA	Pastore *et al.*, 2020
Punica granatum L.	*Penicillium pinophilum*	Potassium solubilization, higher dehydrogenase activity, alkaline activity, acid phosphatase activity, microbial biomass carbon content.	Pomegranate rhizosphere in semi-arid ecosystem	Maity *et al.*, 2019
Camellia sinensis (L.) O. Kuntze	*Bacillus subtilis, Burkholderia cepacian, Pseudomonas putida.*	Higher K solubilization, followed by production of plant growth promoting substances.	Southern Indian tea plantation soils	Bagyalakshmi *et al.*, 2017
Oryza sativa L.	*Bacillus pumilus strain JPVS11*	High indole-3-acetic acid (IAA),1-aminocyclo propane-1-carboxylicacid (ACC) deaminase activity, P-solubilization, proline accumulation and exopolysaccharides (EPS) production, conferring salinity stress.	Sodic soil of eastern Uttar Pradesh, India	Kumar *et al.*, 2020b

(Continued)

TABLE 13.1 (Continued)
Plant–Microbe Interaction and Factors Regulated in Different Studies

Plant Studied	Microbes Used	Factors Regulated	Isolation Source	References
Phaseolus vulgaris	Rhizobium etli	Pest protection and plant growth promotion. Specific antibiotic, oxalate oxides, HCN solubilization, and secrete lytic enzymes, which forms a protective shield in its rhizosphere.	NA	Kumar et al., 2020a; Kumar et al., 2020c
Coriandrum sativum L.	Not identified	Lead resistant microbes, Improved growth, photosynthesis, and antioxidant enzyme activities of the plants under Pb stress	NA	Fatemi et al., 2020
Triticum aestivum L.	Bacillus megaterium, Arthrobacter chlorophenolicus and Enterobacter sp.	Improved plant growth, nutrient acquisition.	Soil samples from rice–wheat, vegetables (Cabbage, Spinach, Ladyfinger, and Bottle guard), agro-forestry (Mango, Papaya, and Guava) and grassland rhizosphere from Indo-Gangetic plains of India.	Kumar et al., 2014
Perennial ryegrass and tall fescue	-	metal-resistant PGPR microbes, enhanced plant growth by secreting plant growth enzymes such as IAA, dissolved phosphorous in presence of metal stress with heavy contamination of Copper and Cadmium.	Metal contaminated mines.	Ke et al., 2020
Avena sativa, Medicago sativa, and Cucumis sativus	Providencia rettgeri P2, Advenella incenata P4, Acinetobacter calcoaceticus P19, and Serratia plymuthica P35	Microbes increased plant growth properties, soil enzyme activities, available nutrients activity, plant hormones which ultimately resulted in increased plant yield.	Rhizosphere soil of Trifolium pratense and Polygonum viviparum	Li et al., 2020

(Continued)

TABLE 13.1 (Continued)
Plant–Microbe Interaction and Factors Regulated in Different Studies

Plant Studied	Microbes Used	Factors Regulated	Isolation Source	References
Arachis hypogaea	EX-1 (*Acinetobacter baumannii stain HAMBI 1846*); EX-3 (*Pseudomonas aeruginosa strain A1K319*); EX-5 (*Bacillus subterraneus strain CF1.9*); KNL-1 (*Bacillus subtilis strain JMP-B*); CTR-4 (*Enterobacter cloacae strain VITKJ1*); ANT-4 (*Bacillus subtilis strain SBMP4*) and Group-2 includes EX-4 (*Pseudomonas otitidis strain SLC8*); KDP-4 (*Pseudomonas aeruginosa strain Kasamber 11*); NLR-4 (*Bacillus species ADMK68*); ANT-6 (*Bacillus subtilis subsp. inaquosorum strain KCTC 13429*)	Multiple plant growth characteristics, including phosphate solubilization, production of HCN and Indole acetic acid along with broad antagonism against Aspergillus niger; A. flavus; Fusarium oxysporum	Rhizosphere and Non-rhizosphere zones of groundnut fields	Syed et al., 2020
Oryza sativa L.	*Pseudomonas and Bacillus*	Arsenic tolerance as well as promotion of plant growth	Rhizosphere soil from As-contaminated paddy field near Baoshan mine	Xiao et al., 2020
Foeniculum vulgare L. and Phaseolus vulgaris L.	*Azotobacter vinelandii + Rhizobium phaseoli, Pseudomonas putida and Pantoea agglomerans, Pseudomonas koreensis and P.vancouverensis*	Plant growth promotion as well as essential oil and fatty acid composition content increment. Seed yield increased by 24%.	NA	Rezaei-Chiyaneh et al., 2020

(Continued)

TABLE 13.1 (Continued)
Plant–Microbe Interaction and Factors Regulated in Different Studies

Plant Studied	Microbes Used	Factors Regulated	Isolation Source	References
Mentha pulegium L.	*Azotobacter chroococcum, Azospirillum brasilense,*	Improved physiological and phytochemical parameters, plant growth promotion activity was observed under drought stress.	NA	Asghari *et al.*, 2020
Oryza sativa L.	*Enterobacter aerogenes MCC 3092*	Cadmium resistance, plant growth promotion.	Soil of metal contaminated field	Pramanik *et al.*, 2018
Arabidopsis thaliana	*Bacillus amyloliquefaciens (SN13 strain)*	Plant growth promotion by inducing OsASR6 gene, altering root auxin sensitivity, and the xylem structure in transgenic plants.	NA	Agarwal *et al.*, 2019
Vigna radiata	*Pseudomonas*	Induced plant growth promotion by secretion of different plant hormones. Potent antifungal activity against Rhizoctonia solani.	Rhizosphere of mung bean	Kumari *et al.*, 2018
Scirpus grossus	*Aeromonas taiwanensis isolate 5E, Bacillus sp. Isolate 7G, Bacillus cereus isolate 8H and 3C, Bacillus velezensis isolate 9I, Bacillus proteolyticus isolate 4D, Bacillus stratosphericus isolate 14N, Bacillus megaterium isolate 11K, Pseudomonas sp. Isolate 12L, Enterobacter cloacae isolate 13M and isolate 16P, Bacillus aerius isolate 15O and Lysinibacillus sp. isolate 10J*	Heavy metal Pb resistance, plant growth promotion.	Rhizosphere region of Scirpus grossus.	Kamaruzzaman *et al.*, 2020

(Continued)

TABLE 13.1 (Continued)
Plant–Microbe Interaction and Factors Regulated in Different Studies

Plant Studied	Microbes Used	Factors Regulated	Isolation Source	References
Arachis hypogea	*Bacillus licheniformis A2*	Plant growth promotion, salinity stress tolerance, phosphate solubilization, and IAA production.	Saline desert of Kutch	Goswami *et al.*, 2014
Solanum tuberosum L.	*Bacillus amyloliquefaciens, Pseudomonas brassicacearum, Lysinibacillus boronitolerans*	Phosphate solubilization, IAA, ammonia, HCN and siderophore production. Plant growth promotion. Showed antifungal activity against Pythium sp. and Fusarium sp.	Rhizospheric and non-rhizospheric soil (bulk) samples from Central Potato Research Station, of three different potato varieties cv. Kufri sindhuri, Kufri chipsona-3 and Kufri lauvkar.	Pathak *et al.*, 2019
Zea mays L.	*Pseudomonas sp. SUT 19 and Brevibacillus sp. SUT 47*	ACC-deaminase, P-solubilization, and plant growth promotion.	Roots of forage *Zea mays L.*	Piromyou *et al.*, 2011
Arachis hypogaea	*Ochrobactrum intermedium*	Temperature and salinity stress resistant, plant growth promotion, lipid component modification.	Rhizosphere of peanut plants, of central and south Córdoba, Argentina	Paulucci *et al.*, 2015

biocontrol process. Several bacterial microbes have been used in the form of consortia to control plant disease and pests such as *Bacillus thuringiensis*, *Pseudomonas* sp., *Streptomyces* and fungi such *as Gliocladium*, *Fusarium*, and *Trichoderma* (Bakker *et al.*, 2014; Kumar *et al.*, 2017; Nielsen *et al.* 2000; Mukherjee et al., 2020). In general, biocontrol agents protect the plant by a variety of mechanisms such as by secretion of secondary metabolites, hydrolytic enzymes, cell wall-degrading enzymes, niche exclusion, and competition for nutrients. By use of microbial consortia as seed inoculant such as *Bacillus* sp., *Trichoderma*, and *Pseudomonas* sp. can protect the cucumber and tomato plant from a variety of pathogenic microbes such as *Alternaria*, *fusarium*, and *Pythium* (Kumar *et al.*, 2014; Xu *et al.*, 2014; Singh and Siddiqui, 2015). Therefore, microbial consortia have been paid attention by researchers to protect the plant from various diseases and pests as compared to synthetic chemicals.

13.3 DRIVING FACTORS OF THE PLANT–MICROBE (PM) INTERACTIONS

Interaction of plants with their hosts and the response after the interaction is mainly based on a number of stimuli factors such as physical, chemical, and environmental. The successful interaction phenomena between plant and their host microbes are mainly due to the commutative effect of all these three factors. Any physical signal deploys susceptible mechanosensory cells of *Mimosa pudica* and *Dionaea* species of plants to respond in various ways (Johnson *et al.*, 2014). During physical stimulation, the plant triggers selective pathways such as the calmodulin pathway that mimic the microbial infestation and thereafter plant develops resistance (Pieterse *et al.*, 2014). Some microbes use some physical add for infection, for example, peg or hyphae help pathogens to utilize sheer mechanical force to breach further into plant tissue (Wang and Qi, 2015). These stimuli cause various responses from a plant such as cytoplasmic streaming, microtubule re-aggregation, and membrane reorganization with the molecular reaction in response to various microbial infections (Ueda *et al.*, 2015). The mechanism underlying the chemical signal perception and response between the plant and its associated microbes is unclear. Either the association of microbes with their host work positively or negatively in which the chemical signal determines the plant and soil community's dynamics (Kumar et al., 2021). These signals can be exploited by the use of genetic engineering technologies for the higher production of crops (Babar *et al.*, 2016; Singh *et al.*, 2018). Environmental factors conclude the biotic and abiotic factors, which determine the interaction and response between the plant and microbes. They can directly modulate the molecular mechanism within microbes that lead to interactive patterns between microbes and their host. For example, in nutrient-deficient conditions, soil favour the growth of siderophore-producing microbes that show the beneficial effect for both partners in stress conditions. In addition, they also show a negative effect (bactericidal role) for pathogenic microbes (Ahmed and Holmstro¨m, 2014; Wackett, 2014). Environmental factors such as light, temperature, carbon flux, agricultural practices and cropping systems are the main drivers that significantly influence the interactions of plants and their associated microbes (Valentı´n-Vargas *et al.*, 2014). Schematic representation of factors associated with plant and microbial interactions is given in Figure 13.2.

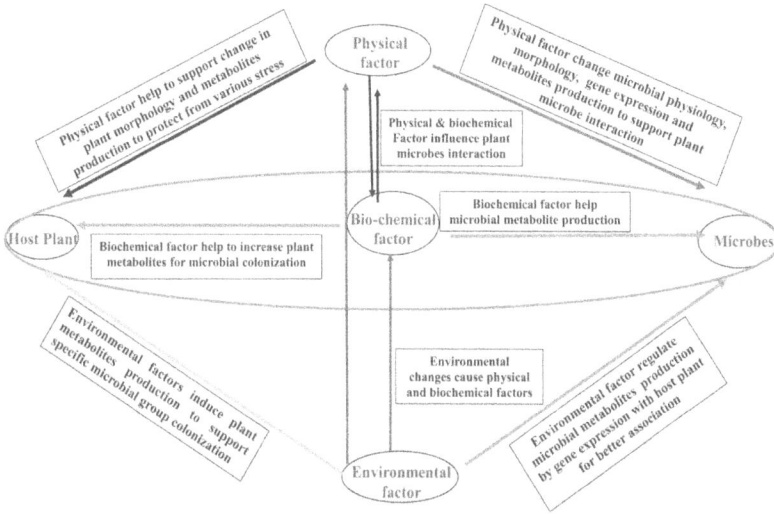

Physical factor help to support change in plant morphology and metabolites production to protect from various stress

Physical factor change microbial physiology, gene expression and microbe interaction to support plant

Physical factor

Physical & biochemical Factor influence plant microbes interaction

Biochemical factor help microbial metabolite production

Bio-chemical factor

Host Plant

Biochemical factor help to increase plant metabolites for microbial colonization

Microbes

Environmental factors induce plant specific microbial group colonization to support metabolites production

Environmental changes cause physical and biochemical factors

Environmental factor regulate microbial metabolites production by gene expression with host plant for better association

Environmental factor

FIGURE 13.2   Diagrammatic representation of different factors (Physical, biochemical and environmental factor) and their impacts associated with plant-microbe interaction.

13.4   BIOTECHNOLOGICAL APPLICATIONS OF PLANT–MICROBE INTERACTIONS

Although there is wide use of plant–microbe interactions to enhance agriculture production, still there are widespread concerns regarding the use of chemical signalling between plant and their associated microbes, and their interactive coordination to overcome the problems of soil pollution. A number of studies have been done in molecular biology such as genomics, metagenomics, proteomics, metabolomics and transcriptomics that can be used to identify traits that maximize the benefits of modern agricultural practices and manipulate the plant–microbe relation towards sustainable agriculture production. Microbes can either act beneficial interactions such as biofertilization, bioremediation, and biocontrol, or act as pathogenic interactions that can cause diseases in the plant. However, the pathogenesis or microbial infection is just a functional phase of microbes that occurs at the propagating phase of their lifecycle. Hence, the use of modern technology in agriculture practice and awareness about plant and microbe interaction tends to focus on more crop production as well as eliminate the pathogenic microbes making it more effective under sustainable agriculture (Li *et al.*, 2013; Newton *et al.*, 2010). Treatment of roots with a variety of PGPR in terms of biofertilizer, inoculation of the plant with a number of bacterial and fungal species in the form of biocontrol agents and plant–microbe interactive consumption of soil pollutants in the form of rhizoremediation are some of the recent progressions under sustainable agricultural production having low-input biotechnology prospects. The application of whole microbes such as bacteria and fungus for crop production becomes old technology while the recent methodology lends to the use of manipulated microbes with the help of biotechnological means (Bourras *et al.*, 2015). Along with better growth and development of crops under sustainable agriculture, humankind may also get benefitted by getting the number of medical and economic benefits that is coming from higher expression secondary metabolites such as alkaloids, volatile oils, and antibiotics by the use of molecular pharming technology (Maag *et al.*, 2015). Hence, the molecular insights of plant and associated microbes show advancement in agriculture practice. Moreover, manipulation of the desired plant–microbe interaction conquest with the help of molecular techniques should facilitate more advances in farming practice.

13.5   CONCLUSION AND FUTURE PERSPECTIVES

Plant–microbe interactions significantly influence the microbial community associated with their host, plant health and development, and biogeochemical cycling in terrestrial ecosystems. Moreover, plant–microbe interaction also plays an important role in the formation of biofertilizer, biocontrol, and bioremediation agents to enhance crop production under sustainable farming practices. However, a number of studies have been carried out on plant–microbe interaction but the exact molecular mechanism between plant and associated microbes for the transportation of signal molecules are still unknown, especially transportation of signal molecules across the cell membrane during the beneficial and pathogenic association. Therefore, understanding the mechanism between microbes and their host plant may provide an overall idea about the biological phenomenon to boost plant health and overcome plant disease and risk management problems. In addition, climatic factors along with various biotic and abiotic factors play a crucial role in the plant–microbe interaction. Thus, future research should be focused on deciphering the mechanism of plant–microbe interactions with changing environmental conditions. The understanding of the interaction mechanism between plant and their associated microbes could help minimize the use of chemical input in the agriculture field, plant disease, and pathogen outbreaks enhancing crop production under sustainable agriculture.

ACKNOWLEDGEMENT

Authors thankful to DST (DST/INT/SL/P-31/2021; DST/SEED/SCSP/STI/2020/426/G), SERB (EEQ/2021/0001083) and Banaras Hindu University-IoE (6031) for providing financial assistance to start research work on plant–microbe interaction study for sustainable agriculture.

REFERENCES

Agarwal, P., Singh, P. C., Chaudhry, V., Shirke, P. A., Chakrabarty, D., Farooqui, A., Nautiyal, C. S., Sane, A. P., & Sane, V. A. (2019). PGPR-induced OsASR6 improves plant growth and yield by altering root auxin sensitivity and the xylem structure in transgenic Arabidopsis thaliana. Journal of Plant Physiology, 240, 153010.

Aggani S. L. (2013). Development of bio-fertilizers and its future perspective. Sch Acad J Pharm, 4, 327–332.

Ahmed, E., & Holmstro¨m, S. J. (2014). Siderophores in environmental research: roles and applications. Microb Biotechnol, 7 (3), 196–208.

Asghari, B., Khademian, R., & Sedaghati, B. (2020). Plant growth promoting rhizobacteria (PGPR) confer drought resistance and stimulate biosynthesis of secondary metabolites in pennyroyal (Mentha pulegium L.) under water shortage condition. Scientia Horticulturae, 263, 109132.

Babar, M. M., Khan, S. F., Zargaham, M. K., & Gul, A. (2016). Plant-microbe interactions: a molecular approach. In: Hakeem, K., Akhtar, M. (Eds.), Plant, Soil and Microbes. Springer, Cham.

Badri, D. V., Weir, T. L., van der Lelie, D., & Vivanco, J. M. (2009). Rhizosphere chemical dialogues: Plant-microbe interactions. Current Opinion in Biotechnology, 20(6), 642650.

Bagyalakshmi, B., Ponmurugan, P., & Balamurugan, A. (2017). Potassium solubilization, plant growth promoting substances by potassium solubilizing bacteria (KSB) from Southern Indian tea plantation soil. Biocatalysis and Agricultural Biotechnology, 12, 116–124.

Bakker, P. A., Ran, L., & Mercado-Blanco, J. (2014). Rhizobacterial salicylate production provokes headaches!. Plant Soil, 382 (1–2), 1–16.

Block, A., Li, G., Fu, Z. Q., & Alfano, J. R. (2008). Phytopathogen type III effector weaponry and their plant targets. Curr Opin Plant Biol, 11 (4), 396403.

Bourras, S., Rouxel, T., & Meyer, M. (2015). Agrobacterium tumefaciens gene transfer: how a plant pathogen hacks the nuclei of plant and nonplant organisms. Phytopathology, 105 (10), 1288–1301.

Bulgarelli, D., Schlaeppi, K., Spaepen, S., Van Themaat, E. V. L., & Schulze-Lefert, P. (2013). Structure and functions of the bacterial microbiota of plants. Annual Review of Plant Biology, 64, 807–838.

Chaudhry, Q., Blom-Zandstra, M., Gupta, S., & Joner, E. J. (2005). Utilising the synergy between plants and rhizosphere microorganisms to enhance breakdown of organic pollutants in the environment. Environ Sci Pollut Res Int, 12 (1), 34–48.

Chen, Y.-h., Yang, X.-z., Zhuang, L., An, X.-h., Li, Y.-q., & Cheng, C.-g. (2020). Efficiency of potassium-solubilizing Paenibacillus mucilaginosus for the growth of apple seedling. Journal of Integrative Agriculture, 19, 2458–2469.

Chisholm, S. T., Coaker, G., Day, B., & Staskawicz, B. J. (2006). Host-microbe interactions: shaping the evolution of the plant immune response. Cell, 124 (4), 803814.

Chouhan, G. K., Mukherjee, A., Gaurav, A. K., Jaiswal, D. K., & Verma, J. P. (2021a). Plant-specific microbiome for environmental stress management: issues and challenges. New and Future Developments in Microbial Biotechnology and Bioengineering, 69–89.

Chouhan, G. K., Verma, J. P., Jaiswal, D. K., Mukherjee, A., Singh, S., de Araujo Pereira, A. P., … & Singh, B. K. (2021b). Phytomicrobiome for promoting sustainable agriculture and food security: opportunities, challenges, and solutions. Microbiological Research, 126763.

Chouhan, G. K., Jaiswal, D. K., Gaurav, A. K., Mukherjee, A., & Verma, J. P. (2021c). PGPM as a potential bioinoculant for enhancing crop productivity under sustainable agriculture. In Biofertilizers (pp. 221–237). Woodhead Publishing.

Das, S., Singh, S., Verma, J. P., & Mukherjee (2021). A. Fungi: a potential candidate for sustainable agriculture and agroecosystem. In New and Future Developments in Microbial Biotechnology and Bioengineering (pp. 159–164). Elsevier.

Dai, W. J., Zeng, Y., Xie, Z. P., & Staehelin, C. (2008). Symbiosis-promoting and deleterious effects of NopT, a novel type 3 effector of Rhizobium sp. strain NGR234. Journal of Bacteriology, 190, 5101–5110. doi: 10.1128/JB.00306-08

Deakin, W. J., & Broughton, W. J. (2009). Symbiotic use of pathogenic strategies: rhizobial protein secretion systems. Nat Rev Microbiol, 7 (4), 312.

Deguchi, Y., Banba, M., Shimoda, Y., Chechetka, S.A., Suzuri, R., Okusako, Y. et al. (2007). Transcriptome profiling of Lotus japonicus roots during arbuscular mycorrhiza development and comparison with that of nodulation. DNA Res, 14 (3), 117–133.

De La Fuente, L., & Burdman, S. (2011). Pathogenic and beneficial plant-associated bacteria. *Encycl. Life Support Syst.*

Djordjevic, M. A., Gabriel, D. W., & Rolfe, B. G. (1987). Rhizobium-the refined parasite of legumes. Annu Rev Phytopathol, 25 (1), 145168.

Dodds, P. N., & Rathjen, J. P. (2010). Plant immunity: towards an integrated view of plant-pathogen interactions. Nat Rev Genet, 11 (8), 539.

Fatemi, H., Pour, B. E., & Rizwan, M. (2020). Isolation and characterization of lead (Pb) resistant microbes and their combined use with silicon nanoparticles improved the growth, photosynthesis and antioxidant capacity of coriander (*Coriandrum sativum* L.) under Pb stress. Environmental Pollution, 266, 114982.

Glick, B. R. (2014). Bacteria with ACC deaminase can promote plant growth and help to feed the world. Microbiological Research, 169, 30–39.

Goswami, D., Dhandhukia, P., Patel, P., & Thakker, J. N. (2014). Screening of PGPR from saline desert of Kutch: growth promotion in Arachis hypogea by *Bacillus licheniformis* A2. Microbiological Research, 169, 66–75.

Hao, K., Liu, J. Y., Ling, F., Liu, X. L., Lu, L., Xia, L. et al. (2014). Effects of dietary administration of Shewanella haliotis D4, Bacillus cereus D7 and Aeromonas bivalvium D15, single or combined, on the growth, innate immunity and disease resistance of shrimp, Litopenaeus vannamei. Aquaculture, 428, 141149.

Hubber, A., Vergunst, A. C., Sullivan, J. T., Hooykaas, P. J., & Ronson, C. W. (2004). Symbiotic phenotypes and translocated effector proteins of the Mesorhizobium loti strain R7A VirB/D4 type IV secretion system. Mol Microbiol, 54 (2), 561574.

Hua, J. (2013). Modulation of plant immunity by light, circadian rhythm, and temperature. Current Opinion in Plant Biology, 16(4), 406–413.

Hu¨ckelhoven, R. (2007). Transport and secretion in plant-microbe interactions. Current Opinion in Plant Biology, 10 (6), 573–579.

Huettermann, A., Orikiriza, L. J., & Agaba, H. (2009). Application of superabsorbent polymers for improving the ecological chemistry of degraded or polluted lands. CLEANSoil, Air, Water, 37 (7), 517526.

Jing, Y., He, Z., & Yang, X. (2007). Role of soil rhizobacteria in phytoremediation of heavy metal contaminated soils. J Zhejiang Univ Sci B, 8, 192–207.

Jones, J. D. G, & Dangl, J. L. (2006). The plant immune system. Nature, 444, 323–329.

Johnson, K., Narasimhan, G., & Krishnan, C. (2014). Mimosa pudica Linn-a shyness princess: a review of its plant movement, active constituents, uses and pharmacological activity. Int J Pharm Sci Res, 5 (12), 5104.

Kafle, A., Garcia, K., Peta, V., Yakha, J., Soupir, A., & Bücking, H. (2018). Beneficial plant -microbe interactions and their effect on nutrient uptake, yield, and stress resistance of soybeans. In Soybean-Biomass, Yield and Productivity. IntechOpen.

Kamaruzzaman, M., Abdullah, S., Hasan, H. A., Hassan, M., Othman, A., & Idris, M. (2020). Characterisation of Pb-resistant plant growth-promoting rhizobacteria (PGPR) from Scirpus grossus. Biocatalysis and Agricultural Biotechnology, 23, 101456.

Kambara, K., Ardissone, S., Kobayashi, H., Saad, M. M., Schumpp, O., Broughton, W. J. *et al.* (2009). Rhizobia utilize pathogen-like effector proteins during symbiosis. Molecular Microbiology, 71, 92–106. doi: 10.1111/j.1365-2958.2008.06507.x

Kaul, S., Sharma, T., & Dhar, M. K. (2016). "Omics" tools for better understanding the plant-Endophyte interactions. Frontiers in Plant Science, 7, 955.

Kayalvizhi, K., & Kathiresan, K. (2019). Microbes from wastewater treated mangrove soil and their heavy metal accumulation and Zn solubilization. Biocatalysis and Agricultural Biotechnology, 22, 101379.

Ke, T., Guo, G., Liu, J., Zhang, C., Tao, Y., Wang, P., Xu, Y., & Chen, L. (2020). Improvement of the Cu and Cd phytostabilization efficiency of perennial ryegrass through the inoculation of three metal-resistant PGPR strains. Environmental Pollution, 271, 116314.

Kim, S., Lim, H., & Lee, I. (2010). Enhanced heavy metal phytoextraction by Echinochloa crus-galli using root exudates. J Biosci Bioeng, 109 (1), 4750.

Kirjavainen, P. V., Karvonen, A. M., Adams, R. I., Täubel, M. Roponen, M., Tuoresmäki, P., Loss, G., Jayaprakash, B., Depner, M., Ege, M. J. *et al.* (2019). Farm-like indoor microbiota in non-farm homes protects children from asthma development. Nature Medicine, 25, 1089–1095.

Krishna, R., Singh, S., Gaurav, A. K., Jaiswal, D. K., Singh, M., & Verma, J. P. (2021). Rhizosphere soil microbiomes: as driver of agriculture commodity and industrial application. In New and Future Developments in Microbial Biotechnology and Bioengineering (pp. 183–195). Elsevier.

Kumar, A., Maurya, B., & Raghuwanshi, R. (2014). Isolation and characterization of PGPR and their effect on growth, yield and nutrient content in wheat (Triticum aestivum L.). Biocatalysis and Agricultural Biotechnology 3, 121–128.

Kumar, A., Singh, S., Gaurav, A. K., Srivastava, S., & Verma, J. P. (2020a). Plant growth-promoting bacteria: biological tools for the mitigation of salinity stress in plants. Frontiers in Microbiology, 11.

Kumar, A., Singh, S., Mukherjee, A., Rastogi, R. P., & Verma, J. P. (2020b). Salt-tolerant plant growth-promoting Bacillus pumilus strain JPVS11 to enhance plant growth attributes of rice and improve soil health under salinity stress. Microbiological Research, 242, 126616.

Kumar, A., & Verma, J. P. (2019). The role of microbes to improve crop productivity and soil health. In Ecological Wisdom Inspired Restoration Engineering (pp. 249–265). Springer, Singapore.

Kumar, A., Verma, H., Singh, V. K., Singh, P. P., Singh, S. K., Ansari, W. A. et al. (2017). Role of Pseudomonas sp. in sustainable agriculture and disease management. In Agriculturally Important Microbes for Sustainable Agriculture (pp. 195–215). Springer, Singapore.

Kumar, C. A., Narinder, S., & Daljeet, S. (2014). Exploitation of indigenous strains of Trichoderma and Pseudomonas fluorescens for the control of damping-off in chilli. Plant Disease Research, 30(1), 6–10.

Kumar, S. M., Reddy, G. C., Phogat, M., & Korav, S. (2018). Role of bio-fertilizers towards sustainable agricultural development: a review. J Pharmacogn Phytochem, 7, 1915–1921.

Kumar, V., Kumar, P., & Khan, A. (2020c). Optimization of PGPR and silicon fertilization using response surface methodology for enhanced growth, yield and biochemical parameters of French bean (Phaseolus vulgaris L.) under saline stress. Biocatalysis and Agricultural Biotechnology, 23, 101463.

Kumar, A., Sharma, P. K., Singh, S., & Verma, J. P. (2021). Impact of engineered nanoparticles on microbial communities, soil health and plants. Plant-Microbes-Engineered Nano-particles (PM-ENPs) Nexus in Agro-Ecosystems: Understanding the Interaction of Plant, Microbes and Engineered Nano-particles (ENPS), 201.

Kumar, S., Pandey, P., & Maheshwari, D. K. (2009). Reduction in dose of chemical fertilizers and growth enhancement of sesame (Sesamum indicum L.) with application of rhizospheric competent Pseudomonas aeruginosa LES4. Eur J Soil Biol, 45 (4), 334340.

Kumari, P., Meena, M., & Upadhyay, R. (2018). Characterization of plant growth promoting rhizobacteria (PGPR) isolated from the rhizosphere of Vigna radiata (mung bean). Biocatalysis and Agricultural Biotechnology, 16, 155–162.

Li, H., Qiu, Y., Yao, T., Ma, Y., Zhang, H., & Yang, X. (2020). Effects of PGPR microbial inoculants on the growth and soil properties of Avena sativa, Medicago sativa, and Cucumis sativus seedlings. Soil and Tillage Research, 199, 104577.

Li, Y., Huang, F., Lu, Y., Shi, Y., Zhang, M., Fan, J. *et al.* (2013). Mechanism of plant-microbe interaction and its utilization in disease-resistance breeding for modern agriculture. Physiological and Molecular Plant Pathology, 83, 51–58.

Links, M. G., Demeke, T., Gräfenhan, T., Hill, J. E., Hemmingsen, S. M., & Dumonceaux, T. J. (2014). Simultaneous profiling of seed-associated bacteria and fungi reveals antagonistic interactions between microorganisms within a shared epiphytic microbiome on Triticum and Brassica seeds. The New Phytologist, 202, 542–553.

Liu, H., Li, J., Carvalhais, L. C., Percy, C. D., Prakash Verma, J., Schenk, P. M., & Singh, B. K. (2021). Evidence for the plant recruitment of beneficial microbes to suppress soil-borne pathogens. The New Phytologist, 229(5), 2873–2885.

Lohar, D. P., Sharopova, N., Endre, G., Penuela, S., Samac, D., Town, C. et al. (2006). Transcript analysis of early nodulation events in Medicago truncatula. Plant Physiol, 140 (1), 221234.

Luo, S., Xu, T., Chen, L., Chen, J., Rao, C., Xiao, X. et al. (2012). Endophyte-assisted promotion of biomass production and metal-uptake of energy crop sweet sorghum by plant-growth-promoting endophyte Bacillus sp. SLS18. Appl Microbiol Biotechnol, 93 (4), 17451753.

Maag, D., Erb, M., Ko"llner, T. G., & Gershenzon, J. (2015). Defensive weapons and defense signals in plants: some metabolites serve both roles. Bioessays, 37 (2), 167–174.

Maity, A., Sharma, J., & Pal, R. (2019). Novel potassium solubilizing bio-formulation improves nutrient availability, fruit yield and quality of pomegranate (Punica granatum L.) in semi-arid ecosystem. Scientia Horticulturae, 255, 14–20.

Mukherjee, A., Chouhan, G. K., Gaurav, A. K., Jaiswal, D. K., & Verma, J. P. (2021). Development of indigenous microbial consortium for biocontrol management. In New and Future Developments in Microbial Biotechnology and Bioengineering (pp. 91–104). Elsevier.

Mukherjee, A., Gaurav, A. K., Patel, A. K., Singh, S., Chouhan, G. K., Lepcha, A., ... & Verma, J. P. (2021). Unlocking the potential plant growth-promoting properties of chickpea (Cicer arietinum L.) seed endophytes bio-inoculants for improving the soil health and crop production. Land Degradation & Development.

Mukherjee, A., Bhowmick, S., Yadav, S., Rashid, M. M., Chouhan, G. K., Vaishya, J. K., & Verma J. P. (2021) Revitalizing of endophytic microbes for soil health management and plant protection. 3 Biotech, 11, 399.

Mukherjee, A., Gaurav, A. K., Singh, S., Chouhan, G. K., Kumar, A., & Das, S. (2019). Role of potassium (K) solubilising microbes (KSM) in growth and induction of resistance against biotic and abiotic stress in plant: a book review. Climate Change and Environmental Sustainability, 7(2), 212–214.

Mukherjee, A., Verma, J. P., Gaurav, A. K., Chouhan, G. K., Patel, J. S., & Hesham, A. E. L. (2020). Yeast a potential bio-agent: future for plant growth and postharvest disease management for sustainable agriculture. Applied Microbiology and Biotechnology, 104(4), 1497–1510.

Newton, A. C., Fitt, B. D., Atkins, S. D., Walters, D. R., & Daniell, T. J. (2010). Pathogenesis, parasitism and mutualism in the trophic space of microbe-plant interactions. Trends in Microbiology, 18(8), 365–373.

Nielsen, T. H., Thrane, C., Christophersen, C., Anthoni, U., & Sorensen, J. (2000). Structure, production characteristics and fungal antagonism of tensin – a new antifungal cyclic lipopeptide from Pseudomonas fluorescens strain 96.578. J Appl Microbiol, 89, 992–1001.

Pastore, G., Kernchen, S., & Spohn, M. (2020). Microbial solubilization of silicon and phosphorus from bedrock in relation to abundance of phosphorus-solubilizing bacteria in temperate forest soils. Soil Biology and Biochemistry, 151, 108050.

Pathak, D., Lone, R., Khan, S., & Koul, K. (2019). Isolation, screening and molecular characterization of free-living bacteria of potato (Solanum tuberosum L.) and their interplay impact on growth and production of potato plant under mycorrhizal association. Scientia Horticulturae, 252, 388–397.

Paulucci, N. S., Gallarato, L. A., Reguera, Y. B., Vicario, J. C., Cesari, A. B., de Lema, M. B. G., & Dardanelli, M. S. (2015). Arachis hypogaea PGPR isolated from Argentine soil modifies its lipids components in response to temperature and salinity. Microbiological Research, 173, 1–9.

Pieterse, C. M. J., Zamioudis, C., Does, D. V., & Van Wees, S. C. M. (2014). Signalling networks involved in induced resistance. In Induced Resistance for Plant Defense [Internet] (pp. 58–80). John Wiley & Sons, Ltd.

Piromyou, P., Buranabanyat, B., Tantasawat, P., Tittabutr, P., Boonkerd, N., & Teaumroong, N. (2011). Effect of plant growth promoting rhizobacteria (PGPR) inoculation on microbial community structure in rhizosphere of forage corn cultivated in Thailand. European Journal of Soil Biology, 47, 44–54.

Pramanik, K., Mitra, S., Sarkar, A., & Maiti, T. K. (2018). Alleviation of phytotoxic effects of cadmium on rice seedlings by cadmium resistant PGPR strain Enterobacter aerogenes MCC 3092. Journal of Hazardous Materials, 351, 317–329.

Prasad, R., Kumar, M., & Varma, A. (2014). Role of PGPR in soil fertility and plant health. Plant-Growth-Promoting Rhizobacteria (PGPR) and Medicinal Plants [Internet] (pp. 247260). Springer International Publishing.

Pritchard, L., & Birch, P. (2011). A systems biology perspective on plant-microbe interactions: Biochemical and structural targets of pathogen effectors. Plant Science, 180(4), 584–603.

Pueppke, S. G., & Broughton, W. J. (1999). Rhizobium sp. strain NGR234 and R. fredii USDA257 share exceptionally broad, nested host ranges. Mol Plant Microbe Interact, 12 (4), 293318.

Rezaei-Chiyaneh, E., Amirnia, R., Machiani, M. A., Javanmard, A., Maggi, F., & Morshedloo, M. R. (2020). Intercropping fennel (Foeniculum vulgare L.) with common bean (Phaseolus vulgaris L.) as affected by PGPR inoculation: a strategy for improving yield, essential oil and fatty acid composition. Scientia Horticulturae, 261, 108951.

Santillo, D., & Johnston, P. (2003). Playing with fire: the global threat presented by brominated flame retardants justifies urgent substitution. Environment International, 29(6), 725–734.

Savary, S., Ficke, A., Aubertot, J. N., & Hollier, C. (2012). Crop losses due to diseases and their implications for global food production losses and food security. Food Security, 4 (4), 519–537.

Singh, B. K., Trivedi, P., Singh, S., Macdonald, C. A., & Verma, J. P. (2018). Emerging microbiome technologies for sustainable increase in farm productivity and environmental security. Microbiology Australia, 39, 17–23.

Singh, N., & Siddiqui, Z. A. (2015). Effects of Bacillus subtilis, Pseudomonas fluorescens and Aspergillus awamori on the wilt-leaf spot disease complex of tomato. Phytoparasitica, 43 (1), 61–75.

Singh, V. K., Singh, A. K., & Kumar, A. (2017). Disease management of tomato through PGPB: current trends and future perspective. 3 Biotech, 7(4), 255.

Singh, S., Jaiswal, D. K., Krishna, R., Mukherjee, A., & Verma, J. P. (2020). Restoration of degraded lands through bioenergy plantations. Restoration Ecology, 28(2), 263–266.

Soto, M. J., Domı́nguez-Ferreras, A., Pe´rez-Mendoza, D., Sanjua´n, J., & Olivares, J. (2009). Mutualism versus pathogenesis: the give-and-take in plantbacteria interactions. Cell Microbiol, 11 (3), 381388.

Soto, M. J., Sanjuan, J., & Olivares, J. (2006). Rhizobia and plant-pathogenic bacteria: common infection weapons. Microbiology, 152 (11), 3167–3174.

Sun, F., Ou, Q., Wang, N., xuan Guo, Z., Ou, Y., Li, N., & Peng, C. (2020). Isolation and identification of potassium-solubilizing bacteria from Mikania micrantha rhizospheric soil and their effect on M. micrantha plants. Global Ecology and Conservation, 23, e01141.

Syed, S., Tollamadugu, N. P., & Lian, B. (2020). Aspergillus and Fusarium control in the early stages of Arachis hypogaea (groundnut crop) by plant growth-promoting rhizobacteria (PGPR) consortium. Microbiological Research, 240, 126562.

Ueda, H., Tamura, K., & Hara-Nishimura, I. (2015). Functions of plant-specific myosin XI: from intracellular motility to plant postures. Curr Opin Plant Biol, 28, 30–38.

Valentı́n-Vargas, A., Root, R. A., Neilson, J. W., Chorover, J., & Maier, R. M. (2014). Environmental factors influencing the structural dynamics of soil microbial communities during assisted phytostabilization of acid-generating mine tailings: a mesocosm experiment. Sci Total Environ, 500, 314–324.

Verma, J. P., Jaiswal, D. K., Singh, S., Kumar, A., Prakash, S., & Curá, J. A. (2017). Consequence of phosphate solubilising microbes in sustainable agriculture as efficient microbial consortium: a review. Climate Change and Environmental Sustainability, 5(1), 1–19.

Wackett, L. P. (2014). Antibiosis in the environment. Environ Microbiol Rep, 6 (5), 532–533.

Wang, L. J., & Qi, X. (2015). Metabolomics research of quantitative disease resistance against barley leaf rust. In: Hakeem, K., & Akhtar, M. (Eds.), Plant Metabolomics. Springer, Dordrecht.

Xiao, A., Li, Z., Li, W. C., & Ye, Z. (2020). The effect of plant growth-promoting rhizobacteria (PGPR) on arsenic accumulation and the growth of rice plants (Oryza sativa L.). Chemosphere, 242, 125136.

Xu, Z., Zhang, R., Wang, D., Qiu, M., Feng, H., Zhang, N. et al. (2014). Enhanced control of cucumber wilt disease by Bacillus amyloliquefaciens SQR9 by altering the regulation of its DegU phosphorylation. Appl Environ Microbiol, 80 (9), 2941–2950.

Youssef, M. M. A., & Eissa, M. F. M. (2014). Biofertilizers and their role in management of plant parasitic nematodes: a review. J Biotechnol Pharm Res, 5, 1–6.

Zehner, S., Schober, G., Wenzel, M., Lang, K., & Göttfert, M. (2008). Expression of the Bradyrhizobium japonicum type III secretion system in legume nodules and analysis of the associated tts box promoter. Mol Plant Microbe Interact, 21 (8), 10871093.

14 Green Synthesized Nanoparticles for Sustainable Agriculture

Divya Mittal[1]*, Reena V Saini*[1]*, Rahul Thakur*[1]*,*
Soumya Pal[1]*, Joydeep Das*[2]*, Samarjeet*
Singh Siwal[1]*, and Adesh K Saini*[1]
[1]Maharishi Markandeshwar (Deemed to be University)
Mullana, Haryana, India
[2]Mizoram University, Mizoram, India

CONTENTS

14.1 INTRODUCTION

The demand for pesticides has increased all over the globe in the last few years. A large quantity of chemical-based pesticides is produced, and thus it is readily available in the market. Globally, around two million tons of chemical pesticides are utilized per year (De et al. 2014). The excessive use of these pesticides leads to health-related problems like various types of cancer, reproductive toxicity, and endocrine disruption (Nicolopoulou-Stamati et al. 2016). Apart from it, these pesticides badly affect the soil structure and its composition (Randall et al. 2013; Lade et al. 2017). So, it is necessary to find a safer and eco-friendly method to substitute chemical-based pesticides. One of the tactics is to utilize biopesticides to control the pests (Devi et al. 2015; Gupta et al. 2016; Mittal et al. 2019). These biocontrol agents contain various genes and their products to control plant pathogens and pests (Devi et al. 2013a; Devi et al. 2013b; Khatri et al. 2013; Usta et al. 2013). As we know, nanoscience is a new field with a vast scope in various fields of sciences like antimicrobial, anticancer (Thakur et al. 2020; Sharma et al. 2021; Kumari et al. 2021), and is also beneficial in agriculture for crop protection to control the pest. In the preparations of nano-based pesticides and insecticides, nanoparticles are synthesized using different types of

DOI: 10.1201/9781003110477-14

materials. In this regard, the nano-pesticides with the biological composition are synthesized and termed as green-nanoparticles (G-NPs) or nano-biopesticides. However, a broad range of research and studies are required to better understand the interaction among nanoparticles, microbes, plants, soil, and humans (Shang et al. 2019).

It has been found that crops are infected by around 67,000 species of organisms, leading to 10–16% of estimated global agricultural loss every year, and out of these, insects and mites are the most common causative agents (Bradshaw et al. 2016). Every year worldwide, around 10,000 species of insect can damage the agricultural crop, and the estimated cost was 13.6% globally (Benedict and Ring 2004). In India, around 17.5% loss of significant field crops is due to insects every year (Dhaliwal et al. 2010). Moreover, farmers trust chemical-based insecticides to control pests, but due to the excessive use of highly toxic pesticides, several problems like environmental imbalance, loss of highly beneficial insects, insecticidal resistance, and crop failures are reported (Dhaliwal et al. 2010). Therefore, novel methods are discovered to prevent these losses and G-NPs are emerging as new tools. Nanomaterials in crop fields aim to minimize the consumption of pesticides and reduce their impact on soil's biotic and abiotic structure.

G-NPs have higher surface area/volume ratio, mechanical properties, adsorption ability, and so on. In addition, the green nano-biopesticides attract global attention because they are a safer agent for the human and environment than the chemical agent for pest control. G-NPs are mainly derived from plants, animals, bacteria, nematodes-based biomaterials. However, for the last two decades, only 2% of biopesticides are globally used in agriculture. Nano-pesticides are the formulation where the size of the particle is in the nano-range harboring traits like releasing the active compound slowly and effectively delivering it for more extended periods. Researchers have developed different classes of nano-pesticide via nanospheres, nano-emulsions, nano-encapsulated formulations, nanogels, metals or their oxides-based nanoparticles. However, the formulation of biobased nano-pesticide is still in the early stage of research (Nuruzzaman et al. 2016; Chippa 2017).

Nano-biopesticides have various benefits over chemical pesticides. These include their small size, high mobility and stability, high surface area/volume ratio, high solubility, effectiveness, and less toxic (Sasson et al. 2007). However, there is a dire need to alter the chemical-based nano-pesticide because the component present in nano-structures could be toxic when runoff in the soil and groundwater. In addition, another environmental and health concern is the accumulation of them in the food chain. The solution to this problem is the use of micelles, composites, or biopolymers during their synthesis, which improves the pesticide activity and prevents its assimilation in the ecosystem. In this chapter, we discussed how microbes can be utilized for the synthesis of bio-nano-formulation using bioactive components and their applications in improving agricultural productivity.

14.2 MICROBES AND THEIR METABOLITES AS BIOPESTICIDES

Many microbes harbor pesticide activity. These biocontrol agents have low toxicity and are eco-friendly (Devi et al. 2013a; Khatri et al. 2013; Usta et al. 2013). However, due to the excessive use of pesticides, the contaminated soils give a tough time to these microbes to survive. Many beneficial microbes are not resistant to pesticides that have contaminated the soil. It was reported that excessive use of chemical-based pesticides depletes the population of beneficial microbes (Milošević, and Govedarica 2002; Meena et al. 2015). In addition to this, sporulation of arbuscular mycorrhizae decreases due to use of herbicides (Pasaribu et al. 2013). Recently, microbes-based pesticides from actinomycetes were extracted which include milbemycin (fermentation product of *Streptomyces hygroscopicus*), polynactins (*S. aureus*-based secondary metabolites), avermectins (abamectin and emamectin from *S. avermitilis fermentation*), spinosad (*Saccharopolyspora spinosa* fermentation) were highly effective against pests. Moreover, they are required at a lower amount and are less harmful to non-target organisms in the agro-ecosystem (Devi et al. 2019). Some national and international certifications now approve these bioformulations for use in fields. In addition to this, the new advanced formulation has been developed for biopesticides including nano-formulation to reduce the number of active components for crop protection because biopesticide exists with certain pesticidal activity drawbacks such as instability under physical conditions like high temperature, aridity, and UV rays. Also, they are not so effective during the heavy attack of pest (Singh et al. 2017).

14.3 GREEN-NANOPARTICLES AND THEIR ROLE IN PEST CONTROL

The limitations imposed by various alternatives of chemical pesticides, there is a need to utilize another alternative which could be used for crop improvement. For this, the G-NPs synthesized from various bio-based sources provide a solution. The synthesis process of G-NPs utilizes safer solvents and biomaterials than the manufacturing of chemical compounds. These G-NPs are made from diverse microorganisms and metal ions which control many pests in agriculture farms and stored grains (Table 14.1).

TABLE 14.1

Biological Component and Metal Used for the Synthesis of Green Nanoparticles (G-NPs)

S. No.	Biological Component	Nanoparticle	Pest	Reference
1	*Pseudomonas fluorescens* MAL2	Copper	*Tribolium castaneum* (red flour beetle)	El-Saadony et al. 2020
2	*Aristolochia indica leaf extract*	Silver	*Helicoverpa armigera* (ballworm)	Siva and Kumar 2015
3	*Avivennia marina*	Silver and lead	*Sitophilus oryzae* (rice weevil)	Sankar and Abideen 2015
4	*Nomuraea rileyi*	Chitosan	*Spodoptera litura*	Chandra et al. 2013
5	DNA-tagged nanogold	Gold	*Spodoptera litura*	Chandrashekharaiah et al. 2015
6	*Euphorbia prostate*	Silver	*Sitophilus oryzae*	Zahir et al. 2012

El-Saadony et al. prepared bioformulation of copper nanoparticles by using cell free suspension of copper resistant bacteria *Pseudomonas fluorescens MAL2*. The bacterial strain was isolated from metal contaminated soil of the industrial region. The G-NPs of *P. fluorescence* showed insecticidal properties against *Tribolium castaneum* (El-Saadony et al. 2020). In another study, the derivative of chitin and metabolite of *Nomuraea rileyi* has been reported to show insecticidal properties against plant pests (Zheng et al. 2005; Rabea et al. 2005). They compared the Uncoated Fungal Metabolite, fungal spores and nanoparticles derived from chitosan and metabolites derived from *Nomuraea rileyi* and named it CNPCFM. Among all the three, the nanoparticles (CNPCFM) showed insecticidal activity against *Spodoptera litura* (Chandra et al. 2013).

Some researchers also developed NPs by using DNA as the green part. For example, Chandrashekharaiah et al. worked on DNA-tagged CdS, nano-TiO2, DNA-tagged Nanogold, and nano-Ag. He examined these NPs against larvae of *S. litura*. Its exposure revealed that DNA-tagged nanogold caused 75% mortality on the fifth stadium larvae of the insect (Chakravarthy et al. 2012; Figure 14.1). In contrast, highest larval mortality was shown by CdS nanoparticles, that is, 79% at 2400 ppm. Similar to Cds, the nano-TiO2 and nano-Ag showed 73% and 56% mortality of *S. litura* larvae, respectively. They also prepared nanoparticles layered with ecdysteroid analogs like halofenozide and tebufenozide, which showed pesticide activity against rice moth *Corcyra cephalonica* (Chandrashekharaiah et al. 2015). These two ecdysteroids reduced the fertility and fecundity of the pests.

In another study, the silver nanoparticles (AgNPs) were formulated with *Aristolochia indica* (leaf extract) studied against HeLa cell lines and larvae of *Helicoverpa armigera*. Results revealed that these bio-nanomaterials are

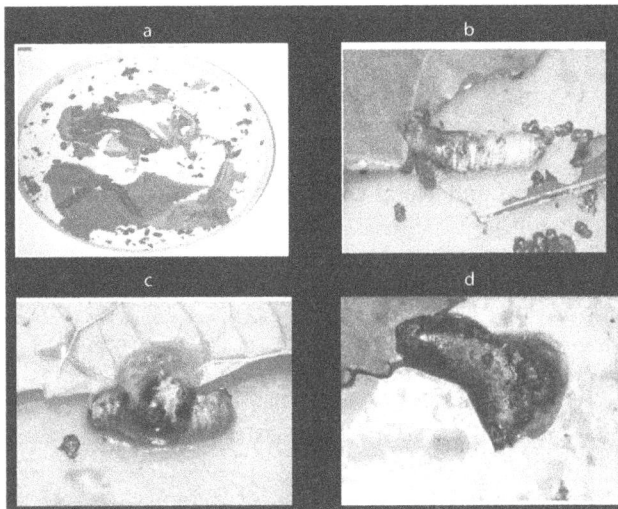

FIGURE 14.1 Larval mortality in *Spodoptera litura* after treatment with DNA-tagged nanogold (a) control (b) *Spodoptera* larva fifth day after treatment, (c) on sixth day, and (d) on the seventh day. (Adapted from Chakravarthy et al. 2012.)

larvicidal, antifeedant, and cytotoxic (Siva and Kumar 2015). In addition, lead and AgNPs were produced using the extract of *Avivennia marina*, which is a mangrove plant, exhibited 100% mortality within 4 days against pest of stored grains *Sitophilus oryzae* (Sankar and Abideen 2015). Nanoparticles using prostrate sandmat *Euphorbia prostrata* were also generated by green synthesis and were found to be significant in controlling adult *S. oryzae* (Zahir et al. 2012). Some nanoparticles were also found to be effective against the red flour beetle *Tribolium castaneum*. Garlic essential oil infused with PEG-coated nanoparticles showed insecticidal activity against *Tribolium castaneum* (Yang et al. 2009). Nano-Diatomaceous Earth (Nano-DE) were also tested against the confused flour beetle *Tribolium confusum* and it turned out to be a better pesticidal and insecticidal material than natural Diatomaceous Earth. They were able to tackle both *T. confusum* and *T. castaneum*. Therefore, nano-pesticides could be utilized on a larger scale to help manage insects, pests, and pathogens (Sabbour and El-Aziz 2015).

Greener synthesis of AgNPs using an extract of *Euphorbia prostate* and *Avicennia marinas* revealed the pesticide activity against *Sitophilus oryzae* (Sankar and Abideen 2015). In addition, AgNPs capped with metabolites of *Aristolochia indica* leaves showed pesticide activity towards *Helicoverpa armigera* larvae (Siva and Kumar 2015).

14.4 EXTRACTION OF METABOLITES FOR G-NPs SYNTHESIS

Extraction of the components from biological material is a crucial step for the production of G-NPs. For the extraction process, we need to observe some important points because it could affect the quality of an extract (Pandey and Tripathi 2014). The collection and separation of target components from plants depend on the selected part from the plant material. In addition, the extraction method exhibits some crucial factors of consideration, that is, the temperature, solvent-to-sample ratio, length of extraction, solvent's pH, and size of the particle of raw materials (Belokurov et al. 2019).

The solvent-based extraction method is the most widely used to obtain natural products. In this method, plant soluble metabolites are separated from insoluble materials by using a suitable solvent. Another critical method is microwave-assisted extraction in which separation of analytes is done by using microwave energy. As in microwave, due to the interaction between polarizable and polar compounds like water and plant materials, a large amount of heat is produced (Azwanida 2015; Bhat et al. 2020). Apart from this, maceration extraction is also used for the extraction of metabolites, and it involves three steps: (a) crushing the plant parts into small pieces, (b) selection of appropriate solvent based on the polarity of the compound and its addition to the plant part in a closed vessel and soaking it at room temperature for three days, and (c) separation of the liquid phase by filtration (Trusheva et al. 2007; Oroian et al. 2020).

Moreover, ultrasound-assisted extraction or ultrasonication shows various benefits over other methods, such as the low quantity of solvent required for extraction, lower temperatures, fewer extraction times and low energy consumption. This is the most straightforward plant extraction technique because it only needs a crushed sample with a suitable solvent placed in an ultrasonic bath (Altemimi et al. 2017).

14.5 METHOD FOR THE SYNTHESIS G-NPs

The plant-based metabolites or the microbe-based metabolites having pesticidal activity are produced and commercialized all over the world. These metabolites are extracted by various approaches and used to produce the metal nanoparticle because metabolite activity is retained at a high metal concentration. However, chemical methods for the production of nanoparticles are highly expensive and use harmful chemicals. So in the past, researchers started utilizing NPs of biological origin like microbes or plant extracts (Narayanan and Sakthivel 2010a). Also, the biological processes are cost-effective, reusable, require less energy and run at room temperature (Ingale and Chaudhari 2013; Narayanan and Sakthivel 2010b; Benelli 2016). The multi-functional NPs formulated with various essential oils and organic polymers are produced which harbor traits like antioxidant, antibiotic, and odor-masking. The potential of nano-biopesticides or nanoparticles can be predicted by comparative studies of raw extract, metal-based NPs from plant extract, biopolymers, and nano-biopesticides. This could be further described by minus the potential of the crude extract from the nano-biopesticides (Lade et al. 2017).

Gold and silver nanoparticles have caught special consideration among all green nanoparticles due to their unique biological and physicochemical properties. These two metals are resistant to oxidation, corrosion and their NPs have attractive qualities, such as high chemical stability, high electrical and thermal conductivity, and the most helpful property for agriculture, that is, these are antimicrobial against a wide variety of microbes (Dauthal and Mukhopadhyay 2016). There are two types of synthesis which are previously reported for obtaining gold and silver G-NPs, which depend on the material used at the starting of preparation (Jamkhande et al. 2019). First is the top-bottom synthesis path that is used when required raw material is available in large quantity and is further crushed into small particles by thermal ablation, lithographic techniques, or grinding. However, the drawback of this method is the consumption of a huge amount of energy and can also lead to surface defects in NPs, which ultimately affects their physical and chemical properties (Yadi et al. 2018; Jamkhande et al. 2019).

The second method is the bottom-up synthesis path or "self-assembly approach," which includes biological and chemical methods where the growth of atoms takes place inside the nucleation centers to produce NPs. The bottom-up synthesis method exhibits more advantages over the top-bottom synthesis, such as low cost, lower time consumption, and NPs produced through this method are of good quality and have uniform chemical composition. Biosynthesis of G-NPs follows a bottom-up approach (Mazhar et al. 2017; Rafique et al. 2017; Gour and Jain 2019; Jamkhande et al. 2019; Castillo-Henríquez et al. 2020). Illustrations of both methods are shown in Figure 14.2. Using this method, Jayaseelan et al. conducted a synthesis of AgNPs using *Tinospora cordifolia* and obtained 55–80 nm G-NPs. These G-NPs showed larvicidal activity against malaria vector, *Anopheles subpictus* Grassi and filariasis vector, *Culex quinquefasciatus* (Jayaseelan et al. 2011). In another study, Logeswari et al. synthesized AgNPs by using commercially available plant powders of *Solanum tricobatum*, *Citrus sinensis.*, *Centella asiatica*, and *Syzygium* cumini. The resulted G-NPs showed antimicrobial activity against pathogenic bacterial strains *S. aureus*, *P. aeruginosa*, *Klebsiella pneumoniae*, and *E. coli* (Logeswari et al. 2013). We strongly

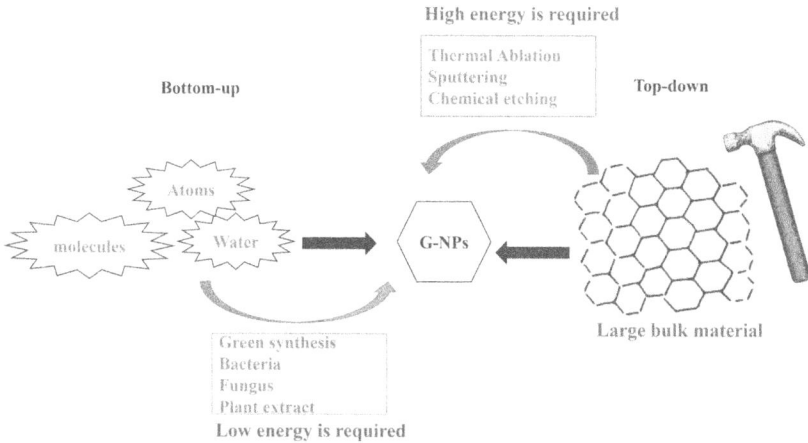

FIGURE 14.2   Illustration of the bottom-up and top-down method.

believe that using this approach< G-NPs can be efficiently synthesized against plant pests for pesticide activities. The basic method of G-NPs is shown in Figure 14.3.

In another approach, the bioformulation of copper nanoparticles by using culture filtrate of *Pseudomonas fluorescens MAL2* was studied (El-Saadony et al. 2020). For this, the metal-contaminated soil samples were collected from sites nearby metal industries for bacterial isolation. Then, the bacteria from soil samples were grown on nutrient agar plates supplemented with 100 ppm of $CuSO_4 \cdot 5H_2O$ (El-Saadony et al. 2020). Bacterial isolates were then grown in Luria–Bertani Broth (LB) at pH 7, 37°C for 24 hours and after this the bacterial suspension was centrifuged for the separation of filtrate which is required for the synthesis of nanoparticles. Further, 10 mL of cell free filtrate was mixed with 40 mL of 100 ppm sterilized $CuSO_4 \cdot 5H_2O$ solution in a 250 mL conical flask and then placed in 120 rpm incubator shaker at 37°C for 24 hours. A color change indicates the presence of G-NPs. The G-NPs of *P. fluorescence* showed insecticidal properties against *Tribolium castaneum* (El-Saadony et al. 2020). A similar approach was used earlier by Lakshmi et al., where silver G-NPs were synthesized by using *Bacillus* strain CS11 of size 42–92 nm

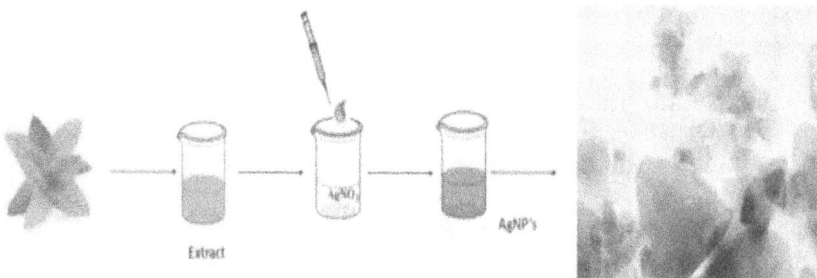

FIGURE 14.3   G-NPs synthesis using silver nitrate and plant extract. (Adapted from Castillo-Henríquez et al. 2020.)

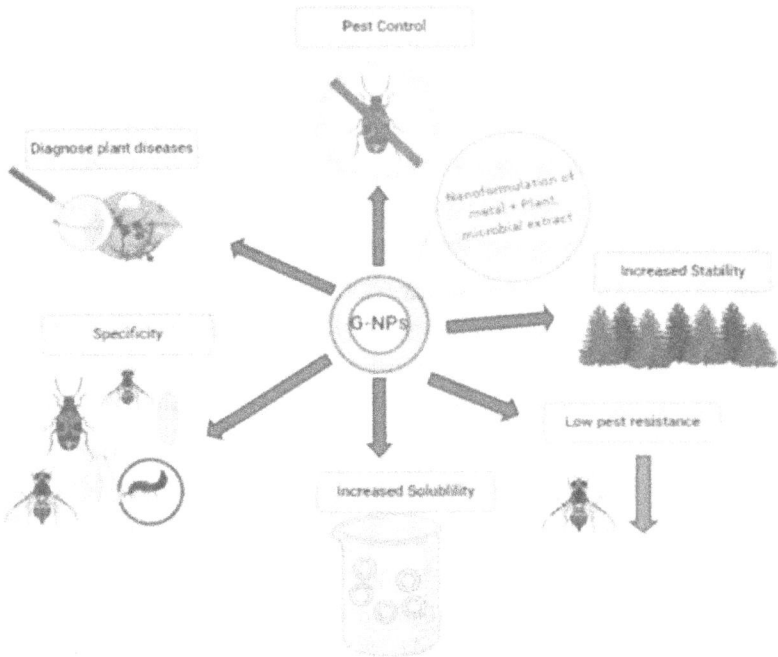

FIGURE 14.4   The possible benefits of G-NPs.

(Lakshmi et al. 2014). In another study, gold and silver nanoparticles were developed by using *Sporosarcina koreensis DC4 strain*. These G-NPs enhance the antimicrobial activity of commercially available antibiotics lincomycin, oleandomycin, vancomycin, novobiocin, penicillin G, and rifampicin against biofilm of pathogenic bacterial strains *S. aureus*, *P. aeruginosa*, and *E. coli* (Singh et al. 2016). Syed et al. conducted green synthesis of gold nanoparticles using *Pseudomonas fluorescens*, resulting in G-NPs of size 5–50 nm that exhibited antibacterial activity against pathogens (Syed et al. 2016). In addition to this, Vijayabharathi et al. worked on the synthesis of AgNPs by using an extracellular extract of actinomycetes *Streptomyces griseoplanus* SAI-25, which showed antifungal activity of charcoal rot pathogen *Marcophomina phaseolina* (Vijayabharathi et al. 2018). Similarly, Sankar and Abideen used the plant extract of *Avivennia marina* for the green synthesis of silver and lead nanoparticles against grain storage pests *Sitophilus oryzae* (Sankar and Abideen 2015). Thus, G-NPs are the appropriate substitute of chemical-based pesticides as with high specificity and solubility, G-NPs control pest population and enhance crop productivity (Figure 14.4).

14.6   SYNTHESIS OF G-NPs BY FUNGI AND BLUE-GREEN ALGAE

It is reported that the enzyme level produced by fungal strains is comparatively better than actinomycetes and bacteria, so fungal-based enzymes can be utilized for synthesizing G-NPs (Kamaraj et al. 2018). These are detoxifying enzymes include beta-glucosidase, carboxylesterase, and glutathione S-transferase. Kamaraj et al. (2018) found that *Trichoderma viride*-based titanium dioxide G-NPs showed larvicidal, anti-feedant, and pupicidal activity against *Helicoverpa armigera*. They have also observed the effect of these TDNPs on detoxifying enzymes, and they found an increase in the concentration of glutathione S- transferase whereas reduction in beta-glucosidase and carboxylesterase during the treatment G-NPs (Chinnaperumal et al. 2018).

Similarly, Mishra et al. (2014) prepared gold nanoparticles by *Trichoderma viride* of size ranges 20–30 nm that inhibits the growth of plant pathogens and parasites. In addition to this, the Au-NPs were synthesized by using blue-green algae *Spirulina platensis* and obtained NPs with an average size of ~5 nm. These G-NPs showed antibacterial activity against Gram-positive bacteria *Bacillus subtilis* and *Staphylococcus aureus* (Suganya et al. 2015). Besides, Salunkhe et al. used the fungus *cochliobolus lunatus* for the eco-friendly synthesis of AgNPs. In this approach, the obtained 3–21 nm G-NPs exhibited larvicidal activity against *Aedes aegypti* and *Anopheles stephensi* (Salunkhe et al. 2011).

14.7   CONCLUSION

Although G-NPs play a vital role in pest control and crop yield, there are certain drawbacks. Consequently, thorough studies into optimization and screening of G-NPs for various plant species are necessary to make these commercially available. Furthermore, the role of G-NPs should also be surveyed in bioremediation to develop its integrated approaches. Most research on bio-nanomaterials in agriculture depends on trials performed under controlled laboratory conditions, whereas inadequate data is available regarding their application in fields. Therefore, additional knowledge related to fieldwork would be highly beneficial for the large-scale application of G-NPs.

REFERENCES

Altemimi, A., Lakhssassi, N., Baharlouei, A., Watson, D. G., and D. A. Lightfoot. 2017. Phytochemicals: Extraction, isolation, and identification of bioactive compounds from plant extracts. *Plants* 6: 42.

Azwanida, N. N. 2015. A review on the extraction methods use in medicinal plants, principle, strength and limitation. *Medicinal and Aromatic Plants Research Journal* 4 (196): 2167–0412.

Belokurov, S. S., Narkevich, I. A., Flisyuk, E. V., Kaukhova, I. E., and M. V. Aroyan. 2019. Modern extraction methods for medicinal plant raw material (Review). *Pharmaceutical Chemistry Journal* 53: 559–563.

Benedict, J. H., and D. R. Ring. 2004. *Transgenic Crops Expressing Bt Proteins: Current Status, Challenges and Outlook.* Enfield, NH: Science Publishers.

Benelli, G. 2016. Plant-mediated biosynthesis of nanoparticles as an emerging tool against mosquitoes of medical and veterinary importance: A review. *Parasitology Research* 115(1): 23–34.

Bhat, A. R., Najar, M. H., Dongre, R. S., and M. S. Akhter. 2020. Microwave assisted synthesis of Knoevenagel Derivatives using water as green solvent. *Current Research in Green Sustainable Chemistry* 3: 100008.

Bradshaw, C. J., Leroy, B., Bellard, C., Roiz, D., Albert, C., Fournier, A., Barbet-Massin, M., Salles, J. M., Simard, F., and F. Courchamp. 2016. Massive yet grossly underestimated global costs of invasive insects. *Nature communications* 7(1): 1–8.

Castillo-Henríquez, L., Alfaro-Aguilar, K., Ugalde-Álvarez, J., Vega-Fernandez, L., Montes de Oca-Vásquez, G., and J. R. Vega-Baudrit. 2020. Green synthesis of Gold and Silver Nanoparticles from plant extracts and their possible applications as antimicrobial agents in the agricultural area. *Nanomaterials* 10(9): 1763.

Chakravarthy, A. K., Bhattacharyya, A., Shashank, P. R., Epidi, T. T., Doddabasappa, B., and S. K. Mandal. 2012. DNA-tagged nano gold: A new tool for the control of the armyworm, *Spodoptera litura* Fab. (Lepidoptera: Noctuidae). *African Journal of Biotechnology* 11: 9295–9301.

Chandra, J. H., Raj, L. F. A. A., Namasivayam, S. K. R., and R. S. A. Bharani. 2013. Improved pesticidal activity of fungal metabolite from *Nomureae rileyi* with chitosan nanoparticles. *Proceedings of the International Conference on Advanced Nanomaterials and Emerging Engineering Technologies* 387–390.

Chandrashekharaiah, M., Kandakoor, S. B., Gowda, G. B., Kammar, V., and A. K. Chakravarthy. 2015. Nanomaterials: A review of their action and application in pest management and evaluation of DNA-Tagged particles. *New Horizons in Insect Science: Towards Sustainable Pest Management* 13: 113–126.

Chhipa, H. 2017. Nanofertilizers and nanopesticides for agriculture. *Environmental Chemistry Letters* 15(1): 15–22.

Chinnaperumal, K., Govindasamy, B., Paramasivam, D., Dilipkumar, A., Dhayalan, A., Vadivel, A., Sengodan, K., and P. Pachiappan. 2018. Bio-pesticidal effects of *Trichoderma viride* formulated titanium dioxide nanoparticle and their physiological and biochemical changes on *Helicoverpa armigera* (Hub.). *Pesticide Biochemistry and Physiology* 149: 26–36.

Dauthal, P., and M. Mukhopadhyay. 2016. Noble metal nanoparticles: Plant-mediated synthesis, Mechanistic aspects of synthesis, and applications. *Industrial and Engineering Chemistry Research* 55: 9557–9577.

De, A., Bose, R., Kumar, A., and S. Mozumdar. 2014. Targeted delivery of pesticides using biodegradable polymeric nanoparticles. In *Springer Briefs in Molecular Science*, Bose, R. (ed.), pp. 5–6. New Delhi: Springer.

Devi, P. V., Duraimurugan, P., Chandrika, K. S., Gayatri, B., and R. D. Prasad. 2019. Nanobiopesticides for crop protection. In *Nanobiotechnology Applications in Plant Protection, Nanotechnology in the Life Sciences*, Abd-Elsalam, K. A. (eds.), pp. 145–168. Cham: Springer.

Devi, U., Khatri, I., Kumar, N., Kumar, L., Sharma, D., Subramanian, S., and A. K. Saini. 2013a. Draft genome of plant-growth-promoting-Rhizobacteria *Serratia fonticola* strain AU-P3(3). *Genome Announcement.* 1: e00946–13.

Devi, U., Khatri, I., Kumar, N., Sharma, D., Subramanian, S., and A. K. Saini. 2013b. Draft genome sequence of Plant-Growth-Promoting Rhizobacterium *Serratia fonticola* strain AU-AP2C, isolated from the pea rhizosphere. *Genome Announcement* 1: e01022–13.

Devi, U., Khatri, I., Saini, R. V., Kumar, L., Singh, D., Kumar, N., Gárriz, A., Subramanian, S., Sharma, D., and A. K. Saini. 2015. Genomic and functional characterization of a novel *Burkholderia* sp. strain AU4i from pea rhizosphere conferring plant growth promoting activities. *Advancement in Genetic Engineering* 4: 129.

Dhaliwal, G. S., Jindal, V., and A. K. Dhawan. 2010. Insect pest problems and crop losses: Changing trends. *Indian Journal of Ecology* 37(1): 1–7.

El-Saadony, M. T., El-Hack, A., Mohamed, E., Taha, A. E., Fouda, M. M., Ajarem, J. S., N. Maodaa, S., Allam, A. A., and N. Elshaer. 2020. Ecofriendly synthesis and insecticidal application of copper nanoparticles against the storage pest *Tribolium castaneum*. *Nanomaterials* 10(3): 587.

Gour, A., and N. K. Jain. 2019. Advances in green synthesis of nanoparticles. *Artificial Cells, Nanomedicine and Biotechnology* 47: 844–851.

Gupta, H., Saini, R. V., Pagadala, V., Kumar, N., Sharma, D. K., and A. K. Saini. 2016. Analysis of plant growth promoting potential of endophytes isolated from *Echinacea purpurea* and *Lonicera japonica*. *Journal of Soil Science and Plant Nutrition* 16: 558–577.

Ingale, A. G., and A. N. Chaudhari. 2013. Biogenic synthesis of nanoparticles and potential applications: An eco-friendly approach. *Journal of Nanomedicine and Nanotechnology* 4(165): 1–7.

Jamkhande, P. G., Ghule, N. W., Bamer, A. H., and M. G. Kalaskar. 2019. Metal nanoparticles synthesis: An overview on methods of preparation, advantages and disadvantages, and applications. *Journal of Drug Delivery Science and Technology* 53: 101–174.

Jayaseelan, C., Rahuman, A. A., Rajakumar, G., Vishnu, Kirthi. A., Santhoshkumar, T., Marimuthu, S., Bagavan, A., Kamaraj, C., Zahir, A. A., and G. Elango. 2011. Synthesis of pediculocidal and larvicidal silver nanoparticles by leaf extract from heartleaf moonseed plant, *Tinospora cordifolia Miers*. *Parasitology Research* 109(1): 185–194.

Kamaraj, C., Gandhi, P. R., Elango, G., Karthi, S., Chung, I. M., and G. Rajakumar. 2018. Novel and environmental friendly approach; Impact of Neem (Azadirachta indica) gum nano formulation (NGNF) on Helicoverpa armigera (Hub.) and Spodoptera litura (Fab.). *International Journal of Biological Macromolecules* 107:59–69.

Khatri, I., Kaur, S., Devi, U., Kumar, N., Sharma, D., Subramanian, S., and A. K. Saini. 2013. Draft genome sequence of plant Growth-promoting rhizobacterium *Pantoea* sp. strain AS-PWVM4. *Genome Announcement* 1: e00947-13.

Lade, B. D., Gogle, D. P., and S. B. Nandeshwar. 2017. Nano biopesticide to constraint plant destructive pests. *Journal of Nanomedicine Research* 6(3): 1–9.

Kumari, R., Saini, A. K., Chhillar, A. K., Saini, V., and R. V. Saini. 2021. Antitumor effect of bio-fabricated silver nanoparticles towards ehrlich ascites carcinoma. *Biointerface Research in Applied Chemistry* 11: 12958–12972.

Lakshmi, D. V., Thomas, R., Varghese, R. T., Soniya, E. V., Mathew, J., and E. K. Radhakrishnan. 2014. Extracellular synthesis of silver nanoparticles by the *Bacillus* strain CS 11 isolated from industrialized area. *3 Biotech* 4: 121–126.

Logeswari, P., Silambarasan, S., and J. Abraham. 2013. Ecofriendly synthesis of silver nanoparticles from commercially available plant powders and their antibacterial properties. *Scientia Iranica*. 20(3): 1049–1054.

Mazhar, T., Shrivastava, V., and R. S. Tomar. 2017. Green synthesis of bimetallic nanoparticles and its applications: A review. *Journal of Pharmaceutical Science and Research* 9: 102.

Meena, R. S., Meena, V. S., Meena, S. K., and J. P. Verma. 2015. The needs of healthy soils for a healthy world. *Journal of Cleaner Production* 102: 560–561.

Milošević, N., and M. Govedarica. 2002. Effect of herbicides on microbiological properties of soil. *Zbornik Matice Srpske za Prirodne Nauke* 102: 5–21.

Mishra, A., Kumari, M., Pandey, S., Chaudhry, V., Gupta, K. C., and C. S. Nautiyal. 2014. Biocatalytic and antimicrobial activities of gold nanoparticles synthesized by *Trichoderma* sp. *Bioresource Technology* 166: 235–242.

Mittal, D., Shukla, R., Verma, S., Sagar, A., Verma, K. S., Pandey, A., Negi, Y. S., Saini, R. V., and A. K. Saini. 2019. Fire in pine grown regions of Himalayas depletes plant growth promoting beneficial microbes in the soil. *Applied Soil Ecology* 139: 117–124.

Narayanan, K. B., and N. Sakthivel. 2010a. Biological synthesis of metal nanoparticles by microbes. *Advances in Colloid and Interface Science* 156(1–2): 1–3.

Narayanan, K. B., and N. Sakthivel. 2010b. Photosynthesis of gold nanoparticles using leaf extract of *Coleus amboinicus* Lour. *Materials characterization* 61(11): 1232–1238.

Nicolopoulou-Stamati, P., Maipas, S., Kotampasi, C., Stamatis, P., and L. Hens. 2016. Chemical pesticides and human health: The urgent need for a new concept in agriculture. *Front Public Health* 4: 148–8.

Nuruzzaman, M. D., Rahman, M. M., Liu, Y., and R. Naidu. 2016. Nano encapsulation, nanoguard for pesticides: A new window for safe application. *Journal of Agricultural and Food Chemistry* 64(7): 1447–1483.

Oroian, M., Dranca, F., and F. Ursachi. 2020. Comparative evaluation of maceration, microwave and ultrasonic-assisted extraction of phenolic compounds from propolis. *Journal of Food Science and Technology* 57: 70–78.

Pandey, A., and S. Tripathi. 2014. Concept of standardization, extraction and pre phytochemical screening strategies for herbal drug. *Journal of Pharmacognosy and Phytochemistry* 2: 115–119.

Pasaribu, A., Mohamad, R. B., Hashim, A., Rahman, Z. A., Omar, D., Morshed, M. M., and D. E. Selangor. 2013. Effect of herbicide on sporulation and infectivity of vesicular arbuscular mycorrhizal (*Glomus mosseae*) symbiosis with peanut plant. *Journal of Animal and Plant Sciences* 23: 1671–1678.

Rabea, E. I., Badawy, M. E. I., Rogge, T. M., Stevens, C. V., Hofte, M., Steurbaut, W., and G. Smagghe. 2005. Insecticidal and fungicidal activity of new synthesized chitosan derivatives. *Pest Management Science* 61: 951–960.

Rafique, M., Sadaf, I., Rafique, M. S., and M. B. Tahir. 2017. A review on green synthesis of silver nanoparticles and their applications. *Artificial Cells, Nanomedicine and Biotechnology* 45: 1272–1291.

Randall, C., Hock, W., Crow, E., Hudak-Wise, C., and J. Kasai. 2013. *National Pesticide Applicator Certification Core Manual*. National Association of State Departments of Agriculture Research Foundation, Washington, DC. https://www.hsdl.org/?view&did=751552 (accessed March 10, 2014).

Sabbour, M. M., and S. E. Abd El-Aziz. 2015. Efficacy of nano-diatomaceous earth against red flour beetle, *Tribolium castaneum* and confused flour beetle, *Tribolium confusum* (Coleoptera: Tenebrionidae) under laboratory and storage conditions. *Bulletin of Environment, Pharmacology and Life Sciences* 4: 54–59.

Salunkhe, R. B., Patil, S. V., Patil, C. D., and B. K. Salunke. 2011. Larvicidal potential of silver nanoparticles synthesized using fungus *Cochliobolus lunatus* against *Aedes aegypti* (Linnaeus, 1762) and *Anopheles stephensi* Liston (Diptera; Culicidae). *Parasitology Research* 109(3): 823–831.

Sankar, M. V., and S. Abideen. 2015. Pesticidal effect of green synthesized silver and lead nanoparticles using *Avicenna marina* against grain storage pest *Sitophilus oryzae*. *International Journal of Nanomaterials and Biostructures* 5: 32–39.

Sasson, Y., Levy-Ruso, G., Toledano, O., and I. Ishaaya. 2007. *Insecticides Design Using Advanced Technologies*. Berlin: Springer.

Shang, Y., Hasan, M., Ahammed. G. J., Li, M., Yin, H., and J. Zhou. 2019. Applications of nanotechnology in plant growth and crop protection. *Molecules* 24(14): 2558.

Sharma, D., Shandilya, P., Saini, N. K., Singh, P., Thakur, V. K., Saini, R. V., Mittal D., Chandan, G., Saini, V., and A. K. Saini. 2021. Insights into the synthesis and mechanism of green synthesized antimicrobial nanoparticles, answer to the multidrug resistance. *Materials Today Chemistry* 19: 100391.

Singh, N., Dotasara, S. K., Kherwa, B., and S. Singh. 2017. Management of tomato fruit borer by incorporating newer and biorationals insecticides. *Journal of Entomology and Zoology Studies* 5(2): 1403–1408.

Singh, P., Singh, H., Kim, Y. J., Mathiyalagan, R., Wang, C., and D. C. Yang. 2016. Extracellular synthesis of silver and gold nanoparticles by *Sporosarcina koreensis* DC4 and their biological applications. *Enzyme and Microbial Technology* 86: 75–83.

Siva, C., and M. S. Kumar. 2015. Pesticidal activity of eco-friendly synthesized silver nanoparticles using *Aristolochia indica* extract against *Helicoverpa armigera* Hubner (Lepidoptera: Noctuidae). *International Journal of Advance Scientific Technological Research* 2: 197–226.

Suganya, K. U., Govindaraju, K., Kumar, V. G., Dhas, T. S., Karthick, V., Singaravelu, G., and M. Elanchezhiyan. 2015. Blue green alga mediated synthesis of gold nanoparticles and its antibacterial efficacy against Gram positive organisms. *Materials Science and Engineering* 47: 351–356.

Syed, B., Prasad, N. M., and S. Satish. 2016. Endogenic mediated synthesis of gold nanoparticles bearing bactericidal activity. *Journal of Microscopy and Ultrastructure* 4(3): 162–166.

Thakur, S., Saini, R. V., Singh, P., Raizada, P., Thakur, V., and A. K. Saini. 2020. Nanoparticles as an emerging tool to alter the gene expression: Preparation and conjugation methods. *Materials Today Chemistry* 17: 100295.

Trusheva, B., Trunkova, D., and V. Bankova. 2007. Different extraction methods of biologically active components from propolis: A preliminary study. *Chemistry Central Journal* 1: 13.

Usta, C. 2013. Microorganisms in biological pest control—A review (bacterial toxin application and effect of environmental factors). *Current Progress in Biological Research* 13: 287–317.

Vijayabharathi, R., Sathya, A., and S. Gopalakrishnan. 2018. Extracellular biosynthesis of silver nanoparticles using *Streptomyces griseoplanus* SAI-25 and its antifungal activity against *Macrophomina phaseolina*, the charcoal rot pathogen of sorghum. *Biocatalysis and Agricultural Biotechnology* 14: 166–171.

Yadi, M., Mostafavi, E., Saleh, B., Davaran, S., Aliyeva, I., Khalilov, R., Nikzamir, M., Nikzamir, N., Akbarzadeh, A., and Y. Panahi. 2018. Current developments in green synthesis of metallic nanoparticles using plant extracts: A review. *Artificial Cells, Nanomedicine and Biotechnology* 46: 336–S343.

Yang, F. L., Li, X. G., Zhu, F., and C. L. Lei. 2009. Structural characterization of nanoparticles loaded with garlic essential oil and their insecticidal activity against *Tribolium castaneum* (Herbst) (Coleoptera: Tenebrionidae). *Journal of Agricultural and Food Chemistry* 57: 10156–10162.

Zahir, A. A., Bagavan, A., Kamaraj, C., Elango, G., and A. A. Rahuman. 2012. Efficacy of plant-mediated synthesized silver nanoparticles against *Sitophilus oryzae*. *Journal of Biopesticides* 5: 95–102.

Zheng, L., Hong, F., Lu, S., and C. Liu. 2005. Effect of nano-TiO2 on strength of naturally aged seeds and growth of spinach. *Biological Trace Element Research* 104: 83–91.

15 Role of Microorganisms in Improving Quality of Medicinal Aromatic Plants (MAPs)

Nirali Desai[1], Helly Shah[1], and Manan Shah[2]
[1]Ahmedabad University, Gujarat, India
[2]Pandit Deendayal Petroleum University, Gujarat, India

CONTENTS

15.1 INTRODUCTION

Treating illnesses with botanical resources is quite ancient worldwide. Traditional medicines are synthesized from plants since time immemorial. According to the World Health Organization (WHO) statistics, 80% of the worldwide population

DOI: 10.1201/9781003110477-15

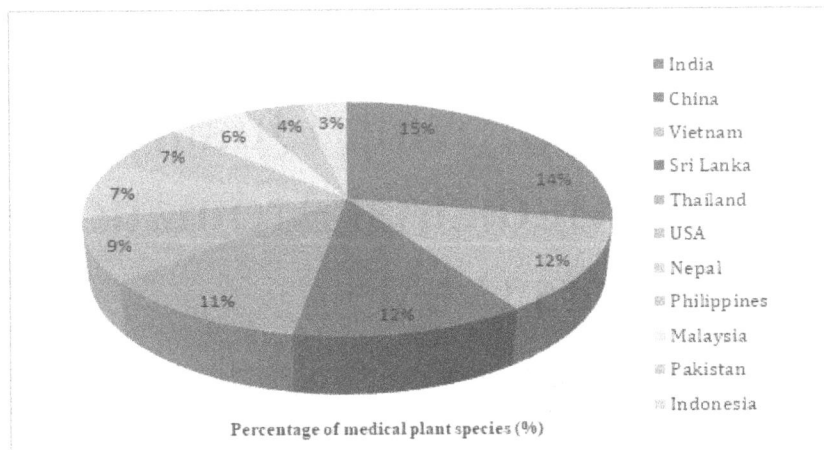

Percentage of medical plant species (%)

FIGURE 15.1   Distribution of medicinal plant species worldwide (Chen et al. 2016).

depends on herbal resources for their routine healthcare and 25% of worldwide drugs are produced from plants (Hamilton 2004). The distribution of medicinal plant species worldwide is shown in Figure 15.1 (Chen et al. 2016). Single or multiple parts of a plant have the property of preventing or treating any kind of illness, that is, physical, mental, or social, which can be characterized as medicinal plants. Examples of which include, ginger-like food plants, Dioscorea yams like plants which are used to synthesize hormones, cotton and jute-like plants for the preparation of surgical dressings, and so on (Sofowora, Ogunbodede, and Onayade 2013). Aromatic plants are the special category of plants that possess aroma or flavor of volatile substances in one or more parts of a plant. Many medicinal plants also exhibit characteristics of aromatic plants. Hence, they are together called medicinal and aromatic products (MAPs) (Rao, Palada, and Becker 2004).

In recent years, natural products are getting recognized for their lesser or no side effect characteristics. Due to this mindset, the demand for MAPs and their importance in industries have increased. MAPs are used in the treatment of many common diseases and as an anti-diabetic, a blood purifier, a worm killer, and so forth (Patel 2016). Plants secondary metabolites like alkaloids, terpenes, flavonoids are used in several industries worldwide. Phytochemicals like glycosides, which are a type of terpenes, can be used for the treatment of cardiac failure. Famous tablets available in the market containing these glycosides are digoxin and digitoxin (Oerther 2011). Another plant secondary metabolite extract, taxol, and artemisia, which is extracted from Taxusbaccata and Artemisia annua respectively, play an important role as anti-neoplastic and antimalarial agents (Croteau et al. 2006; Pollier, Moses, and Goossens 2011). Many known plant products like coriander, a known carminative and important spice; Catharanthusroseus (Rose) which shows potent pharmacological activity; aloe vera, a strong purgative which has important content for daily skin-care and hair-care products are used commonly (Gurib-Fakim 2006). Likewise, hundreds and thousands of products are extracted from these plants which are very much useful for the treatment of different organs of the body. MAPs are the major source of the herbal cosmetic industry. Herbal preparation contains many chemicals and products isolated from plants because of their antioxidant as well as anti-inflammatory properties (Glaser 2004; Draelos 2016). In another half of the twentieth century, demand for processed food was increased and so the safety issues related to it. Therefore, scientists are now trying to identify more naturally occurring food-additives for processed food. MAPs possess antimicrobial (bactericidal and fungicidal) properties, which made them one of the most important ingredients in the processed food and the food industry worldwide (Michael Davidson, Sofos, and Larry 2005). In addition, medicinal and aromatic plants are crucial for the socioeconomic status of the countries. They help the system by bettering the rural parts of the community (Tabuti, Dhillion, and Lye 2003; Pattanaik, Reddy, and Murthy 2008). According to the WHO fact sheet published in 2003, the global market for plants-derived medicines stands over US$60 billion annually (Ghorbanpour et al. 2017). One study by BCC research in 2009 stated that the global market for plant-derived drugs will increase with an 11.0% annual growth rate (Global Market for Botanical and Plant-Derived Drugs Worth $32.9 Billion in 2013 n.d.). Moreover, there are reports available which stated the risk of extinction of MAPs and loss of its biodiversity because of habitat loss and overconsumption (Aslam et al. 2017; Iqbal et al. 2017; Jeelani et al. 2018). Overall, it supports the increasing demand for medicinal and aromatic plants and, thus, plants with better yield and quality are desirable.

To match the demand for medicinal and aromatic plants, it is important to increase the cultivation of them. However, various factors hinder the production of MAPs in terms of quantity as well as quality. These factors include environmental parameters (like temperature, humidity, rain, CO_2 level, etc.), availability of nutrients, chemicals present in the surrounding environment, pesticides or plant diseases caused by bacteria or viruses (Shaikh and Mokat 2018). For any medicinal plants which grow in a particular geographical region, the climate of that region works as an optimum condition for that particular plant. Change in that environment will lead to a decrease in the yield or quality of the plants (Gupta et al. 2019; Yuan et al. 2020). Numbers of medicinal and aromatic plants based on their geographical locations are shown in Table 15.1. The availability of nutrients can also be one of the important

TABLE 15.1
Number of Medicinal and Aromatics Plants (in Range) according to Some Countries

Countries	Number of Medicinal and Aromatic Plants
Hungary, Jordan	100–500
Philippines, Nepal, France, Bulgaria, Sri Lanka, South Korea	500–1,000
Malaysia	1,000–1,500
Thailand, Vietnam, Pakistan	1,500–2,000
USA, India	2,000–3,000
China, Turkey	Above 3,000

factors which regulate the production of medicinal plants. Access nutrients or insufficient nutrients affect the growth of the plant, and it is reported that they may affect the quality of the products obtained from the MAPs (Máthé 2015). Also, pesticides are known to enhance the production of plants by killing insects and factors that affect production. However, their debris can affect plants as well as human health (Igbedioh 1991). Furthermore, it can badly affect the commodities developed from the plants, the surrounding environment, soil fertility, water retention, groundwater, and many non-targeted organisms (Aktar, Sengupta, and Chowdhury 2009). These reports show the need for safe methods that can enhance the production of the MAPs and do little or no harm to the quality and quantity of the plants.

There are various methods available at present that use different approaches to enhance the cultivation of plants. Some of them are widely accepted worldwide and very much known. Some microorganisms like bacteria and fungi aid in improving the quality and growth of the plants. For example, Rhizobium bacteria which are soil bacteria that help plants to fix atmospheric nitrogen in the soil. By doing this, it helps to increase the soil's nutritional efficiency and help the plants to grow healthy. It is a very known example of the symbiosis relationship between plants and bacteria (van Rhijn and Vanderleyden 1995). Similarly, Rhizosphere fungi help medicinal and aromatic plants in several ways like it enhances the nutrition in the soil, helps in increased production of secondary metabolites, also involved in increasing aroma and quality of essential oil obtained from aromatic plants (Shaikh and Mokat 2018). Likewise, there are many other methods used to improve the quality of MAPs which are discussed further in detail.

15.2 METHODS FOR IMPROVING THE QUALITY OF MEDICINAL AND AROMATIC PLANTS

There are various approaches like molecular approach, technological approach, agronomical approach, chemical approach, and plant-breeding approach for improving the quality of medicinal and aromatic plants. Further, in this chapter, these approaches are explained, and a new bacterial approach is described in detail.

To improve the yield and quality of medicinal plants and their important secondary metabolites, transgenic plants can be developed using the molecular or transgenic approach. Mainly molecular markers can be used to evaluate breeding, genetic diversity, and genotype potential at the initial growth stage which overall will help in the improvement of medicinal cultivars. To increase medicinal value, aromatic polyketides should be more in number than the complex ones. Besides, these content and composition information of the plant can be discovered by doing QTL analyses which can be further used for the development of improved medicinal plant species (Gupta et al. 2008). Technological or chemical approaches include various lab techniques or the invention of new tools to increase yield, health, quality traits, and growth of plants (Prisa Domenico 2020). Biotechnological approaches use many chemical markers while unconventional approaches use various in-vitro and ex-situ techniques to manipulate or combine genes to create species with desired traits which overall help in protecting endangered species (Khan, Al-Qurainy, and Nadeem 2012). Agroforestry systems and multiple strategies are also used for encouraging cultivation, to modify potency, and for the conservation of medicinal and aromatic plants. In this approach, artificial arrangements are used to maintain ideal conditions as quality and biomass yield of plants are dependent on edaphic factors, geographical distribution, duration as well as the intensity of shade effects, and so on (Rao, Palada, and Becker 2004). For example, integration of shade-tolerant plants, in situ, and ex-situ conservation methods are used in which wild nurseries, natural habitats and reserves, botanic gardens, and seed banks are developed to protect and conserve species of high demand, endangered medicinal and aromatic plants. Cultivation and good agricultural practices are also used to enhance the production of medicinal plants as well as to boost yields of active compounds (Chen et al. 2016). Furthermore, copious plant-breeding approaches like hybrid, combinational, synthetic cross, and variation methods are used for increasing variability and improving the quality of plants (Chandra 2017).

However, these approaches have their disadvantages and haven't always been successful. Thus, to inhibit increased risk, pollution, and toxicity caused using outdated approaches, it is necessary to re-evaluate some existing approaches and to find alternatives for the ones which use artificial chemicals (Shameer and Prasad 2018). In addition, to increase agricultural productivity and by looking at various other problems like disturbed biodiversity, increased hunger due to population growth and climate changes, it is mandatory to develop or shift towards the use of sustainable and environmentally friendly methods (Glick 2012). As a result, nowadays, environmentally sustainable bacterial approaches are used to increase plant production, quality, and growth. The most common approach includes the use of growth-promoting bacteria (PGPR) (Roriz et al. 2020). These approaches are further described in-depth.

There are several bacteria commercially used to enhance the growth and quality of medicinal and aromatic plants while only a few types are used quite often. Plants interact with soil by their roots or root hairs. Generally, in soil, several microorganisms like bacteria, fungi, and algae reside near the plant roots which are known as the Rhizosphere. Rhizospheric bacteria that induce enhancement in the production and quality of the plants are also called plant growth-promoting rhizobacteria (PGPR) (Dojima and Craker 2016). PGPR can be further classified into extracellular PGPR (ePGPR) which reside between the cells of the outer part of the roots (i.e., cortex) and intracellular PGPR (iPGPR) which usually live in nodules present in the root. Bacteria genera like Bacillus, Erwinia, Pseudomonas, Caulobacter, Micrococcus, Chromobacterium, and Arthrobacter fall under the ePGPR category whereas, iPGPR include endophytic microbes like Rhizobium, Bradyrhizobium, Allorhizobium (Martínez-Viveros et al. 2010; Bhattacharyya and Jha 2012; Gouda et al. 2018). They mainly regulate the plant hormones like cytokinins, ethylene, auxins, and gibberellins, and enhance the accessibility of Fe, P, and N in the soil which help in increasing the plant growth (Tahir et al. 2017). PGPR uses different mechanisms to promote the growth of plants, which are described below.

15.2.1 SOLUBILIZATION OF PHOSPHORUS

Phosphorus in their soluble forms, that is, orthophosphate anions (HPO_4^- & $H_2PO_4^-$) are essential for plant growth. From the total phosphorus, only 1% of these necessary phosphorus anions are freely available in the soil (Dojima and Craker 2016). Thus, to access the remaining and to solubilize phosphorus, PGPRs help in different ways: (i) by releasing organic acids, which may lower the pH, or enhance chelation; and (ii) by mineralizing organic phosphate. Mineralization of organic phosphate by bacteria can be done by releasing non-specific acid phosphatases or phytases (Sharma et al. 2013). Bacterial species that dominantly help in solubilizing phosphorus are Pseudomonas species (*Pseudomonas putida, Pseudomonas calcis, Pseudomonas fluorescens*, etc.) and Bacillus species (*Bacillus* sp., *B. polymyxa, B. coagulans, B. cereus, B. subtilis, B. fusiformis*, etc.) (Sharma et al. 2013). Other bacteria like *Xanthomonas* sp. (Vazquez et al. 2000), *Arthrobacter* sp., *Serratia phosphoticum, Thiobacillus ferrooxidans, Delfia* sp., *Rhodococcus* (Chen et al. 2006), *Klebsiella, Vibrio, Enterobacter* (Chung et al. 2005) are also involved in solubilizing phosphate.

15.2.2 FIXATION OF NITROGEN

Nitrogen is one of the major nutrients involved in plant growth and development. Plants cannot take up atmospheric nitrogen from the surroundings thus, using the nitrogen-fixation cycle, plants assimilate necessary ions from the soil via their root cells. The nitrogen-fixation cycle includes processes like ammonification, nitrification, and denitrification which convert atmospheric nitrogen into ammonium (NH_4+), Nitrite (NO_2-), and Nitrate (NO_3-). These processes are carried out by bacteria residing in the nearby soil and ions are further used in various biological processes. Biofertilizers form of PGPRs are involved in these processes and which interact with plants either using symbiotic mode or non-symbiotically (free-living). Symbiotic PGPR usually includes *Rhizobium* sp., *Beijerinckia* sp., *K. pneumoniae, Azoarcus* sp. (Ahemad and Kibret 2014; Gouda et al. 2018), while there are also free-living diazotrophs like *Azospirillum, Azotobacter*, and *Pseudomonas* sp. which are involved in nitrogen fixation for medicinal plants like Catharanthus roseus, Ocimum sanctum, Coleus, and aloe vera (Karthikeyan et al. 2008; Dojima and Craker 2016). Moreover, other free-living bacteria near the rhizosphere like *Azoarcus, Cyanobacteria, Gluconacetobacter, Burkholderia* are also involved in nitrogen fixation (Bürgmann et al. 2004; Orr et al. 2011; Dojima and Craker 2016).

15.2.3 POTASSIUM SOLUBILIZATION

Being the third most important nutrients after nitrogen and phosphate, potassium is essential for the healthy growth and development of plants (Armengaud, Breitling, and Amtmann 2010; Kour et al. 2020). Potassium regulates nearly 60 different enzymes that are involved in growth hormone synthesis pathways. It is also important for osmotic regulation, stomata opening, and maintenance of neutralization of pH in plants. Insufficient absorption of potassium leads to a slow growth rate, poorly developed root, and lower yield. In soil, 90% of potassium is available in different forms like, in soil solution, in minerals, as insoluble matter, and freely present in soil which can be taken up by the ion exchange process (Kour et al. 2020). Potassium solubilizing PGPR usually produces organic acids like citric acid, lactic acid, malic acid, and oxalic acid (Liu, Lian, and Dong 2012) which acidify the microbial cells and its environment. This process leads to the release of K+ from minerals (Ullman et al. 1996;

Shrivastava, Srivastava, and D'Souza 2016). *Bacillus mucilaginosus, Paenibacillus glucanolyticus*, Pseudomonas sp., *Penicillium frequentans, Enterobacter hormaechei*, Cladosporium sp., produce organic acid for K-solubilization (Kour et al. 2020). *Bacillus pseudomycoides* isolated from tea-growing soil solubilizes potassium in great amounts (Pramanik et al. 2019). The potassium release rate is affected by the *Cenococcum geophilum* fr. (Xue et al. 2019). Additionally, *Rhizobium rhizogenes, Bacillus megaterium, Rhizobium leguminosarum, Thiobacillus thiooxidans* (El-barbary and El-Badry 2018), *Mesorhizobium sp., Arthrobacter sp.* (Xiao et al. 2017), *Bacillus licheniformis*, and *Pseudomonas azotoformans* (Saha et al. 2016) are K-solubilizer bacteria and which are isolated from different plant soils (Ahmad and Zargar 2017; Kour et al. 2020). Furthermore, K-solubilizing bacteria aid in plant growth and development using different mechanisms like nitrogen fixation, zinc solubilization, production of hydrolytic enzymes, siderophore, and phytohormone (Kumar et al. 2019; Yadav 2020). For instance, *Bacillus megaterium, Duganella violaceusniger, Psychrobacter fozii, Pseudomonas monteilii, Stenotrophomonas maltophilia, Pseudomonas lini, Pseudomonas thivervalensis, Paenibacillus amylolyticus, Paenibacillus dendritiformis* (Kumar, Dubey, and Maheshwari 2012), *Bacillus horikoshii, Bacillus amyloliquefaciens, Achromobacter piechaudii, Klebsiella sp., Exiguobacterium antarcticum, Stenotrophomonas maltophilia, Bacillus atrophaeus, Bacillus pumilus, Planococcus salinarum, Paenibacillus polymyxa, Bacillus mojavensis* promote the production of IAA, GA-like phytohormones, siderophores, hydrogen cyanide (HCN), ammonia, 1-aminocyclopropane 1-carboxylic acid (ACC), beta-glucanase, lipase, and chitinase (Verma et al. 2015, 2016).

15.2.4 PRODUCTION OF SIDEROPHORES

Siderophores are iron-chelating low molecular weight compounds produced by some microorganisms and plant species. Siderophores promote plant growth in iron-deficient environments by solubilizing iron in ferric complexes obtained from the minerals. Bacteria like *Sphingobacterium* sp., *Pseudomonas poae, Delftia acidovorans, Achromobacter xylosoxidans, Enterobacter endosymbiont, Bacillus* sp., and *Rhodococcus* sp., efficiently produce siderophores and rescue plant growth from iron-stress conditions (Tian et al. 2009; Dojima and Craker 2016). *Rhizobium meliloti*, root-nodulating bacteria isolated from the *Mucuna pruriens* (a medicinal plant) produces siderophores to promote plant growth (Arora, Kang, and Maheshwari 2001).

Fluorescent *Pseudomonas* produces yellow-green pigments (pyoverdines) which function as siderophores (Demange et al. 1986; Kloepper et al. 2004).

15.2.5 PRODUCTION OF PHYTOHORMONES

Phytohormones like auxin, cytokine, gibberellic acid, abscisic acid, ethylene, jasmonic acid, brassinosteroids are very important for different processes which are essential for plant growth and development and are produced by specific PGPRs known as phyto stimulators. Amongst them, auxin, cytokine, and gibberellic acid are very much important for root formation, shoot formation, and its elongation, and seed germination, respectively. Additionally, they inhibit the senescence process in plants. On the other hand, ethylene and abscisic acid induce ripening and senescence in the plants. Therefore, microbes help in inducing or protecting plant growth by increasing the production of growth hormones like auxin, cytokine, and gibberellic acid or by decreasing the production or effect of negative growth regulators ethylene and abscisic acid (Tsavkelova et al. 2006; Bhattacharyya and Jha 2012). As mentioned above, other than for root initiation, auxins are very important for cell division, cell differentiation, increase the rate of reaction, synthesis of pigment, and so on. Indole acetic acid (IAA) is one of the auxins present in plants, which exhibits the greatest beneficial activity (Tsavkelova et al. 2006). Synthesis of IAA is observed by many kinds of bacteria like rhizospheric, free-living, epiphytic, and symbiotic bacteria and these different bacteria use different biosynthesis methods. Many bacteria synthesize IAA by indole-3-pyruvic acid and indole-3-acetic aldehyde whereas some use tryptophan-dependent and tryptophan independent pathways. IAA biosynthesis can be increased by Agrobacterium tumefaciens, phytopathogenic bacteria (Tsavkelova et al. 2006). Major genera involved in IAA formation are Pseudomonas, Xanthomonas, Bacillus (Suzuki, He, and Oyaizu 2003), Methylobacterium, Amino Bacteria, Paracoccus, and Methylovorus (Ivanova, Doronina, and Trotsenko 2001). Bacterial species that synthesize IAA are Agrobacterium sp., Azotobacter sp., Enterobacter sp., Azospirillum sp., Alcaligenes sp., Bradyrhizobium sp., Rhizobium sp., Herbaspirillum sp. (Tien, Gaskins, and Hubbell 1979; Cacciari 1989; Costacurta and Vanderleyden 1995; Patten 1996; Basu 2000; Mohite 2013), Klebsiella, Archromobacter, Flavobacterium, Sphingomonas, Corynebacterium, Stenofrophomonas, Mycobacterium, Microbacterium, Arthrobacter, Rhodococcus (Tsavkelova et al. 2006) and symbiotic and free-living cyanobacteria like Anabaena, Anabaenopsis, Calothrix, Plectonema, Cylindrospermum, and Chlorogloeopsis (Sergeeva, Liaimer, and Bergman 2002).

Cytokine phytohormone is derived from adenine and it regulates shoot initiation, plant cell division, promotion of branching, chloroplast formation, and so on. Different cytokines like kinetin, zeatin, isopentenyl adenine can be found in plants and are synthesized by different microorganisms (Tsavkelova et al. 2006). Bacteria that belong to genera Rhizobium, Arthrobacter, Azospirillum, Bacillus, Azotobacter, Pseudomonas form cytokine (Cacciari 1989; Salamone 2001; Tsavkelova et al. 2006). Additionally, bacterial from genera Methylopolia, Methylomonas, Xanthobacter, Aminobacter, Methylosinus, Hyphomicrobium, Paracoccus, Blastobacter, and Methyloarcula have an inherent ability to synthesize cytokinins (Ivanova 2000; Koenig, Morris, and Polacco 2002). Furthermore, phytopathogenic bacteria like Erwiniaherbicola, A. rhizogenes, Agrobacterium tumefaciens, and PGHPR strains like Pseudomonas fluorescens also synthesize cytokinins (Salamone 2001).

Gibberellins are the largest class of phytohormones and can be synthesized in both plants and microorganisms using different methods or pathways (Saifuddin et al. 2009). They are important for seed germination primarily and in breakage of the seed dormancy. Mutant gibberellin synthesis pathway protein was used during the Green Revolution to increase the cultivation of plants for agriculture. They are also very important for meristematic cell division and elongation of intercalary regions. Additionally, they stimulate fluorescence, activate the synthesis of the membrane and amylolytic enzymes, and so on (Michael 1983). Bacteria which synthesize gibberellins are Azospirillum, Azotobacter, Pseudomonas,

Acinetobacter, Agrobacterium, Clostridium, Xanthomonas, Flavobacterium, Rhizobium, Micrococcus (Barea, Navarro, and Montoya 1976; Tien, Gaskins, and Hubbell 1979; Cacciari 1989), Bacillus (B. pumilus, B. macroides) (Joo et al. 2004) and Cyanobacteria (Cylindrospermum sp., Anabaenopsis sp.). Likewise, ethylene is also a very important gaseous hormone for plant growth and development. It is initially known as ripening hormone, as it is involved in the ripening of fruits. Later, it was found to be involved in various processes like flower induction, and the release of dormancy. However, excessive production of ethylene adversely affects the plant in various ways like the shedding of leaves, abnormal growth of roots, etiolated seedling, and so on. Accelerated production of ethylene takes place in response to stress (Abeles, Morgan, and Saltveit 1993). To increase plant growth and development, it is important to decrease the synthesis of ethylene. Ethylene is a methionine-derived hormone and the precursor of it is ACC. Some bacteria synthesize enzyme ACC deaminase which can hydrolyze ACC and so ethylene production. This way they help to regulate the concentration of ethylene (Glick, Penrose, and Li 1998; Saleem et al. 2007; Egamberdieva, Shrivastava, and Varma 2015). For the treatment of plant, for the growth of roots by inhibiting ethylene and for resistance against various stresses like salinity stress, waterlogging stress, drought stress, temperature stress, pathogenicity stress, contaminants like heavy metals, organic and air pollutants stress, bacteria which contain ACC deaminase can be used and examples of which are *Methylobacterium fujisawaense, Bacillus (B. circulans* DUC1, *B. Wrmus* DUC2, *B. globisporus* DUC3, *B. pumilus, B. lentus), Pseudomonas sp.* (*P. cepacia, P. putida), Alcaligenes sp., Bradyrhizobium sp., Enterobacter cloacae, Klebsiellaoxytoca* 10MKR7, *Azospirillum brasilense* Cd1843 (Saleem et al. 2007; Egamberdieva, Shrivastava, and Varma 2015), and *Achromobacter xylosoxidants* (Dojima and Craker 2016).

15.2.6 PRODUCTION OF VOLATILE ORGANIC COMPOUNDS (VOCS)

Bacterial-produced VOCs, especially by rhizobacteria constitute a significant role in the regulation of metabolism or synthesis of phytohormones and overall, in enhancing plant growth and health. Bacterial VOCs are low molecular weight lipophilic compounds with high vapor pressure and low boiling point and can be used as eco-friendly alternatives because of their high efficiency, effectiveness, and low cost (Fincheira and Quiroz 2018). Plant growth surges when VOCs produced by PGPR interact with phytohormones during morphogenetic processes. Acetoin and 2,3-butanediol VOCs are the first to get reported which induces the growth promotion. Pseudomonas fluorescens SS101 produces 2-methyl-n-1-tridecene, 2-butanone, and 13-tetradecadien-1-ol which increase the growth of *Nicotiana tabacum* used for treating skin orders and some cancers (Charlton 2004; Tahir et al. 2017). It also produces some VOCs which can improve the yield of oil constituents like pulegone and menthone of peppermint as well as of basil. Also, Bacillus subtilis GB03 aids in increasing the growth of shoot-root biomass, leaf area, and concentration of eugenol and alpha-terpineol of basil (Dojima and Craker 2016). Some Gram-positive species like Arthrobacteragilis UMCV2 and *Paenibacillus polymyxa* E681 are known to produce VOCs which help in enhancing growth, weight, stem length, lateral root density of Medicago Sativa and in increasing fresh weight, surface leaf area, respectively. There are Gram-negative species like *Pseudomonas simiae*, *Proteus vulgaris*, and some belonging to Chromobacterium, *Serratia*, *Pandoraea*, and *Burkholderiagenera* have been reported to enhance the primary root-shoot length, biomass, and the number of lateral roots significantly (Fincheira and Quiroz 2018). When abiotic or biotic stress is induced, VOCs protect or stabilize cell membranes and also enhance the plant resistance by quenching toxic accumulation of reactive oxygen species (ROS) (Brilli, Loreto, and Baccelli 2019). Moreover, VOCs help to improve iron and sulfur nutrition in plants by protecting them against salinity and drought stress. Some bacterial species like *Bacillus amyloliquefaciens* GB03, *Pseudomonas simiae*, *Pseudomonas chlororaphis* strain O6, *Bacillus safensis* strain W10, *Ochrobactrum pseudogregnonense* strain IP8, and *Bacillus sp.* strain B55 help in increasing salt tolerance, greater biomass production, protection from water loss; pathogens; inducing diseases, heavy metal, and system tolerance; optimizing homeostasis of iron in plants which have medicinal uses (Liu and Zhang 2015).

15.2.7 DISRUPTING QUORUM SENSING SIGNALS IN PATHOGENS

The quorum-sensing signal is a phenomenon used for bacterial communication in which bacteria use signal molecules called autoinducers to assess their population density. Oligopeptides and N-acyl-L-homoserine lactones (AHLs) are used by gram-negative and gram-positive bacteria, respectively. By sensing these autoinducers in the surroundings, bacteria regulate their physiological activities like virulence, conjugation, symbiosis, antibiotic production which may negatively affect plant growth. For example, bacterium *Serratia plymuthica* HRO-C48, root-associated bacteria negatively regulate IAA production by quorum sensing (Müller et al. 2009). *Agrobacterium tumefaciens*, *Pseudomonas aeruginosa*, *Erwinia chrysanthemi*, *Pantoea stewartii*, and *Burkholderia cepacia* induce pathogenesis by quorum sensing (Dong et al. 2000; Dong et al. 2001). On the contrary, there are some bacteria identified which produce enzymes like lactonase, decarboxylase, oxidoreductase, deaminase, and acylase that can degrade quorum-sensing molecules associated with negative regulation such as N-AHLs (Chen et al. 2013). These types of enzymes are sometimes released by quorum-sensing bacteria like Agrobacterium tumefaciens to degrade quorum sensing (Zhang, Wang, and Zhang 2002). Gram-negative and gram-positive bacteria involved in disrupting quorum-sensing signals and in protecting plants mainly include species like Bacillus (B. subtilis, B. thuringiensis, B. mycoides, B. marcorestinctum) (Dong et al. 2000; Dong et al. 2001), Arthobacter (Park et al. 2003), Pseudomonas (Sio et al. 2006), Streptomyces (Park et al. 2005), Klebsiella Ralstonia (Lin et al. 2003), Shewanella (Morohoshi et al. 2008), and Comanomas (Uroz et al. 2007; Chen et al. 2013).

15.2.8 RESISTANCE TOWARDS PATHOGENS

Stress occurs due to pests like bacteria, viruses, fungi, insects, and nematodes, and plant-specific pathogens are usually referred to as biotic stress. PGPRs help in enhancing plant immunity against pathogens. Pattern recognition receptors (PRRs) help plant's immunity system by identifying essential survival structures of microbes called microbe-associated molecular patterns (MAMPs) and by commencing immune signaling. Pathogenic and nonpathogenic bacteria induce various defense responses like systemic acquired resistance and induced resistance which are used for treating plants against different pathogens and infections (Meena et al. 2020). Bacteria like *Bacillus*, *Bradyrhizobium*, and *Pseudomonas* help plants to protect against pathogen attacks by activating immunity and strength. Additionally, some bacterial species like *Trichoderma spp.*, *Streptomyces spp.*, *Ampelomyces quisqualis*, and *Bacillus subtilis* (Dojima and Craker 2016) can be used as biopesticides which promote the growth of plants by directly removing pathogens via the production of antibiotics, increasing nutritional availability (mostly N, P, K), and other important factors (Bhattacharyya and Jha 2012). The bacterial species Pseudomonas helps in inhibiting phytopathogens using different mechanisms like by producing cyanide which is used as a biocontrol agent and by producing metabolites, lytic enzymes like chitinase and glucanase, which help in inducing antifungal and antibiotic properties and in destroying cell walls of pathogens (Vejan et al. 2016). Other than Pseudomonas, Bacillus, Azospirillum, Paenibacillus, Azotobacter, Streptomyces, and Enterobacter are used as biocontrol agents as they secrete compounds having disease-repressing activity. Different bacterial strains like Rhizobium (*R. leguminosarumbv. Viciae*, *R. leguminosarumbv. Trifolii*, *R. meliloti*) and Bacillus (B. subtilis, B. japonicum) have also been known for producing enzymes, antibiotics, and for helping plants by inhibiting various pathogenic microorganisms as well as inducing resistance (Meena et al. 2020).

15.2.9 RESISTANCE TOWARDS ABIOTIC STRESS

Stress occurred due to different factors like temperature, salinity, drought, nutrient deficiency, and high amounts of heavy metals in soil are called abiotic stress. Among all the abiotic stresses, salt stress is the most prominent one and it occurs when there is an excessive salt concentration in the soil near the root (EC > 4 Sm-1) and high exchangeable sodium (>15%) (Akhtar 2019). It is estimated that approximately 20% of cultivated land and 33% of irrigated lands are affected with high salt (Jamil et al. 2011). For any plant to grow and develop, it requires optimum temperature (Fitter and Hay 2012) as increased or decreased temperature leads to abnormal cellular metabolism (Somerville 1995; Kour et al. 2020). Drought stress is a natural calamity that arises due to less or no rain or water deficiency in the surrounding of plants which often lead to physiological deteriorations like photosynthesis, flowering, fruiting, and so forth (Osakabe et al. 2014; Xu et al. 2014). Stresses alter the physiological activities of plants such as vegetation, seed germination, nutrient, and water uptake (Akbarimoghaddam et al. 2011). To ameliorate stress conditions, bacteria release some chemicals or signaling molecules which further give signals to plants to synthesize growth hormones like auxin, gibberellins, cytokinin. These growth hormones will make changes in plant physiology in such a way that they can uptake enough nutrients to survive in the stress condition. Some bacteria also release enzymes like ACC deaminase, which controls the synthesis of ethylene (Prasad, Kumar, and Varma 2015) and abscisic acid (Laloi, Apel, and

Danon 2004). For example, in salt stress conditions, Azospirillum sp. controls the uptake of Na^+ and increases the uptake of the K^+ and Ca^+. It also alleviates the activity of nitrogen solubilizing enzymes in maize plants (Hamdia, Shaddad, and Doaa et al. 2004). Similarly, Pseudomonas putidaabsorbs produces Mg^{+2}, K^+, and Ca^+ in an increased amount in cotton plants (Yao et al. 2010). In mung beans, Rhizobium sp. and Pseudomonas sp. (Ahmad et al. 2013); in barley and oats, Acinetobacter sp., Pseudomonas sp. (Chang et al. 2014); and in maize plant, Pseudomonas syringae, P. fluorescens, and Enterobacteraero genes (Nadeem et al. 2007) increase the IAA production and activity of ACC deaminase (Akhtar 2019). In Arachishypogaea (groundnut), Bacillus licheniformis A2 produces NH3, chitinase, HCN, IAA, and P-solubilizing enzymes due to which root length and biomass increases (Nautiyal, Christian, and Parker 2013). In Silybummarianum, Pseudomonas extermorientalis produces auxin, exopolysaccharides, and form biofilm which lead to an increase in the elongation of root and shoot under salt stress conditions (Egamberdieva, Jabborova, and Mamadalieva 2013). Under heat stress conditions, Burkholeriaphy to firmans showed ACC-deaminase activity, releases phenylalanine ammonia, and siderophores in Vitisvinifera (grapevine), and maintain the growth of potato plants (Bensalim, Nowak, and Asiedu 1998). This increases antioxidant, color pigment, and catalase activity (Barka, Nowak, and Clément 2006). ACC deaminase at low temperature and under salt stress promotes the growth of canola plants (Cheikh and Jones 1994). In drought or dry soil conditions, Azospirillumbrasilense, Azotobacterchroococcum, and Bacillus sp. produce P-solubilizers, siderophores, IAA in Ociumumbasilium L. (Basil) (Heidari and Golpayegani 2012), Trigonella foenum graecum L. (fenugreek) (Tank and Saraf 2003), and Lactuca sativa L. (lettuce) (Arkhipova et al. 2007), respectively. Inoculation of Arthrobacter sp. and Bacillus sp. significantly reduced stress-inducible genes in capsicum annuum plant under osmotic stress (Sziderics et al. 2007). Also, Pseudomonas tolaasii, P. fluorescens, Alcaligenes sp., and Mycobacterium sp. release siderophores, IAA, exopolysaccharides to protect Brassica napus (rapsi) plant against cadmium (Dell'Amico, Cavalca, and Andreoni 2008).

15.3   CHALLENGES AND FUTURE SCOPE

This chapter highlights the beneficial effects and different mechanisms of plant-associated microbes which are useful for enhancing plant growth and for improvement of its quality. However, they produce some compounds which have adverse outcomes on the growth of the plant (Dojima and Craker 2016). There is a wide range of PGPR strains available and selection of relevant microbial strains for maximized advantages in growth and yield can be very difficult. Also, the positive and negative effects of certain bacterial products are dependent upon the selection, compatibility, environmental conditions, and combination of plants; thus, one needs to be careful (Meena et al. 2020). For example, biocontrol agent cyanide produced by Pseudomonas species has beneficial as well as inhibitory effects and so knowledge about the same is very important (Vejan et al. 2016). Moreover, fewer experiments have been carried out especially for medicinal and aromatic plants; thus, more studies can be carried out to know about all the possible mechanisms in detail at all levels by which they enhance the growth and essential constituents (Egamberdieva and Teixeira da Silva 2015). To achieve minimal adverse effects, various techniques for selection of strains can be generated, to carry out research about PGPRs mechanisms in detail and colonization of the rhizosphere can be promoted more, and PGPRs can be combined with transgenic techniques (Bhattacharyya and Jha 2012).

15.4 CONCLUSIONS

It can be concluded that the usage of microbiome or combination of bacteria is more effective in increasing plant growth and quality than the single bacterium. Most commonly, different strains and species of rhizobacteria like Bacillus, Stenotrophomonas, Arthrobacter, Pseudomonas, Acinetobacter, Rhizobium, Flavobacterium, Enterobacter, Azotobacter, and Serratia are used in combinations or as formulations for enhancing the growth activity of medicinal and aromatic plants. This bacterial approach replaces the use of harmful pesticides, artificial supplements, and chemical fertilizers and helps in enhancing the yield, quality, and growth of plants without doing any harm. As mentioned in the chapter, PGPR strains use copious mechanisms to increase growth of plants as well as for protecting them against various stresses, pathogens, and other factors.

DECLARATION

AUTHORS' CONTRIBUTION

All the authors make a substantial contribution to this chapter. ND, HS and MS participated in drafting the manuscript. ND, HS and MS wrote the main chapter. All the authors discussed the results and implication on the chapter at all stages.

ACKNOWLEDGMENTS

The authors are grateful to Department of Chemical Engineering, School of Technology, Pandit Deendayal Petroleum University for the permission to publish this research.

AVAILABILITY OF DATA AND MATERIAL

All relevant data and material are presented in the main paper.

COMPETING INTERESTS

The authors declare that they have no competing interests.

FUNDING

Not applicable.

CONSENT FOR PUBLICATION

Not applicable.

ETHICS APPROVAL AND CONSENT TO PARTICIPATE

Not applicable.

REFERENCES

Ahemad, Munees, and Mulugeta Kibret. 2014. "Mechanisms and Applications of Plant Growth Promoting Rhizobacteria: Current Perspective." Journal of King Saud University – Science. https://doi.org/10.1016/j.jksus.2013.05.001.

Ahmad, Malik, and M. Zargar. 2017. "Characterization of Potassium Solubilizing Bacteria (KSB) in Rhizospheric Soils of Apple (Malus Domestica Borkh.) in Temperate Kashmir." Journal of Applied Life Sciences International. https://doi.org/10.9734/jalsi/2017/36848.

Ahmad, Maqshoof, Zahir A. Zahir, Muhammad Khalid, Farheen Nazli, and Muhammad Arshad. 2013. "Efficacy of Rhizobium and Pseudomonas Strains to Improve Physiology, Ionic Balance and Quality of Mung Bean under Salt-Affected Conditions on Farmer's Fields." Plant Physiology and Biochemistry. https://doi.org/10.1016/j.plaphy.2012.11.024.

Akbarimoghaddam, H., M. Galavi, A. Ghanbari, N. Panjehkeh et al. 2011. "Salinity Effects on Seed Germination and Seedling Growth of Bread Wheat Cultivars." Trakia Journal of Sciences 9 (1): 43–50.

Akhtar, Mohd Sayeed. 2019. Salt Stress, Microbes, and Plant Interactions: Mechanisms and Molecular Approaches. Singapore: Springer Nature.

Aktar, MdWasim, Dwaipayan Sengupta, and Ashim Chowdhury. 2009. "Impact of Pesticides Use in Agriculture: Their Benefits and Hazards." Interdisciplinary Toxicology 2 (1): 1–12.

Arkhipova, T. N., E. Prinsen, S. U. Veselov, E. V. Martinenko, A. I. Melentiev, and G. R. Kudoyarova. 2007. "Cytokinin Producing Bacteria Enhance Plant Growth in Drying Soil." Plant and Soil. https://doi.org/10.1007/s11104-007-9233-5.

Armengaud, Patrick, Rainer Breitling, and Anna Amtmann. 2010. "Coronatine-Insensitive 1 (COI1) Mediates Transcriptional Responses of Arabidopsis Thaliana to External Potassium Supply." Molecular Plant 3 (2): 390–405.

Arora, N. K., S. C. Kang, and D. K. Maheshwari. 2001. "Isolation of Siderophore-Producing Strains of Rhizobium Meliloti and Their Biocontrol Potential against Macrophomina Phaseolina that Causes Charcoal Rot of Groundnut." Current Science 81 (6): 673–77.

Barea, J. M., E. Navarro, and E. Montoya. 1976. "Production of Plant Growth Regulators by Rhizosphere Phosphate-solubilizing Bacteria." Journal of Applied Bacteriology 40 (2): 129–34. https://doi.org/10.1111/j.1365-2672.1976.tb04161.x.

Barka, Essaid Ait, Jerzy Nowak, and Christophe Clément. 2006. "Enhancement of Chilling Resistance of Inoculated Grapevine Plantlets with a Plant Growth-Promoting Rhizobacterium, Burkholderia Phytofirmans Strain PsJN." Applied and Environmental Microbiology. https://doi.org/10.1128/aem.01047-06.

Basu, Datta. 2000. "Indole Acetic Acid Production by a Rhizobium Species from Root Nodules of a Leguminous Shrub, Cajanus Cajan." Microbiol Res. 2000 Jul;155(2):123-7. https://doi.org/10.1016/S0944-5013(00)80047-6.

Bensalim, Salah, Jerzy Nowak, and Samuel K. Asiedu. 1998. "A Plant Growth Promoting Rhizobacterium and Temperature Effects on Performance of 18 Clones of Potato." American Journal of Potato Research. https://doi.org/10.1007/bf02895849.

Bhattacharyya, P. N., and D. K. Jha. 2012. "Plant Growth-Promoting Rhizobacteria (PGPR): Emergence in Agriculture." World Journal of Microbiology and Biotechnology. https://doi.org/10.1007/s11274-011-0979-9.

Brilli, Federico, Francesco Loreto, and Ivan Baccelli. 2019. "Exploiting Plant Volatile Organic Compounds (VOCs) in Agriculture to Improve Sustainable Defense Strategies and Productivity of Crops." Frontiers in Plant Science 10 (March): 264.

Bürgmann, Helmut, Franco Widmer, William Von Sigler, and Josef Zeyer. 2004. "New Molecular Screening Tools for Analysis of Free-Living Diazotrophs in Soil." Applied and Environmental Microbiology. https://doi.org/10.1128/aem.70.1.240-247.2004.

Cacciari, I., Lippi, D., Pietrosanti, T. and Pietrosanti, W., 1989. Phytohormone-like substances produced by single and mixed diazotrophic cultures of Azospirillum and Arthrobacter. Plant and soil, 115(1), pp.151–153.

Chandra, Lakshman. 2017. "Breeding of Medicinal and Aromatic Plants-an Overview." International Journal of Botany and Research 7 (2): 25–34.

Chang, Pearl, Karen E. Gerhardt, Xiao-Dong Huang, Xiao-Ming Yu, Bernard R. Glick, Perry D. Gerwing, and Bruce M. Greenberg. 2014. "Plant Growth-Promoting Bacteria Facilitate the Growth of Barley and Oats in Salt-Impacted Soil: Implications for Phytoremediation of Saline Soils." International Journal of Phytoremediation. https://doi.org/10.1080/15226514.2013.821447.

Charlton, Anne. 2004. "Medicinal Uses of Tobacco in History." Journal of the Royal Society of Medicine 97 (6): 292–96.

Cheikh, N., and R. J. Jones. 1994. "Disruption of Maize Kernel Growth and Development by Heat Stress (Role of Cytokinin/Abscisic Acid Balance)." Plant Physiology. https://doi.org/10.1104/pp.106.1.45.

Chen, Shi-Lin, Hua Yu, Hong-Mei Luo, Qiong Wu, Chun-Fang Li, and André Steinmetz. 2016. "Conservation and Sustainable Use of Medicinal Plants: Problems, Progress, and Prospects." Chinese Medicine. https://doi.org/10.1186/s13020-016-0108-7.

Chen, Fang, Yuxin Gao, Xiaoyi Chen, Zhimin Yu, and Xianzhen Li. 2013. "Quorum Quenching Enzymes and Their Application in Degrading Signal Molecules to Block Quorum Sensing-Dependent Infection." International Journal of Molecular Sciences 14 (9): 17477–500. https://doi.org/10.3390/ijms140917477.

Chen, Y. P., P. D. Rekha, A. B. Arun, F. T. Shen, W-A Lai, and C. C. Young. 2006. "Phosphate Solubilizing Bacteria from Subtropical Soil and Their Tricalcium Phosphate Solubilizing Abilities." Applied Soil Ecology. https://doi.org/10.1016/j.apsoil.2005.12.002.

Chung, Heekyung, Myoungsu Park, Munusamy Madhaiyan, Sundaram Seshadri, Jaekyeong Song, Hyunsuk Cho, and Tongmin Sa. 2005. "Isolation and Characterization of Phosphate Solubilizing Bacteria from the Rhizosphere of Crop Plants of Korea." Soil Biology and Biochemistry. https://doi.org/10.1016/j.soilbio.2005.02.025.

Costacurta A. and Vanderleyden J. 1995. "Synthesis of Phytohormones by Plant-Associated Bacteria." Crit Rev Microbiol. 1995; 21(1): 1–18. doi: 10.3109/10408419509113531. PMID: 7576148.

Croteau, Rodney, Raymond E. B. Ketchum, Robert M. Long, Rüdiger Kaspera, and Mark R. Wildung. 2006. "Taxol Biosynthesis and Molecular Genetics." Phytochemistry Reviews: Proceedings of the Phytochemical Society of Europe 5 (1): 75–97.

Dell'Amico, Elena, Lucia Cavalca, and Vincenza Andreoni. 2008. "Improvement of Brassica Napus Growth under Cadmium Stress by Cadmium-Resistant Rhizobacteria." Soil Biology and Biochemistry. https://doi.org/10.1016/j.soilbio.2007.06.024.

Demange, P., S. Wendenbaum, A. Bateman, A. Dell, J. M. Meyer, and M. A. Abdallah. 1986. "Bacterial Siderophores: Structure of Pyoverdins and Related Compounds." Iron, Siderophores, and Plant Diseases. https://doi.org/10.1007/978-1-4615-9480-2_15.

Dojima, Tomoko, and Lyle E. Craker. 2016. "Potential Benefits of Soil Microorganisms on Medicinal and Aromatic Plants." ACS Symposium Series. https://doi.org/10.1021/bk-2016-1218.ch006.

Dong, Yi Hu, Lian Hui Wang, Jin Ling Xu, Hai Bao Zhang, Xi Fen Zhang, and Lian Hui Zhang. 2001. "Quenching Quorum-Sensing-Dependent Bacterial Infection by an N-Acyl Homoserine Lactonase." Nature 411 (6839): 813–817. https://doi.org/10.1038/35081101.

Dong, Yi Hu, Jin Ling Xu, Xian Zhen Li, and Lian Hui Zhang. 2000. "AiiA, an Enzyme that Inactivates the Acylhomoserine Lactone Quorum-Sensing Signal and Attenuates the Virulence of Erwinia Carotovora." Proceedings of the National Academy of Sciences of the United States of America 97 (7): 3526–31. https://doi.org/10.1073/pnas.97.7.3526.

Draelos, Zoe Diana. 2016. "A Pilot Study Investigating the Efficacy of Botanical Anti-Inflammatory Agents in an OTC Eczema Therapy." Journal of Cosmetic Dermatology. https://doi.org/10.1111/jocd.12199.

Egamberdieva, Dilfuza, Dilfuza Jabborova, and Nilufar Mamadalieva. 2013. "Salt Tolerant Pseudomonas Extremorientalis Able to Stimulate Growth of SilybumMarianum under Salt Stress." Medicinal and Aromatic Plant Science and Biotechnology 7: 7–10.

Egamberdieva, Dilfuza, and Jaime A. Teixeira da Silva. 2015. "Medicinal Plants and PGPR: A New Frontier for Phytochemicals." In Plant-Growth-Promoting Rhizobacteria (PGPR) and Medicinal Plants (Vol. 42, pp. 287–303). Springer International Publishing. https://doi.org/10.1007/978-3-319-13401-7_14.

Egamberdieva, Dilfuza, Smriti Shrivastava, and Ajit Varma. 2015. Plant-Growth-Promoting Rhizobacteria (PGPR) and Medicinal Plants. Switzerland: Springer.

El-barbary, T. A. A., and M. A. El-Badry. 2018. "Solubilization of Potassium from Gluconite by Microorganisms." Journal of Basic and Environmental Sciences 5: 240–44.

Fincheira, Paola, and Andrés Quiroz. 2018. "Microbial Volatiles as Plant Growth Inducers." Microbiological Research 208 (March): 63–75.

Fitter, Alastair H., and Robert K. M. Hay. 2012. Environmental Physiology of Plants. San Diego, CA: Academic Press.

Frederick Abeles, Page Morgan, and Mikal Saltveit. 1993. "Ethylene in Plant Biology." Cell 72 (1): 11–2. https://doi.org/10.1016/0092-8674(93)90043-p.

Ghorbanpour, Mansour, Javad Hadian, Shahab Nikabadi, and Ajit Varma. 2017. "Importance of Medicinal and Aromatic Plants in Human Life." Medicinal Plants and Environmental Challenges. https://doi.org/10.1007/978-3-319-68717-9_1.

Glaser, Dee Anna. 2004. "Anti-Aging Products and Cosmeceuticals." Facial Plastic Surgery Clinics of North America. https://doi.org/10.1016/j.fsc.2004.03.004.

Glick, Bernard R. 2012. "Plant Growth-Promoting Bacteria: Mechanisms and Applications." Scientifica, 2012, 1–15. https://doi.org/10.6064/2012/963401.

Glick, Bernard R., Donna M. Penrose, and Jiping Li. 1998. "A Model for the Lowering of Plant Ethylene Concentrations by Plant Growth-Promoting Bacteria." Journal of Theoretical Biology 190 (1): 63–68. https://doi.org/10.1006/jtbi.1997.0532.

Global Market for Botanical and Plant-Derived Drugs Worth $32.9 Billion in 2013. n.d. https://www.bccresearch.com/pressroom/bio/global-market-botanical-plant-derived-drugs-worth-$32.9-billion-2013.

Gouda, Sushanto, Rout George Kerry, Gitishree Das, Spiros Paramithiotis, Han-Seung Shin, and Jayanta Kumar Patra. 2018. "Revitalization of Plant Growth Promoting Rhizobacteria for Sustainable Development in Agriculture." Microbiological Research 206 (January): 131–40.

Gupta, Akanksha, Prem Pratap Singh, Pardeep Singh, Kalpna Singh, Anand Vikram Singh, Sandeep Kumar Singh, and Ajay Kumar. 2019. "Medicinal Plants Under Climate Change: Impacts on Pharmaceutical Properties of Plants." Climate Change and Agricultural Ecosystems. https://doi.org/10.1016/b978-0-12-816483-9.00008-6.

Gupta, Rahul, Kumar Gaurav Bajpai, Samta Johri, and A. M. Saxena. 2008. "An Overview of Indian Novel Traditional Medicinal Plants with Anti-Diabetic Potentials." African Journal of Traditional, Complementary and Alternative Medicines 5 (1): 1–17.

Gurib-Fakim, Ameenah. 2006. "Medicinal Plants: Traditions of Yesterday and Drugs of Tomorrow." Molecular Aspects of Medicine. https://doi.org/10.1016/j.mam.2005. 07.008.

Hamdia, M. Abd El-Samad, M. A. K. Shaddad, and M. M. Doaa. 2004. "Mechanisms of Salt Tolerance and Interactive Effects of Azospirillum Brasilense Inoculation on Maize Cultivars Grown under Salt Stress Conditions." Plant Growth Regulation. https://doi. org/10.1023/b:grow.0000049414.03099.9b.

Hamilton, Alan C. 2004. "Medicinal Plants, Conservation and Livelihoods." Biodiversity and Conservation. https://doi.org/10.1023/b:bioc.0000021333.23413.42.

Heidari, Mostafa, and Amir Golpayegani. 2012. "Effects of Water Stress and Inoculation with Plant Growth Promoting Rhizobacteria (PGPR) on Antioxidant Status and Photosynthetic Pigments in Basil (Ocimum Basilicum L.)." Journal of the Saudi Society of Agricultural Sciences. https://doi.org/10.1016/j.jssas.2011.09.001.

Igbedioh, S. O. 1991. "Effects of Agricultural Pesticides on Humans, Animals, and Higher Plants in Developing Countries." Archives of Environmental Health 46 (4): 218–24.

Iqbal, Muhammad, Yamin Bibi, Naveed Iqbal Raja, Muhammad Ejaz, Mubashir Hussain, Farhat Yasmeen, Hafiza Saira, and MuhammadImran. 2017. "Review on Therapeutic and Pharmaceutically Important Medicinal Plant Asparagus Officinalis L." Journal of Plant Biochemistry & Physiology. https://doi.org/10.4172/2329-9029.1000180.

Ivanova. 2000. "Facultative and Obligate Aerobic Methylobacteria Synthesize Cytokinins". Microbiology., 69(6), 646–651. https://doi.org/10.1023/A:1026693805653.

Ivanova, E. G., N. V. Doronina, and Yu A. Trotsenko. 2001. "Aerobic Methylobacteria Are Capable of Synthesizing Auxins." Microbiology 70 (4): 392–97. https://doi.org/10.1023/ A:1010469708107.

Jamil, A., S. Riaz, M. Ashraf, and M. R. Foolad. 2011. "Gene Expression Profiling of Plants under Salt Stress." Critical Reviews in Plant Sciences. https://doi.org/10.1080/0735268 9.2011.605739.

Jeelani, Syed Mudassir, Gulzar A. Rather, Arti Sharma, and Surrinder K. Lattoo. 2018. "In Perspective: Potential Medicinal Plant Resources of Kashmir Himalayas, Their Domestication and Cultivation for Commercial Exploitation." Journal of Applied Research on Medicinal and Aromatic Plants 8 (March): 10–25. https://doi.org/10.1016/ j.jarmap.2017.11.001.

Joo, Gil Jae, Young Mog Kim, In Jung Lee, Kyung Sik Song, and In Koo Rhee. 2004. "Growth Promotion of Red Pepper Plug Seedlings and the Production of Gibberellins by Bacillus Cereus, Bacillus Macroides and Bacillus Pumilus." Biotechnology Letters 26 (6): 487–91. https://doi.org/10.1023/B:BILE.0000019555.87121.34.

Karthikeyan, B., C. Abdul Jaleel, G. M. A. Lakshmanan, and M. Deiveekasundaram. 2008. "Studies on Rhizosphere Microbial Diversity of Some Commercially Important Medicinal Plants." Colloids and Surfaces B: Biointerfaces. https://doi.org/10.1016/ j.colsurfb.2007.09.004.

Khan, Salim, Fahad Al-Qurainy, and Mohammad Nadeem. 2012. "Biotechnological Approaches for Conservation and Improvement of Rare and Endangered Plants of Saudi Arabia." Saudi Journal of Biological Sciences 19 (1): 1–11. https://doi.org/10.1016/j.sjbs. 2011.11.001.

Kloepper, J. W., M. S. Reddy, R. Rodríguez-Kabana, D. S. Kenney, N. Kokalis-Burelle, and N. Martinez-Ochoa. 2004. "Application for Rhizobacteria in Transplant Production and Yield Enhancement." Acta Horticulturae. https://doi.org/10.17660/actahortic.2004. 631.28.

Koenig, Robbin L., Roy O. Morris, and Joe C. Polacco. 2002. "TRNA Is the Source of Low-Level Trans-Zeatin Production in Methylobacterium Spp." *Journal of Bacteriology* 184 (7): 1832–42. https://doi.org/10.1128/JB.184.7.1832-1842.2002.

Kour, Divjot, Kusam Lata Rana, Tanvir Kaur, Neelam Yadav, Suman Kumar Halder, Ajar Nath Yadav, Shashwati Ghosh Sachan, and Anil Kumar Saxena. 2020. "Potassium Solubilizing and Mobilizing Microbes: Biodiversity, Mechanisms of Solubilization, and Biotechnological Implication for Alleviations of Abiotic Stress." New and Future Developments in Microbial Biotechnology and Bioengineering. https://doi.org/10.1016/b978-0-12-820526-6.00012-9.

Kumar, Pankaj, R. C. Dubey, and D. K. Maheshwari. 2012. "Bacillus Strains Isolated from Rhizosphere Showed Plant Growth Promoting and Antagonistic Activity against Phytopathogens." Microbiological Research 167 (8): 493–99.

Kumar, Vinod, Shourabh Joshi, Naveen C. Pant, Punesh Sangwan, Ajar Nath Yadav, Abhishake Saxena, and Dharmendra Singh. 2019. "Molecular Approaches for Combating Multiple Abiotic Stresses in Crops of Arid and Semi-Arid Region." Energy, Environment, and Sustainability. https://doi.org/10.1007/978-981-15-0690-1_8.

Laloi, Christophe, Klaus Apel, and Antoine Danon. 2004. "Reactive Oxygen Signalling: The Latest News." Current Opinion in Plant Biology 7 (3): 323–28.

Lin, Yi Han, Jin Ling Xu, Jiangyong Hu, Lian Hui Wang, Say Leong Ong, Jared Renton Leadbetter, and Lian Hui Zhang. 2003. "Acyl-Homoserine Lactone Acylase from Ralstonia Strain XJ12B Represents a Novel and Potent Class of Quorum-Quenching Enzymes." Molecular Microbiology 47 (3): 849–60. https://doi.org/10.1046/j.1365-2958.2003.03351.x.

Liu, Dianfeng, Bin Lian, and Hailiang Dong. 2012. "Isolation of Paenibacillussp. and Assessment of Its Potential for Enhancing Mineral Weathering." Geomicrobiology Journal. https://doi.org/10.1080/01490451.2011.576602.

Liu, Xiao-Min, and Huiming Zhang. 2015. "The Effects of Bacterial Volatile Emissions on Plant Abiotic Stress Tolerance." Frontiers in Plant Science 6 (September): 774.

Martínez-Viveros, O., M. A. Jorquera, D. E. Crowley, G. Gajardo, and M. L. Mora. 2010. "Mechanisms and Practical Considerations Involved In Plant Growth Promotion by Rhizobacteria." Journal of Soil Science and Plant Nutrition. https://doi.org/10.4067/s0718-95162010000100006.

Máthé, Ákos. 2015. Medicinal and Aromatic Plants of the World: Scientific, Production, Commercial and Utilization Aspects. Switzerland, Springer.

Michael Davidson, P., John N. Sofos, and Branen A. Larry. 2005. Nitrite. Antimicrobials in Food, 3rd edition. Boca Raton, FL: CRC Press. https://doi.org/10.1201/9780429058196-7.

Mohite. 2013. "Isolation and Characterization of Indole Aceticacid (IAA) Producing Bacteria" from *Rhizosphericsoil and Its Effect on Plant Growth* 15 (3): 2012–14.

Mukesh, Meena, Prashant Swapnil, Kumari Divyanshu, Sunil Kumar, Harish, Yashoda Nandan Tripathi, Andleeb Zehra, Avinash Marwal, and Ram Sanmukh Upadhyay. 2020. "PGPR-Mediated Induction of Systemic Resistance and Physiochemical Alterations in Plants against the Pathogens: Current Perspectives." Journal of Basic Microbiology 60 (10): 828–61.

Morohoshi, Tomohiro, Masashi Kato, Katsumasa Fukamachi, Norihiro Kato, and Tsukasa Ikeda. 2008. "N-Acylhomoserine Lactone Regulates Violacein Production in Chromobacterium Violaceum Type Strain ATCC 12472." FEMS microbiology letters, 279(1), pp. 124-130.

Müller, Henry, Christian Westendorf, Erich Leitner, Leonid Chernin, Kathrin Riedel, Silvia Schmidt, Leo Eberl, and Gabriele Berg. 2009. "Quorum-Sensing Effects in the Antagonistic Rhizosphere Bacterium Serratia Plymuthica HRO-C48." FEMS Microbiology Ecology 67 (3): 468–78. https://doi.org/10.1111/j.1574-6941.2008.00635.x.

Nadeem, Sajid Mahmood, Zahir Ahmad Zahir, Muhammad Naveed, and Muhammad Arshad. 2007. "Preliminary Investigations on Inducing Salt Tolerance in Maize through Inoculation with Rhizobacteria Containing ACC Deaminase Activity." Canadian Journal of Microbiology. https://doi.org/10.1139/w07-081.

Nautiyal, Jaya, Mark Christian, and Malcolm G. Parker. 2013. "Distinct Functions for RIP140 in Development, Inflammation, and Metabolism." Trends in Endocrinology and Metabolism: TEM 24 (9): 451–59.

Oerther, Sarah E. 2011. "Plant Poisonings: Common Plants That Contain Cardiac Glycosides." Journal of Emergency Nursing. https://doi.org/10.1016/j.jen.2010.09.008.

Orr, Caroline H., Angela James, Carlo Leifert, Julia M. Cooper, and Stephen P. Cummings. 2011. "Diversity and Activity of Free-Living Nitrogen-Fixing Bacteria and Total Bacteria in Organic and Conventionally Managed Soils." Applied and Environmental Microbiology. https://doi.org/10.1128/aem.01250-10.

Osakabe, Yuriko, Keishi Osakabe, Kazuo Shinozaki, and Lam-Son P. Tran. 2014. "Response of Plants to Water Stress." Frontiers in Plant Science. https://doi.org/10.3389/fpls.2014.00086.

Park, Sun Yang, Sang Jun Lee, Tae Kwang Oh, Jong Won Oh, Bon Tag Koo, Do Young Yum, and Jung Kee Lee. 2003. "AhlD, an N-Acylhomoserine Lactonase in Arthrobacter Sp., and Predicted Homologues in Other Bacteria." Microbiology 149 (6): 1541–50. https://doi.org/10.1099/mic.0.26269-0.

Park, Sun-yang, Hye-ok Kang, Hak-sun Jang, Jung-kee Lee, Bon-tag Koo, and Do-young Yum. 2005. "Identification of Extracellular N-Acylhomoserine Lactone Acylase from a Streptomyces Sp. and Its Application to Quorum Quenching." Applied and Environmental Microbiology., 71(5), 2632–2641. https://doi.org/10.1128/AEM.71.5.2632-2641.2005.

Patel, D. K. 2016. "Medicinal and Aromatic Plants: Role in Human Society." Medicinal & Aromatic Plants. https://doi.org/10.4172/2167-0412.1000e175.

Pattanaik, Chiranjibi, C. Sudhakar Reddy, and M. S. R. Murthy. 2008. "An Ethnobotanical Survey of Medicinal Plants Used by the Didayi Tribe of Malkangiri District of Orissa, India." Fitoterapia 79 (1): 67–71. https://doi.org/10.1016/j.fitote.2007.07.015.

Patten, Glick. 1996. "Bacterial Biosynthesis of Indole-3-Acetic Acid." Canadian Journal of Microbiology., 42(3), 207–220. https://doi.org/10.1139/m96-032.

Pollier, Jacob, Tessa Moses, and Alain Goossens. 2011. "Combinatorial Biosynthesis in Plants: A (p)review on Its Potential and Future Exploitation." Natural Product Reports 28 (12): 1897–1916.

Pramanik, P., A. J. Goswami, S. Ghosh, and C. Kalita. 2019. "An Indigenous Strain of Potassium-Solubilizing Bacteria Bacillus Pseudomycoides Enhanced Potassium Uptake in Tea Plants by Increasing Potassium Availability in the Mica Waste-Treated Soil of North-East India." Journal of Applied Microbiology 126 (1): 215–22.

Prasad, Ram, Manoj Kumar, and Ajit Varma. 2015. "Role of PGPR in Soil Fertility and Plant Health." Soil Biology. https://doi.org/10.1007/978-3-319-13401-7_12.

Prisa Domenico. 2020. "Plant Growth Promoting Rhizobacteria: Increase of Vegetative and Roots Biomass in Portulacaria Afra." GSC Advanced Research and Reviews 2 (2): 001–007. https://doi.org/10.30574/gscarr.2020.2.2.0005.

Rao, M. R., M. C. Palada, and B. N. Becker. 2004. "Medicinal and Aromatic Plants in Agroforestry Systems." Agroforestry Systems. https://doi.org/10.1023/b:agfo.0000028993.83007.4b.

Rhijn, P. van, and J. Vanderleyden. 1995. "The Rhizobium-Plant Symbiosis." Microbiological Reviews. https://doi.org/10.1128/mmbr.59.1.124-142.1995.

Roriz, Mariana, Susana M. P. Carvalho, Paula M. L. Castro, and Marta W. Vasconcelos. 2020. "Legume Biofortification and the Role of Plant Growth-Promoting Bacteria in a Sustainable Agricultural Era." Agronomy 10 (3). https://doi.org/10.3390/agronomy10030435.

Saha, Madhumonti, Bihari Ram Maurya, Vijay Singh Meena, Indra Bahadur, and Ashok
    Kumar. 2016. "Identification and Characterization of Potassium Solubilizing
    Bacteria (KSB) from Indo-Gangetic Plains of India." Biocatalysis and Agricultural
    Biotechnology. https://doi.org/10.1016/j.bcab.2016.06.007.
Salamone, Hynes, and Nelson L.M. 2001. "Cytokinin Production by Plant Growth Promoting
    Rhizobacteria and Selected Mutants." Canadian Journal of microbiology, 47(5),
    pp. 404–411. https://doi.org/10.1139/w01-029.
Saifuddin, Mohammed, A. B.M.S. Hossain, O. Normaiuza, A. Nasrulhaq Boyce, and K. M.
    Mofieruzzaman. 2009. "The Effects of Naphthaleneacetic Acid and Gibbmllic Acid in
    Prolonging Bract Longevity and Delaying Discoloration of Bougainvillea Spectabilis."
    Biotechnology 8 (3): 343–50. https://doi.org/10.3923/biotech.2009.343.350.
Saleem, Muhammad, Muhammad Arshad, Sarfraz Hussain, and Ahmad Saeed Bhatti. 2007.
    "Perspective of Plant Growth Promoting Rhizobacteria (PGPR) Containing ACC
    Deaminase in Stress Agriculture." Journal of Industrial Microbiology & Biotechnology
    34 (10): 635–48.
Sergeeva, Elena, Anton Liaimer, and Birgitta Bergman. 2002. "Evidence for Production of the
    Phytohormone Indole-3-Acetic Acid by Cyanobacteria." Planta, 215(2), pp. 229–238.
    https://doi.org/10.1007/s00425-002-0749-x.
Shaikh, Mosma Nadim, and Digambar Nabhu Mokat. 2018. "Role of Rhizosphere Fungi
    Associated with Commercially Explored Medicinal and Aromatic Plants: A Review."
    Current Agriculture Research Journal. https://doi.org/10.12944/carj.6.1.09.
Shameer, Syed, and T. N. V. K. V. Prasad. 2018. "Plant Growth Promoting Rhizobacteria
    for Sustainable Agricultural Practices with Special Reference to Biotic and Abiotic
    Stresses." Plant Growth Regulation 84 (3): 603–15. https://doi.org/10.1007/s10725-
    017-0365-1.
Sharma, Seema B., Riyaz Z. Sayyed, Mrugesh H. Trivedi, and Thivakaran A. Gobi. 2013.
    "Phosphate Solubilizing Microbes: Sustainable Approach for Managing Phosphorus
    Deficiency in Agricultural Soils." SpringerPlus 2 (October): 587.
Shrivastava, Manoj, P. C. Srivastava, and S. F. D'Souza. 2016. "KSM Soil Diversity and
    Mineral Solubilization, in Relation to Crop Production and Molecular Mechanism."
    Potassium Solubilizing Microorganisms for Sustainable Agriculture. https://doi.org/
    10.1007/978-81-322-2776-2_16.
Sio, Charles F., Linda G. Otten, Robbert H. Cool, Stephen P. Diggle, Peter G. Braun,
    Rein Bos, Mavis Daykin, Miguel Cámara, Paul Williams, and Wim J. Quax. 2006.
    "Quorum Quenching by an N-Acyl-Homoserine Lactone Acylase from Pseudomonas
    Aeruginosa PAO1." Infection and Immunity 74 (3): 1673–82. https://doi.org/10.1128/
    IAI.74.3.1673-1682.2006.
Sofowora, Abayomi, Eyitope Ogunbodede, and Adedeji Onayade. 2013. "The Role and
    Place of Medicinal Plants in the Strategies for Disease Prevention." African Journal of
    Traditional, Complementary, and Alternative Medicines: AJTCAM/African Networks
    on Ethnomedicines 10 (5): 210–29.
Somerville, C. 1995. "Direct Tests of the Role of Membrane Lipid Composition in Low-
    Temperature-Induced Photoinhibition and Chilling Sensitivity in Plants and
    Cyanobacteria." Proceedings of the National Academy of Sciences of the United States
    of America 92 (14): 6215–18.
Sumaira, Aslam, Naveed Iqbal Raja, Mubashir Hussain, Muhammad Iqbal, Muhammad
    Ejaz, Danish Ashfaq, Hira Fatima, Muhammad Ali Shah, Abd-Ur-Rehman, and Maria
    Ehsan. 2017. "Current Status of Withania Somnifera (L.) Dunal: An Endangered
    Medicinal Plant from Himalaya." American Journal of Plant Sciences. https://doi.org/
    10.4236/ajps.2017.85076.
Suzuki, Shino, Yuxi He, and Hiroshi Oyaizu. 2003. "Indole-3-Acetic Acid Production in
    Pseudomonas Fluorescens HP72 and Its Association with Suppression of Creeping

Bentgrass Brown Patch." Current Microbiology, 47(2), 138–143. https://doi.org/10.1007/s00284-002-3968-2.

Sziderics, A. H., F. Rasche, F. Trognitz, A. Sessitsch, and E. Wilhelm. 2007. "Bacterial Endophytes Contribute to Abiotic Stress Adaptation in Pepper Plants (Capsicum annuumL.)." Canadian Journal of Microbiology. https://doi.org/10.1139/w07-082.

Tabuti, J. R. S., S. S. Dhillion, and K. A. Lye. 2003. "Traditional Medicine in Bulamogi County, Uganda: Its Practitioners, Users and Viability." Journal of Ethnopharmacology. https://doi.org/10.1016/s0378-8741(02)00378-1.

Tahir, Hafiz A. S., Qin Gu, Huijun Wu, Waseem Raza, Alwina Hanif, Liming Wu, Massawe V. Colman, and Xuewen Gao. 2017. "Plant Growth Promotion by Volatile Organic Compounds Produced by SYST2." Frontiers in Microbiology 8 (February): 171.

Tank, Neelam, and Meenu Saraf. 2003. "Phosphate Solubilization, Exopolysaccharide Production and Indole Acetic Acid Secretion by Rhizobacteria Isolated from TrigonellaFoenum-Graecum." Indian Journal of Microbiology 43 (1): 37–40.

Tian, Fang, Yanqin Ding, Hui Zhu, Liangtong Yao, and Binghai Du. 2009. "Genetic Diversity of Siderophore-Producing Bacteria of Tobacco Rhizosphere." Brazilian Journal of Microbiology: [publication of the Brazilian Society for Microbiology] 40 (2): 276–84.

Tien, TM, MH Gaskins, and DH Hubbell. 1979. "Plant Growth Substances Produced by Azospirillum Brasilense and Their Effect on the Growth of Pearl Millet (Pennisetum americanum L.)" *Applied and Environmental Microbiology* 37 (5): 1016–24.

Tsavkelova, E. A., S. I. Klimova, T. A. Cherdyntseva, and A. I. Netrusov. 2006. "Microbial Producers of Plant Growth Stimulators and Their Practical Use: A Review." Prikladnaia Biokhimiia i Mikrobiologiia. 42 (2): 133–43.

Ullman, William J., David L. Kirchman, Susan A. Welch, and Philippe Vandevivere. 1996. "Laboratory Evidence for Microbially Mediated Silicate Mineral Dissolution in Nature." Chemical Geology. https://doi.org/10.1016/s0009-2541(96)00036-8.

Uroz, Stéphane, Phil Oger, Siri Ram Chhabra, Miguel Cámara, Paul Williams, and Yves Dessaux. 2007. "N-Acyl Homoserine Lactones Are Degraded via an Amidolytic Activity in Comamonas Sp. Strain D1." 187 (3): 17136382. https://doi.org/10.1007/s00203-006-0186-5.

Vazquez, P., G. Holguin, M. E. Puente, A. Lopez-Cortes, and Y. Bashan. 2000. "Phosphate-Solubilizing Microorganisms Associated with the Rhizosphere of Mangroves in a Semiarid Coastal Lagoon." Biology and Fertility of Soils. https://doi.org/10.1007/s003740050024.

Vejan, Pravin, Rosazlin Abdullah, Tumirah Khadiran, Salmah Ismail, and AmruNasrulhaq Boyce. 2016. "Role of Plant Growth Promoting Rhizobacteria in Agricultural Sustainability-A Review." Molecules 21 (5). https://doi.org/10.3390/molecules21050573.

Verma, Priyanka, Ajar Nath Yadav, Kazy Sufia Khannam, Sanjay Kumar, Anil Kumar Saxena, and Archna Suman. 2016. "Molecular Diversity and Multifarious Plant Growth Promoting Attributes of Bacilli Associated with Wheat (Triticum aestivumL.) Rhizosphere from Six Diverse Agro-Ecological Zones of India." Journal of Basic Microbiology. https://doi.org/10.1002/jobm.201500459.

Verma, Priyanka, Ajar Nath Yadav, Kazy Sufia Khannam, Neha Panjiar, Sanjay Kumar, Anil Kumar Saxena, and Archna Suman. 2015. "Assessment of Genetic Diversity and Plant Growth Promoting Attributes of Psychrotolerant Bacteria Allied with Wheat (Triticum Aestivum) from the Northern Hills Zone of India." Annals of Microbiology. https://doi.org/10.1007/s13213-014-1027-4.

Xiao, Yujie, Xunjue Wang, Wenli Chen, and Qiaoyun Huang. 2017. "Isolation and Identification of Three Potassium-Solubilizing Bacteria from Rape Rhizospheric Soil and Their Effects on Ryegrass." Geomicrobiology Journal. https://doi.org/10.1080/01490451.2017.1286416.

Xue, Xiaopeng, Li Zhang, Yunxiang Peng, Ping Li, and Jianguo Yu. 2019. "Effects of Mineral Structure and Microenvironment on K Release from Potassium Aluminosilicate Minerals by Cenococcum Geophilum Fr." Geomicrobiology Journal. https://doi.org/10.1080/01490451.2018.1485064.

Xu, Zhenzhu, Hideyuki Shimizu, Shoko Ito, Yasumi Yagasaki, Chunjing Zou, Guangsheng Zhou, and Yuanrun Zheng. 2014. "Effects of Elevated CO2, Warming and Precipitation Change on Plant Growth, Photosynthesis and Peroxidation in Dominant Species from North China Grassland." Planta. https://doi.org/10.1007/s00425-013-1987-9.

Yadav, Ajar Nath. 2020. Microbiomes of Extreme Environments: Biodiversity and Biotechnological Applications. Boca Raton, USA: CRC Press, Taylor and Francis Group.

Yao, Lixia, Zhansheng Wu, Yuanyuan Zheng, Imdad Kaleem, and Chun Li. 2010. "Growth Promotion and Protection against Salt Stress by Pseudomonas Putida Rs-198 on Cotton." European Journal of Soil Biology. https://doi.org/10.1016/j.ejsobi.2009.11.002.

Yuan, Yingdan, Xinggang Tang, Zhaohui Jia, Chong Li, Jieyi Ma, and Jinchi Zhang. 2020. "The Effects of Ecological Factors on the Main Medicinal Components of Dendrobium Officinale under Different Cultivation Modes." Forests. https://doi.org/10.3390/f11010094.

Zhang, Hai Bao, Lian Hui Wang, and Lian Hui Zhang. 2002. "Genetic Control of Quorum-Sensing Signal Turnover in Agrobacterium Tumefaciens." Proceedings of the National Academy of Sciences of the United States of America 99 (7): 4638–43. https://doi.org/10.1073/pnas.022056699.

16 The Importance of Microbes in Organic Matter Composting

Zimin Wei, Junqiu Wu, Xiaomeng Chen,
Haishi Qi, Mingzi Shi, Yufeng Chen,
Yue Zhao, Xu Zhang, and Xinyu Xie
Northeast Agricultural University, Harbin, China

CONTENTS

DOI: 10.1201/9781003110477-16

16.1　INTRODUCTION

Composting is one of the most important ways to decrease organic waste pollution (Wu et al., 2020a). During composting, organic matter will be transformed into mature products which can fix heavy metal, adsorb organic contaminant, improve soil fertility and so on (Chen et al., 2019a; Cui et al., 2019; Zhu et al., 2020). Meanwhile, compared with landfilling and burning for managing organic waste, composting has the advantages of high conversion efficiency and low risk of secondary pollution. However, it is the microorganisms to drive the composting process, any microenvironment unsuitable for microbial metabolism will lead to composting failure. For example, the intermediate products derived from incomplete transformation of organic matter are not conducive to seed germination (Cui et al., 2017). Meanwhile, soil organic matter will be further destroyed by the failure composting products in the process of soil application. Therefore, the important role of microorganism in composting process should not be ignored.

Microorganisms perform two contrary roles in transforming organic carbon (C) during composting: promoting release of CO_2 into the atmosphere by catabolism or fixing C into compounds that with high humification degree and not easy to be degraded by metabolism (Liang et al., 2017). During the process of organic matter transformation, both microbial activities and microorganisms' cells play an important role in the fixation of organic C. This has stimulated research to consider the direct incorporation of microbial residues (cellular components from both living and senesced biomass) into the stable organic C pool (Benner, 2011; Miltner et al., 2012; Ludwig et al., 2015; Schaeffer et al., 2015). In addition, in the process of fixing organic C, microorganisms can also promote the transformation of other plant nutrients, for the nitrogen or phosphorus. During composting, NH_4^+-N will be transformed into NO_2^--N by ammonia-oxidizing bacteria (AOB), which might be conducive in reducing NH_3 emission, decreasing organic nitrogen loss (Wang et al., 2018a; Zeng et al., 2018; Chen et al., 2019a; An et al., 2020). Previous study of Zhang et al. (2016a) had confirmed that NO_2^--N concentration was increased with decreased NH_3 emission and NH_4^+-N concentration by inoculating enriched AOB. In addition, NH_3 recycling, biochar addition can also alter bacteria diversity to promote nitrogen transformation (Wang et al., 2020). Therefore, bacteria community plays an important role in nitrogen transformation. Similarly, during composting, microbe drive the hydrolysis of complex organic components produces a large number of small molecular organic acids which transform insoluble inorganic phosphorus (IP) into soluble ones (Wei et al., 2018). The soluble phosphorus contributes to the absorption of phosphorus by plants. Meanwhile, microorganisms can convert IP into organic phosphorus (OP) and store it in the cell through metabolism (Zeng et al., 2019). These conversion ways not only increase the available phosphorus concentration, but also decrease the loss of phosphorus nutrients. On the other hand, microbial activities can also promote the degradation of antibiotics and the passivation of heavy metals during composting (Chen et al., 2020; Zhu et al., 2019). At present, compost has been used as a medium to treat heavy metals, antibiotics, micro-plastics and other pollutants (Zhu et al., 2019; Chen et al., 2020; Shi et al., 2020). Accordingly, the function of microorganism in composting process is diversified. Thus, the specific goals can be achieved by regulating microbial activity such as C sequestration, pollution reduction and so on. Microorganisms are the main drivers of organic matter transformation in composting and play critical roles in the C balance through the decomposition and anabolism of organic waste of different origin. Here, microbial transformation of C, nitrogen, phosphorus and pollutants and its influencing factors have been discussed, aiming to understand the regulation of composting microenvironment on microorganisms. It is significant for achieving specific goals of composting, for C fixation, nitrogen conservation, phosphorus increases and pollution removal.

16.2 DRIVEN EFFECTS OF MICROBE IN CARBON CONVERSION DURING COMPOSTING

16.2.1 RELATIONSHIPS BETWEEN MICROBIAL ACTIVITIES AND C METABOLISM

Generally speaking, various organic solid wastes, including domestic waste, straw, weeds, litter, fruit and vegetable waste can be used as raw materials into the composting system. The common characteristic of these materials is that they are rich in organic C. Organic C metabolism is a critical process in composting. Organic C metabolism in composting is not only related to humus formation but also provides energy for microbial growth (Wu et al., 2017). During composting, the flow of organic C can be divided into two pathways, that is, "degradation and synthesis pathway" (Tan, 2014). The "degradation pathway" is complete decompose of organic C into CO_2, waters and minerals, which not only causes the C loss but also enhances greenhouse gas emission (Lu et al., 2018; Wang et al., 2019; Zhang et al., 2019). The "synthesis pathway" is the re-bonding of C metabolic intermediates to form humus (Wu et al., 2017; Yu et al., 2019; Zhao et al., 2019). As final compost by-products, humus, is widely imposed to repair environmental pollution and improve soil fertility, because its strong adsorption ability and high C content (Hao et al., 2019; Yang et al., 2019. Therefore, improving the "synthesis pathway" and restraining the "degradation pathway" is meaningful for the sequestration of C and development of the green ecological environment. The "degradation pathway" and "synthesis pathway" are not proceeding spontaneously rather than are closely related to the microbial activity (Wang et al., 2018; Zhao et al., 2018). Besides, some researchers reported that microbial community succession and changes in abundance were key to regulate the composting process (Wu et al., 2020b; Zhu et al., 2020). Wu et al. (2020b) reported that inoculation of bacterial compound accelerated the degradation coarse fiber content by 8.82%, which were due to the fact that key enzymes and core microbes were stimulated during rice straw composting. Liu et al. (2020)

illustrated that inoculated microbial communities had a significant impact on both bacterial and fungal community structure. In addition, microbial inoculation could accelerate the composting process by stimulating key resident microbes in the initial stage (Liu et al., 2020). Depending on the temperature change, the composting can be generally divided into four phases: heating phase, thermophilic phase, cooling phase and maturity phase (Wu et al., 2017). The microbial activity is the main cause of composting temperature changes (Wu et al., 2017). Based on this, the microorganisms have different activities at varied composting periods, which may result in the different C metabolism and diverse precursors formation rules during composting. Meanwhile, the category and quantity of organic C have been shown to affect the microbial community composition and functionality. Qi et al. (2020) found that low-molecular-weight organic C was easily consumed by microbes eventually producing CO_2, while highly polymerized organic C molecules were resistant to microbial degradation. Therefore, the sources of organic C in different raw materials are the determinant factors of microbial community structure, richness and function during composting.

16.2.2 EFFECTS OF MICROBE IN CARBON DEGRADATION

Cellulosic-like materials as one of the main sources of composting raw materials. Cellulosic-like materials with high C content, mainly consisting of a three-dimensional network of cellulose (40–50%), hemicellulose (25–30%) and lignin (15–20%). The highly ordered microfiber structure formed by the orderly interlaced arrangement of cellulose and hemicellulose is the main component of plant cell wall. The outer layer of the microstructure is covered by highly complex aromatic polymer lignin, which makes the degradation of lignocellulose have strong resistance (Alessi et al., 2018; Bredon et al., 2018). Lignin, cellulose and hemicellulose are important sources for the formation of humus C skeleton (Tan, 2014). Lignin, hemicellulose and cellulose can be collectively referred to as lignocellulose. The biopolymers of importance in humus synthesis is called precursors of humus, like lignocellulose. In addition, lignocellulose degradation products are also important precursors in the synthesis of humic acid, such as polysaccharides, quinones, polyphenols and so on (Wu et al., 2017). However, due to the complexity of lignocellulose structure, it is difficult to be degraded by microorganisms in nature (McCann & Carpita, 2015; Larisa et al., 2017). Because of its inherent complexity and heterogeneity, a large number of organisms (producing diverse enzymes) is currently needed to efficiently degrade lignocellulose into its monomer compounds.

Cellulase is a mixed enzyme composed of a variety of glycoside hydrolases, mainly including endoglucanases (EG), cellobiohydrolases (CBH) and β-glucosaccharase (BG) (Payne et al., 2015). Among them, EG acts on the amorphous region inside the cellulose polysaccharide chain to produce oligosaccharides with different lengths and the end of the new chain. CBH acts on the end of the reducing and non-reducing cellulose polysaccharide chain to release glucose or cellobiose. BG is used to hydrolyze cellobiose to produce glucose (Beckham et al., 2014; Harris et al., 2014). The degraders are thus supposed to show dynamic responses to the lignocellulose, reaching higher efficiency when acting together than working alone (Larisa et al., 2017).

This process is known as synergistic effects. In addition, polysaccharide monooxygenase (LPMO) is a kind of nonhydrolase that destroys the β-1,4-glycosidic bond of cellulose polysaccharide chain by catalytic oxidation. LPMO can preferentially act on the crystalline region of cellulose and accelerate the whole process of cellulose degradation (Agger et al., 2014; Müller et al., 2014; Johansen, 2016). Carbohydrate binding module is a kind of non-catalytic polysaccharide hydrolysis module that assists in the hydrolysis of cellulose. Carbohydrate binding module is generally connected with the catalytic structure of cellulase. Carbohydrate binding module is mainly used to anchor on the hydrolyzed substrate to strengthen the binding rate of cellulose hydrolase and substrate, thereby enhancing the hydrolysis efficiency (Várnai et al., 2014).

For lignin, white-rot, brown-rot and soft-rot fungi have widely been reported as the major organisms responsible for the partial decomposition of lignin (Leonowicz et al., 1999; Sugimoto et al., 2017; Pertile et al., 2018). The white-rot fungi are reported to decompose lignin into CO_2 and H_2O. Patches of a white substance are often formed in the residue, hence the name white-rot. These white patches have been identified as pure forms of cellulose. After inoculation with white rot fungi, the activities of lignin peroxidase, manganese peroxidase and laccase increased. (Yang et al., 2020). The brown-rot fungi are useful for the removal of the methoxyl, $-OCH_3$, group from lignin, leaving the hydroxy phenols behind, which upon oxidation in the air produce brown colors. The soft-rot fungi are most active in wet environments and are specifically adapted to decomposing lignin. Because the fact is that composting is a complex system, a wide variety of microorganisms and their secretion of enzymes often work together. During composting, the degradation of lignocellulose could be limited by imbalances in moisture content, C/N ratio and ventilation rate (Rashed et al., 2010). This is due to the inhibition of microbial activity, thereby affecting the lignocellulose hydrolysis enzyme secretion, reducing the lignocellulose hydrolysis efficiency. Therefore, adjusting appropriate moisture content, C/N ratio and ventilation rate can provide an appropriate environment for lignocellulose degradation during composting. Meanwhile, increasing the activity of functional microorganisms by inoculation is also a very effective method.

16.2.3 Effects of Microbes in Carbon Fixation

During composting, partly organic macromolecules (e.g., lignocellulose, protein) are first degraded to humus precursors (e.g., polysaccharide, amino acid). These precursors will form humus by abiotic and biotic ways. The process in which precursors form humus is called humification. The formation of precursors in compost involves a variety of hypotheses. One group of theories is based on the direct conversion of biopolymers into humus due to the depolymerization of biopolymers. The other group assumes that during the formation of humus, small molecules liberated by the complete decomposition of biopolymers polymerization (Qi et al., 2020). The lignin theory is considered the example of the biopolymer degradation theory. In this pathway, phenolic compounds are obtained from lignin hydrolysis. Then, phenolic hydroxyl will be oxidized to quinone under the action of polyphenol oxidase, forming semiquinone free radicals, which helps in condensation with simple protein molecules, peptides and free amino acids to form the core carbon skeleton of humus. There is a significant negative correlation between the concentration of phenolic compounds and humus production during composting, that is, the composting humus production increased with the concentration of phenolic compounds decreased (Wu et al., 2017). The phenolic compound could be produced by two ways: the biodegradation of lignocellulose and the synthesis with microorganisms. Microbial secondary metabolism of non-aromatic substances produces a lot of phenols and hydroxyl aromatic substances (e.g., shikimic acid pathway). The second of theories suggests that monophenols, quinones, monomeric lignin, amino acids and even NH_3 are the basic units of humus formation, not lignin. The reaction between the small molecules can take place by chemical condensation or polymerization (Jokic et al., 2004; Hardie et al., 2009; Wu et al., 2018). Maillard reaction is a typical reaction of low-weight substances polymerization. In this reaction, reducing sugars and amino acids obtained from hydrolysis of organic matter and microbial synthesis will be automatically dehydrated and condensed to form humus by chemical reactions. But it should be viewed that the phenols, quinones and amino acids were derived biologically. Not only can they be formed from the degradation of large macromolecular substance such as lignocellulose, but they can also be derived from secondary metabolites of microorganisms. For biotic pathway of humus formation, it emphasizes the important role of microbial assimilation and metabolism in the formation of humus. The cellular autolysis pathway is similar to the microbial synthesis pathway, and precursors for humus synthesis are derived from microbial synthesis. However, the main difference is that humus precursors in the microbial synthesis pathway are mainly microbial metabolites, but humus precursors in the cellular autolysis pathway include microbial cell residues and other structural compounds. Therefore, microorganisms can not only act as decomposers to break down organic macromolecules to provide substrates for humus formation, but can also act as "producers" to synthesize humus precursors, for example, polyphenols, carboxylic acids, amino acids and proteins during composting.

16.2.4 REGULATE THE RELATIONSHIP BETWEEN MICROBE AND CARBON METABOLISM

In recent years, a variety of methods have been used to regulate the production of humus during composting. Some studies showed that mineral materials (e.g., biochar, ilmenite, montmorillonite) could help microorganism to synthesize humus precursors (e.g., phenols, lipids, polysaccharides) in multiple environmental systems (Kallenbach et al., 2017; Yu et al., 2019). Montmorillonite is conducive to the production of polyphenolic aromatic compounds, and even lignin derivatives can be detected. The ilmenite is more likely to synthesize alkane compounds such as fatty acids and proteins. Thus, different types of minerals have great differences in the pathways of microbial synthesis and metabolism. The addition of minerals may not directly affect the metabolic activity of microorganisms. The presence of minerals changes the porosity, pH, conductivity and even the organic matter carried by the minerals on the microbial activity microenvironment, which indirectly affects the microbial activity. The higher the content of clay minerals, gibbsite and other minerals in the system, the greater the pH, the richer the microbial diversity, thus, the more conducive to the formation and retention of organic C in the system (Razanamalala et al., 2017). Metal oxide has gradually become a potential additive to promote the formation of humus in compost. The interaction between metal oxides and humus can also be regarded as the interaction between metal oxides and organic matter. There are two main types of interaction: one is a chemical catalyst to promote the condensation or polymerization of humic acid precursors to form humus; the other acts as a protective agent to form a protective film on the outer layer of the humic structure/organic matter, preventing them from being hydrolyzed by microorganisms (Nakayama et al., 2010; Johnson et al., 2015). In addition, researchers show that adding metal oxides significantly alters microbial population structure and abundance during composting, which is more conducive to the formation of humus. The addition of metal oxides can accelerate the process of abiotic pathway in humus formation during composting. Meanwhile, researchers have conducted numerous studies on improving compost reaction processes by regulating microhabitats in composting (Zhang et al., 2020). Microbial activity in compost is influenced by a variety of physical and chemical factors, such as temperature, moisture, pH and ventilation rate. Despite numerous studies on microhabitat regulation in composting, its nature can be classified into three categories; one is to explore the response relationship between compost humic acid components and functional microorganisms. Then determine the priority regulatory components closely related to the formation of humus and the functional microorganism involved in the synthesis of precursors. The second is the study of the metabolic mechanism of key functional microorganisms in the related pathways of humic acid precursors. Thereby, identification of regulatory nodes in different metabolic pathways with the functional microbial groups and key enzyme. The third is the study of the regulation between functional microorganisms of humus and environmental factors, which ultimately affect microorganisms by regulating environmental factors associated with them. Based on the above view, microorganisms play an irreplaceable role in the transformation of organic C during composting, which can degrade polymer organic C and synthesize humus. However, the mechanism of how microorganisms produce humus through their own metabolism still needs further study.

16.3 THE MICROBIAL NITROGEN METABOLISM OF COMPOSTING

Nitrogen is one of the basic ingredients in composting environment. The nitrogen cycle, a fundamental biochemical process, is a redox reaction primarily mediated by microbial communities with metabolic diversity during composting (Wang et al., 2013; Kuypers et al., 2018). In general, the major nitrogen cycle reactions include ammonification, nitrification, nitrogen fixation and denitrification. Ammonifying, nitrifying, and nitrogen-fixating and denitrifying microbes coordinately form the intact microbial networks and play the key role in nitrogen cycle (Wang et al., 2016).

16.3.1 Ammonifying Microbes

Ammoniation emerges as a degradation process of organic nitrogen, which typically occurs in heating phase and is accompanied by a large amount of NH_3 during composting (Jones & Kielland, 2012). Ammonification during composting is the initial step of nitrogen mineralization, which is mainly influenced by microbes (Jones & Kielland, 2012). Ammonifying microbes are one kind of microbial populations that participate in ammonification during composting (Jones & Kielland, 2012). The most important ammonifying microbes are bacterial species during composting, which can release large amount of basic metalloprotease (Bach & Munch, 2000). The main bacterial species that produce metalloprotease, including *Bacillus cereus*, *Pseudomonas fluorescens*, *Bacillus. mycoides* and *Flavobacterium*. In addition, the *Bacillus* spp. secretes peptidases to degrade polypeptides. In particular, the large amount of basic metalloprotease produced by *B. cereus* or *B. mycoides*, and basic serine proteases produced by *B. subtilis* make the most contributions to polypeptides degradation (Bach et al., 2000).

Composting environment is an ecosystem with complicated and unstable physicochemical conditions that deeply influences ammoniation accomplished by ammonifying bacteria (Bach et al., 2000; Watanabe & Hayano, 1994). It has been reported that ammonification was influenced by oxygen concentration, with ammonifying effects varying with different oxygen concentration at different depths. In addition, nitrogen contents can affect the interactions between ammonifying bacteria during composting (Alef & Kleiner, 1987; Cárdenasgonzález et al., 1999). The ammonifying bacterial number may decrease as the nitrogen contents reduce. Furthermore, ammonifying bacteria tend to utilize C nutrition rather than nitrogen nutrition in environments with high C/N ratio, then high C/N ratio inhibits ammonification (Alef & Kleiner, 1987; Cárdenasgonzález et al., 1999; Stark et al., 2008). Arginine can promote the ammonia accumulation, a lower C/N ratio caused by additional amino acids is benefits ammoniation. What's more, ammonifying bacteria prefer environments with basic or neutral conditions (Alef & Kleiner, 1987; Cárdenasgonzález et al., 1999; Stark et al., 2008).

The natural organic nitric source involved in ammonification mainly consists of proteins. Under various physical, chemical and biological effects, these proteins are first denatured and degraded into polypeptides with the help of microbial extracellular proteases (Jones et al., 2012). The polypeptides then break the peptide bonds and further degrade polypeptides in dipeptides and amino acids, which can be directly absorbed and assimilated as nitric source by microbes. After a series of intracellular biochemical processes, microbes will discharge extra nitrogen to the extracellular environment as ammonia (Cárdenasgonzález et al., 1999). At least 95% organic nitric sources are provided by compost, of which only 5% is available to microbes. Most natural dissolved organic nitric components with high molecular weight (HMW-DON) are stable and hardly degradable, while microbes can utilize only dissolved organic nitrogen with low molecular weight (LMW-DON) (Watanabe & Hayano, 1994; Cárdenasgonzález et al., 1999; Bach et al., 2000). Thus, extracellular transformation of HMW-DON to LMW-DON is probably the rate-limiting step of ammonification during composting. In general, ammonification is an enzymatic biochemical process during composting. Degradation of HMW-DON is mainly proceeded by microbial enzyme (Smith et al., 2017), among which proteases play a vital role in nitrogen utilization by microbes in soil. Proteases that participate in ammonification mainly consist of fungal cysteine proteases and aspartic proteases, as well as bacterial serine proteases and basic or neutral metalloprotease (Jones et al., 2012). Catalyzed by deaminase, some LMW-DON can be converted into gaseous ammonia (NH_3), resulting in nitrogen loss during composting. Once LMW-DONs like amino acids enter microbial cellular environment, their amino groups will be transferred into α-ketoglutarate under transamination, producing glutamic acids that further participate in other amino acids and protein synthesis (Bach et al., 2000: Jones et al., 2012). When cells lack carbohydrate, the other products, organic acids, can be synthesized with ketoglutarate into pyruvic acids during transamination. Pyruvic acids then enter in the cellular TCA cycles, providing energy to the cell. In essence, like all biological processes, ammonification is basically controlled by various functional genes, including extracellular ammonification-related genes and intracellular ones that encode deaminase. It is well understood that proteases play an important role in nitrogen ammonification during composting, and many protease-encoding genes have been found and characterized. In bacteria, these genes include *arp*, *npr* and *sub*, which individually encodes a basic metalloenzyme, a neutral metalloprotease or a serine peptidase. In fungus, these genes include a aspartic protease-conding gene *pepA* from *Aspergillus awamori*, a basic protease-encoding gene *alp* from *A.oryzae* and a metalloenzyme-coding gene *mep* from *A. fumigatus* (Bach et al., 2000; Jones et al., 2012; Smith et al., 2017).

16.3.2 AMMONIA-OXIDIZING MICROBES

Nitrification is a biological process that oxidates ammonia to nitrate under aerobic conditions, including autotrophic nitrification and heterotrophic nitrification. Microbes with nitrifying function include ammonia oxidizing bacteria (AOB) and ammonia oxidizing archaea (AOA) (Francis et al., 2005). AOB can be divided into *Nitrosomonas*, *Nitrosospira*, *Nitrosolobus*, *Nitrosovibrio* and *Nitrosococcus*, while AOA belongs to another evolutionary branch independent to AOB. Both AOA and AOB greatly influence the nitrification at different stages during composting (Zeng et al., 2011). Generally, AOA exists in all stages of composting, while AOB exists only in some stages. Diversity of AOA is consistent with potential nitrification rate at very significant level and reached the peak at high temperature stage. Diversity of AOB is consistent with potential nitrification rate at significant level in heating stage and maturing stage (Maeda et al., 2010). The main AOB in agricultural waste compost were *Nitrosospira* and *Nitrosomonas*, with *Nitrosomonas* being the dominant species at all composting stages. Temperature, nitrate content and ammonium content influence the AOB community at a very significant level. Many studies demonstrated that *Nitrosospira* and *Nitrosomonas* are the dominant AOB species. While the functions of AOB during composting are well studied and received, functions of AOA are still controversial (Maeda et al., 2010).

Classical autotrophic nitrification contains ammonia oxidation and nitrite oxidation. During autotrophic nitrification, ammonia (NH_4^+) is first oxidized to hydroxylamine (NH_2OH) by ammonia monooxygenase (Amo), followed with NH_2OH being oxidized to nitrite by hydroxylamine oxidase (HAOs) (Yamamoto et al., 2010; Xie et al., 2012). Amo is a membrane protein which consists of three subunits α, β and γ. Subunits α, β and γ are individually encoded by *amoA*, *amoB* and *amoC* gene. *AmoA* gene encodes the active site of Amo, and is well studied as the target gene for molecular ecology of AOA and AOB during autotrophic nitrification (Santoro et al., 2008; Tourna et al., 2008). However, the *amoA* diversity of AOA is usually higher than AOB during composting. Enriched AOA into the composting also accelerated composing rate and increased nitrogen content. On the contrary, there are studies demonstrating that *amoA* of AOA cannot be detected at all stages of composting. Considering that pH and ammonium content show great influence on AOA activity, differences of AOA community in various composting system may result from different biological and environmental factors. Heterotrophic nitrification has a wide ranges of function matrix, including inorganic nitrogen like ammonia in composts and organic nitrogen like amine or acidamide (Nicol et al., 2008; Santoro et al., 2008; Tourna et al., 2008). The two known enzymes potentially involved in microbial heterotrophic nitrification are Amo and hydroxylamine cytochrome reductase, which have been purified and characterized in *Paracoccus*, *Pseudomonas* and *Arthrobacter* strains. Notably, all three species are widely distributed in composts and play an important role during nitrification.

16.3.3 Nitrogen-Fixating Microbes

Nitrogen fixation is beneficial for reducing nitrogen loss and promoting composting. Nitrogen fixation is a part of nitrogen cycles that coverts atmospheric dinitrogen (N_2) to ammonia (NH_4^+) (Eady, 1996). Biological nitrogen fixation is accomplished by microbes that contain nitrogenase-related genes. Nitrogenases are widely distributed in both archaea and bacteria, indicating that nitrogen fixation is an important and ubiquitous biochemical process for microbes (Zehr et al., 2003). All nitrogenases, with high similarities in amino acid sequence, protein structure and biological function, consist of two component metalloproteins. They can be divided into three types including molybdenum iron (Mo-Fe) protein, iron (Fe-Fe) protein and vanadium-iron (V-Fe) protein (Vitousek, 1991; Zehr et al., 2003). Though nitrogen-fixing genes, or nitrogenase-encoding genes (*nifH*), are large in both number and distribution among microbes, the primary sequence of *nifH* genes as well as nitrogenases are high conserved, which facilitates cloning and analysis of *nifH* genes with polymerase chain reaction (PCR) techniques, especially for *nifH* genes that originates from uncultured microorganisms (Bothe et al., 2010). *NifH* is commonly used as a biomarker for the presence of microbes with nitrogen-fixating functions during composting. With an increasing number of studies on phylogenetic analysis of *nifH* genes appear, understanding of diversity and evolutionary relationship of *nifH* genes becomes progressive deepening (Robson & Postgate 1980). Comparison of *nifH* phylogenetic trees to *16S* RNA gene phylogenetic trees shows that these two genes, to some degree, are one-to-one correspondence, indicating a corresponding relationship between nitrogen-fixing genes and nitrogen-fixing microbes (Berman-Frank et al., 2003; Inomura et al., 2017). Early studies have shown that *nifH* genes can be divided into four clusters. Cluster I includes *nifH* from α-Proteobacteria, β-Proteobacteria and Cyanobacteria, and *vnfH* from γ-Proteobacteria. Cluster II includes *nifH* from methanogenic microbes and *anfH* from bacterias. Cluster III contains *nifH* from anaerobia microbes, such as Clostridium and sulfur-reducing bacterias of δ-Proteobacteria (MacKellar et al., 2016; Martinez-Perez et al., 2016). Cluster IV includes all kinds of *nifH* from archaea. In addtion, *anfH*, *vnfH* and *nifH* encode Fe-containing electron-transport proteins. Expect for some soil nitrogen-fixating bacteria like *Vinelandii* that contain all three types nitrogenases, most nitrogen-fixating microbes may only contain Mo-Fe nitrogenase (Burris & Roberts, 1993; Lechene et al., 2007; Brune, 2014). *Bradyrhizobium* spp. lives in specific root nodules, providing leguminous plants with ammonium nitrogen nutrients, However, it can be also detected in composting.

16.3.4 DENITRIFYING MICROBES

Furthermore, composting systems rely on heterotrophic denitrification for nitrate reduction. Denitrification leads to a great loss of nitrogen and serious environment pollution during composting. Denitrification is a four-step biochemical process: (1) NO_3^- is reduced to NO_2^-, which is catalyzed by nitrate reductase encoded by *napG* and *napA*; (2) NO_2^- is reduced to NO, which is catalyzed by nitrite reductase encoded by *nirS*; (3) NO is reduced to N_2O, which is catalyzed by nitric oxide reductase encoded by *nor*; and (4) N_2O is reduced to N_2, which is catalyzed by nitrous oxide reductase encoded by *nosZ* (Smith et al., 2007; Sánchez & Minamisawa, 2008). Among these key enzymes of denitrification, the *nosZ*-encoding nitrous oxide reductase is the most important one for its ability to reduce air pollution during denitrification. Denitrifying microbes may carry one or more types of denitrification functional enzymes. Complete denitrifying microbes can successively reduce nitrate to N_2 with transient accumulation of intermediates (nitrite, NO and N_2O), which have the ability to synthesize all fully functional enzymes. Relatively, denitrifying microbes lack one or more functional enzymes that are considered truncated denitrifying microbe (Moreno-Vivian et al., 1999). Truncated denitrifying microbes have been frequently observed in the composting, which are physiological and regulatory characteristics diverse. Therefore, partial denitrification may occur during composting containing truncated denitrifying microbes. However, significant accumulation of nitrite or N_2O may also be conducted by some complete denitrifying microbes during composting (Lycus et al., 2018). It may be due to different denitrification functional enzymes with different reducing abilities under specific conditions. Besides denitrification, the denitrifying microbes can participate in various metabolic pathways (Lycus et al., 2017). For instance, denitrifying bacteria *Azoarcus*, *Georgfuchaia*, *Rhodoferax* and *Sulfuritalea* play pivotal roles in degrading polycyclic aromatic hydrocarbons (PAHs). *Paracoccus* and *Pseudomonas*, the typical aerobic denitrifying microbes, can aerobically reduce nitrate, nitrite to nitrogen or N_2O. Recent studies show that some autotrophic microbes can perform denitrification with inorganic C sources (CO_2, CO_3^{2-}), such as *Thiobacillus*. Denitrifying polyphosphate bacteria is metabolically diverse, which perform nitrate reduction while accumulating phosphorus. In general, the real-time quantification of nitrite reductase functional genes *nirS* or *nirK* is the gold standard for the presence of denitrifying microorganisms (Kraft et al., 2014; Lycus et al., 2018).

Currently, most studies on denitrification functional genes are focused on composting. The abundance of denitrifying gene is higher in samples with higher nitrite concentration, while the abundance of denitrifying gene is lower in samples with lower nitrite concentration. Researchers also found the presence of denitrifying bacteria in high oxygen content environments. Most works today in understanding denitrification during composting are converged on isolation and quantification of denitrifying bacteria. Further studies on denitrification during composting include diversity and abundance of denitrifying bacteria; the interactions between environmental factors and denitrifying bacterial community; and relationship between denitrifying gene and inorganic nitrogen transformation in genetic level.

For identifying the factors that shape nitrogen-transforming networks, greater insight into the physiology of the involved microorganisms, as well as deeper understanding of their ecology and evolution are needed. In addition, the matrix's physical structure and characteristics should be studied and controlled to gain a deeper insight into the processes during composting. Undoubtedly, to understand the regular transformation of different forms of nitrogen and properly regulate nitrogen cycle is essential for effective control of nitrogen loss and quality improvement during composting.

16.4   RELATIONSHIPS BETWEEN MICROBE AND PHOSPHOROUS

16.4.1   PHOSPHORUS: A MAJOR NUTRIENT COMPOSITION IN ORGANIC SOLID WASTE COMPOSTING

Phosphorus (P) is one of the most important mineral nutrients for all the plant, which can enhance the resistance of plant to drought, cold and the change of pH. Generally, the materials for compost always depends on the area's environment: difference of the environment with different materials often have different environmental effects. Given that, the phosphorus composition from different compost with seven different sources were studied: pig manure (PM), chicken manure (CM), municipal solid waste (MSW), kitchen waste (KW), green waste (GW), straw waste (SW), fruits and vegetables waste (FVW) (Wei et al., 2015).

It could be seen that there were significant differences in total phosphorus (TP) concentration of composts from different sources, among which livestock manure (include PM and CM) contains the highest phosphorus concentration. It might be due to the excessive mineral phosphorus added in the feed during the livestock breeding process (Figure 16.1a). The trend of Olsen P (Olsen P is the most easily used form of phosphorus by plants, which was mobile and easily mineralizable, including water soluble P and $NaHCO_3$ soluble P) is similar to the distribution of TP (Figure 16.1c). It is worth noting that the concentration of Olsen P in GW also has a quantitative advantage over other groups. It might be due to the fact that GW was mostly composed of varies plants and the plants must contain a lot of phosphorus in different forms. The contribution rate of OP and IP concentration to TP concentration showed that there were three kinds of distribution of OP and IP (Figure 16.1b). The IP concentration

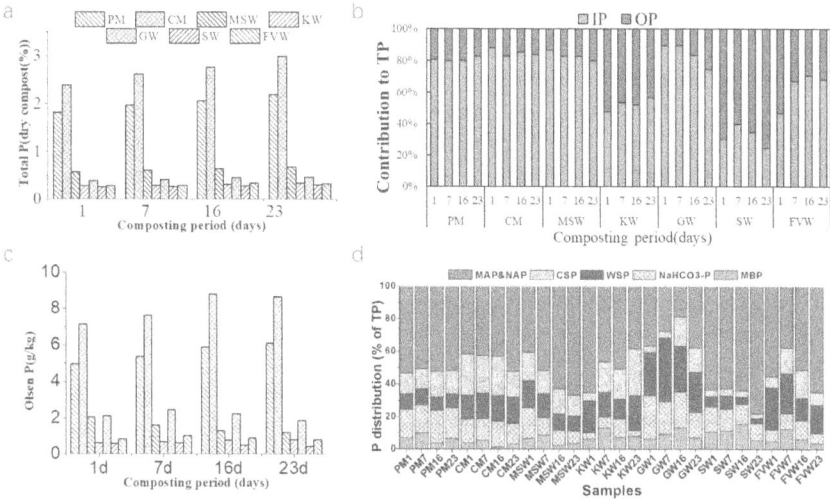

FIGURE 16.1 Different phosphorus composition of composts from seven different sources during composting: (a) total phosphorus content of composts from seven different sources at four different phases of composting; (b) the contribution to TP of IP and OP during composting derived from different materials; (c) changes in concentration of Olsen phosphorus (Olsen P) during composting; and (d) the contribution to TP of different P fractions during composting derived from different materials (Wei et al., 2015).

was absolutely dominant in TP concentration, which accounts for about 80% in PM, CM, MSW and GW; the concentration of OP and IP was roughly evenly distributed, which was more obvious in KW and FVW; the OP concentration was dominant in TP concentration, which accounted for more than 60% in SW.

Another phosphorus-grading method showed the contribution of different P fractions to TP during different materials composting (Figure 16.1d): the concentration of Olsen P only taken a little part of TP concentration, while the most absolutely dominant part in TP concentration was the concentration of moderately available phosphorus (MAP) and non-available phosphorus (NAP). As a result, there still had some stubborn problems that prevent people from successfully trying to make phosphate fertilizer by using the high phosphorus concentration of compost products since the MAP&NAP cannot be used directly by plants.

In view of the increasing global resource shortages, people are beginning to consider more methods for resource reuse. Then compost undoubtedly becomes the main source of resource reuse due to its environmental protection and high efficiency characteristics accompanied by various rich nutrients. However, combined with the discussion above on the phosphorus composition during different materials composting, it is not difficult to speculate that compost products can be used as organic fertilizers for agriculture, but it must not supplement the land with enough available phosphorus. Only a little part of IP such as Olsen P and citrate-soluble P (CSP) can provide phosphate fertilizer for plants, most of the phosphorus will be quickly fixed with various metal ions or leached into the water system. In other words, the phosphorus cannot be effectively reused inherent in compost products because of the high concentration of MAP&NAP in it. Therefore, some methods need to be provided to solve this "stubborn ingredient."

Organic P includes phytic acid, nucleic acid, phospholipid, phosphoprotein, glycolipid and many other phosphates. In addition to natural organic phosphates, most OP exist in the form of micro biomass P (MBP) which circulates between microorganisms and the environment as they grow and live, and acts as a backup source of most available phosphorus. Thanks to the high flexibility of MBP, people began to try to change the phosphorus fractions in the regional environment by means of microbial phosphorus metabolic transformation pathway.

16.4.2 MICROBIAL ACTION ON PHOSPHORUS FRACTION CONVERSION

16.4.2.1 Microbial Phosphorus Metabolism Pathway

Phosphorus is one of the indispensable elements in organisms, and all known organisms contain phosphorus (Sun et al., 2012). Phosphorus is an important component of biological genetic material such as nucleic acid, and it is also an important component of adenosine triphosphate (ATP), which is a key substance of biological energy metabolism. Phosphorus is also a component of many enzymes, biofilm phospholipids, nicotinamide adenine nucleotide (NAD), flavin adenine dinucleotide (FAD), phosphati-dylcholine, phosphatidylethano-lamine, phosphatidylglycerol, phosphati-dylserine and so on. These phosphorus fractions are involved in regulating acid-base balance in the organism, contributing to the metabolism of glucose, protein, fat and playing a very important role in various life activities.

When microorganisms die for various reasons, the phosphorus-containing compounds in the microbial will be released as litter and become free phosphorus components in the environment. Although the death of a single microorganism is insignificant, it is well known that microorganisms are a diverse and large family with extremely short life cycles compared to animals and plants. The continuous succession of microbial communities provides a part of stable phosphorus conversion pathways for the phosphorus pool in the environment: microorganisms absorb the available phosphorus to satisfy their own growth. Once death, they will release the compounds composed of phosphorus and return them to the environmental phosphorus pool. When the phosphorus metabolism pathways of microorganisms were studied in detail, it was discovered that some microbial communities with special functions could bring unexpected changes to the phosphorus pool in the environment.

16.4.2.2 Phosphate Solubilizing Bacteria and Its Phosphorus-Solubilizing Function

People discovered the relationship between phosphorus-solubilizing bacteria and phosphorus in the environment in the early 20th century. Gerretsen screened bacteria that can dissolve phosphate rock powder, which can help plants absorb insoluble phosphorus for normal growth (Gerretsen, 1948). Since then, more people have studied and reported on microorganisms with the ability to dissolve phosphorus.

Up to now, there are many kinds of microorganisms that can mineralize and decompose insoluble phosphate, and most of them are isolated from common soil samples or rhizosphere soil of plants. However, in general, the environmental adaptation range of the bacteria is relatively limited (Oliveira & Alves, 2009; Mander et al., 2012; Acevedo et al., 2014; Vassilev et al., 2014). According to the type of poorly soluble or hard-to-use phosphorus and the strength of phosphorus-solubilizing capacity, phosphorus-solubilizing microorganisms can be divided into two types: one are the IP solubilizing microorganisms that can secrete organic acids to decompose IP compounds into soluble IP and secretory them into the ex-environment; the other are the organophosphate-dissolving microorganisms that decompose phospholipids, nucleic acids and phytochemicals into soluble IP. Meanwhile, they can also be divided into phosphate-dissolving bacteria, phosphate-dissolving actinomycetes and phosphate-dissolving fungi based on the type of phosphate-dissolving microorganisms. In addition to the most common species of *Pseudomonas* and *Bacillus*, phosphate-solubilizing bacteria mainly include some bacteria in the genus of *Escherichia*, *Agrobacterium*, *Arthrobacter*, *Serratia*, *Proteus*, *Micrococcus*, *Rhodococcus*, *Chryseobacterium*, *Gordonia*, *Phyllobacterium*, *Xanthomonas*, *Vibrio proteolyticus*, *Klebsiella*, *Azotobacter* and many others. Phosphorus-dissolving fungi mainly include *Penicillium*, *Aspergillus*, *Rhizopus*, *Fusarium*, *Sclerotium* and so on. And phosphate-solubilizing actinomycetes are mainly *Streptomyces* (Molla & Chowdhury, 1984; Freitas et al., 1997). In soil research, the phosphate solubilization mechanism of phosphate-solubilizing microorganisms is generally considered to be the secretion of small molecular organic acids, proton exchange and complexation to dissolve and release poorly soluble phosphate. At the same time, the phosphatases and phytase of microorganisms play an important role in the release of soil OP.

The phosphate solubilization mechanism of IP phosphate-solubilizing microorganisms is very complicated with many mechanisms, which are also different for different strains. The following mechanisms are generally believed to exist. (1) Some microorganisms can secrete small molecular organic acids for acidolysis of insoluble phosphides. These organic acids can not only reduce the microenvironment pH but also combined with the cations such as iron, aluminum, calcium, magnesium and so on, thus releasing the phosphate bound to the cations. In this process, the type and quantity of the organic acids secreted by phosphorus-soluble microorganisms will also affect the phosphorus-soluble effect (Hoberg et al., 2005; Patel et al., 2008; Patel et al., 2010). (2) Some microorganisms can secrete protons (H^+). During the process of ingesting cations (such as NH_4^+), some microorganisms use the energy generated during the conversion of ATP to release H^+ to the outside of the cell membrane through the proton pump, which in turn causes the pH of the medium to drop to produce phosphorus solubilization. On the other hand, the assimilation of NH_4^+ is accompanied by the production of organic or inorganic acids, but some studies have shown that the main role of organic acids at this time is to provide protons (Whitelaw et al., 1999; Lin et al., 2006; Acevedo et al., 2014). (3) The CO_2 released by part of the microbial population through respiration can reduce soil pH within a certain range, which will lead to the dissolution of some phosphates. (4) Some microorganisms can decompose plant residues to produce organic compounds with aromatic polyphenol functional groups such as humic acid and fulvic acid which can complexing with metal cations such as calcium, iron and aluminum in the environment and release the phosphate bonded to the metal cations. At present, scholars at home and abroad generally believe that the process of microbial phosphorus dissolution is most likely to be related to the reduction of organic acid, pH and acidity.

16.4.3 EFFECT OF PHOSPHATE-SOLUBILIZING BACTERIA IN ORGANIC SOLID WASTE COMPOST

Phosphorus-solubilizing microorganisms are also widely present in solid waste compost. The microorganisms selected from the compost have been reported to be *Bacillus sp.*, *Pseudomonas sp.*, *Arthrobater sp.*, *Flarobacterium sp.*, *Alcaligenes sp.* and *Serratia sp.* Several yeasts and molds have also been reported as phosphorus-solubilizing microorganisms, which are similar to the species in the soil, but different microbial strains have different phosphate solubilization capabilities (Oliveira& Alves, 2009; Mander et al., 2012; Acevedo et al., 2014; Vassilev et al., 2014).

Chang & Yang (2009) screened three strains of phosphate-solubilizing bacteria, one strain of phosphate-solubilizing actinomycetes and one strain of phosphate-solubilizing fungi from different compost plants and biological fertilizers. These phosphate solubilizing strains not only have strong phosphate solubilizing ability (they can secrete amylase, carboxymethyl cellulase, chitinase, pectinase, protease, lipase and nitrogenase), but also have temperature tolerance. After inoculation of compost, the microorganisms can significantly accelerate compost maturity and increase the soluble phosphorus content of compost products number of phosphate-dissolving microorganisms (Chang & Yang, 2009). Mupambwa's research results showed that inoculation with two strains of phosphate-dissolving bacteria can promote the degradation of organic matter and the process of phosphorus mineralization in earthworm compost using cow dung–paper waste mixture as the main material (Mupambwa et al., 2016). Wei et al. (2008) showed that the inoculation of thermophilic IP-dissolving bacteria could significantly promote the conversion efficiency of the insoluble phosphate rock powder added in the compost, and at the same time increased the plant-available phosphorus content in the compost product. While the planting phosphorus-solubilizing bacteria at different stages of composting not only increased the diversity and abundance of phosphorus-solubilizing bacteria and increase the bacterial diversity during the compost, but also significantly increased the content of potentially available phosphorus (Wei et al., 2017).

The microbial changes in the compost have great volatility and unpredictability due to the different materials, composting environment and many other external factors. Microbes and the compost interact and achieve each other. Here, only one of the functional microorganisms related to phosphorus conversion is briefly described. The role of more microorganisms in the compost needs to be further explored.

16.5 MICROBIAL CONTRIBUTION TO POLLUTANT CONTROL DURING ORGANIC MATTERS COMPOSTING

16.5.1 MICROBIAL FUNCTIONAL ROLES UNDER HEAVY METALS STRESS

16.5.1.1 Differences of Microbial Community in Different Materials Composting

There are many kinds of organic solid wastes, including crop straw, livestock manures, domestic wastes, KW and so on. Among them, livestock manures are mainly rich in heavy metals, antibiotics and other pollutants. These pollutants will affect the composting process, especially microorganisms (Zhang et al., 2018; Zhao et al., 2019). Therefore, the response of microorganisms to stress is an important topic. It plays a decisive role in composting. It is worth noting that different dietary habits of livestock contribute to differences in the main components of their manures, leading to differences in the original microbial community in the manures. This results in different responses to environmental pressure. For example, in our previous study, CMs and bovine manures were employed as research objects (Chen et al., 2019b). The NMDS and PCoA analysis showed the bacterial community of CMs was completely separated from that of bovine manures, suggesting that the bacterial community structures of the two raw materials composting were different. The results of bacterial community β diversity showed that the bacterial diversity of CMs and bovine manures composting was significantly different ($P < 0.05$), and the bacterial diversity of CMs composting was higher than that of bovine manures composting. Besides, there were more specific operational taxonomic units in CMs composting than in bovine manures composting, especially 17 specific OTUs in CMs from breeding farm, indicating that the bacteria in CMs had more specific functions than bovine manures. Based on these, it can be concluded that the bacterial communities of CMs and bovine manures are significantly different during composting. In addition, the responses of microorganisms from different raw materials to the driving of basic nutrients (proteins and polysaccharides) in manures are different. Our previous study suggested that microbial degradation of nutrients is "specific" or "universal" (Chen et al., 2019c). "Specificity" means that proteins and polysaccharides can only be degraded and transformed by certain bacteria, while "universality" means that different kinds of bacteria can degrade and transform nutrients such as proteins and polysaccharides. Some specific bacteria can effectively transform polysaccharides during the thermophilic phase of CMs composting, while other related bacteria can transform proteins. However, during bovine manures composting, bacteria can degrade nutrients universally. They can transform both proteins and polysaccharides. The difference may be caused by animal feed. The feed compositions of chicken and bovine are obviously different. For example, bovine feed contains more proteins and crude fibers, and a large amount of rumen buffer is added to the feed to help digestion (Chibisa et al., 2016). However, the content of calcium and antibiotics in chicken feed is higher (Sapkota et al., 2007). The difference of nutrient substrates between CMs and bovine manures will lead to different responses of bacterial community to nutrient substrates, resulting in differences in microbial community composition and succession.

16.5.1.2   Effect of Microbial Community on Heavy Metals Passivation during Different Manures Composting

In our previous study, the results of structural equation models showed that the driving factors of reducing the bioavailability of heavy metals were different during CMs and bovine manures composting (Chen et al., 2019a). However, the specific ways to reduce the bioavailability of heavy metals in the same type of manure from farms and domestic sources were the same. During bovine manures composting, the main reason for reducing the bioavailability of heavy metals is the passivation of heavy metals caused by physicochemical properties (e.g. organic matters content, pH value, temperature, moisture content) (Yamada et al., 2008; Xu et al., 2017). However, during CMs composting, in addition to the physicochemical properties, the inactivation of bacteria is also an important factor. Therefore, the contribution of microbial community to heavy metals passivation is different during different animal manures composting. The difference in the ways of reducing the toxicity of heavy metals between CMs and bovine manures are mainly due to the bacterial community structure. Furthermore, variance partitioning analysis was used to reveal the difference of contribution of bacterial community and physicochemical properties to heavy metal content changes. The results showed that 44.1% of heavy metals content variables were caused by physicochemical properties during CMs composting, while 46.9% of heavy metals content variables were caused by physicochemical properties during bovine manures composting, which indicated that physiochemical properties play a dominant role in affecting heavy metals content. During CMs composting, bacterial community can also explain 29.0% of the heavy metals content variables, and physicochemical properties and bacterial community can explain 6.0% of the variables. However, the contribution of bacterial community to heavy metal variables was not significant ($P > 0.05$) during bovine manures composting. Therefore, the effect of microbial communities on heavy metals passivation is discrepant during different animal manures composting, which is due to the differences of microbial community structure from different raw materials. The difference of microbial community between CMs and bovine manures caused by nutrient difference will lead to different responses to environmental stress.

16.5.1.3   The Functional Role of Microorganisms

Resistance and tolerance are the most direct way for microorganisms to survive in heavy metal environment, but they are essentially different (Brauner et al., 2016). Resistance is mainly used to characterize the ability of microorganisms to resist heavy metals stress. The minimum inhibitory concentration (MIC) is often used to define the resistance of microorganisms. However, tolerance can be used to characterize the adaptability of microorganisms to heavy metal stress. The tolerance is usually defined by the length of delay period of microbial growth. Our previous study has also confirmed that microbial resistance and tolerance are two independent parameters, and the resistance and tolerance of microorganisms are universal under heavy metals stress during composting. Furthermore, the results of enzymes activity test showed that the catalase activity of microorganisms decreased gradually with the composting, however, the protease activity of microorganisms was gradually increased. This indicated that microorganisms in the later stage of composting would be more adaptable to the environment of heavy metals, and the microbial metabolic capacity would also become stronger. Therefore, the microorganisms that are sensitive to heavy metals exist in livestock manures, but with the extinction of sensitive strains during composting, the resistant microorganisms will become the dominant strains. In addition, the experimental results on the removal rate of heavy metals showed that microorganisms in composting could act on heavy metals under heavy metal stress; however, the effects of different composting materials and periods on ionic and organic heavy metals were different. Based on this, the roles of microorganisms can be defined as "sensitive," "resister" and "actor." The relationship among the sensitive, resister and actor microorganisms is often like the game of "scissors, stone and paper." The resister microorganisms can defeat the sensitive microorganisms, the actor microorganisms can control the resister microorganisms and the sensitive microorganisms will threaten the actor microorganisms. In theory, the sensitive microorganisms will die when the sensitive and resister microorganisms are exposed to heavy metals stress, while the resister microorganisms can survive due to some resistance mechanisms. However, both the resister and actor microorganisms can coexist under heavy metal stress, and the roles of actor microorganisms can promote their growth. In addition, when the sensitive and actor microorganisms are exposed to heavy metals stress, the growth and metabolism of sensitive microorganisms will be inhibited due to the toxicity of heavy metals at first. However, the actor microorganisms can survive and reduce the toxicity of heavy metals through retention, adsorption and transformation. The heavy metals pressure of the environment is reduced so that the sensitive microorganisms can continue to grow. In the actual environment, the three functional roles exist at the same time, so they continue to evolve and eventually form a relatively stable microbial community under the heavy metals pressure. This is the reason for the change results of bacterial community under heavy metal stress during composting.

16.5.2 EFFECTS OF MICROORGANISMS ON THE TOXICITY OF ANTIBIOTIC RESISTANCE GENES

16.5.2.1 Distribution Characteristics of Host Bacteria with Resistance Genes

The long-term abuse of antibiotics in livestock breeding leads to gene mutation, and large amount of antibiotic resistance genes produced by intestinal microorganisms of livestock under the action of antibiotics selection pressure (Yazdankhah et al., 2014). The resistance bacteria carrying antibiotic resistance genes are discharged out of the body with manures, and the spread of antibiotic resistance genes in the environment is an important factor which leads to the antibiotic resistance pollution (Ma et al., 2017). Many studies have shown that animal manures are an important repository of antibiotic resistance genes (Qian et al., 2017). Fortunately, composting can significantly reduce the abundance of resistant bacteria in livestock manures. On the one hand, the high temperature of composting at 50–65°C can kill pathogens, including some antibiotic-resistant bacteria. On the other hand, high temperature of composting will lead to the decomposition of antibiotics, which will weaken the selection pressure of microorganisms and reduce the frequency of mutation, so as to reduce the production of resistant bacteria. Some studies have shown that different kinds of microorganisms have different ability to capture antibiotics resistance genes or to mutate their own resistance (Andersson & Hughes, 2014). Some bacteria carrying resistance genes usually have a significant positive correlation with antibiotic resistance genes abundance, which are called potential host bacteria of antibiotic resistance genes. The change of these potential host bacteria is an important factor, which leading to the change of resistance genes abundance during composting. As a powerful tool to study the interactions between microorganisms and antibiotic resistance genes in complex environment, the network analysis is widely used to mine potential host bacteria of resistance genes in environmental samples (Zhang et al., 2018). The network analysis found that the host bacteria with tetracycline resistance genes were the most during different materials composting. However, a few antibiotics resistance genes have high abundance, but no potential host bacteria are found during composting. This indicates that these antibiotic resistance genes may be more easily transmitted by horizontal transfer during composting, thus it is impossible to determine the fixed host bacteria.

16.5.2.2   Analysis of Potential Host Bacteria of Mobile Genetic Elements

Clinically, mobile genetic elements can gather various antibiotic resistance genes and spread to bacteria and environment, thus the integron gene is usually regarded as a pollutant from the DNA perspective (Gillings, 2018). The mobile genetic elements cannot transfer horizontally directly among microorganisms, but they can transfer among microorganisms (within or among species) with other genetic elements as carriers, such as plasmids, transposons and insertion sequences (Wang et al., 2015; Zhang et al., 2016b; Zhang et al., 2016c). Our previous study showed the removal rate of IntI1 and IntI2 was significantly higher during CMs composting than that of bovine manures composting (P < 0.05), which was due to the difference between food and intestinal bacteria of animals (Zhu et al., 2019). As intI1 and intI2 are common gene capture systems in bacterial genome (Gillings, 2014), the removal of intI1 and intI2 is closely related to the change of host bacteria during composting (Su et al., 2015). In general, the variation of potential host bacteria of mobile genetic elements can be divided into three categories during composting. The first type is eliminated during the thermophilic period of composting, which indicated that high temperature could kill or inhibit such host bacteria. Previous studies have also shown that high temperature composting helps to control antibiotic resistance genes and mobile genetic elements by eliminating their host bacteria (Liao et al., 2017). The second type of host bacteria begin to disappear during the cooling or maturity stage of composting. The third group of host bacteria cannot be eliminated by composting, which indicates that this kind of host bacteria has strong tolerance to high temperature, such as the phylum Proteobacteria. Many studies have shown that the Proteobacteria is the host of resistance genes. The Proteobacteria is a phylum mainly existing in the later stage of composting. The presence of these stubborn host bacteria leads to the incomplete removal of mobile genetic elements during composting. Therefore, the eradication of host bacteria is still an important way to control mobile genetic elements.

In our previous study, projection pursuit models were used to evaluate the horizontal transfer risk of antibiotic resistance genes during composting (Zhu et al., 2019). The results showed that, in general, composting process can reduce the risk of antibiotic resistance gene transfer. The transfer risk decreased more sharply during CMs composting; however, the risk change of bovine manures composting was relatively slow, which was related to the change trend of host bacteria.

REFERENCES

Acevedo, E., Galindo-Castañeda, T., Prada, F., Navia, M., Romero, H. M., 2014. Phosphate-solubilizing microorganisms associated with the rhizosphere of oil palm (Elaeis guineensis Jacq.) in Colombia. *Applied Soil Ecology* 80: 26–33.

Agger, J. W., Isaksen, T., Várnai, A., et al., 2014. Discovery of LPMO activity on hemicelluloses shows the importance of oxidative processes in plant cell wall degradation. *Proceedings of the National Academy of Sciences of the United States of America* 111: 6287–6292.

Alef, K., Kleiner, D., 1987. Applicability of arginine ammonification as indicator of microbial activity in different soils. *Biology and Fertility of Soils* 5(2): 148–151.

Alessi, A. M., Bird, S. M., Oates, N. C., et al., 2018. Defining functional diversity for lignocellulose degradation in a microbial community using multi-omics studies. *Biotechnology for Biofuels* 11(1): 166.

An, X., Cheng, Y., Miao, L., et al., 2020. Characterization and genome functional analysis of an efficient nitrile-degrading bacterium, Rhodococcus rhodochrous BX2, to lay the foundation for potential bioaugmentation for remediation of nitrile-contaminated environments. *Journal of Hazardous Materials* 389: 121906.

Andersson, D. I., Hughes, D., 2014. Microbiological effects of sublethal levels of antibiotics. *Nature Reviews Microbiology* 12: 465–478.

Bach, H. J., Munch, J. C., 2000. Identification of bacterial sources of soil peptidases. *Biology and Fertility of Soils* 31(3): 219–224.

Beckham, G. T., Ståhlberg, J., Knott, B. C., et al., 2014. Towards a molecular-level theory of carbohydrate processivity in glycoside hydrolases. *Current Opinion in Biotechnology* 27: 96–106.

Benner, R., 2011. Biosequestration of C by heterotrophic microorganisms. *Nature Reviews Microbiology* 9: 75–75.

Berman-Frank, I., Lundgren, P., Falkowski, P., 2003. Nitrogen fixation and photosynthetic oxygen evolution in cyanobacteria. *Research in Microbiology* 154(3): 157–164.

Bothe, H., Schmitz, O., Yates, M. G., et al., 2010. Nitrogen fixation and hydrogen metabolism in cyanobacteria. *Microbiology and Molecular Biology Reviews* 74: 529–551.

Brauner, A., Fridman, O., Gefen, O., et al., 2016. Distinguishing between resistance, tolerance and persistence to antibiotic treatment. *Nature Reviews Microbiology* 14: 320–330.

Bredon, M., Dittmer, J., Noël, C., et al., 2018. Lignocellulose degradation at the holobiont level: teamwork in a keystone soil invertebrate. *Microbiome* 6(1): 162.

Brune, A., 2014. Symbiotic digestion of lignocellulose in termite guts. *Nature Reviews Microbiology* 12: 168–180.

Burris, R. H., Roberts, G., 1993. Biological nitrogen fixation. *Annual Review of Nutrition* 13: 317–335.

Cárdenasgonzález, B., Ergas, S. J., Switzenbaum, M. S., 1999. Characterization of compost biofiltration media. *Journal of the Air & Waste Management Association* 49(7): 784.

Chang, C., Yang, S. J., 2009. Thermo-tolerant phosphate-solubilizing microbes for multi-functional biofertilizer preparation. *Bioresource Technology* 100(4): 1648–1658.

Chen, X. M., Liu, R., Hao, J. K., et al., 2019c. Protein and carbohydrate drive microbial responses in diverse ways during different animal manures composting. *Bioresource Technology* 271: 482–486.

Chen, X. M., Zhao, X. Y., Ge, J. P., et al., 2019b. Recognition of the neutral sugars conversion induced by bacterial community during lignocellulose wastes composting. *Bioresource Technology* 294: 122153.

Chen, X. M., Zhao, Y., Zeng, C. C., et al., 2019a. Assessment contributions of physicochemical properties and bacterial community to mitigate the bioavailability of heavy metals during composting based on structural equation models. *Bioresource Technology* 289: 121657.

Chen, X. M., Zhao, Y, Zhao, X. Y., et al., 2020. Selective pressures of heavy metals on microbial community determine microbial functional roles during composting: Sensitive, resistant and actor. *Journal of Hazardous Materials* 398: 122858.

Chibisa, G. E., Beauchemin, K. A., Penner, G. B., 2016. Relative contribution of ruminal buffering systems to pH regulation in feedlot cattle fed either low- or high-forage diets. *Animal* 10(7): 1164–1172.

Cui, H. Y., Zhang, S. B., Zhao, M. Y., Zhao, Y., Wei, Z. M., 2019. Parallel faction analysis combined with two-dimensional correlation spectroscopy reveal the characteristics of mercury-composting-derived dissolved organic matter interactions. *Journal of Hazardous Materials* 384: 121395.

Cui, H. Y., Zhao, Y., Chen, Y. N., et al., 2017. Assessment of phytotoxicity grade during composting based on EEM/PARAFAC combined with projection pursuit regression. *Journal of Hazardous Materials* 326: 10–17.

Eady, R. R., 1996. Structure-function relationships of alternative nitrogenases. *Chemical Reviews* 96: 3013–3030.

Francis, C. A., Roberts, K. J., Beman, J. M., et al., 2005. Ubiquity and diversity of ammonia-oxidizing archaea in water columns and sediments of the ocean. *National Academy of Sciences of the United States of America* 102 (41): 14683–14688.

Gerretsen, F. C. J. 1948. The influence of microorganisms on the phosphate intake by plant. *Plant and Soil* 1: 51–81.

Gillings, M. R., 2014. Integrons: Past, present, and future. *Microbiology & Molecular Biology Reviews* 78: 257–277.

Gillings, M. R., 2018. DNA as a pollutant: The clinical class 1 integron. *Current Pollution Reports* 4: 49–55.

Hao, J. K., Wei, Z. M., Wei, D., et al., 2019. Roles of adding biochar and montmorillonite alone on reducing the bioavailability of heavy metals during chicken manure composting. *Bioresoure Technology* 294: 122199.

Hardie, A. G., Dynes, J. J., Kozak, L., et al., 2009. The role of glucose in abiotic humification pathways as catalyzed by birnessite. *Journal of Molecular Catalysis A-Chemical* 308: 114–126.

Harris, P. V., Xu, F., Kreel, N. E., et al., 2014. New enzyme insights drive advances in commercial ethanol production. *Current Opinion in Chemical Biology* 19(1): 162–178.

Hoberg, E., Marschner, P., Lieberei, R., 2005. Organic acid exudation and pH changes by Gordonia sp and Pseudomonas fluorescens grown with P adsorbed to goethite. *Microbilogical Research* 160(2): 177–187.

Inomura, K., Bragg, J., Follows, M. J., 2017. A quantitative analysis of the direct and indirect costs of nitrogen fixation: A model based on Azotobacter vinelandii. *The ISME Journal* 11: 166–175.

Johnson, K., Purvis, G., Lopez-Capel, E., et al., 2015. Towards a mechanistic understanding of C stabilization in manganese oxides. *Nature Communications* 6: 7628.

Jokic, A., Wang, M. C., Liu, C., et al., 2004. Integration of the polyphenol and Maillard reactions into a unified abiotic pathway for humification in nature: the role of delta-MnO(2). *Organic Geochemistry* 35: 747–762.

Jones, D. L., Kielland, K., 2012. Amino acid, peptide and protein mineralization dynamics in a taiga forest soil. *Soil Biology & Biochemistry* 55(2): 60–69.

De Freitas, J. R. Banerjee, M. R. Germida, J. J. 1997. Phosphate-solubillizing rhizobactera enchance the growth and yield but not phosphorus uptake of canola (Brassica napus L.). *Biology and Fertility of Soils* 24: 358–364.

Freitas, J., Banerjee, M. R., Germida, J. J. 1997. Phosphate-solubilizing rhizobacteria enhance the growth and yield but not phosphorus uptake of canola (Brassica napus L.). *Biology and Fertility of Soils* 24(4), 358–364.

Johansen, K. S., 2016. Discovery and industrial applications of lytic polysaccharide monooxygenases. *Biochemical Society Transactions* 44(1): 143–149.

Kallenbach, C. M., Frey, S. D., Grandy, A. S., 2017. Direct evidence for microbial-derived soil organic matter formation and its ecophysiological controls. *Nature Communications* 7: 13630.

Kraft, B., Tegetmeyer, H. E., Sharma, R., et al., 2014. The environmental controls that govern the end product of bacterial nitrate respiration. *Science* 345: 676–679.

Kuypers, M. M. M., Marchant, H. K., Kartal, B., 2018. The microbial nitrogen-cycling network. *Nature Reviews Microbiology* 16(5): 263–276.

Larisa, C. T., Salles, J. F., Dirk, V. E. J., 2017. Bacterial synergism in lignocellulose biomass degradation–complementary roles of degraders as influenced by complexity of the C source. *Frontiers in Microbiology* 8: 1628.

Lechene, C. P., Luyten, Y., McMahon, G., et al., 2007. Quantitative imaging of nitrogen fixation by individual bacteria within animal cells. *Science* 317: 1563–1566.

Leonowicz, A., Matuszewska, A., Luterek, J., et al., 1999. Biodegradation of lignin by white rot fungi. *Fungal Genet Biology* 27(2–3): 175–185.

Liang, C., Schimel, J. P., Jastrow, J. D., 2017. The importance of anabolism in microbial control over soil C storage. *Nature Microbiology* 2(8): 1–6.

Liao, H. P., Lu, X. M., Rensing, C., et al., 2017. Hyperthermophilic composting accelerates the removal of antibiotic resistance genes and mobile genetic elements in sewage sludge. *Environmental Science & Technology* 52(1): 266–276.

Lin, T. F., Huang, H. I., Shen, F. T., Young, C. C., 2006. The protons of gluconic acid are the major factor responsible for the dissolution of tricalcium phosphate by Burkholderia cepacia CC-A174. *Bioresource Technology* 97(7): 957–960.

Liu, H. J., Huang, Y., Duan, W. D., et al., 2020. Microbial community composition turnover and function in the mesophilic phase predetermine chicken manure composting efficiency. *Bioresoure Technology* 313: 123658.

Lu, Q., Zhao, Y., Gao, X. T., et al., 2018. Effect of tricarboxylic acid cycle regulator on C retention and organic component transformation during food waste composting. *Bioresoure Technology* 256: 128–136.

Ludwig, M., Achtenhagen, J., Miltner, A., et al., 2015. Microbial contribution to SOM quantity and quality in density fractions of temperate arable soils. *Soil Biology and Biochemistry* 81: 311–322.

Lycus, P., Bøthun, K. L., Bergaust, L., et al., 2017. Phenotypic and genotypic richness of denitrifiers revealed by a novel isolation strategy. *The ISME Journal* 11(10): 2219.

Lycus, P., Soriano-Laguna, M. J., Kjos, M., et al., 2018. A bet-hedging strategy for denitrifying bacteria curtails their release of N_2O. *Proceedings of the National Academy of Sciences of the United States of America* 115(46): 11820–11825.

Ma, L., Li, A. D., Yin, X. L., et al., 2017. The Prevalence of integrons as the carrier of antibiotic resistance genes in natural and man-made environments. *Environmental Science & Technology* 51: 5721–5728.

MacKellar, D., Lieber, L., Norman, J. S., et al., 2016. *Streptomyces thermoautotrophicus* does not fix nitrogen. *Scientific Reports* 6: 20086.

Maeda, K., Toyoda, S., Shimojima, R., et al., 2010. Source of nitrous oxide emissions during the cow manure composting process as revealed by isotopomer analysis of and *amoA* abundance in betaproteobacterial ammonia-oxidizing bacteria. *Applied & Environmental Microbiology* 76(5): 1555–1562.

Mander, C., Wakelin, S., Young, S., Condron, L., O'Callaghan, M., 2012. Incidence and diversity of phosphate-solubilising bacteria are linked to phosphorus status in grassland soils. *Soil Biology and Biochemistry* 44(1): 93–101.

Martinez-Perez, C., Mohr, W., Löscher, C. R., et al., 2016. The small unicellular diazotrophic symbiont, UCYN-A, is a key player in the marine nitrogen cycle. *Nature Microbiology* 1: 16163.

Molla, M. A. Z., Chowdhury, A. A. 1984. Microbial mineralization of organic phosphate in soil. *Plant and Soil* 78: 393–399.

McCann, M. C., Carpita, N. C., 2015. Biomass recalcitrance: a multi-scale, multi-factor, and conversion-specific property. *Journal of Experimental Botany* 66(14): 4109–4118.

Miltner, A., Bombach, P., Schmidt-Brücken, B., Kästner, M., 2012. SOM genesis: microbial biomass as a significant source. *Biogeochemistry* 111: 41–55.

Moreno-Vivian, C., Cabello, P., Martinez-Luque, M., et al., 1999. Prokaryotic nitrate reduction: molecular properties and functional distinction among bacterial nitrate reductases. *Journal of Bacteriology* 181: 6573–6584.

Müller, G., Varnai, A., Johansen, K. S., et al., 2014. Harnessing the potential of LPMO-containing cellulase cocktails poses new demands on processing conditions. *Biotechnology for Biofuels* 8(1): 187.

Mupambwa, H. A., Ravindran, B., Mnkeni, P. N. S., 2016. Potential of effective microorganisms and Eisenia fetida in enhancing vermi-degradation and nutrient release of fly ash incorporated into cow dung–paper waste mixture. *Waste Management* 48: 165–173.

Nakayama, M., Shamoto, M., Kamimura, A., 2010. Surfactant-induced electrodeposition of layered manganese oxide with large interlayer space for catalytic oxidation of phenol. *Chemistry of Materials* 22(21): 5887–5894.

Nicol, G. W., Leininger, S., Schleper, C., et al., 2008. The influence of soil ph on the diversity, abundance and transcriptional activity of ammonia oxidizing archaea and bacteria. *Environmental Microbiology* 10(11): 2966.

Oliveira, C. A., Alves, J., 2009. Phosphate solubilizing microorganisms isolated from rhizosphere of maize cultivated in an oxisol of the Brazilian Cerrado Biome. *Soil Biology and Biochemistry* 41(9): 1782–1787.

Patel, D. K., Archana, G., Kumar, G. N., 2008. Variation in the nature of organic acid secretion and mineral phosphate solubilization by Citrobacter sp DHRSS in the presence of different sugars. *Current Microbiology* 56(2): 168–174.

Patel, K. J., Singh, A. K., Nareshkumar, G., Archana, G., 2010. Organic-acid-producing, phytate-mineralizing rhizobacteria and their effect on growth of pigeon pea (Cajanus cajan). *Applied Soil Ecology* 44(3): 252–261.

Payne, C. M., Knott, B. C., Mayes, H. B., et al., 2015. Fungal cellulases. *Chemical Reviews* 115(3): 130448.

Pertile, G., Panek, J., Oszust, K., et al., 2018. Intraspecific functional and genetic diversity of Petriella setifera. *PeerJ* 2018: 6.

Qi, H. S., Zhao, Y., Zhao, X. Y., et al., 2020. Effect of manganese dioxide on the formation of humin during different agricultural organic wastes compostable environments: It is meaningful C sequestration. *Bioresoure Technology* 299: 122596.

Qian, X., Gu, J., Sun, W., et al., 2017. Diversity, abundance, and persistence of antibiotic resistance genes in various types of animal manure following industrial composting. *Journal of Hazardous Materials* 344: 716–722.

Rashed, F. M., Saleh, W. D., Moselhy, M. A., 2010. Bioconversion of rice straw and certain agro-industrial wastes to amendments for organic farming systems: Composting, quality, stability and maturity indices. *Bioresource Technology* 101: 5952–5960.

Razanamalala, K., Razafimbelo, T., Maron, P. A., et al., 2017. Soil microbial diversity drives the priming effect along climate gradients: a case study in Madagascar. *The ISME Journal* 12(2): 451.

Robson, R. L., Postgate, J. R., 1980. Oxygen and hydrogen in biological nitrogen fixation. *Annual Review of Microbiology* 34(1): 183–207.

Sánchez, C., Minamisawa, K., 2018. Redundant roles of Bradyrhizobium oligotrophicum Cu-type (*nirK*) and cd 1-type (*nirS*) nitrite reductase genes under denitrifying conditions. *FEMS Microbiology Letters* (5): 5.

Santoro, A. E., Francis, C. A., Sieyes, N. R. D., et al., 2008. Shifts in the relative abundance of ammonia-oxidizing bacteria and archaea across physicochemical gradients in a subterranean estuary. *Environmental Microbiology* 10(4): 1068–1079.

Sapkota, A. R., Lefferts, L. Y., Mckenzie, S., et al., 2007. What do we feed to foodproduction animals? A review of animal feed ingredients and their potential impacts on human health. *Environmental Health Perspectives* 115(5): 663–670.

Schaeffer, A., Nannipieri, P., Kästner, M., Schmidt, B., Botterweck, J., 2015. From humus to soil organic matter–microbial contributions. In honour of Konrad Haider and James P. Martin for their outstanding research contribution to soil science. *Journal of Soils and Sediments* 15: 1865–1881.

Shi, J., Wu, D., Su, Y., Xie, B., 2020. (Nano)microplastics promote the propagation of antibiotic resistance genes in landfill leachate. *Environmental Science: Nano.* 10.1039/d0en00511h.

Smith, C. J., Mckew, B. A., Coggan, A., et al., 2017. Primers: Functional genes for nitrogen-cyclingmicrobes in oil reservoirs. In Terry J, McGenity, Kenneth N. Timmis, Balbina Nogales (Eds), Hydrocarbon and Lipid Microbiology Protocols.Springer, Berlin, Heidelberg, 207–241.

Smith, C. J., Nedwell, D. B., Dong, L. F., et al., 2007. Diversity and abundance of nitrate reductase genes (*narG* and *napA*), nitrite reductase genes (*nirS* and *nrfA*), and their transcripts in estuarine sediments. *Applied Microbiology and Biotechnology* 73: 3612–3622.

Stark, C. H., Condron, L. M., O'Callaghan, M., et al., 2008. Differences in soil enzyme activities, microbial community structure and short-term nitrogen mineralisation resulting from farm management history and organic matter amendments. *Soil Biology & Biochemistry* 40(6): 1352–1363.

Su, J. Q., Wei, B., Ou-Yang, W. Y., et al., 2015. Antibiotic resistome and its association with bacterial communities during sewage sludge composting. *Environmental Science & Technology* 49(12): 7356–7363.

Sugimoto, T., Hosoya, S., Yamamoto, K., et al., 2017. Effect of ozonation on composting Japanese cedar wood meal. *Holzforschung* 71(11): 913–918.

Sun, S., Gu, M., Cao, Y., et al., 2012. A constitutive expressed phosphate transporter, OsPht1;1, modulates phosphate uptake and translocation in phosphate-replete rice. *Plant Physiology* 159(4): 1571–1581.

Tan, K. H., 2014. *Humic Matter in Soil and the Environment: Principles and Controversies*, Second Edition. CRC Press, Boca Raton.

Tourna, M., Freitag, T. E., Nicol, G. W., et al., 2008. Growth, activity and temperature responses of ammonia-oxidizing archaea and bacteria in soil microcosms. *Environmental Microbiology* 10(5): 1357–1364.

Várnai, A., Mäkelä, M. R., et al., 2014. Chapter Four–Carbohydrate-Binding modules of fungal cellulases: Occurrence in nature, function, and relevance in industrial biomass conversion. *Advances in Applied Microbiology* 88: 103–165.

Várnai, A., Mkel, M. R., Djajadi, D. T., Rahikainen, J., Viikari, L., 2014. Carbohydrate-binding modules of fungal cellulases: Occurrence in nature, function, and relevance in industrial biomass conversion. *Advances in Applied Microbiology* 88, 103–165.

Vassilev, N., Mendes, G., Costa, M., Vassileva, M., 2014. Biotechnological tools for enhancing microbial solubilization of insoluble inorganic phosphates. *Geomicrobiology Journal* 31(9): 751–763.

Vitousek, P. M., Howarth, R. W., 1991. Nitrogen limitation on land and in the sea: How can it occur? *Biogeochemistry* 13: 87–115.

Wang, K., Wu, Y., Li, W., Wu, C., Chen, Z., 2018a. Insight into effects of mature compost recycling on N_2O emission and denitrification genes in sludge composting. *Bioresource Technology* 251: 320–326.

Wang, K., Mao, H. L., Wang, Z., et al., 2018. Succession of organics metabolic function of bacterial community in swine manure composting. *Journal of Hazardous Materials* 360: 471–480.

Wang, L. Q., Zhao, Y., Ge, J. P., et al., 2019. Effect of tricarboxylic acid cycle regulators on the formation of humic substance during composting: The performance in labile and refractory materials. *Bioresoure Technology* 292: 235–242.

Wang, Q., Qian, L., Mao, D., et al., 2015. The horizontal transfer of antibiotic resistance genes is enhanced by ionic liquid with different structure of varying alkyl chain length. *Frontiers in Microbiology* 6: 864.

Wang, R., Zhao, Y., Xie, Y., et al., 2020. Role of NH_3 recycling on nitrogen fractions during sludge composting. *Bioresource Technology* 295: 122175.

Wang, X. Q., Zhao, Y., Wang H, et al., 2016. Reducing nitrogen loss and phytotoxicity during beer vinasse composting with biochar addition. *Waste Management* 61: 150–156.

Wang, X., Selvam, A., Chan, M. T., et al., 2013. Nitrogen conservation and acidity control during food wastes composting through struvite formation. *Bioresource Technology* 147(8): 17–22.

Watanabe, K., Hayano, K., 1994. Source of soil protease based on the splitting sites of a polypeptide. *Soil Science and Plant Nutrition* 40(4): 697–701.

Wei, Y., Zhao, Y., Xi, B., et al., 2015. Changes in phosphorus fractions during organic wastes composting from different sources. *Bioresource Technology* 89: 349–56.

Wei, Z., Xi, B., Wang, S., et al., 2008. Phosphate transform of composting with pre-mixing insoluble phosphate using high temperature dissolved phosphorus microbes inoculation. *Environmental Science* 07: 2073–2076.

Wei, Y., Zhao, Y., Fan, Y., 2017. Impact of phosphate-solubilizing bacteria inoculation methods on phosphorus transformation and long-term utilization in composting. *Bioresource Technology* 241: 134–141.

Wei, Y., Zhao, Y., Shi, M., et al., 2018. Effect of organic acids production and bacterial community on the possible mechanism of phosphorus solubilization during composting with enriched phosphate-solubilizing bacteria inoculation. *Bioresource Technology* 247: 190–199.

Whitelaw, M. A., J., 1999. Growth promotion of plants inoculated with phosphate-solubilizing fungi. *Advances in Agronomy* 69: 99–151.

Wu, D., Wei, Z. M., Gao, X. Z., et al., 2020b. Reconstruction of core microbes based on producing lignocellulolytic enzymes causing by bacterial inoculation during rice straw composting. *Bioresoure Technology* 315: 123849.

Wu, J. Q., Zhao, Y., Wang, F., et al., 2020a. Identifying the action ways of function materials in catalyzing organic waste transformation into humus during chicken manure composting. *Bioresource Technology* 303: 122927.

Wu, J. Q., Zhao, Y., Qi, H. S., et al., 2017. Identifying the key factors that affect the formation of humic substance during different materials composting. *Bioresoure Technology* 244: 1193–1196.

Wu, J. Q., Qi, H. S., Huang, X. N., et al., 2018. How does manganese dioxide affect humus formation during bio-composting of chicken manure and corn straw?. *Bioresoure Technology* 269: 169–178.

Xie, K. Z., Jia X. S., Xu, P. Z., et al., 2012. Improved composting of poultry feces via supplementation with ammonia oxidizing archaea. *Bioresource Technology* 120(3): 70–77.

Xu, Q. X., Li, X. M., Ding, R. R., et al., 2017. Understanding and mitigating the toxicity of cadmium to the anaerobic fermentation of waste activated sludge. *Water Research* 124(1): 269–279.

Yamada, T., Suzuki, A., Ueda, H., et al., 2008. Successions of bacterial community in composting cow dung wastes with or without hyperthermophilic pre-treatment. *Applied Microbiology and Biotechnology* 81(4): 771–781.

Yamamoto, N., Otawa, K., Nakai, Y., et al., 2010. Diversity and abundance of ammonia-oxidizing bacteria and ammonia-oxidizing archaea during cattle manure composting. *Microbial Ecology* 60(4): 807.

Yang, K. J., Zhu, L. J., Zhao, Y., et al., 2019. A novel method for removing heavy metals from composting system: the combination of functional bacteria and adsorbent materials. *Bioresoure Technology* 293: 122095.

Yang, Y., Yu, C. L., Wang, X. M., 2020. Inoculation with xanthomonas oryzae pv. oryzae induces thylakoid membrane association of rubisco activase in oryza meyeriana. *Journal of Plant Physiology* 168(14): 170–174.

Yazdankhah, S., Rudi, K. Bernhoft, A., 2014. Zinc and copper in animal feed - development of resistance and co-resistance to antimicrobial agents in bacteria of animal origin. *Microbial Ecology in Health and Disease* 25: 25862.

Yu, H. M., Zhao, Y; Zhang, C., et al., 2019. Driving effects of minerals on humic acid formation during chicken manure composting: Emphasis on the carrier role of bacterial community. *Bioresoure Technology* 294: 122239.

Zehr, J. P., Jenkins, B. D., Short, S. M., et al., 2003.Nitrogenase gene diversity and microbial community structure: a cross-system comparison. *Environmental Microbiology* 5: 539–554.

Zeng, F., Jin, W., Zhao, Q., 2019. Temperature effect on extracellular polymeric substances (EPS) and phosphorus accumulating organisms (PAOs) for phosphorus release of anaerobic sludge. *RSC Advances* 9(4): 2162–2171.

Zeng, G. M., Zhang, J. C., Chen, Y. N., et al., 2011. Relative contributions of archaea and bacteria to microbial ammonia oxidation differ under different conditions during agricultural waste composting. *Bioresource Technology* 102(19): 9026–9032.

Zeng, G., Zhang, L., Dong, H., et al., 2018. Pathway and mechanism of nitrogen transformation during composting: Functional enzymes and genes under different concentrations of PVP-AgNPs. *Bioresource Technology* 253: 112–120.

Zhang, J., Sui, Q., Tong, J., et al., 2018. Soil types influence the fate of antibiotic-resistant bacteria and antibiotic resistance genes following the land application of sludge composts. *Environment International* 118: 34–43.

Zhang, J. Y., Chen, M. X., Sui, Q. W., et al., 2016a. Impacts of addition of natural zeolite or a nitrification inhibitor on antibiotic resistance genes during sludge composting. *Water Research* 91: 339–349.

Zhang, J. Y., Chen M. X., Sui Q. W., et al., 2016b. Fate of antibiotic resistance genes and its drivers during anaerobic co-digestion of food waste and sewage sludge based on microwave pretreatment. *Bioresource Technology* 217: 28–36.

Zhang, S. C., Wang, J. L., Chen, X., et al., 2020. Industrial-scale food waste composting: effects of aeration frequencies on oxygen consumption, enzymatic activities and bacterial community succession. *Bioresource Technology* 320: 124357.

Zhang, Y., Zhao, Y., Chen, Y., et al., 2016c. A regulating method for reducing nitrogen loss based on enriched ammonia-oxidizing bacteria during composting. *Bioresource Technology* 221: 276–283.

Zhang, Z. C., Zhao, Y., Yang, T. X., et al., 2019. Effects of exogenous protein-like precursors on humification process during lignocellulose-like biomass composting: amino acids as the key linker to promote humification process. *Bioresoure Technology* 291: 121882.

Zhao, X. Y., Tan, W. B., Dang, Q. L., et al., 2019. Enhanced biotic contributions to the dechlorination of pentachlorophenol by humus respiration from different compostable environments. *Chemical Engineering Journal* 361: 1565–1575.

Zhao, X. Y., Wei, Y. Q., Fan, Y. Y., et al., 2018. Roles of bacterial community in the transformation of dissolved organic matter for the stability and safety of material during sludge composting. *Bioresoure Technology* 267: 378–385.

Zhu, L. J., Zhao, Y., Yang, K. J., et al., 2019. Host bacterial community of MGEs determines the risk of horizontal gene transfer during composting of different animal manures. *Environmental Pollution* 250: 166–174.

Zhu, L. J., Wei, Z. M., Yang, T. X., et al., 2020. Core microorganisms promote the transformation of DOM fractions with different molecular weights to improve the stability during composting. *Bioresource Technology* 229: 122575.

17 Acyl Homoserine Lactone Regulated Quorum Sensing in Plant Co-Habitating Microorganisms and Their Agricultural Significance

Himani Meena and Busi Siddhardha
Pondicherry University, Puducherry, India

CONTENTS

17.1 INTRODUCTION

The global threat in scarcity of food supply to the human race in the current situation has increased the tension among scientists and economists. The demand for food has enhanced chemical fertilizers, affecting plant health and consumer health. The prolonged use of chemical fertilizers has escalated the chances of genetic manipulation in the plant genome, leading to pathogenic strains resistance. The chemical inputs have also increased the abundance of toxic substances in the soil via bioaccumulation of metal ions and bioaugmentation. This needs urgent attention to rectify the public concerns related to stomach cancer, methemoglobinemia, and other diseases

DOI: 10.1201/9781003110477-17

caused by harmful substances. Chemical fertilizers have caused numerous detrimental effects on soil fertility, salinity, nutrient uptake, and soil porosity, directly
affecting soil health and surrounding environment. To overcome these severe issues,
authorities should take measures to come forward to enhance the soil fertility and
food supply for the whole world without causing lethal side effects on plants and
consumers. Microorganisms are present in a harmonious manner known as a symbiotic and non-symbiotic relationship with plants. The microbial community harbors
within plant tissues possess specific phenomenon which helps them vigorously grow
in the plant habitat. The rhizospheric microorganism plays a vital role in plant health
and boosts the host defense system against phytopathogens. These microorganisms
associated with plants can be categorized based on their beneficial and non-beneficial
relationship; obligate, facultative, and parasitic. The microbial consortia contribute to nutrition supply, crop quality, crop yield, plant growth-promoting parameters,
nitrogen (N) fixation, phosphate solubilization, biotic, abiotic stress adaptation,
phytohormones, siderophore, and root exudates secretions as beneficial elements.
Microorganisms are essential assets to fight phytopathogens. The concept of a complicated microbial relationship with the host can be defined as "friends with benefits"
leading to beneficial but harmless collaboration. To better understand the complexity
of the relationship, we need to explore specific aspects of plant–microbe interaction (Zhang et al. 2017; Rodriguez et al. 2019; Uroz et al. 2019; Papik et al 2020;
Hartmann 2021). Several molecular mechanisms involved in the phyllosphere and
rhizosphere species mutualistic process explained in the following sections.

17.2   QUORUM-SENSING SYSTEM

The communication system that recruits many bacterial populations and regulates
gene expression is called quorum sensing (QS). The QS system initiates via the production of a signal molecule, secretion in the surrounding environment, accumulation of the signaling molecules, reaching a threshold level and modulates the genetic
expression of virulence factors, secondary metabolite production, biofilm development, and antibiotic resistance. The signal molecule is a pheromone type compound
that attracts the microbial population and modifies specific physiochemical parameters for microbial consortia extraordinary growth. The QS system in microorganisms
describes one of the defense mechanisms to fight against harsh conditions and adapt
to the unfavorable environment without altering their growth pattern (Gonzalez and
Venturi 2013; Kanchiswamy et al. 2015; Phour et al. 2020).

17.3   ACYL HOMOSERINE LACTONE

Acyl homoserine lactone (AHL) is an autoinducer molecule that control a two-
component regulatory system in Gram-negative bacteria. The AHL molecule synthesized by the autoinducer synthase gene binds with the cognate transcriptional receptor,
stabilizing the AHL concentration. The system activates virulence factor production,
type III secretory system, and biofilm formation contributing to antibiotic resistance
and persister cells at a threshold level. Briefly, the AHL molecules' chemical structure reveals much more about the synthesis of the particular molecules. The AHL

molecule comprises an acyl side chain derived from S-adenosyl methionine (SAM) attached to the lactone ring acquired from the acyl-acyl carrier protein (ACP). The chemical reaction initiates with cyclization of home-cysteine residue from SAM, forming an amide bond with hexanoyl group from ACP followed by the release of 5′-S-methyl-5′-thioadenosine and concurrent lactonization-producing AHL molecule. A variety of AHL molecules are present based on their carbon chain length and carbon substitute positioned on the third position. The two-component regulatory system comprises AHL molecule, a LuxI-type synthase protein, and LuxR regulatory receptor. At low cell density, signal molecule diffuses across the cell membrane and acts as chemo-pheromones attracting more bacterial cells, further escalating bacterial multiplication leading to a higher AHL molecule concentration in the environment. This condition activates target gene expression essential for establishing the infection, spread, and ultimate death of the lost (Yajima 2014; Verbon and Liberman 2016; Bettenworth et al. 2019; Prescott and Decho 2020; Ke et al. 2021).

A group of bacterial species, *Rhizobacterium, Pseudomonas, Enterobacter, Flavobacterium, Alcaligenes, Serratia, Yersinia, Acinetobacter, Achromobacter, Azospirillum, Methylobacterium*, and *Agrobacterium* spp. are contributing to the Gram-negative community of the phyllosphere and rhizosphere microbial population. *Pseudomonas aeruginosa* possesses a wide range of host systems, including plants, humans, and insects. A sophisticated two-component regulatory system controls *P. aeruginosa* pathogenicity and is composed of four individual operons. These four operons are governed by a specific set of AHL molecules differing in their carbon chain lengths. *P. aeruginosa* communicates via LasI/LasR, RhlI/RhlR, PQS, and IQS system, and causes infection in immune-compromised patients, patients suffering from acute and chronic respiratory disease, burn wounds, and is responsible for urinary tract infections. A hierarchy of operon is initiated by production of C_{12}-HSL molecule AHL synthase (LasI), binding of C_{12}-HSL to its cognate receptor, LasR leading to accumulation of AHL molecules in the surrounding environment. This ligand-protein binding activates RhlI synthase, which is responsible for producing the C4-HSL molecule, a ligand molecule required to activate RhlR, a transcriptional regulator. This cascade further activates the expression of multiple genes essential for virulence and survival in non-favorable conditions. The PQS is regulated via both LasI/LasR and RhlI/RhlR and activated by 2-heptyl-3-hydroxy-4-quinolone (PQS) signal molecule binds to MvfR receptor protein. The PQS signaling system causes a negative feedback loop on the LasI/LasR system and has a positive feedback loop effect on the RhlI/Rhlr system. The IQS system in *P. aeruginosa* is linked to phosphate depletion where lack of phosphate molecules in the environment leads to stimulation of IQS molecule production and binding to IqsR receptor (El-Sayed et al. 2001; Smith and Iglewski 2003, Kostylev et al. 2019; Yan and Wu 2019; Thi et al. 2020). Another Pseudomonas group member, *Pseudomonas syringae*, possesses a one-component system known as AhlI/AhlR system, a typical QS system. The AhlI synthase initiates C_6-HSL molecule production and catalyzes binding with AhlR receptor protein further controlling bacterial motility and virulence factor production. AhlI/AhlR can be achieved via induction of AefR and GacA activators, compelled for type III secretion system activation via *hrpL* gene (Xie et al. 2020). In *Pseudomonas chlororaphis* PA23, two new QS systems were identified: PhzI/PhzR (N-(3-OH-Hexanoyl)-L-HSL) and CsaI/CsaR that attributes to pigment production, siderophore and HCN production, biofilm development, and helps in root colonization (Maddula et al. 2006; Selin et al. 2012; Shah et al. 2020). Another root-colonizing bacteria, *Pseudomonas fluorescens* 2P24, revealed the presence of QS system (PcoI/PcoR) under Mg^{2+} starvation conditions (Wei and Zhang 2006; Yan et al. 2009).

Many reports have been cited for the QS mechanism of the nodulating root bacteria, *Bradyrhizobium japonicum*, and explored the importance of QS system during root colonization. *B. japonicum* harbors a symbiotic relationship with the soybean plant, significantly impacts nutritional uptake, and induces root propagation. QS system root colonizing bacteria is essential for root surface attachment and xylem invasion with the help of toxin molecules that create an infection site that helps establish a mutualistic relationship within host microbes. QS regulatory proteins, nodD1 and nodAD2, are involved and regulated via QS-signaling mechanism during the root nodulation process. The NodD1 is a positive regulatory factor triggers by genistein and daidzein, plant isoflavonoids, whereas the NodD2 has negative feedback that suppresses *nodD1* gene expression. Several studies have shown that isovaleryl-homoserine lactone modulates bjaI/bjaR system as a signaling molecule and acts as a triggering agent for the secretion of surface modulation, biosurfactant production, and biofilm formation that leads to root colonization (Jitacksorn and Sadowsky 2008; Lindemann et al. 2011; Bogino et al. 2015; Fagotti et al. 2019).

Enterobacter asburiae is an epiphytic bacterium that produces AHL molecule known as N-butanoyl homoserine lactone (C_4-HSL) and N-hexanoyl homoserine lactone (C_6-HSL) that controls QS mechanism in the microorganism. The binding of the AHL molecule to the respective receptors modulates further secretion of exoenzyme and toxins, helps in combating pathogenic microorganisms (Lau et al. 2013; Shastry et al. 2018; Martins et al. 2018).

Suppiger et al. (2013) reported AHL-based QS regulatory system in *Burkholderia cepacia* complex (BCC) (including *B. cenocepacia* and *B. multivorans*) designated as CepI/CepR and CciI/CciR systems. The CepI produces N-octanoyl-homoserine lactone (C_8-HSL) and N-hexanoyl-homoserine lactone (C_6-HSL) that directly binds to CepR, transcriptional regulatory protein and synchronize genetic expression of downstream genes responsible for exoenzymes production, siderophore production, conjugation, sporulation, bacterial motility, and biofilm formation (Huber et al. 2001; McKeon et al. 2011; Slinger et al. 2019). The second QS system, CciI/CciR prefers C_6-HSL over C_8-HSL to express bacterial pathogenic phenotypes and biofilm development within the host plant. Biofilm forming BCC harbors cyanide synthase with homologous to the *hcn* gene in *P. aeruginosa,* affecting cellular respiration in Cystic fibrosis patients. Regulation of QS system in *B. cepacia* complex is an integral part of the host-microbes interaction, essential for plant growth-promoting activities (Ryall et al. 2008, Chen et al. 2010; O'Grady et al. 2012).

Li et al. (2017) and Hayashi et al. (2019) described *PhcBSR* operon in *Ralstonia solanacearum*, a known QS mechanism causing agent for "bacterial wilt" in the plants. A *phcB* gene is responsible for 3-hydroxypalmitic acid methyl ester (3-OH-PAME) or (R)-methyl 3-hydroxymyristate (3-OHMAME) production that acts as a quorum-sensing signal molecule for the *PhcS* and *PhcR* operon. The *PhcBSR* operon is responsible for biosurfactant production and bacterial adhesion to the plant tissues (Kumar et al. 2016a). Kai et al. (2015) suggested the role of *phcA* gene in bacterial pathogenesis via modulating the secretion of endoglucanase and ralfurnanone needed for cell wall degradation for smooth invasion through wounds. A strain with mutant *phcA* gene lacks EPS and endoglucanase production but is highly active during bacterial movement (Khokhani et al. 2017). Lowe-Power et al. (2018) described bacterial dynamics changes along with metabolic rate when the population shifts from individual to an involved community. An increment in the level of QS molecules, followed by cell wall degrading enzymes and effector proteins via Type III secretory system, biofilm formation inhibits the flow rate of xylem from root to leaves and leads to plant disease by causing nutrient-deprived conditions.

Rhizobacteria is among the most studied soil-dwelling microbial consortia associated with root division, nutrient absorption, secretion of root exudates, root proliferation, and xylem phloem transportation service involved in plant growth and development (Sanchez-Contreras et al. 2007). Multiple QS systems have been reported in *Rhizobium leguminosarum*a and *Rhizobium etli* needed for acculturation in the surrounding conditions. The *cinI* gene (AHL synthase) produces N-(3-hydroxy-7-cis tetradecenoyl)-L-homoserine lactone (3OH-C(14:1)-HSL) modulates gene expression of an adjacent gene known as *cinR* gene, a transcriptional regulator protein. The cinI/cinR operon controls the bacterial transition from log phase to stationary phase and leads to secondary metabolites production responsible for the adaption of bacterial cells to the environmental conditions. Another QS system found in *R. leguminosarum* is raiI/raiR operon, where the raiR protein respond to the AHL molecules N-3-hydroxyoctanoyl-L-homoserine lactone (3OH-C_8-HSL) or C_8-HSL produced by *raiI* gene. A sophisticated gene regulatory system associated with legume-microbe interactions, where *rhiI* synthase gene releases a range of C_6-HSL, C_7-HSL, and C_8-HSL utilized by rhiR protein and are necessary for activation of *rhiA* gene and *rhi*ABC genes. The activated genes are further required for root nodulation and influence the legume bacterial cell multiplication (Lithgow et al. 2001; Daniels et al. 2002; Daniels et al. 2006; McAnulla et al. 2007; Edwards et al. 2009; Frederix et al. 2011; Dixit et al. 2017).

Moreover, *Sinorhizobium meliloti* and *Sinorhizobium fredii* reported with sinI/sinR and expR operon a requisite system for biosurfactant production, flagellar motility, and nodule production. The *sinI* gene initiates long-chain AHL molecules production ranging from C_{12}-HSL, C_{14}-HSL, 3-oxo-C_{14}-HSL, C_{16}-HSL, C(16:1)-HSL, 3-oxo-C_{16}-HSL, 3-oxo-C(16:1)-HSL, and C_{18}-HSL that controls gene expression of the downstream gene, *sinR*. The expR regulatory system influences the root nodulation process and acclimation via EPS production triggered by an unknown signal molecule (Marketon and Gonzalez 2002; Gao et al. 2012; Perez-Montano et al. 2014; Calatrava-Morales et al. 2018; Xu et al. 2018; Acosta-Jurado et al. 2020).

Serratia marcescens, *Serratia proteamaculans*, and *Serratia plymuthica* are opportunistic phytopathogens and are causing agents for cucurbit yellow vine disease and leaf spot disease. There is a diversified range of chemical moieties for the QS system regulation in *S. marcescens* for bacterial swarming motility, serrawattin production, and biofilm development. Studies show that the binding of C_4-HSL/C_6-HSL molecules (*smaI and swrI*) to their respective receptors, smaR and swrR initiates

QS-mediated pathogenicity, modulating various virulence phenotypes. Another set of QS system is SpnI/SpnR operon which operates via circulation of C_6-HSL, 3-oxo-C_6-HSL,C_7-HSL, C_8-HSL that binds with SpnR, a regulatory protein. The specific QS system controls EPS production, prodigiosin synthesis and enzymatic secretion. *Serratia plymuthica* has been reported for the SpiI/SpiR system required for secondary metabolite production and butendiol fermentation process. The signaling molecules are 3-OH-C_6-HSL, 3-OH-C_8-HSL, C_4-HSL, 3-oxo-C_6-HSL, and C_6-HSL recognized by the receptor based on their different structural moieties. *Serratia proteamaculans* identified C_6-HSL and 3-oxo-C_6-HSL as AHL molecules secreted by SprI which interacts with SprR protein to control the pathogenic phenotype production such as protease, nuclease, and chitinase enzymes found to be accountable for bacterial pathogenicity and movement (Liu et al. 2007; Van Houdt et al. 2007; Wevers et al. 2009; Zaitseva et al 2019). An illustration of QS system in microorganisms has been demonstrated in Figures 17.1 and 17.2. A tabulated form of AHL molecules with respective cognate receptor protein in different microorganisms has been included and enlisted in Table 17.1.

17.4   AHL MOLECULES IN AGRICULTURAL IMPORTANT MICROORGANISMS AND SIGNIFICANCE

17.4.1   ROOT-SHOOT GROWTH AND DEVELOPMENT

Microbial communities have specific dynamics with the host plant and QS system. They are among the beneficial mechanisms that can influence secondary metabolite production, shoot-root growth, seed germination, and improve plant defense against pathogens via siderophore production, antibiotic production, and Phyto-stimulatory molecules (Basu et al. 2017; Ortiz-Castro and Lopez-Bucio 2019). Several reports have been published regarding the plant growth-promoting phenotype displayed by microbial consortia residing in plant systems. In Pakistan, Hanif et al. (2020) reported an abundance of *Serratia* spp. in the soil, enhancing plant growth via AHL production. *Serratia* spp. (KPS-14 and KPS-10) produce structurally diversified chemical AHL molecules (C_6-HSL, 3oxo-C_{10}-HSL, 3oxo-C_{12}-HSL, 3OH-C_5-HSL, 3OH-C_6-HSL, C_{10}-HSL, 3OH-C_6-HSL, C_8-HSL, 3oxo-C_9-HSL, 3OH-C_9-HSL) in the surrounding environment of *Solanum tuberosum* L. AHL-producing KPS-10 and KPS-14 displayed anti-fungal activity against *F. oxysporum*. Both KPS-10 and KPS-14 solubilized phosphate at a higher rate and produced indole-acetic acid (IAA) in the medium, which could mediate the plant growth-promoting characteristics. Bacterial species colonize and form biofilm in the root region, simultaneously improving root division, seed germination, and enhancing shoot length: weigh ratio. Miao et al. (2012) reported that adding exogenous AHL molecules enhances the root formation and division in the *Arabidopsis thaliana* plant. Plant growth medium supplemented with 3-oxo-C_8-HSL molecule positively impacts seed germination, root formation, and plant defense mechanism. Rankl et al. (2016) suggested that accumulation of AHL molecules in the root zone can enhance root division, improve plant quality, and increase health conditions. The group supplemented *Hordeum vulgare* L. growth medium with three AHL molecules; C_6-HSL, C_8-HSL, and C_{12}-HSL

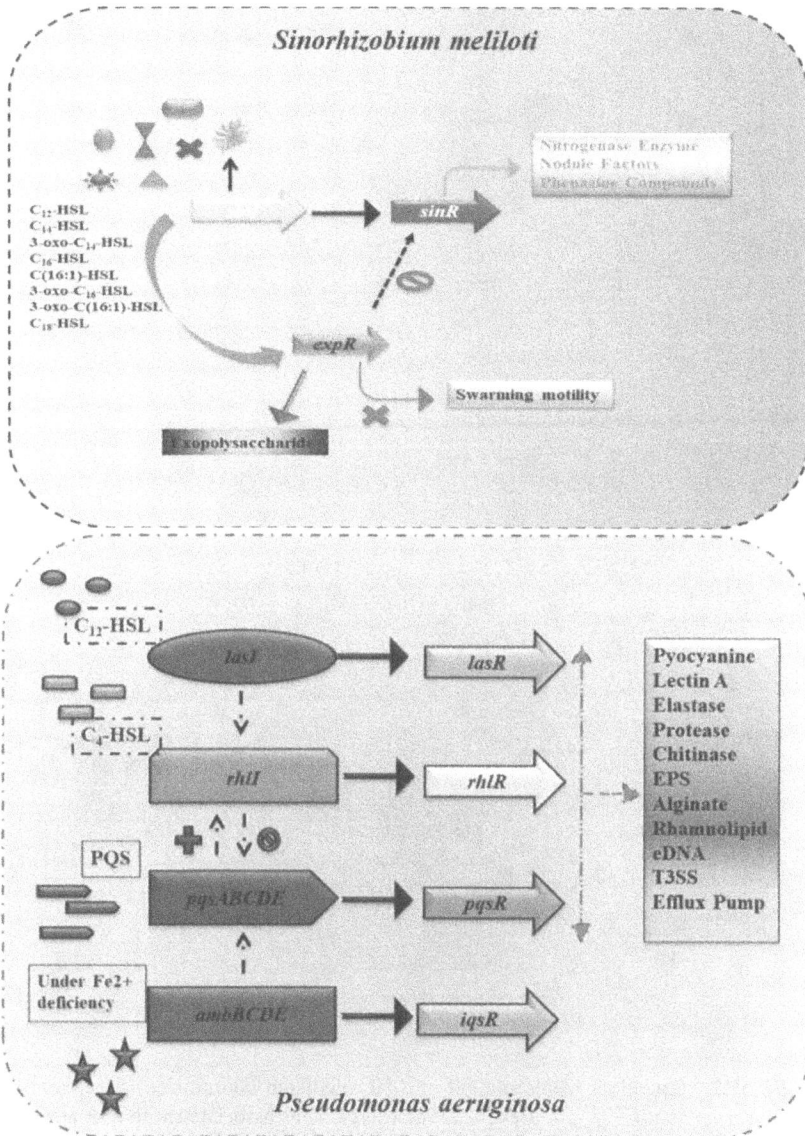

FIGURE 17.1 Quorum-sensing system in *S. meliloti* and *P. aeruginosa*. Microbial secretion of secondary metabolites and extracellular proteins in the presence of AHL molecules binding with cognate receptor proteins.

that altered physiochemical parameters related to the plant growth patterns. After the AHL molecules treatment, the root transition from proliferation to differentiation stage supported phenotypic and genotypic changes. The AHL molecules have boosted potassium (K^+) uptake in the root region, accumulation of nitric oxide (NO) influences root meristem via lateral root differentiation, and creates a hyperpolarized

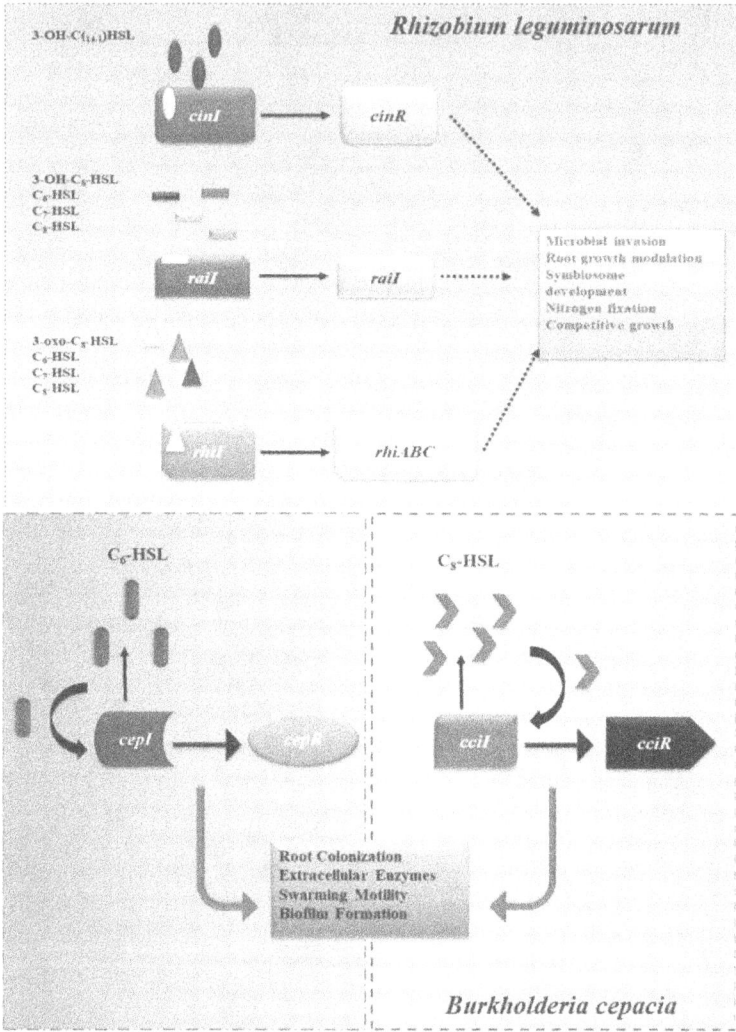

FIGURE 17.2 Graphical representation of AHL-mediated quorum-sensing system in *R. leguminosarum* and *B. cepacia*. Beneficial microorganisms associated with root nodulation, elongation and shoot length development.

condition alterations in root architecture. Microbial colonization in the root site stimulates root hair development and lateral meristem cell differentiation in *Crocus sativus* L. plant (Saffron). A group of researchers conducted an *in-vivo* experiment and recorded the plant growth-promoting activities of indigenous microbial species that effectively control the growth of *Sclerotium rolfsii* and *F. oxysporum*. *Brevibacterium frigoritolerans* (AIS-3), *Alcaligenes faecalis* subsp. *Phenolicus* (AIS-8) and *Bacillus aryabhattai* (AIS-10) secretes microbial metabolites, that is, HCN, IAA, and siderophore, with extracellular enzymes such as lipase and protease elastase to establish a

TABLE 17.1

Catalogue of Various AHL Molecules Based on Their Molecular Structure, Transcriptional Regulators, and Functional Phenotypic Traits

S. No.	Microorganisms	QS System	AHL Molecules	Phenotypes	Reference
1	Pseudomonas aeruginosa	LasI/LasR RhlI/RhlR PQSIQS	C_{12}-HSL C_4-HSL 2-heptyl-3-hydroxy-4-quinolone	Pyocyanine, Pyoverdine, Motility, Siderophore, Cellulolytic Enzymes, T3SS Effector Proteins, Extracellular DNA, Exopolysaccharide, Rhamnolipid, Alginate, Hydrogen Cyanide	El-Sayed et al. 2001; Smith and Iglewski 2003; Kostylev et al. 2019; Yan and Wu 2019; Thi et al. 2020
2	Pseudomonas syringae	AhlI/AhlR	C_6-HSL	T3SS Effector Proteins, Exopolysaccharide, Hormone Mimics	Xie et al. 2020
3	Pseudomonas chlororaphis	PhzI/PhzR CsaI/CsaR	N-(3-OH-Hexanoyl)-L-HSL	Phenazine-1-carboxamide	Maddula et al. 2006; Selin et al. 2012; Shah et al. 2020
4	Pseudomonas fluorescens	PcoI/PcoR	Mg^{2+} starvation	Mupirocin, Hemolysin, Phenazine Compounds	Wei and Zhang 2006; Yan et al. 2009
5	Bradyrhizobium japonicum	NodD1/NodD2	Unknown	Nodule Factors, Nitrogen-Fixing Enzymes	Jitacksorn and Sadowsky 2008; Lindemann et al. 2011; Bogino et al. 2015; Fagotti et al. 2019
6	Enterobacter asburiae	EasR	C_4-HSL C_6-HSL	Hydrogen Cyanide, Phenazines, Antibiotics	Lau et al. 2013; Martins et al. 2018; Shastry et al. 2018
7	Burkholderia cepacia complex (BCC)	CepI/CepR	C_6-HSL	Swarming Motility, Biofilm Formation	Huber et al. 2001; Ryall et al. 2008; Chen et al. 2010; McKeon et al. 2011; O'Grady et al. 2012; Suppiger et al. 2013; Slinger et al. 2019
		CciI/CciR	C_8-HSL	Protease, Swarming Motility	

(Continued)

TABLE 17.1 (Continued)
Catalogue of Various AHL Molecules Based on Their Molecular Structure, Transcriptional Regulators, and Functional Phenotypic Traits

S. No.	Microorganisms	QS System	AHL Molecules	Phenotypes	Reference
8	*Ralstonia solanacearum*	PhcS/ PhcR	3-OH-PAME 3-OHMAME	Endoglucanase	Kai et al. 2015; Kumar et al. 2016a; Khokhani et al. 2017; Li et al. 2017; Lowe-Power et al. 2018; Hayashi et al. 2019
9	*Rhizobium leguminosarum* *Rhizobium etli*	CinI/ CinR RaiI/RaiR RhiA/RhiABC	3OH-C(14:1)-HSL 3OH-C$_8$-HSL C$_8$-HSL C$_6$-HSL C$_7$-HSL C$_8$-HSL	Nitrogenase Enzyme, Nodule Factors, Phenazine Compounds	Lithgow et al. 2001; Daniels et al. 2002; Daniels et al. 2006; McAnulla et al. 2007; Sanchez-Contreras et al. 2007; Edwards et al. 2009; Frederix et al. 2011; Dixit et al. 2017
10	*Sinorhizobium meliloti* *Sinorhizobium fredii*	SinI/SinR ExpR	C$_{12}$-HSL C$_{14}$-HSL 3-oxo-C$_{14}$-HSL C$_{16}$-HSL C(16:1)-HSL 3-oxo-C$_{16}$-HSL 3-oxo-C(16:1)-HSL C$_{18}$-HSL	Nitrogenase Enzyme, Nodule Factors, Phenazine Compounds	Marketon and Gonzalez 2002; Gao et al. 2012; Perez-Montano et al. 2014; Calatrava-Morales et al. 2018; Xu et al. 2018; Acosta-Jurado et al. 2020

(Continued)

TABLE 17.1 *(Continued)*

Catalogue of Various AHL Molecules Based on Their Molecular Structure, Transcriptional Regulators, and Functional Phenotypic Traits

S. No.	Microorganisms	QS System	AHL Molecules	Phenotypes	Reference
11	*Serratia marcescens*	SmaI/SmaR	C_4-HSL	Prodigiosin	Liu et al. 2007; Van Houdt et al. 2007; Wevers et al. 2009; Zaitseva et al 2019
	Serratia proteamaculans	SwrI/SwrR	C_6-HSL		
	Serratia plymuthica	SpnI/SpnR	C_6-HSL		
			3-oxo-C_6-HSL		
			C_7-HSL		
			C_8-HSL		
		SpiI/SpiR	3-OH-C_6-HSL		
			3-OH-C_8-HSL		
			C_4-HSL		
			3-oxo-C_6-HSL,		
			C_6-HSL		
		SprI/SprR	C_6-HSL		
			3-oxo-C_6-HSL		

strong bacterial community in the root zone. The bacterial colonies stimulated plant growth factor production, phosphate solubilization, and increased crop yield percentage. The microbial community displayed anti-fungal properties against *Sclerotium rolfsii* and *F. oxysporum* via sporulation and mycelial formation inhibition (Rasool et al. 2021). You et al. (2021) suggested that *B. contaminans* secretes exopolysaccharide molecules stimulated by cadmium metal presence. The morphological changes in the root architecture were observed after treatment with *B. contaminans* under an escalated Cd^{2+} level. Inoculation with the *B. contaminans* has improved plant growth via the production of HCN, siderophore, IAA, 1-aminocyclo propane-1-carboxylate (ACC) deaminase enzymes and phytohormones. In a recent study, Maqsood et al. (2021) reported the plant growth-promoting efficacy of *Burkholderia* sp. AQ12 via the release of IAA, ACC deminase enzyme, and solubilizing the phosphate helps root formation and development. The rice plant lateral root development has improved the crop productivity and proved *Burkholderia* sp. AQ12 and showed visible increase in the shoot formation (tillering). Jin et al. (2012) documented the effect of extracellular AHL molecules on the G-coupled receptor protein regulation, which directly involves root differentiation. The study focused on the genetic expression of two genes, *cand2* and *cand7* from *Arabidopsis thaliana*. They reported that elevated AHL molecules have up-regulated the downstream gene responsible for root development. The AHL molecules promoted root-site differentiation and primary root elongation, hence displaying a wide range of plant growth-promoting activities. Li et al. (2019) conducted an in-vivo experiment to isolate three bacterial strains from *Arachis hypogaea* root nodules that belonged to the *Bradyrhizobium* genera. The *Bradyrhizobium* sp. demonstrated induced root nodulation in *Lablab purpureus* and *Vigna radiate* plant effectively. A significant change in the nitrogen fixation level was observed with a higher root elongation rate after inoculation with *Bradyrhizobium guangzhouense* sp. nov., *Bradyrhizobium nanningense* sp. nov., and *Bradyrhizobium zhanjiangense* sp. nov. In another article, *Bradyrhizobium yuanmingense* strain P10 130 reported for the efficacy of root nodulation and regulation of nitrogen cycle (nitrification and denitrification) in soybean (*Desmodium incanum*) plant (Toniutti et al. 2021). Bromfield et al. (2017) suggested that *Amphicarpaea* bracteata (S1), *Desmodium canadense* (S2), *Desmodium glutinosum* and *Agave Americana* have an abundance of *Bradyrhizobium sp.* responsible for overall plant growth via phosphate solubilization, nitrogen fixation, triggers root nodulation, providing the nutrient source, and promotes root development.

17.4.2 BIOCONTROL

Biocontrol agents are naturally occurring microorganisms that provide protection against disease-causing agents and suppress the multiplication and virulence factor production in habitat plant. Biocontrol agents are also known as biological suppression agents due to their biostatic nature instead of biocidal quality. Many reports in the scientific forum that supports use of bacterial species as a biocontrol agent under a limited bacterial concentration. In China, Marian et al. (2018) screened soil samples for the isolation of endophytic microorganisms. They found that 18 isolates belong to the soil-dwelling Rhizospheric bacteria known as *Ralstonia* (TCR112) and *Mitsuaria* (TWR114) genera out of 441 isolates and could restrict growth of pathogenic bacteria. In one pot experiment, both *Ralstonia* and *Mitsuaria* showed anti-pathogenic efficacy against *Ralstonia pseudosolanacearum* via productin of secondary metabolites, that is, indole-acetic acid, protease, siderophore and polyga-lacturonase. The antagonistic approach against tomato bacterial wilt disease using the habitat bacterial species could help to avoid chemical agents for a better future.

Kurabachew and Wydra (2013) reported the efficacy of AHL-producing bacte-rial species as antagonistic agents against bacterial wilt on tomato plant caused by *Ralstonia solanacearum*. A total number of 150 isolates were screened and 13 AHL producing *Bacillus cereus* (BC1AW), *P. putida* (PP3WT) and *S. marcescens* noted down for their ability to minimize pathogen growth. The AHL producers (BC1AW and PP3WT) were competent enough to reduce diseased bacterial conditions such as split roots, brown stems, patched leaves, and overall plant growth. A group of researchers from China has designed an experimental set up that used mutant strains of soil-borne *R. solanacearum* lacking virulence factor production and the ability to cause disease in the tomato plants. Mutant strain PRS-84-4-49 demonstrated anti-virulence activity against the bacterial wilt in *Pogostemon cablin* plant. The results showed a positive effect on the plant growth rate, minimized bacterial growth, and biofilm formation when treated with mutant *R. solanacearum* PRS-84-4-49 (Zhang et al. 2020). In another report, Ge et al. (2017) demonstrated impairment of disease-causing capability of bacterial wilt agent, *R. solanacearum*, after treatment with AHL-producing *P. aeruginosa* VIH2 strain. The AHL-producing bacteria, *P. aerugi-nosa* VIH2, displayed a wide range of toxic substance production via type III secre-tion system (T3SS) to manage the bacterial wilt disease. The T3SS releases effector proteins to the pathogen cell membrane and causes cell membrane damage resulting in cell leakage and cell death. Morohoshi et al. (2013) reported a strain of *P. chlo-roraphis* subsp. *aurantiaca* StFRB508 producing a wide range of AHL molecules, that is, C_4-HSL, C_6-HSL, and 3-OH-C_6-HSL. These AHL molecules contribute to the anti-fungal efficacy of StFRB508 against *Fusarium oxysporum* f. sp. *congluti-nans*. The study reveals the importance of the *phzI* gene that regulates phenazine compound production, an important bioactive compound responsible for anti-fungal activity. In another research, Agaras et al. (2015) demonstrated the involvement of QS system-mediated secondary metabolites production, that is, biosurfactant, phenazine compound, protease, HCN and siderophore production in *Pseudomonas* sp. isolated from soil. The majority of the isolates belonged to *P. putida* and *P. chlororaphis* group and can reduce fungal contamination. These species also showed plant growth-promoting capabilities to induce nutrient uptake and improved health conditions. Nakahara et al. (2021) recorded the inhibition of fungal wilt in eggplant by *R. sola-nacearum*, which is further used to develop mutant strains. The mutant strains were deficient in phenotypic conversion, responsible for the transition from disease-causing pathogen to non-pathogenic strain. These mutants showed inhibition of *Verticillium* wilt infected by *Verticillium dahliae* and reduced the fungal growth without showing any plant growth-promoting properties. An experimental study showed that *Bacillus subtilis* V26 colonizes potato root and acts as a potential candidate for the biocontrol of *Fusarium* wilt and tuber dry rot disease caused by *Fusarium* spp. (*F. oxysporum*, *F. gramineaurum Fusarium solani*, *Fusarium sambucinum*). The *B. subtilis* V26 demonstrated secretion of secondary metabolites (HCN, siderophore, IAA, bacilysin, macrolactin, surfactin, fengycin, bacyllomicin, iturin) that indirectly affect root division via inhibition of the primary roots leading to lateral root formation along with enhanced shoot biomass. These physiological changes are noted as a significant parameter in plant growth-promoting traits (Khedher et al. 2021). Furthermore, the bacterial colonization in plants via biofilm formation by the beneficial microbial community evolves as a new strategy to induce plant defense system and minimize pathovars multiplication as an efficient biocontrol agent.

17.4.3 CROP PRODUCTIVITY

Valetti et al. (2018) suggested the availability of *Bacillus*, *Serratia*, *Arthrobacter*, and *Pantoea* can enhance *Brassica napus* plant growth in the soil via molecular interaction in favorable conditions. The experimental set up demonstrated that inoculation of phosphate solubilizing bacteria in the root area and noted the significant changes during the process. The group reported considerable alterations in the seed germination and rapeseed oil yield in a positive manner. The P solubilizers have improved soil fertility phosphate groups for metabolic pathways, phosphate groups for metabolic pathways, and promoted plant growth and yield percentage in *B. napus*. Jayamohan et al. (2020) documented 11 species of *Pseudomonas* that produce AHL molecules and secrete extracellular enzymes, pigments, nitrogen fixation, and ACC deaminase enzyme. These extracellular substances were involved in enhancing crop productivity via the seed priming method. Briefly, the *Pseudomonas putida* spread over tomato seeds and placed under favorable conditions in the greenhouse. The strain has increased the seed germination rate, cellular metabolism, shoot elongation and has ultimately increased the crop yield while inhibiting growth of *F. oxysporum* MTCC1755. Kumar et al. (2014) reported *Bacillus megaterium*, *Arthrobacter chlorophenolicus*, and *Enterobacter* sp. for phosphate solubilization, nitrogen fixation, HCN, siderophore production mediated by the QS system that influences seed germination and plant growth under pot and field conditions. The plants demonstrated a significant increase in nutrient uptake contributing to plant growth, measured by grain yield, straw yield, plant height, and root reconfiguration in *Triticum aestivum* L.

Another research group, Kumar et al. (2016b) documented *Bacillus* sp., *Pseudomonas* sp., and *R. leguminosarum* for their plant growth-promoting abilities on *Phaseolus vulgaris* in pot and field experimental set up, simultaneously. The microbial consortia enabled seed germination, seed yield, shoot-root elongation, and crop production. Moshynets et al. (2019) conducted an experiment using exogenous C_6-HSL and observed for effect on seeds from *Triticum aestivum* L. (Volodarka and Yatran 60). They performed seed priming with C_6-HSL molecules and monitored the seed for a certain period. The experimental data displayed significant changes in seed germination rate affecting coleoptile and radicle development and seed yield. Two strains, *Pseudomonas thivervalensis* (STF3) and *S. marcesens* (STJ5) found to be potential candidates for enhancing plant growth and crop productivity via siderophore production and ACC deaminase activity in *Zea mays* L. (Shahzad et al. 2013). The microorganisms are enlisted in Table 17.2 based on their AHL structures and significance on agriculturally important crops.

TABLE 17.2

List of AHL-Producing Microorganism and Their Significance on Agricultural Important Crops

S. No.	Microorganisms	AHL Molecules	Plant	Beneficial Activities	Reference
		Plant Growth Promoting Bacteria			
1	*Serratia* spp. (KPS-14 & KPS-10)	C_6-HSL, $3oxo$-C_{10}-HSL, $3oxo$-C_{12}-HSL, $3OH$-C_5-HSL, $3OH$-C_6-HSL, C_{10}-HSL, $3OH$-C_6-HSL, C_8-HSL, $3oxo$-C_9-HSL, and $3OH$-C_9-HSL	*Solanum tuberosum* L.	Root division, seed germination, enhancing shoot length: weight ratio, antagonistic agents against *F. oxysporum*	Hanif et al. 2020
2	-	Exogenous 3-oxo-C_8-HSL	*Arabidopsis thaliana*	Seed germination, root formation, and plant defense mechanism	Miao et al. 2012
3	-	Exogenous C_6-HSL, C_8-HSL, and C_{12}-HSL	*Hordeum vulgare* L.	Root division, lateral root differentiation	Rankl et al. 2016
4	*Brevibacterium frigoritolerans* (AIS-3), *Alcaligenes faecalis* subsp. *Phenolicus* (AIS-8), and *Bacillus aryabhattai* (AIS-10)	-	*Crocus sativus* L.	Antifungal activity against *Sclerotium rolfsii* and *F. oxysporum*	Rasool et al. 2021
5	*Burkholderia contaminans*	-	*Glycine max*	Plant growth promotion	You et al. 2021
6	*Burkholderia* sp. AQ12	-	*Oryza sativa*	Tillering, crop productivity, and root formation and development	Maqsood et al. 2021
7		Exogenous AHL	*Arabidopsis thaliana*	Root development	Jin et al. 2012
8	*Bradyrhizobium guangzhouense* sp. nov., *Bradyrhizobium nanningense* sp. nov., and *Bradyrhizobium zhanjiangense* sp. nov.		*Lablab purpureus, Vigna radiate*	Nitrogen fixation	Li et al. 2019

(Continued)

TABLE 17.2 (Continued)
List of AHL-Producing Microorganism and Their Significance on Agricultural Important Crops

S. No.	Microorganisms	AHL Molecules	Plant	Beneficial Activities	Reference
9	*Bradyrhizobium yuanmingense* strain P10 130		*Desmodium incanum*	Nitrification and denitrification	Toniutti et al. 2021
10	*Bradyrhizobium sp.*		*Amphicarpaea bracteata* (S1), *Desmodium canadense* (S2), and *Desmodium glutinosum* *Agave Americana*	Phosphate solubilization, nitrogen fixation, and root nodulation and development	Bromfield et al. 2017
Biocontrol					
11	*Ralstonia* (TCR112) *Mitsuaria* (TWR114)	-	*Solanum lycopersicum*	Tomato bacterial wilt	Marian et al. 2018
12	*Bacillus cereus* (BC1AW), *Pseudomonas putida* (PP3WT), and *S. marcescens*	-	*Solanum lycopersicum*	*Ralstonia solanacearum* wilt	Kurabachew and Wydra 2013
13	*Ralstonia solanacearum* PRS-84-4-49	-	*Pogostemon cablin*	Patchouli bacterial wilt	Zhang et al. 2020
14	*Pseudomonas aeruginosa* VIH2	-	*Solanum lycopersicum*	*Ralstonia solanacearum* wilt	Ge et al. 2017
15	*Pseudomonas chlororaphis* subsp. *aurantiaca* StFRB508	C_4-HSL, C_6-HSL, and 3-OH-C_6-HSL	*Fusarium oxysporum* f. sp. *conglutinans.*		Morohoshi et al. 2013
16	*P. putida* and *P. chlororaphis*	C_{10}-HSL,	*Colletotrichum truncatum* *Fusarium oxysporum* *Fusarium graminearum* *Macrophomina phaseolina*	Biocontrol of fungal pathogens, Direct plant growth promotion,	Agaras et al. 2015

(Continued)

TABLE 17.2 (Continued)
List of AHL-Producing Microorganism and Their Significance on Agricultural Important Crops

S. No.	Microorganisms	AHL Molecules	Plant	Beneficial Activities	Reference
17	R. solanacearum	-	Solanum melongena	Verticillium wilt	Nakahara et al. 2021
18	Bacillus subtilis V26	-	Solanum tuberosum	Fusarium wilt and tuber dry rot disease	Khedher et al. 2021
Crop Productivity					
19	Bacillus, Serratia, Arthrobacter, and Pantoea	-	Brassica napus	Seed germination Rapeseed oil yield	Valetti et al. 2018
20	Pseudomonas putida	-	Solanum lycopersicum	Seed germination, cellular metabolism, shoot elongation, and inhibiting growth of F. oxysporum MTCC1755	Jayamohan et al. 2020
21	Bacillus megaterium, Arthrobacter chlorophenolicus, and Enterobacter sp.	-	Triticum aestivum L.	Increase grain yield, straw yield, plant height, and root re-configuration	Kumar et al. 2014
22	Bacillus sp., Pseudomonas sp., and R. leguminosarum	-	Phaseolus vulgaris	Seed germination, seed yield, shoot-root elongation, and crop production	Kumar et al. 2016b
23	-	Exogenous C_6-HSL	Triticum aestivum L.	Seed germination rate, coleoptile and radicle development, and seed yield	Moshynets et al. 2019
24	Pseudomonas thivervalensis (STF3) and S. marcesens (STJ5)	-	Zea mays L.	Enhancing plant growth and crop productivity	Shahzad et al. 2013

17.5 CONCLUSION AND FUTURE PERSPECTIVE

The chapter has tried to explain the importance of plant–microbial interaction according to the molecular mechanism between Gram-negative bacteria and QS-mediated phenotypic secretion. In Gram-negative bacteria, QS system plays an essential role in host interaction as it releases several bioactive molecules for plant growth and development. The AHL molecules are diffusible molecules that activates a cascade of genes, which further regulates bacterial pathogenicity. The emergence of biofertilizer can be fulfilled by using different exogenous AHL molecules or introducing AHL producers and AHL mutants' strains. AHL-mediated molecular dynamics within the plant–microbe complex regulate root and development, shoot development, biocontrol, and crop productivity, directly or indirectly.

REFERENCES

Acosta-Jurado S., Alías-Villegas C., Almozara A. et al. (2020). Deciphering the symbiotic significance of quorum sensing systems of *Sinorhizobium fredii* HH103. *Microorganisms* 8(1), 68.

Agaras B.C., Scandiani M., Luque A. et al. (2015). Quantification of the potential biocontrol and direct plant growth promotion abilities based on multiple biological traits distinguish different groups of *Pseudomonas* spp. isolates. *Biological Control* 90, 173–186.

Basu S., Rabara R., Negi S. (2017). Towards a better greener future – An alternative strategy using biofertilizers. I: Plant growth promoting bacteria. *Plant Gene* 12, 43–49.

Bettenworth V., Steinfeld B., Duin H. et al. (2019). Phenotypic heterogeneity in bacterial quorum sensing systems. *Journal of Molecular Biology* 431(23), 4530–4546.

Bogino P.C., Nievas F.L., Giordano W. (2015). A review: Quorum sensing in *Bradyrhizobium*. *Applied Soil Ecology* 94, 49–58.

Bromfield E.S.P., Cloutier S., Tambong J.T. et al. (2017). Soybeans inoculated with root zone soils of Canadian native legumes harbour diverse and novel *Bradyrhizobium* spp. that possess agricultural potential. *Systematic and Applied Microbiology* 40, 440–447.

Calatrava-Morales N., McIntosh, M. Soto, M.J. (2018). Regulation mediated by N-Acyl homoserine lactone quorum-sensing signals in the Rhizobium-Legume symbiosis. *Genes* 9(5), 263.

Chen X., Buddrus-Schiemann K., Rothballer M. et al. (2010). Detection of quorum sensing molecules in *Burkholderia cepacia* culture supernatants with enzyme-linked immunosorbent assays. *Analytical and Bioanalytical Chemistry* 398(6), 2669–2676.

Daniels R., De Vos D.E., Desair J. et al. (2002). The *cin* quorum sensing locus of *Rhizobium etli* CNPAF512 affects growth and symbiotic nitrogen fixation. *The Journal of Biological Chemistry* 277(1), 462–468.

Daniels R., Reynaert S., Hoekstra H. et al. (2006). Quorum signal molecules as biosurfactants affecting swarming in *Rhizobium etli*. *PNAS* 103 (40), 14965–14970.

Dixit S., Dubey R.C., Maheshwari D.K. et al. (2017). Roles of quorum sensing molecules from *Rhizobium etli* RT1 in bacterial motility and biofilm formation. *Brazilian Journal of Microbiology* 48(4), 815–821.

Edwards A., Frederix M., Wisniewski-Dye F. et al. (2009). The *cin* and *rai* Quorum-sensing regulatory systems in *Rhizobium leguminosarum* are coordinated by *ExpR* and *CinS*, a small regulatory protein coexpressed with CinI. *Journal of Bacteriology* 191 (9), 3059–3067.

El-Sayed, A.K., Hothersall, J., Thomas, C.M. (2001). Quorum-sensing-dependent regulation of biosynthesis of the polyketide antibiotic mupirocin in *Pseudomonas fluorescens* NCIMB 10586. *Microbiology* 147(Pt 8), 2127–2139.

Fagotti D.D.S.L., Abrantes J., Cerezini P. et al. (2019). Quorum sensing communication: *Bradyrhizobium-Azospirillum* interaction via N-acyl-homoserine lactones in the promotion of soybean symbiosis. *Journal of Basic Microbiology* 59(1), 38–53.

Frederix M., Edwards A., McAnulla C. et al. (2011). Co-ordination of quorum-sensing regulation in *Rhizobium leguminosarum* by induction of an anti-repressor. *Molecular Microbiology* 81(4), 994–1007.

Gao M., Coggin A., Yagnik K. et al. (2012). Role of specific quorum-sensing signals in the regulation of exopolysaccharide II production within *Sinorhizobium meliloti* spreading colonies. *PloS One* 7(8), e42611.

Ge X., Wei W., Li G. et al. (2017). Isolated *Pseudomonas aeruginosa* strain VIH2 -and antagonistic properties against Ralstonia solanacearum. *Microbial Pathogenesis* 111, 519–526.

Gonzalez J.F., Venturi V. (2013). A novel widespread interkingdom signaling circuit. *Trends in Plant Science* 18(3), 167–174.

Hanif M.K., Malik K.A., Hameed S. et al. (2020). Growth stimulatory effect of AHL producing *Serratia* spp. from potato on homologous and non-homologous host plants. *Microbiological Research* 238, 126506.

Hartmann A. (2021). Quorum sensing N-acyl-homoserine lactone signal molecules of plant beneficial Gram-negative rhizobacteria support plant growth and resistance to pathogens. *Rhizosphere* 16, 100258.

Hayashi K., Senuma W., Kai K. et al. (2019). Major exopolysaccharide, EPS I, is associated with the feedback loop in the quorum sensing of *Ralstonia solanacearum* strain OE1-1. *Molecular Plant Pathology* 20(12), 1740–1747.

Huber B., Riedel K., Hentzer M. et al. (2001). The cep quorum-sensing system of *Burkholderia cepacia* H111 controls biofilm formation and swarming motility. *Microbiology* 147(Pt 9), 2517–2528.

Jayamohan N.S., Patil S.V., Kumudini B.S. (2020). Seed priming with *Pseudomonas putida* isolated from rhizosphere triggers innate resistance against Fusarium wilt in tomato through pathogenesis-related protein activation and phenylpropanoid pathway. *Pedosphere* 30(5), 651–660.

Jin G., Liu F., Maa H. et al. (2012). Two G-protein-coupled-receptor candidates, Cand2 and Cand7, are involved in *Arabidopsis* root growth mediated by the bacterial quorum-sensing signals N-acyl-homoserine lactones. *Biochemical and Biophysical Research Communications* 417, 991–995.

Jitacksorn, S., Sadowsky, M.J. (2008). Nodulation gene regulation and quorum sensing control density-dependent suppression and restriction of nodulation in the *Bradyrhizobium japonicum*-soybean symbiosis. *Applied and Environmental Microbiology* 74(12), 3749–3756.

Kai K., Ohnishi H., Shimatani M. et al. (2015). Methyl 3-Hydroxymyristate, a diffusible signal mediating phc quorum sensing in *Ralstonia solanacearum*. *Chembiochem: A European Journal of Chemical Biology* 16(16), 2309–2318.

Kanchiswamy C.N., Malnoy M., Maffei M.E. (2015). Bioprospecting bacterial and fungal volatiles for sustainable agriculture. *Trends in Plant Science* 20(4), 206–211.

Ke J., Wang B., Yoshikuni Y. (2021). Microbiome engineering: Synthetic biology of plant-associated microbiomes in sustainable agriculture. *Trends in Biotechnology* 39(3), 244–261.

Khedher S.B., Mejdoub-Trabelsi B., Tounsi S. (2021). Biological potential of Bacillus subtilis V26 for the control of Fusarium wilt and tuber dry rot on potato caused by Fusarium species and the promotion of plant growth. *Biological Control* 152, 104444.

Khokhani D., Lowe-Power T.M., Tran T.M. et al. (2017). A single regulator mediates strategic switching between attachment/spread and growth/virulence in the plant pathogen *Ralstonia solanacearum*. *American Society for Microbiology* 8(5), e00895–17.

Kostylev M., Kim D.Y., Smalley N.E. et al. (2019). Evolution of the *Pseudomonas aeruginosa* quorum-sensing hierarchy. *Proceedings of the National Academy of Sciences* 116(14), 7027–7032.

Kumar J.S., Umesha S., Prasad K.S. et al. (2016a). Detection of quorum sensing molecules and biofilm formation in *Ralstonia solanacearum*. *Current Microbiology* 72(3), 297–305.

Kumar A., Maurya B.R., Raghuwanshi R. (2014). Isolation and characterization of PGPR and their effect on growth, yield and nutrient content in wheat (*Triticum aestivum* L.). *Biocatalysis and Agricultural Biotechnology* 3, 121–128.

Kumar P., Pandey P., Dubey R.C. et al. (2016b). Bacteria consortium optimization improves nutrient uptake, nodulation, disease suppression and growth of the common bean (*Phaseolus vulgaris*) in both pot and field studies. *Rhizosphere* 2, 13–23.

Kurabachew H., Wydra K. (2013). Characterization of plant growth promoting rhizobacteria and their potential as bioprotectant against tomato bacterial wilt caused by *Ralstonia solanacearum*. *Biological Control* 67, 75–83.

Lau, Y.Y., Sulaiman, J., Chen, J.W. et al. (2013). Quorum sensing activity of *Enterobacter asburiae* isolated from lettuce leaves. *Sensors* 13(10), 14189–14199.

Li P., Yin W., Yan J. et al. (2017). Modulation of inter-kingdom communication by PhcBSR quorum sensing system in *Ralstonia solanacearum* Phylotype I Strain GMI1000. *Frontiers in Microbiology* 8, 1172.

Li Y.H., Wanga R., Sui Z.H. et al. (2019). *Bradyrhizobium nanningense* sp. nov., *Bradyrhizobium guangzhouense* sp. nov. and *Bradyrhizobium zhanjiangense* sp. nov., isolated from effective nodules of peanut in Southeast China. *Systematic and Applied Microbiology* 42, 126002.

Lindemann A., Pessi G., Schaefer A.L. et al. (2011). Isovaleryl-homoserine lactone, an unusual branched-chain quorum-sensing signal from the soybean symbiont *Bradyrhizobium japonicum*. *Proceedings of the National Academy of Sciences* 108(40), 16765–16770.

Lithgow J., Danino V.E., Jones J. et al. (2001). Analysis of N-acyl homoserine-lactone quorum-sensing molecules made by different strains and biovars of *Rhizobium leguminosarum* containing different symbiotic plasmids. *Plant and Soil* 232, 3–12.

Liu X., Bimerew M., Ma Y. et al. (2007). Quorum-sensing signaling is required for production of the antibiotic pyrrolnitrin in a rhizospheric biocontrol strain of *Serratia plymuthica*. *FEMS Microbiology Letters* 270(2), 299–305.

Lowe-Power T.M., Khokhani D., Allen C. (2018). How *Ralstonia solanacearum* exploits and Thrives in the flowing plant xylem environment. *Trends in Microbiology* 26(11), 929–942.

Marian M., Nishioka T., Koyama H. et al. (2018). Biocontrol potential of *Ralstonia* sp. TCR112 and *Mitsuaria* sp. TWR114 against tomato bacterial wilt. *Applied Soil Ecology* 128, 71–80.

Maddula V.S., Zhang, Z., Pierson, E.A. et al. (2006). Quorum sensing and phenazines are involved in biofilm formation by *Pseudomonas chlororaphis* (aureofaciens) strain 30–84. *Microbial Ecology* 52(2), 289–301.

Marketon M.M. Gonzalez, J.E. (2002). Identification of two quorum-sensing systems in *Sinorhizobium meliloti*. *Journal of Bacteriology* 184(13), 3466–3475.

Martins M.L., Pinto U.M., Riedel K. et al. (2018). Quorum sensing and spoilage potential of psychrotrophic *Enterobacteriaceae* isolated from milk. *BioMed Research International*. Article ID 2723157, 13.

Maqsood A., Shahid M., Hussain S. et al. (2021). Root colonizing *Burkholderia* sp. AQ12 enhanced rice growth and upregulated tillering-responsive genes in rice. *Applied Soil Ecology* 157, 103769.

McAnulla C., Edwards A., Sanchez-Contreras M. et al. (2007). Quorum-sensing-regulated transcriptional initiation of plasmid transfer and replication genes in *Rhizobium leguminosarum* biovar *viciae*. *Microbiology* 153(Pt 7), 2074–2082.

McKeon S.A., Nguyen D.T., Viteri D.F. et al. (2011). Functional quorum sensing systems are maintained during chronic *Burkholderia cepacia* complex infections in patients with cystic fibrosis. *Journal of Infectious Diseases* 203(3), 383–392.

Miao C., Liu F., Zhao Q. et al. (2012). A proteomic analysis of *Arabidopsis thaliana* seedling responses to 3-oxo-octanoyl-homoserine lactone, a bacterial quorum-sensing signal. *Biochemical and Biophysical Research Communications* 427(2), 293–298.

Morohoshi T., Wang W.Z., Suto T. et al. (2013). Phenazine antibiotic production and antifungal activity are regulated by multiple quorum-sensing systems in *Pseudomonas chlororaphis* subsp. *aurantiaca* StFRB508. *Journal of Bioscience and Bioengineering* 116(5), 580–584.

Moshynets O.V., Babenko L.M., Rogalsky S.P. et al. (2019). Priming winter wheat seeds with the bacterial quorum sensing signal N-hexanoyl-Lhomoserine lactone (C6-HSL) shows potential to improve plant growth and seed yield. *PLoS ONE* 14(2): e0209460.

Nakahara H., Mori T., Matsuzoe N. (2021). Screening of phenotypic conversion mutant strains of *Ralstonia solanacearum* for effective biological control of *Verticillium* wilt in eggplant. *Crop Protection* 142, 105530.

Ortiz-Castro R. Lopez-Bucio J. (2019). Phytostimulation and root architectural responses to quorum sensing signals and related molecules from rhizobacteria. *Plant Science* 284, 135–142.

O'Grady E.P., Viteri D.F., Sokol P.A. (2012). A unique regulator contributes to quorum sensing and virulence in *Burkholderia cenocepacia*. *PLoS One* 7(5), e37611.

Papik J., Folkmanova M., Polivkova-Majorova M. (2020). The invisible life inside plants: Deciphering the riddles of endophytic bacterial diversity. *Biotechnology Advances* 44, 107614.

Perez-Montano F., Jimenez-Guerrero I., Del Cerro P. et al. (2014). The symbiotic biofilm of *Sinorhizobium fredii* SMH12, necessary for successful colonization and symbiosis of *Glycine max cv* Osumi is regulated by quorum sensing systems and inducing flavonoids via NodD1. *PLoS One* 9(8), e105901.

Phour M., Sehrawat A., Sindhu S.S. et al. (2020). Interkingdom signaling in plant-rhizomicrobiome interactions for sustainable agriculture. *Microbiological Research* 241, 126589.

Prescott R.D., Decho A.W. (2020). Flexibility and adaptability of quorum sensing in nature. *Trends in Microbiology* 28(6), 436–444.

Rankl S., Gunse B., Sieper T. et al. (2016). Microbial homoserine lactones (AHLs) are effectors of root morphological changes in barley. *Plant Science* 253, 130–140.

Rasool A., Mir M.I., Zulfajri M. et al. (2021). Plant growth promoting and antifungal asset of indigenous rhizobacteria secluded from saffron (*Crocus sativus* L.) rhizosphere. *Microbial Pathogenesis* 150, 104734.

Rodriguez P.A., Rothballer M., Chowdhury S. et al. (2019). Systems biology of plant-microbiome interactions. *Molecular Plant* 12(6), 804–821.

Ryall B., Lee X., Zlosnik, J.E. et al. (2008). Bacteria of the *Burkholderia cepacia* complex are cyanogenic under biofilm and colonial growth conditions. *BMC Microbiology* 8, 108.

Sanchez-Contreras M., Bauer W.D., Gao M. et al. (2007). Quorum-sensing regulation in rhizobia and its role in symbiotic interactions with legumes. *Philosophical Transactions of the Royal Society of London. Series B, Biological Sciences* 362(1483), 1149–1163.

Selin, C., Fernando, W., de Kievit, T. (2012). The PhzI/PhzR quorum-sensing system is required for pyrrolnitrin and phenazine production, and exhibits cross-regulation with RpoS in *Pseudomonas chlororaphis* PA23. *Microbiology* 158(4), 896–907.

Shah N., Gislason A.S., Becker M. et al. (2020). Investigation of the quorum-sensing regulon of the biocontrol bacterium *Pseudomonas chlororaphis* strain PA23. *PLoS One* 15(2), e0226232.

Shahzad S.M., Arif M.S., Riaz M. et al. (2013). PGPR with varied ACC-deaminase activity induced different growth and yield response in maize (*Zea mays* L.) under fertilized conditions. *European Journal of Soil Biology* 57, 27–34.

Shastry, R.P., Dolan, S.K., Abdelhamid, Y. et al. (2018). Purification and characterisation of a quorum quenching AHL-lactonase from the endophytic bacterium *Enterobacter* sp. CS66. *FEMS Microbiology Letters* 365(9), fny054.

Slinger, B.L., Deay, J.J., Chandler, J.R. et al. (2019). Potent modulation of the CepR quorum sensing receptor and virulence in a *Burkholderia cepacia* complex member using non-native lactone ligands. *Scientific Report* 9, 13449.

Smith, R.S. Iglewski, B.H. (2003). *Pseudomonas aeruginosa* quorum sensing as a potential antimicrobial target. *The Journal of Clinical Investigation* 112(10), 1460–1465.

Suppiger A., Schmid N., Aguilar C. et al. (2013). Two quorum sensing systems control biofilm formation and virulence in members of the *Burkholderia cepacia* complex. *Virulence* 4(5), 400–409.

Thi, M., Wibowo, D., Rehm, B. (2020). *Pseudomonas aeruginosa* Biofilms. *International Journal of Molecular Sciences* 21(22), 8671.

Toniutti M.A., Albicoro F.J., Castellani L.G. et al. (2021). Genome sequence of *Bradyrhizobium yuanmingense* strain P10 130, a highly efficient nitrogen-fixing bacterium that could be used for *Desmodium incanum* inoculation. *Gene* 768, 145267.

Uroz S., Courty P.E., Oger P. (2019). Plant symbionts are engineers of the plant-associated microbiome. *Trends in Plant Science* 24(10), 905–916.

Valetti L., Iriarte L., Fabra A. (2018). Growth promotion of rapeseed (*Brassica napus*) associated with the inoculation of phosphate solubilizing bacteria. *Applied Soil Ecology* 132, 1–10.

Van Houdt R., Givskov M., Michiels C.W. (2007). Quorum sensing in *Serratia*. *FEMS Microbiology Reviews* 31(4), 407–424.

Verbon E.H., Liberman L.M. (2016). Beneficial microbes affect endogenous mechanisms controlling root development. *Trends in Plant Science* 21(3), 218–229.

Wei, H.L. Zhang, L.Q. (2006). Quorum-sensing system influences root colonization and biological control ability in *Pseudomonas fluorescens* 2P24. *Antonie Van Leeuwenhoek* 89(2), 267–280.

Wevers E., Moons P., Van Houdt R. et al. (2009). Quorum sensing and butanediol fermentation affect colonization and spoilage of carrot slices by *Serratia plymuthica*. *International Journal of Food Microbiology* 134, 1–2, 63–69.

Xie, Y., Liu, W., Shao, X. et al. (2020). Signal transduction schemes in *Pseudomonas syringae*. *Computational and Structural Biotechnology Journal* 18, 3415–3424.

Xu Y. Y., Yang J. S., Liu C. et al. (2018). Water-soluble humic materials regulate quorum sensing in *Sinorhizobium meliloti* through a novel repressor of *expR*. *Frontiers in Microbiology* 9, 3194.

Yan, S. Wu, G. (2019). Can biofilm be reversed through quorum sensing in *Pseudomonas aeruginosa*?. *Frontiers in Microbiology* 10, 1582.

Yajima A. (2014). Recent progress in the chemistry and chemical biology of microbial signaling molecules: quorum-sensing pheromones and microbial hormones. *Tetrahedron Letters* 55(17), 2773–2780.

Yan Q., Gao W., Wu X.G. et al. (2009). Regulation of the PcoI/PcoR quorum-sensing system in *Pseudomonas fluorescens* 2P24 by the PhoP/PhoQ two-component system. *Microbiology* 155(Pt 1), 124–133.

You L.X., Zhang R.R., Dai J.X. et al. (2021). Potential of cadmium resistant *Burkholderia contaminans* strain ZCC in promoting growth of soy beans in the presence of cadmium. *Ecotoxicology and Environmental Safety* 211, 111914.

Zaitseva Y.V., Koksharova O.A., Lipasova V.A. et al. (2019). SprI/SprR quorum sensing system of *Serratia proteamaculans* 94. *BioMed Research International* 12:3865780.

Zhang R., Vivanco J.M., Shen Q. (2017). The unseen rhizosphere root-soil-microbe interactions for crop production. *Current Opinion in Microbiology* 37, 8–14.

Zhang Y., Li G., Li Q. et al. (2020). Identification and characterization of virulence-attenuated mutants in *Ralstonia solanacearum* as potential biocontrol agents against bacterial wilt of *Pogostemon cablin*. *Microbial Pathogenesis* 147, 104418.

Index

For Product Safety Concerns and Information please contact our EU
representative GPSR@taylorandfrancis.com
Taylor & Francis Verlag GmbH, Kaufingerstraße 24, 80331 München, Germany

www.ingramcontent.com/pod-product-compliance
Lightning Source LLC
Chambersburg PA
CBHW060752220326
41598CB00022B/2410

* 9 7 8 0 3 6 7 6 2 7 1 3 3 *